Semiclassical Mechanics with Molecular Applications

Second Edition

M. S. Child

Emeritus Professor of Theoretical Chemistry, University of Oxford

OXFORD
UNIVERSITY PRESS

OXFORD
UNIVERSITY PRESS

Great Clarendon Street, Oxford, OX2 6DP,
United Kingdom

Oxford University Press is a department of the University of Oxford.
It furthers the University's objective of excellence in research, scholarship,
and education by publishing worldwide. Oxford is a registered trade mark of
Oxford University Press in the UK and in certain other countries

First Edition published in 1991
Reprinted 2005
Second Edition published in 2014
Impression: 1

Published in the United States of America by Oxford University Press
198 Madison Avenue, New York, NY 10016, United States of America

British Library Cataloguing in Publication Data
Data available

Library of Congress Control Number: 2014932180

ISBN 978–0–19–967298–1

Printed in Great Britain by
Clays Ltd, St Ives plc

Prefaces

The roots of semiclassical mechanics lie in the old quantum theory, but it has seen a remarkable revival in recent years under the stimulus of modern experiments which demand interpretation in a physically intuitive and computationally accurate way. The subject concerns the connection between classical and quantum mechanics; and the present book bears less on the mathematical foundations than on the global influence of classical phase space structures on the quantum mechanical observables. The theory is illustrated by examples from molecular physics, where large quantum numbers are encountered, because the classical skeleton becomes more apparent as the quantum numbers increase; but it should be emphasized that the mathematical approximations involved also prove remarkably accurate even for the lowest quantum states.

The book has been written with graduate students in mind, and sets of problems are included to encourage its use as a graduate text. It is also intended partly as a reference work and each chapter has been written as far as possible to be self-contained. The main text tries to emphasize physical principles and practical applications, leaving certain important aspects to be treated in more mathematical detail in the appendices.

The writing of such a book is not lightly undertaken and it is a particular pleasure to acknowledge the hospitality of the Chemistry Department of the University of Colorado at Boulder and of the Laboratoire de Photophysique Moléculaire, Université de Paris Sud, where early drafts were written. My thanks are also due to those who have read the manuscript and made helpful and critical comments, in particular Drs Sally Chapman, Stephen Gray, Rick Gilbert, Don Noid, Steven Neshyba and Mr Anthony Ducket.

Acknowledgement is also due to the following for permission to reproduce copyright material: Academic Press Inc., American Institute of Physics, CRC Press Inc., Elsevier Science Publishers, Institute of Physics, Royal Society of Chemistry, Taylor and Francis Ltd and John Wiley and Sons Inc.

M. S. Child
Oxford
October 1990

While the early chapters of the first edition lay the foundations of semi-classical theory, the later ones cover extended applications to specific systems. Inevitably the emphasis of research in these latter areas has changed over the intervening twenty years. The present edition brings these applications up to date in two important respects. The entire chapter on reactive scattering has been rewritten, to highlight recent insight into the interpretation of differential cross-sections, geometric phase effects on chemical reactions and the instanton theory of chemical rate constants. The chapter on wave-packets has been moved and a new section is introduced to highlight the widely used Herman–Kluk propagator treatment of non-separable systems, with a related new appendix on properties of the classical trajectory monodromy matrix. Finally a new section on so-called 'quantum monodromy' has been added to the classical angle–action chapter, to reflect important recent work on topological impediments to the existence of a global system of angle–action variables, which have significant implications for the interpretation of highly excited bound molecular states.

At the same time I have become increasingly aware that many of the problems at the ends of the chapters, which are intended to stimulate the enthusiasm of the reader, are actually unreasonably hard for graduate students at the beginning of their careers. This new edition will lighten their task by adding a set of worked solutions.

It is a pleasure to acknowledge comments and advice on the new material by Stuart Althorpe, Jonathan Connor and Bill Miller. In addition, Rick Heller, Maksym Kryvohuz and David Manolopoulos kindly provided copies of their published diagrams. I am also grateful to Jessica White at Oxford University Press for her assistance, advice and encouragement in preparing the text.

Finally, in addition to the publishers listed above, acknowledgement for permission to reproduce copyright material is due to the American Physical Society and Springer Science and Business Media B.V.

M. S. Child
Oxford
November 2013

Contents

1 Introduction

1.1 Classical and quantum mechanical structures

Semiclassical mechanics is a short-wavelength link between classical and quantum mechanics, similar to that between wave and geometrical optics. Its objective is to maintain contact with the structure imposed by classical mechanics, without sacrificing quantum mechanical accuracy, or, in the vivid words of Berry and Mount (1972), to 'sew quantum mechanical flesh onto classical bones'. The overriding principles are that the classical phase space is divided into 'accessible' and 'inaccessible' regions characteristic of the type of motion, and that the boundaries of these regions impose a classical structure, which is evidenced in the quantum mechanics by interference in the bright accessible regions and exponential decay into the shadow regions. The strength of the classical skeleton is that this structure is maintained whether the observations are made in coordinate space, quantum number space, or some mixed representation. In all cases the central quantity is the classical action, whose dimensions are those of Planck's constant; on one hand it guides the classical motion, according to the minimum action principle, and on the other it adds a phase to each classical trajectory, which gives rise both to quantum mechanical interference and, by a mathematical extension, to quantum mechanical tunnelling.

To appreciate the main ideas, notice first that the free-motion wave-function, $\psi(x) = A\exp(ipx/\hbar)$, has a phase determined by the classical action, $\int p\mathrm{d}x$, but no interference is seen (i.e. $|\psi(x)|^2$ is independent of x) because all values of x are classically accessible, each being encountered only once along the classical trajectory. Interference arises from a classical boundary, of which the simplest is a classical turning point at $x = a$ say, where $p(a) = 0$. Regions to one side, say $x < a$, are no longer accessible and the folding back of the classical trajectory causes interference between the incoming and outgoing trajectory branches responsible for the familiar wavefunction oscillations, whose phases depend on the action, $\int_a^x p(x)\mathrm{d}x$. A further important point is that the impenetrable forbidden zone can be made accessible by the introduction of 'non-physical' imaginary momenta and imaginary actions, $\mathrm{i}\int_a^x |p(x)|\mathrm{d}x$, which govern the exponential fall-off of the wavefunction into these shadow regions. The resultant pattern of exponential decay on one side of a turning point and oscillatory behaviour on the other is qualitatively similar for all one-dimensional wavefunctions, with a form described by the function introduced by Airy (1838) to describe the appearance of supernumerary optical rainbows. The remarkable feature is that this Airy functional form is characteristic of any situation involving two interfering trajectories, where the quantity in question may be a wavefunction, a scattering amplitude, or

Semiclassical Mechanics with Molecular Applications. Second Edition. M. S. Child

a distribution of final state quantum numbers after a collision. Similarly, more complicated situations, involving several interfering trajectories, may be categorized by catastrophe theory (Thom 1975; Postern and Stewart 1978) according to the topology of the classical boundaries involved, and each catastrophe type has its own generic diffraction integral (Berry 1976; Connor et al. 1976). All these features—the existence of sharp classical boundaries between bright (classically accessible) and dark (forbidden) regions, penetration into the latter by 'non-physical' complex classical paths, and the dependence of the resulting interference pattern on topological features of the boundary—carry over to quite general contexts, regardless of whether the physical variables in question are coordinates, quantum numbers, or scattering angles.

Other important consequences flow from restrictions on the classical phase volume available to the system. At the simplest level, the existence of two classical turning points, a and b, sets up two counter-propagating wave trains, which can only be reconciled by restricting the available action, $\int_a^b p(x)\mathrm{d}x$, to particular values. In other words the quantum mechanical standing-wave condition leads directly to the Bohr quantization condition. The outcome of collisional or spectroscopic disturbances to such quantized systems is also affected by flux constraints arising from this action restriction. For example, the 'Condon reflection' relationship between the form of the spectral emission intensity in Fig. 1.1(a) and the functional form of the initial vibrational wavefunction in Fig. 1.1(b) may be seen as a direct consequence of the availability of $2.5h$ action units, although the distribution is in one case over a radial variable, R, and in the other over a quantum number v''. Similar 'reflection' behaviour is observed in a variety of collisional contexts. It is also evident from Fig. 1.1 that the accessible range of discrete quantum numbers in molecular systems is sufficiently dense to allow interpolation, in this case, for the dependence of the spectral intensity on the continuous classical action variable—an idea that is exploited in the direct inversion of spectroscopic data (see Chapter 6).

A third important connection between classical and quantum mechanics arises from the existence of a phase space separatrix between two or more types of classical motion. The case of chemical reaction is the most obvious, because it creates a clear separation between 'reactive' and 'non-reactive' classical motions, a division that is shaded in quantum mechanics by the possibility of tunnelling, but which broadly persists at the quantum level. By extension, classical trapping between two 'transition states' (or classical dividing lines) on the potential energy surface is known to give rise to the appearance of quantum mechanical resonances. Similar bifurcations also occur, with dramatic effect, in the coupled motions of bound systems. They correspond at the simplest level to the familiar Fermi resonances of molecular spectroscopy (Herzberg 1945), but they also provide the seeds of classical chaos (Tabor 1989). Possible quantum mechanical consequences to this chaos have been much discussed (Noid et al. 1981; Casati 1985; Berry et al. 1987; Ozorio de Almeida 1988; Gutzwiller 1990), but detailed discussion is taken as beyond the scope of this book. The remarkable connection between 'scars' on the wavefunction and periodic orbits should, however, be noted (Heller 1984).

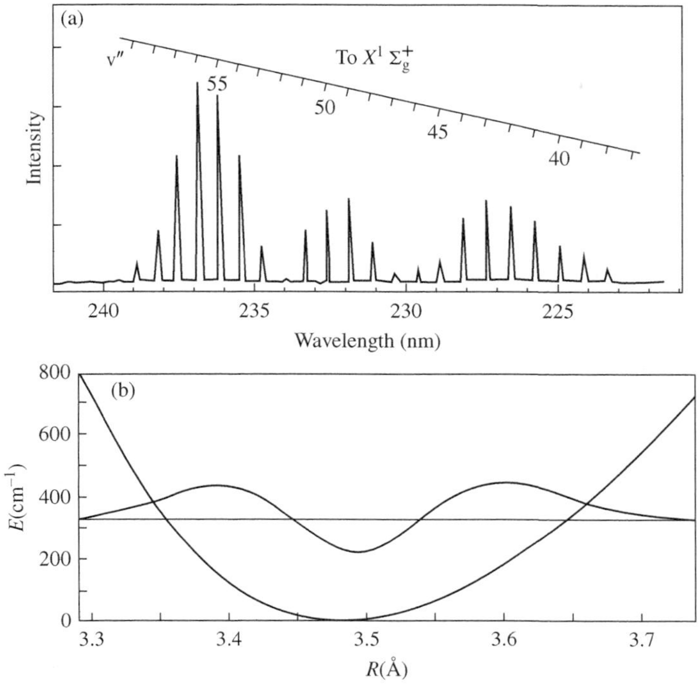

Fig. 1.1 Condon reflection relationship between (a) bands in the $F'(0_u^+) - X(^1\Sigma_g^+)$ emission system of I_2 and (b) the form of the initial $v' = 2$ vibrational wavefunction. (Taken in part from Ishiwata et al. (1984) with permission. Copyright 1984, AIP Publishing LLC.)

1.2 Historical perspectives

Historically, the first ideas date from the recognition by Bohr (1913) that quantization of the classical action (in equal units for all systems) can accommodate the wide variety of discrete energy level patterns shown by simple systems. In other words, the proper classical actions translate directly to equally spaced quantum numbers, although the energy levels themselves are strictly quantized (in the sense of having equal spacings) only for the case of a harmonic oscillator. Extension of the angle–action picture later led to a correspondence between classical Fourier components and quantum mechanical matrix elements, via an angle representation for the wave-function. The book by Born (1960) gives an excellent exposition from this old quantum theory viewpoint. There was also early recognition by Rydberg (1931) and Klein (1932) that the Bohr–Sommerfeld quantization rule offered a direct inversion route from measured energy levels back to the form of the potential energy function. At the same time, analysis of the nature of asymptotic solutions of the Schrödinger equation led, via the generalized de Broglie wavelength, $\lambda(x) = p(x)/h$, to the Jeffreys (1925), Wentzel (1926), Kramers (1926), Brillouin (1926a) (JWKB) connection between the wavefunction and the corresponding classical trajectory. What is now recognized as a powerful uniform improvement to this picture, due to Langer (1937), and the curve-crossing papers of Landau (1932), Zener (1932) and Stückelberg (1932), also date from this period.

Modern developments in the molecular field were stimulated by the famous rainbow scattering papers of Ford and Wheeler (1959) which showed how interfering classical trajectories can explain the intricate scattering patterns arising from perhaps several hundred quantum mechanical partial waves. A now seemingly natural, but then far from obvious, generalization of this interfering trajectory picture was the formulation of the classical S matrix, as was derived from a JWKB viewpoint by Marcus (1970) and by use of path integral methods (Feynman and Hibbs 1965) by Miller (1970, 1976) and Pechukas (1969). Not least among the surprises was the demonstration that 'classically forbidden' processes could become 'allowed' by integration of the classical equations of motion in complex time (Miller and George 1972; Stine and Marcus 1972). There have also been important technical advances in the treatment of singularities at the 'caustic' boundaries between classically allowed and classically forbidden events, which make the difference between a physically intuitive and a quantitatively reliable theory. The review by Berry and Mount (1972) was an important influence, and the ultimate outcome was an analysis of the relation between the topology of caustic boundaries and the nature of the generic interference patterns (Berry 1976; Connor 1976), derived from catastrophe theory (Thom 1975; Postern and Stewart 1978).

Important developments have also occurred in the treatment of bound states and the analysis of molecular spectroscopy. In separable situations a variety of direct inversion methods have exploited the fact that the quantization equation and formulae for relevant matrix elements or scattering cross-sections contain explicit reference to the underlying potential energy functions. The main thrust has been to extend the RKR (Rydberg 1931; Klein 1932; Rees 1946) technique to more complicated situations, but the somewhat different LeRoy and Bernstein (1970) analysis of near dissociation behaviour is also a significant advance.

Particular attention has been given in recent years to the treatment of non-separable bound states. Insights into the nature of possible classical motions come from the astronomical literature, which demonstrated an important distinction between regular and chaotic behaviour (Tabor 1989). This regularity is expressed (Percival 1977) in terms of restriction of the $2n$-dimensional motion to the surface of an $(n + 1)$-dimensional Einstein (1917) invariant torus, through which appropriate topologically distinct cuts yield as many closed phase space areas as there are degrees of freedom—these areas being the proper classical action integrals for the motion. The direct approach to quantization is therefore to adjust the trajectory starting conditions, by one means or another, to yield suitably quantized values of these actions (Einstein 1917; Brillouin 1926b; Keller 1958) (EBK). Complications can, however, occur in resonance situations due to non-classical (or generalized tunnelling) interactions between different modes of motion, in which case classical perturbation approaches have proved fruitful (Sibert et al. 1982; Swimm and Delos 1979). Some success has also been obtained by these methods in weakly chaotic situations (Skodje et al. 1985; Jaffé and Reinhardt 1982) but the most fundamental approach for strongly chaotic systems is to employ the Gutzwiller (1970) sum over periodic orbits, because these orbits remain even after all semblance of order is lost.

1.3 Scope and organization of the text

The text is presented as an introduction to the application of semiclassical methods in practical contexts, with particular reference to molecular physics. Brink (1985) covers complementary applications in nuclear physics. Emphasis is placed on physical principles, and those seeking a more detailed mathematical exposition are referred to Maslov (1972), Maslov and Fedoriouk (1981) and Delos (1986), supplemented by Jeffreys (1962), Heading (1962), Froman and Froman (1967) and Dingle (1973).

The main text includes a fairly full analytical discussion of each topic, as an aid to the student. Some or all of this material may prove suitable as a graduate text. Problem sets are included for most chapters, with worked solutions at the end of the book. Most are of an analytical 'prove that' variety, with reference on occasion to the related literature; others take the form of numerical exercises. Mathematical appendices on some specialized topics are also included. Each chapter has been written, as far as possible, to be self-contained for ease of reference.

Chapter 2 starts with an introduction to the first-order JWKB approximation, which is followed by analysis of behaviour in the neighbourhood of a classical turning point, where the approximation breaks down. The resulting connection between classically allowed and classically forbidden regions includes an important Maslov (1972) phase term due to reflection at the turning point, which plays a role throughout the book. An alternative, more powerful, procedure due to Langer (1937) and later popularized by Miller and Good (1953) is then shown to yield a globally valid uniform approximation in terms of special functions matched to the turning point structure in hand. A particular class of higher-order phase integral approximations (Froman 1966) is also described, whose strengths and weaknesses are tested against the pathological problem of a quartic oscillator.

Chapter 3 deals with quantization, first at a simple level in which the only complications arise from isolated classical turning points and singularities in the potential function. Subsequent sections examine more complicated tunnelling and curve-crossing problems, which are treated by means of JWKB connection formulae derived from the asymptotic properties of the exact solutions for locally valid model problems, whose mathematical properties are discussed in Appendix A. Explicit quantization conditions are derived for double minimum and restricted rotation problems and for the complex energies ($E = E_n - \mathrm{i}\Gamma_n/2$) relevant to resonance or predissociation phenomena, in which E_n are the resonance positions and Γ_n the level widths. The material in Chapter 4 is complementary in that it deals with quantization from Bohr's original angle–action viewpoint. Applications are given to degenerate and non-degenerate harmonic oscillators, angular momentum, the hydrogen atom and symmetric and asymmetric tops. The concept of 'quantum monodromy' as a possible topological impediment to the existence of a global angle–action system is also introduced.

A discussion of semiclassical matrix elements is included in Chapter 5, partly to illustrate particular types of spectroscopically observable interference structures and partly as a preliminary to the semiclassical inversion methods described in Chapter 6. Most of the latter involve variants of the RKR technique but they extend to tunnelling problems, inversion of molecular properties as functions of the quantum numbers, predissociation line-width fluctuations, oscillatory intensity variations and atom–atom scattering cross-sections. The LeRoy and Bernstein (1970) methods for linearizing Birge–Sponer extrapolations (Herzberg 1950) and analysing the near dissociation behaviour of rotational constants are also described.

Chapter 7 completes the discussion of bound states by reference to the quantization of non-separable systems, whose behaviour may be broadly classed as regular or chaotic (Percival 1977; Tabor 1989). Attention is concentrated on the former, because the quantum mechanical consequences of classical chaos are poorly understood at the time of writing, but a descriptive appendix is included on the transition to chaos, and certain of the quantization schemes designed for regular systems have been found to extend at least to weakly chaotic ones. Within the regular regime, as identified by restriction of the motion to the surface of an Einstein (1917) invariant torus, the methods described may be divided into two classes according to whether quantization is to be made at the EBK level (which is equivalent to Bohr–Sommerfeld quantization in one degree of freedom) or whether non-linear resonance effects demand generalized tunnelling corrections. Since the choice depends on the structure of the relevant phase space, Chapter 7 starts with an introduction to the diagnostic tool of the Poincaré (1892) section. Trajectory-based procedures seem to be preferred at the EBK level, some of which give particular insight into the structure of the phase space, while others are particularly valuable for systems with many degrees of freedom. Details are given in the text, but the Gutzwiller (1970) sum over periodic orbits is worth special mention as being applicable in principle to chaotic systems. Classical perturbation methods for the treatment of non-linear resonance complications are also described.

While the foregoing discussion is seen mainly from a time-independent viewpoint, Chapter 8 outlines the time-dependent wavepacket picture, along lines pioneered by Heller (1975) and coworkers. An account of the relationship between certain special wavepackets and those of equivalent classical ensembles is followed by the method of 'spectral quantization', which constructs the stationary state wavefunctions from a swarm of classically guided wavepackets (DeLeon and Heller 1984). The section concludes by outlining the powerful Herman–Kluk propagation technique for the treatment of high-dimensional systems.

The three remaining chapters deal with atomic and molecular collisions. Chapter 9 starts with a résumé of classical and quantum mechanical elastic scattering and goes on to outline the semiclassical treatment of rainbow and glory atom–atom scattering, exchange oscillations and the Stückelberg oscillations associated with non-adiabatic transitions. The extension of this interfering trajectory approach to inelastic scattering and photodissociation is covered by the account of classical S matrix theory in Chapter 10, in which one of the highlights is the use of complex time trajectories to handle classically forbidden events. Secondly, insight into the scattering process as phase space transformation also leads to the formulation of uniform approximations designed to eliminate spurious semiclassical singularities and to a generalization of the spectroscopic Condon reflection principle (Herzberg 1950) to scattering processes. Chapter 11 completes the main text by covering aspects of reactive scattering. This chapter has been completely rewritten for the present edition, to include sections on the interpretation of differential cross-sections, the relevance of geometric phase on reactive scattering and the instanton theory of chemical rate constants in the deep tunnelling regime.

Turning to points of mathematical detail, Appendix A explores the consequences of the so-called Stokes phenomenon associated with abrupt changes in the nature of the asymptotic solutions to Schrödinger-type equations. This offers insight into the origin of Maslov (1972) phase contributions and into limitations in the validity of JWKB representations in various contexts. The cases covered include isolated turning points, potential maxima and minima, curve-crossing situations and Demkov (1964) transitions.

The focus of Appendix B is on one of the most important technical advances in recent years—the 'uniform' elimination of semiclassical singularities arising from coalescence between two or more contributing classical trajectories. In mathematical terms such coalescence corresponds to repeated stationary points of relevant phase functions, which means that their topologies may be classified by catastrophe theory (Thom 1975; Postern and Stewart 1978). Moreover, the algebraic 'unfolding' functions, whose stationary point coalescences cause the catastrophes, also automatically define appropriate diffraction integrals that smooth out the semiclassical singularities. There are, however, certain situations in which the resulting generic class of catastrophe-based uniform approximations does not apply and two other forms are also discussed.

The third special topic, covered in Appendix C, is the semiclassical connection between unitary transformations in quantum mechanics and the canonical transformations of classical mechanics (Goldstein 1980; Percival and Richards 1982). This provides an opportunity to make contact with Feynman path integral methods (Feynman and Hibbs 1965), as used by Miller (1976) in his approach to the classical S matrix. A section on the geometrical origin of semiclassical approximations to $3j$ and $6j$ symbols is also included.

Appendix D charts stages in the transition from order to chaos, by use of the Henon (1969) map, which mimics the behaviour of dynamical systems. Points of particular importance concern island structures associated with non-linear resonances, the destruction of separatrices between different parts of the phase plane by homoclinic oscillations, and factors affecting the stability of different types of separatrix. Two further appendices, E and F, include reference material on specific angle–action transformations and on the properties of the trajectory monodromy matrix.

The book concludes in Appendix G with sets of worked solutions to the problems in the main text.

2
Phase integral approximations

The simplest representation for the quantum mechanical wavefunction in terms of classical quantities is due to Jeffreys (1925), Wentzel (1926), Kramers (1926) and Brillouin (1926a, b). It generalizes the free-motion form, $\psi = \exp(\mathrm{i}px/\hbar)$, by allowing for variation in the momentum $p(x)$ as a function of position, a correction that replaces the exponent by a phase integral, $\int p(x)\mathrm{d}x$, and introduces a normalization term related to the local velocity. A similar description may also be applied in classically forbidden regions by allowing the momentum to become imaginary, $p(x) = \mathrm{i}|p(x)|$, provided that the local de Broglie wavelength, $\lambda(x) = h/p(x)$, varies sufficiently slowly with x. Details are given in Section 2.1 and discussed at greater length by Froman and Froman (1967) and Froman (1980).

The validity criterion on this relatively primitive approximation is shown to imply a catastrophic breakdown around any classical turning point where $p(x) = 0$ and $\lambda(x) \to \infty$; and analysis of the wavefunction in such turning point regions leads, in Sections 2.2 and 2.3, to more powerful so-called uniform phase integral approximations, dependent on the turning point structure in hand. Section 2.4 gives a brief introduction to higher-order phase integral approximations that have been employed in the spectroscopic literature.

These ideas are illustrated below in a way that is intended to bring out some of the central themes of the general theory. In particular the equivalence between *primitive semiclassical* and *stationary phase* descriptions and their relation to the superior *uniform* approximations are of very general relevance. Two specific results, the Bohr quantization condition and the form of the JWKB phase shift, given by eqns (2.57) and (2.54) respectively, play important roles throughout the book.

2.1 The JWKB approximation

The relationship between classical and quantum mechanics is most directly introduced by expressing the Schrodinger equation in the form

$$\hbar^2(\mathrm{d}^2\psi/\mathrm{d}x^2) + p^2(x)\psi = 0 \tag{2.1}$$

where $p(x)$ is the classical momentum,

$$p(x) = \{2m[E - V(x)]\}^{1/2}. \tag{2.2}$$

If p is independent of x, the two independent solutions of (2.1) are obviously

$$\psi(x) = A\exp(\pm\mathrm{i}px/\hbar). \tag{2.3}$$

Semiclassical Mechanics with Molecular Applications. Second Edition. M. S. Child

Hence if $p(x)$ varies slowly with x, it is natural to attempt a solution in the more general exponential form

$$\psi(x) = A \exp[\pm \mathrm{i} S(x)/\hbar], \tag{2.4}$$

whose second derivative is given by

$$\mathrm{d}^2\psi/\mathrm{d}x^2 = [-S'^2/\hbar^2 \pm \mathrm{i}S''/\hbar]\psi, \tag{2.5}$$

where $S'(x)$ denotes $(\mathrm{d}S/\mathrm{d}x)$. An expansion in powers of $\hbar$,

$$S(x) = S_0(x) + \hbar S_1(x) + \hbar^2 S_2(x) + \ldots, \tag{2.6}$$

therefore yields

$$\begin{aligned}[-S_0^2 + p(x)] &+ \hbar[-2S_0'S_1' + \mathrm{i}S_0''] \\ &+ \hbar^2[-2S_0'S_2' - S_1'^2 + \mathrm{i}S_1''] \ldots = 0,\end{aligned} \tag{2.7}$$

after collecting terms according to powers of $\hbar$. The JWKB solution is obtained by equating to zero the successive coefficients of $\hbar^n$ in eqn (2.7). For example, to order $\hbar^0$,

$$S_0'^{\,2} = p^2(x), \tag{2.8}$$

so that

$$S_0(x) = \pm \int p(x)\mathrm{d}x. \tag{2.9}$$

Similarly

$$S_1'(x) = \mathrm{i}(S_0''/2S_0') \tag{2.10}$$

with solution

$$S_1(x) = (\mathrm{i}/2)\ln|S_0'| = (\mathrm{i}/2)\ln[p(x)], \tag{2.11}$$

while from the third term in eqn (2.7)

$$2S_2'(x) = -(S_0')^{-1/2}[\mathrm{d}^2(S_0')^{-1/2}/\mathrm{d}x^2]. \tag{2.12}$$

Two features of the first-order JWKB solution implied by (2.9) and (2.11) may be noted. In the first place the derivation places no requirement that the momentum $p(x)$ should be real. Hence in addition to the general classically allowed solution, applicable where $p^2(x) > 0$,

$$\psi = [p(x)]^{-1/2}\left[A' \exp\left(\frac{\mathrm{i}}{\hbar}\int_{x_0}^{x} p(x')dx'\right) + A'' \exp\left(-\frac{\mathrm{i}}{\hbar}\int_{x_0}^{x} p(x')dx'\right)\right], \tag{2.13}$$

or

$$\psi = A\,[p(x)]^{-1/2} \cos\left(\frac{\mathrm{i}}{\hbar}\int_{x_0}^{x} p(x')dx' + \alpha\right), \tag{2.14}$$

there is also a classically forbidden form

$$\psi = |p(x)|^{-1/2}\left[X\exp\left(\frac{1}{\hbar}\int_{x_0}^{x}|p(x')|dx'\right) + Y\exp\left(-\frac{1}{\hbar}\int_{x_0}^{x}|p(x')|dx'\right)\right], \tag{2.15}$$

applicable in regions where $p^2(x) < 0$, or $p(x) = \pm\mathrm{i}|p(x)|$.

Secondly, the term $[p(x)]^{-1/2}$ in eqn (2.13) normalizes the wavefunction in such a way that the flux density

$$j = -\frac{\mathrm{i}\hbar}{2m}\left(\psi^*\frac{\mathrm{d}\psi}{\mathrm{d}x} - \psi\frac{\mathrm{d}\psi^*}{\mathrm{d}x}\right) \tag{2.16}$$

is independent of x, to the extent that S_1' is ignored compared with S_0', because according to (2.13)

$$j = m^{-1}(|A'|^2 - |A''|^2). \tag{2.17}$$

Looked at in another way, the factor $p^{-1/2}(x)$ relates the amplitude of the wavefunction to the classical velocity in such a way that $\psi^2(x)\mathrm{d}x$ is proportional to the time spent in crossing the element $\mathrm{d}x$. One immediate consequence, which is analysed in detail below, is that the JWKB wavefunction diverges at any classical turning point, although it will be seen in later sections that other semiclassical approximations (e.g. the uniform approximations in Section 2.3 and the angle–action forms in Section 4.1) can be found to circumvent this difficulty.

To assess the validity of the present JWKB form, particularly in the neighbourhood of a turning point, it is important to recognize that eqn (2.6) is in fact an asymptotic approximation (Dingle 1973) whose successive terms typically decrease to a minimum and then increase in such a way that the most accurate approximation is obtained by truncation at the smallest term. Consequently the validity of eqns (2.13)–(2.15) requires that

$$|\hbar S_1'| \ll S_0', \tag{2.18}$$

or equivalently, in the light of (2.9) and (2.10),

$$\hbar|\mathrm{d}p/\mathrm{d}x| \ll |p^2(x)|. \tag{2.19}$$

Alternatively

$$|\mathrm{d}\lambda/\mathrm{d}x| \ll 2\pi \tag{2.20}$$

where $\lambda(x) = h/p(x)$.

Clearly eqn (2.19) breaks down at any classical turning point, where $p(x) = 0$, and the range of this breakdown may be assessed by assuming that

$$p^2(x) = 2mF(x - a), \tag{2.21}$$

consistent with a linear variation in the potential energy function $V(x)$. It is then readily verified that the inequality (2.19) is equivalent to

$$2(2mF/\hbar^2)^{1/2}|x - a|^{3/2} \gg 1. \tag{2.22}$$

One way to interpret this condition is to visualize the term $(x-a)$ as an uncertainty, Δx, in distance from the turning point, a, and the term $2(2mF)^{1/2}|x - a|^{1/2}$ as a classical momentum uncertainty, Δp, associated with the two possible signs of $p(x)$ given by (2.21). Hence (2.22) is equivalent to the uncertainty requirement

$$\Delta p \Delta x \gg \hbar. \tag{2.23}$$

For another view of the inequality, in relation to the form of the wave-function, eqn (2.22) is seen to imply that

$$\frac{1}{\hbar}\int_a^x |p(x)|\mathrm{d}x = \frac{2}{3}(2mF/\hbar^2)|x - a|^{3/2} \gg \frac{1}{3}, \tag{2.24}$$

which means that the primitive JWKB approximation may be expected to apply when the phase integral exceeds, say, 10/3 or approximately π, which corresponds roughly to the first zero beyond the turning point. Figure 2.1, which shows a comparison between the symmetric $[\psi(x) = \psi(-x)]$ JWKB form

$$\psi = A[p(x)]^{-1/2}\cos\left(\frac{1}{\hbar}\int_0^x p(x)\mathrm{d}x\right) \tag{2.25}$$

and the exact quantum mechanical wavefunction, confirms that this is the case for the $n = 0$ and $n = 2$ states of a harmonic oscillator.

One might suppose that the approximation would improve by including higher terms but it may be verified by the use of eqns (2.9)–(2.12) and (2.21) that in the linear potential case

$$\frac{2\hbar S_1'}{S_0'} = \frac{8}{5}\frac{\hbar S_2'}{S_1'} = \frac{\hbar}{(2mF)^{1/2}}(x - a)^{-3/2}. \tag{2.26}$$

Consequently the inequality $|\hbar S_1'| \ll |S_0''|$ ensures that $|\hbar S_2'| \ll |S_1'|$, which means that the breakdown of one strong inequality leads to breakdown of the other. No substantial improvement is therefore expected, in the case of a linear potential, by extending eqn (2.6) to the term in $S_2(x)$.

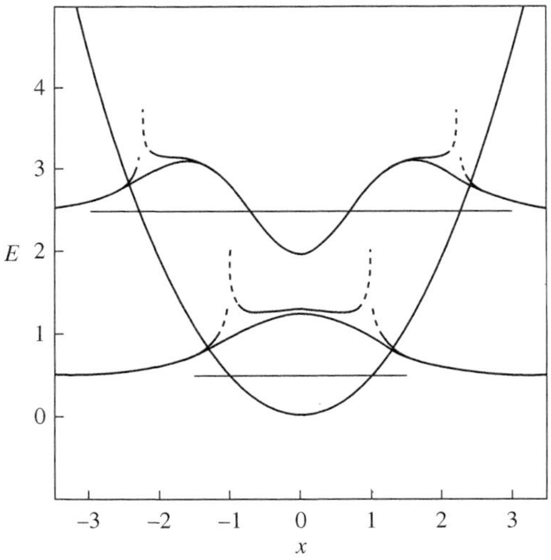

Fig. 2.1 Comparison between the exact (solid line) and JWKB (dashed line) for the $n = 0$ and $n = 2$ states of a harmonic oscillator.

The validity criterion (2.19) may also be used to estimate the energy E at which the wavefunction for motion above the quadratic barrier,

$$V = -\frac{1}{2}kx^2, \tag{2.27}$$

may be represented by a single JWKB term; in other words the energy above which reflection may be neglected. In this case

$$p^2(x) = m\kappa(x_E^2 + x^2), \tag{2.28}$$

where

$$x_E^2 = (2E/\kappa), \tag{2.29}$$

and the validity criterion becomes

$$f(x) = \frac{\hbar}{(m\kappa)^{1/2}} \frac{x}{(x_E^2 + x^2)^{3/2}} \ll 1. \tag{2.30}$$

Since the maximum value of $f(x)$ occurs at $x^2 = x_E^2/2$, it follows that there will be negligible reflection if

$$E \gg (3)^{-3/2}\hbar(\kappa/m)^{1/2}, \tag{2.31}$$

a prediction that accords well with the exact theory because (see eqn (3.54)) the true reflection probability at, say, $E = 10(3)^{-3/2}\hbar(\kappa/m)^{1/2}$ is only 5×10^{-6}.

2.2 Turning point behaviour

The purpose of this section is to relate the nature of the wavefunction in the neighbourhood of a turning point, as shown in Fig. 2.2, to the form of the associated classical phase space orbit and in doing so to determine the phase constant α in eqn (2.14). Put another way, the intention is to establish the relative phases of the two oscillatory terms in eqn (2.13) that connect with the single decreasing exponential in eqn (2.15)—a crucial result for the theory. The simplest model, involving a linear potential function

$$V(x) = E - F(x - a), \tag{2.32}$$

will be adopted, but this limitation will be relaxed in Section 2.3. A mathematically revealing alternative analysis of the phase behaviour in the complex coordinate plane is outlined in Appendix A.

Before turning to the quantum mechanical discussion it is convenient to realize that eqn (2.32) implies a classical orbit in Fig. 2.2 of the form

$$p = \pm p(x) = \pm[2mF(x - a)]^{1/2}, \tag{2.33}$$

and hence that the phase integrals in eqn (2.9) may be expressed as

$$S_0(x) = \pm \int_a^x p(x)\mathrm{d}x = \pm(2/3)(2mF)^{1/2}(x - a)^{3/2}, \tag{2.34}$$

where the turning point, a, has been chosen as the obvious phase reference point, x_0, in eqn (2.13).

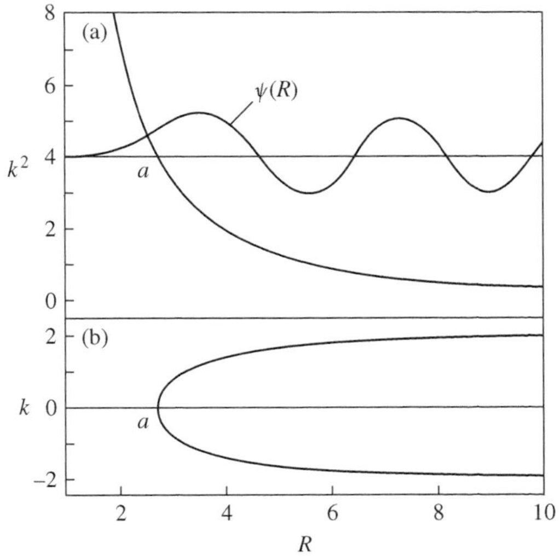

Fig. 2.2 Comparison between the wavefunction $\psi(R)$ and the phase space orbit $k(R)$, in the neighbourhood of a turning point, $R = a$. Note the exponential decrease of $\psi(R)$ into the classically inaccessible region, $R < a$.

The next step is to examine the form of the wavefunction. The usual Schrödinger equation for this problem

$$\left(\frac{1}{2m}\hat{p}^2 - F(x-a)\right)\psi = 0, \tag{2.35}$$

where $\hat{p} = -\mathrm{i}\hbar(\mathrm{d}/\mathrm{d}x)$, cannot be solved in closed form, but one can express $\psi(x)$ as the Fourier transform of its momentum space counterpart $\varphi(p)$ (Landau and Lifshitz 1965)

$$\psi(x) = \int_{-\infty}^{\infty} \varphi(p)\exp(\mathrm{i}px/\hbar)\mathrm{d}p, \tag{2.36}$$

where $\varphi(p)$ satisfies the first-order equation

$$\left(\frac{1}{2m}p^2 - F(\hat{x}-a)\right)\varphi = 0, \tag{2.37}$$

with $\hat{x} = \mathrm{i}\hbar(\mathrm{d}/\mathrm{d}p)$. The solution of eqn (2.37), namely

$$\varphi(p) = C\exp\left[-\frac{\mathrm{i}}{\hbar}\left(\frac{1}{6mF}p^3 + ap\right)\right], \tag{2.38}$$

therefore yields the integral representation

$$\psi(x) = C\int_{-\infty}^{\infty} \exp\left[-\frac{\mathrm{i}}{\hbar}\left(\frac{1}{6mF}p^3 - p(x-a)\right)\right]\mathrm{d}p, \tag{2.39}$$

which may be reduced by suitable scaling to the form of an Airy function (Abramowitz and Stegun 1965),

$$Ai[-\alpha(x-a)] = \frac{1}{2\pi}\int_{-\infty}^{\infty} \exp\left(\frac{iu^3}{3} - \mathrm{i}u\alpha(x-a)\right)\mathrm{d}u. \tag{2.40}$$

It is, however, more convenient for present purposes to retain the physical form in eqn (2.39).

The interesting feature is that $\psi(x)$ appears as an integral over all p, whereas the classical motion allows only two momentum values, $p = \pm p(x)$, as given by (2.33). The semiclassical connection between these two pictures is established by use of a saddle point or stationary phase (Jeffreys and Jeffreys 1956) approximation to $\psi(x)$. The argument is most compactly presented by expressing (2.39) in the form

$$\psi(x) = C\int_{-\infty}^{\infty} \exp[f(p)]\mathrm{d}p, \tag{2.41}$$

where

$$f(p) = -\frac{\mathrm{i}}{\hbar}\left(\frac{1}{6mF}p^3 - p(x-a)\right). \tag{2.42}$$

The mathematical technique is to distort the integration path into the complex p plane in such a way as to concentrate the variation of the integrand into one or more roughly Gaussian-shaped profiles. The associated physical implications are most revealing. Consider first the case $x > a$, and note that $f(p)$ is stationary with respect to p at the 'classical' points given by eqn (2.33)

$$p = p_{\pm} = \pm[2mF(x-a)]^{1/2} = \pm p(x). \tag{2.43}$$

Secondly the values of $f(p)$ at these points are given by the phase integrals in eqn (2.34), because

$$f(p_{\pm}) = f_{\pm} = \pm(\mathrm{i}/\hbar)(2/3)(2mF)^{1/2}(x-a)^{3/2} = \pm\frac{\mathrm{i}}{\hbar}\int_a^x p(x)\mathrm{d}x. \tag{2.44}$$

Finally, the second derivatives are purely imaginary

$$f''(p_{\pm}) = f''_{\pm} = \mp \mathrm{i}p(x)/\hbar mF. \tag{2.45}$$

It is therefore convenient to perform a quadratic expansion about, say, the point p_+ and to substitute

$$p - p_+ = (2/|f''_+|)^{1/2}\exp(-\mathrm{i}\pi/4)q, \tag{2.46}$$

so that

$$\exp[f(p)] \simeq \exp[f(p_+)]\mathrm{e}^{-q^2}. \tag{2.47}$$

Real variations of q, corresponding to motion along the path C_+ in Fig. 2.3(a), therefore give rise to the desired Gaussian form, and the contribution $\psi_+(x)$ from this segment of the path becomes

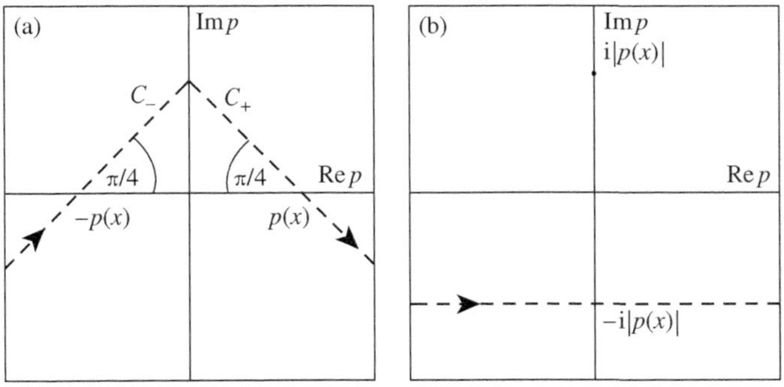

Fig. 2.3 Integration paths in the complex momentum plane appropriate to (a) classically allowed and (b) classically forbidden x values. The points $\pm p(x)$ and $\pm\mathrm{i}|p(x)|$ are saddle points of the integrand in eqn (2.41).

$$
\begin{aligned}
\psi_+(x) &= C\int_{C_+} \exp[f(p)]\mathrm{d}p \\
&\simeq \left[\frac{2C^2}{|f''_+|}\right]^{1/2} \exp\left(f_+ - \mathrm{i}\frac{\pi}{4}\right)\int_{-\infty}^{\infty} \mathrm{e}^{-q^2}\mathrm{d}q \\
&= [2\pi C^2/|f''_+|]^{1/2}\exp\left(f_+ - \mathrm{i}\frac{\pi}{4}\right) \\
&= C'[p(x)]^{-1/2}\exp\left(\frac{\mathrm{i}}{\hbar}\int_a^x p(x)\mathrm{d}x - \mathrm{i}\frac{\pi}{4}\right).
\end{aligned} \tag{2.48}
$$

A similar argument yields a contribution $\psi_-(x)$ equal to the complex conjugate of $\psi_+(x)$, from the path C_-. Finally, the directions of the steepest descent paths in Fig. 2.3(b) valid for $x < a$ show that the appropriate path C_0 should pass only through

$$
p = -\mathrm{i}|p(x)|, \tag{2.49}
$$

so that one obtains the single term

$$
\psi(x) \overset{x \lll a}{\approx} C'|p(x)|^{-1/2}\exp\left(-\frac{1}{\hbar}\int_x^a |p(x)|\mathrm{d}x\right). \tag{2.50}
$$

Overall this asymptotic analysis therefore recovers the forms for $\psi_\pm(x)$ in (2.13) but with the added benefit that their relative phases are determined to connect properly with the classically forbidden form in eqn (2.50):

$$
\begin{aligned}
\psi(x) \overset{x \ggg a}{\sim} \psi_+(x) + \psi_-(x) &= 2C'[p(x)]^{-1/2}\cos\left(\frac{1}{\hbar}\int_a^x p(x)\mathrm{d}x - \frac{\pi}{4}\right) \\
&= A[p(x)]^{-1/2}\sin\left(\frac{1}{\hbar}\int_a^x p(x)\mathrm{d}x + \frac{\pi}{4}\right).
\end{aligned} \tag{2.51}
$$

Similar reasoning leads to the conclusion that

$$
\psi(x) \overset{x \lll b}{\sim} B[p(x)]^{-1/2}\sin\left(\frac{1}{\hbar}\int_x^b p(x)\mathrm{d}x + \frac{\pi}{4}\right), \tag{2.52}
$$

to the left of a right-hand turning point at $x = b$. Note the order of integration limits in eqn (2.52).

Not surprisingly the validity criteria for eqns (2.51) and (2.52) are the same as for the JWKB approximation, but the argument is slightly different. One sees that the approximations for $\psi_\pm(x)$ in terms of the complete Gaussian integrals require that the stationary phase points $p = p(x)$ and $p = -p(x)$ should be sufficiently well separated that $q^2 \gg 1$ at the point $p = \mathrm{i}p(x)$ where the contours $C_\pm$ intersect. In other words, on combining eqns (2.45) and (2.46),

$$
q^2 = p^2(x)/f''(p_+) = p^3(x)/\hbar mF = p^2(x)/\hbar p'(x) \gg 1, \tag{2.53}
$$

where the latter inequality coincides with eqn (2.19).

The importance of eqns (2.51) and (2.52) is that they determine the absolute phase of the wavefunction in regions where eqn (2.53) is satisfied. Hence they immediately lead to an expression for the scattering phase shift, δ, in the standard asymptotic representation

$$\psi(x) \overset{x\to\infty}{\sim} A\sin(kx+\delta), \tag{2.54}$$

namely

$$\delta = \lim_{x\to\infty}\left[\frac{1}{\hbar}\left(\int_a^\infty p(x)\mathrm{d}x - p(\infty)x\right)+\frac{\pi}{4}\right], \tag{2.55}$$

where $p(\infty) = k\hbar$ is the asymptotic momentum. Thus $\delta - \pi/4$ may be interpreted as the negative of the shaded area in Fig. 2.4(a).

A second important implication, applicable to bound states, is that eqns (2.51) and (2.52) can simultaneously apply only at specific energies because their equivalence requires that

$$A\,\cos\left(\frac{1}{\hbar}\int_a^x p(x)\mathrm{d}x - \frac{\pi}{4}\right) = B\,\sin\left(\frac{1}{\hbar}\int_x^b p(x)\mathrm{d}x + \frac{\pi}{4}\right), \tag{2.56}$$

an identity that can only apply if $B = (-1)^v A$ and

$$\frac{1}{\hbar}\int_a^b p(x)\mathrm{d}x = \left(v+\frac{1}{2}\right)\pi. \tag{2.57}$$

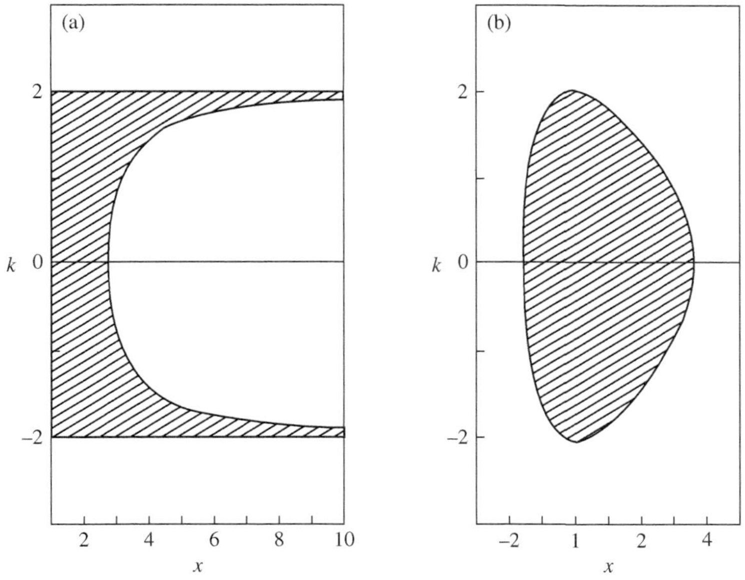

Fig. 2.4 The (shaded) phase space areas relevant to (a) the semiclassical phase shift in eqn (2.55) and (b) the quantization condition (2.58).

Alternatively

$$\oint p(x)\mathrm{d}x = 2\int_a^b p(x)\mathrm{d}x = \left(v + \frac{1}{2}\right)h, \tag{2.58}$$

which is the famous Bohr quantization condition, dependent on the classical action (see Chapter 4), which is equal to the enclosed area in Fig. 2.4(b).

This Bohr–Sommerfeld quantization rule is remarkably robust even for the lowest energy states of many systems, although the JWKB wavefunctions themselves suffer catastrophic failures at the classical turning points, a paradox that is addressed in Sections 2.3, 3.1 and 4.1. It is, for example, exact for all states of the harmonic oscillator, for which eqn (2.58) takes the form

$$\begin{aligned}(v+1/2)\pi &= \frac{1}{\hbar}\int_{-x_0}^{x_0}\left[2m(E-\frac{1}{2}x^2)\right]^{1/2}\mathrm{d}x\\ &= \frac{(mk)^{1/2}}{\hbar}\int_{-x_0}^{x_0}(x_0^2-x^2)^{1/2}\mathrm{d}x\\ &= \frac{(mk)^{1/2}}{\hbar}\frac{\pi x_0^2}{2},\end{aligned} \tag{2.59}$$

where $x_0^2 = (2E/k)$. Consequently, after minor rearrangement,

$$E = (v + 1/2)\hbar\,(m/k)^{1/2}. \tag{2.60}$$

A similarly exact result is obtained for the Morse oscillator (see eqn (3.9)) and for the hydrogen atom after introduction of the Langer (1937) substitution $\ell(\ell+1) \to (\ell + 1/2)^2$, for reasons that are indicated in the following section and discussed at greater length in Section 3.1. There are, however, also cases in which first-order quantization, by eqn (2.58), is seriously inaccurate (see Section 2.4).

2.3 Uniform approximations

The argument in the preceding section, whereby the JWKB form (2.51) emerges as an asymptotic approximation to a certain integral, will now be reversed to obtain a powerful family of uniform approximations, first suggested by Langer (1937) and later extended by Miller and Good (1953).

The underlying idea is that the qualitative shape of the wavefunction is dictated by the disposition of its classical turning points. Thus, Fig. 2.5 compares (a) the true wavefunction and its phase orbit with (b) those of a comparison equation with a single turning point. Classical motion bound from the left gives rise to a wavefunction that oscillates in the classical region and dies away exponentially into the forbidden one; similarly, bound motion between two turning points is represented by a function that oscillates between them and dies away outside. From here it is a short step to speculate that the whole family of possible single turning point wavefunctions might be expressible in terms of a single canonical function taken here as the Airy function shown in Fig. 2.5(b). Figure 2.6 shows a similar mapping between (a) a bound state

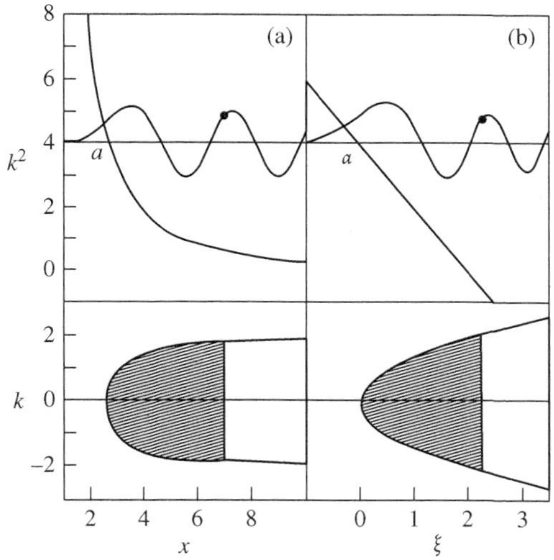

Fig. 2.5 The phase space (equal shaded area) mapping used to relate (a) the solution of (2.61) with $k^2 = 4.0 + (5.5)^2/x^2$ to (b) the solution of the comparison equation (2.63) with $\kappa^2(\xi) = \xi$. The circles indicate corresponding values of $\psi(x)$ and $\Psi(\xi)$.

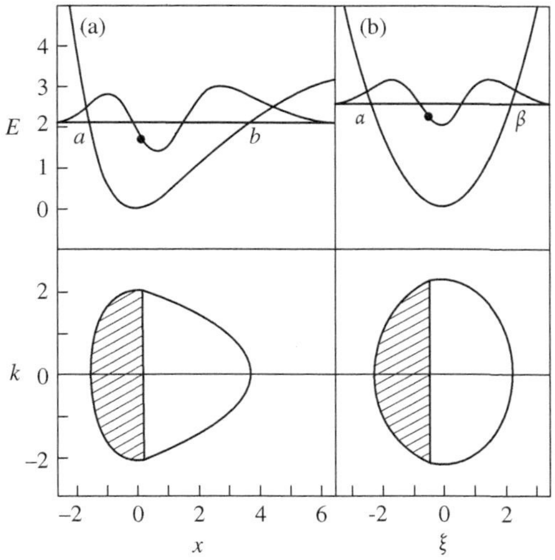

Fig. 2.6 The phase space (equal shaded area) mapping used to relate (a) the $n = 2$ wavefunction for a seven level Morse oscillator to (b) the $n = 2$ scaled harmonic oscillator wavefunction. The circles indicate corresponding values of $\psi(x)$ and $\Psi(\xi)$.

(two turning point) wavefunction and its phase orbit and (b) those of a harmonic oscillator. The nodal structure in panel (a) is well reproduced. By contrast the JWKB approximation attempts to force a sinusoidal or exponential form throughout; and the turning point singularities simply reflect the consequences of this inappropriate choice of canonical form.

To see how this mapping between the true wavefunction $\psi(x)$ and a canonical function $\Psi(\xi)$ is implemented, suppose that the solution of

$$\left(\frac{\mathrm{d}^2}{\mathrm{d}x^2} + k^2(x)\right)\psi = 0, \tag{2.61}$$

where $k^2(x) = p^2(x)/\hbar^2$, is sought in the form

$$\psi_{\text{app}}(x) = A(x)\Psi(\xi(x)) \tag{2.62}$$

where $\Psi(\xi)$ satisfies a suitable comparison equation

$$\left(\frac{\mathrm{d}^2}{\mathrm{d}\xi^2} + \kappa^2(\xi)\right)\Psi = 0 \tag{2.63}$$

and the functions $A(x)$ and $\xi(x)$ remain to be determined.

The forms of the primitive JWKB expressions

$$\begin{aligned}\psi(x) &= k^{-1/2}(x)\exp\left(\mathrm{i}\int_{x_0}^{x} k(x')\mathrm{d}x'\right)\\ \Psi(\xi) &= \kappa^{-1/2}(\xi)\exp\left(\mathrm{i}\int_{\xi_0}^{\xi} \kappa(\xi')\mathrm{d}\xi'\right)\end{aligned} \tag{2.64}$$

suggest a solution such that

$$A(x) = [\kappa(\xi)/k(x)]^{1/2}, \tag{2.65}$$

with $\xi(x)$ implicitly defined by the equation

$$\int_{\xi_0}^{\xi} \kappa(\xi)\mathrm{d}\xi = \int_{x_0}^{x} k(x)\mathrm{d}x. \tag{2.66}$$

In fact the combination of (2.62), (2.63) and (2.65) plus the following identity implied by (2.66),

$$(\mathrm{d}\xi/\mathrm{d}x) = k(x)/\kappa(\xi), \tag{2.67}$$

means that $\psi_{\text{app}}(x)$ satisfies

$$\left(\frac{\mathrm{d}^2}{\mathrm{d}x^2} + k^2(x) + \gamma(x)\right)\psi_{\text{app}} = 0, \tag{2.68}$$

which is identical with eqn (2.61) apart from the error term

$$\gamma(x) = -\left(\frac{\mathrm{d}\xi}{\mathrm{d}x}\right)^{1/2}\frac{\mathrm{d}^2}{\mathrm{d}x^2}\left(\frac{\mathrm{d}\xi}{\mathrm{d}x}\right)^{-1/2} = \left(\frac{k(x)}{\kappa(\xi)}\right)^{1/2}\frac{\mathrm{d}^2}{\mathrm{d}x^2}\left(\frac{k(x)}{\kappa(\xi)}\right)^{-1/2}. \tag{2.69}$$

The validity of this approximation therefore depends on the inequality

$$k^2(x) \gg \gamma(x), \tag{2.70}$$

which can be made much weaker than the JWKB criterion, (2.19), by careful choice of the comparison equation (2.63).

The most obvious requirement is that $k(x)/\kappa(\xi)$ should vary slowly with x, a condition that is well satisfied in any JWKB region where $k(x)$ and $\kappa(\xi)$ are individually slowly varying. A similar slow variation can also be maintained in the turning point region provided that the definition of $\xi(x)$ by means of eqn (2.66) ensures that $k(x)$ and $\kappa(\xi)$ vanish simultaneously. This means that the lower integration limits ξ_0 and x_0 must be taken as corresponding turning points. Similarly, in situations involving several turning points, $(\xi_0, \xi_1, \ldots)$ and $(x_0, x_1, \ldots)$, the function $\kappa(\xi)$ in the comparison equation must contain sufficient adjustable parameters to ensure that

$$\int_{\xi_r}^{\xi_{r+1}} \kappa(\xi) d\xi = \int_{x_r}^{x_{r+1}} k(x)\mathrm{d}x. \tag{2.71}$$

Notice that this turning point correspondence, combined with the necessarily linear variations of $k^2(x)$ and $\kappa^2(\xi)$ in the turning point region, makes their ratio $k(x)/\kappa(\xi)$ constant. Furthermore the first contribution to the error term comes from quadratic variations in $k(x)/\kappa(\xi)$ with respect to x. Hence there are relatively few situations in which $\gamma(x)$ becomes sufficiently large to violate eqn (2.70), although one important example is discussed at the end of the section.

Isolated turning points

The natural comparison equation for an isolated turning point involves the choice $k^2(\xi) = \xi$ with the turning point at $\xi_0 = 0$. The resulting equation

$$\left(\frac{\mathrm{d}^2}{\mathrm{d}x^2} + \xi\right)\Psi = 0 \tag{2.72}$$

is satisfied by the Airy function (Abramowitz and Stegun 1965)

$$\Psi(\xi) = Ai(-\xi), \tag{2.73}$$

and the function $\xi(x)$ is given according to (2.66) by

$$\int_{\xi_0}^{\xi} \kappa(\xi)\mathrm{d}\xi = \int_{\xi_0}^{\xi} \xi^{1/2}\mathrm{d}\xi = \frac{2}{3}\xi^{3/2} = \int_{x_0}^{x} k(x)\mathrm{d}x, \tag{2.74}$$

so that

$$\xi(x) = \left(\frac{3}{2}\int_{x_0}^{x} k(x)\mathrm{d}x\right)^{2/3}. \tag{2.75}$$

The combined result for $\psi_{\text{app}}(x)$ is therefore

$$\psi_{\text{app}}(x) = [\xi(x)]^{1/4} k^{-1/2}(x) Ai[-\xi(x)]. \tag{2.76}$$

Figure 2.7 shows that it gives a strikingly accurate approximation even in situations where the potential $V(x)$ is markedly non-linear.

One also finds from the known asymptotic forms of $Ai(-\xi)$ (see Abramowitz and Stegun (1965) and Section 2.2 above),

$$\begin{aligned} Ai(-\xi) &\overset{\xi>1}{\approx} \pi^{-1/2}\xi^{-1/4}\sin\left(\frac{2}{3}\xi^{3/2} + \pi/4\right) \\ &\overset{\xi<-1}{\approx} \frac{1}{2}\pi^{-1/2}|\xi|^{-1/4}\exp\left[-\frac{2}{3}(-\xi)^{3/2}\right], \end{aligned} \tag{2.77}$$

that $\psi_{\text{app}}(x)$ goes over to the forms in (2.50) and (2.51)

$$\begin{aligned} \psi_{\text{app}} &\overset{x\gg x_0}{\approx} \pi^{1/2}k^{-1/2}(x)\sin\left(\int_{x_0}^{x} k(x)\mathrm{d}x + \frac{\pi}{4}\right) \\ &\overset{x\ll x_0}{\approx} \frac{1}{2}\pi^{1/2}|k^{-1/2}(x)|\exp\left(-\int_{x}^{x_0}|k(x)|\mathrm{d}x\right), \end{aligned} \tag{2.78}$$

although the limitation to a linear potential $V(x)$ has been relaxed.

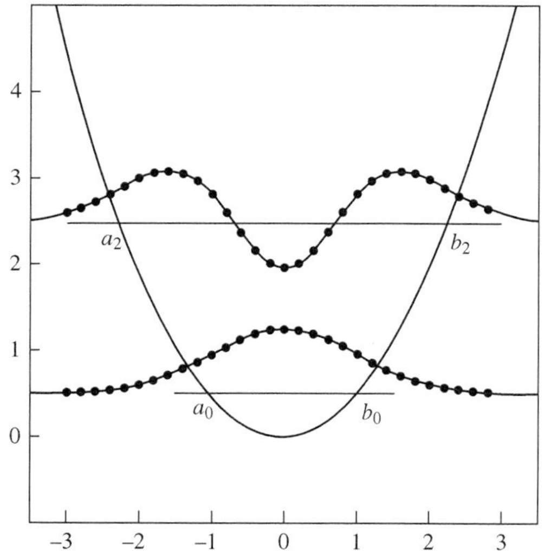

Fig. 2.7 Comparison between the exact (solid line) and uniform Airy (points) wavefunctions for the $n = 0$ and $n = 2$ states of a scaled harmonic oscillator. Note the absence of singularities, such as those in Fig. 2.1.

Notice that care is necessary in handling eqn (2.75) for classically forbidden values, $x < x_0$. $k(x)$ must be written as

$$k(x) = |k(x)| \exp(\mathrm{i}\pi/2) = -|k(x)| \exp(3\mathrm{i}\pi/2) \tag{2.79}$$

in order to ensure that $\xi(x)$ is real and negative; thus

$$\begin{aligned} \xi(x) &= \left[-\frac{3}{2} \int_{x_0}^{x} |k(x)| \mathrm{d}x \exp(3\mathrm{i}\pi/2) \right]^{2/3} \\ &= -\left(\frac{3}{2} \int_{x}^{x_0} |k(x)| \mathrm{d}x \right)^{2/3}, \end{aligned} \tag{2.80}$$

because $\mathrm{e}^{\mathrm{i}\pi} = -1$.

The first of eqns (2.78) is of some importance because it shows that justification for the phase correction term, $\pi/4$, responsible for the $\frac{1}{2}$ in the quantization condition (2.58), is no longer restricted to the linear potential approximation. Equation (2.76) will also prove useful in later chapters by providing an accurate semiclassical approximation for $\psi(x)$ in the turning point region.

Potential well

In the case of a potential well the simplest comparison equation is that for a scaled harmonic oscillator

$$\left(-\frac{1}{2} \frac{\mathrm{d}^2}{\mathrm{d}\xi^2} + \frac{1}{2} \xi^2 \right) \Psi_v = \left(v + \frac{1}{2} \right) \Psi_v, \tag{2.81}$$

for which (Messiah 1961)

$$\begin{aligned} \Psi_v(\xi) &= \pi^{-1/4} (2^v v!)^{-1/2} H_v(\xi) \exp\left(-\frac{1}{2} \xi^2 \right) \\ \kappa(\xi) &= (2v + 1 - \xi^2)^{1/2}. \end{aligned} \tag{2.82}$$

Hence if the turning points for the real problem are denoted a and b, and those for the comparison equation are $\pm\xi_0 = \pm(2v+1)^{1/2}$, solutions of (2.61) are restricted by virtue of eqn (2.71) to energies at which

$$\begin{aligned} \int_a^b \kappa(x) \mathrm{d}x &= \int_{-\xi_0}^{\xi_0} \kappa(\xi) \mathrm{d}\xi = \int_{-\xi_0}^{\xi_0} (\xi_0^2 - \xi^2)^{1/2} \mathrm{d}\xi \\ &= \xi_0^2 \pi/2 = \left(v + \frac{1}{2} \right) \pi, \end{aligned} \tag{2.83}$$

which is a much stronger justification for the Bohr quantization (2.57) than that given in Section 2.2. The physical wavefunction is given, according to eqns (2.62), (2.65) and (2.66), by

$$\psi_v(x) = [2v + 1 - \xi^2(x)]^{1/2} k^{-1/2}(x) \Psi_v[\xi(x)], \tag{2.84}$$

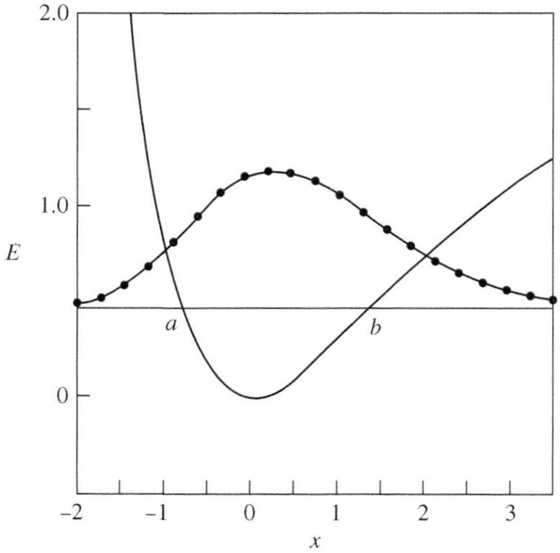

Fig. 2.8 Comparison between the exact (solid line) and uniform harmonic (dots) wavefunctions for the $n = 0$ state of a scaled Morse oscillator with seven bound levels. The uniform approximation has been scaled by a factor of 0.95 because normalization is not preserved in this representation.

with $\xi(x)$ defined by the transcendental equation

$$\int_{-\xi_0}^{\xi} (\xi_0^2 - \xi^2)\mathrm{d}\xi = \frac{1}{2}\xi[\xi_0^2 - \xi^2(x)]^{1/2} + \frac{\xi_0^2}{2}\left[\frac{\pi}{2} + \arcsin\left(\frac{\xi}{\xi_0}\right)\right]$$
$$= \int_a^x k(x)\mathrm{d}x. \quad (2.85)$$

Figure 2.8 shows excellent agreement between the approximation in (2.84) and the exact wavefunction, even for a strongly anharmonic potential.

Potential barrier

The final illustration of this uniform technique concerns barrier penetration, using the scaled quadratic model as a comparison equation:

$$\left(-\frac{1}{2}\frac{\mathrm{d}^2}{\mathrm{d}\xi^2} - \frac{1}{2}\xi^2\right)\Psi = \varepsilon\Psi \quad (2.86)$$

Properties of the exact solution are discussed in Appendix A. The turning points in this case lie at

$$\xi = \pm\xi_0 = \pm(-2\varepsilon)^{1/2} \quad (2.87)$$

which means that they are purely imaginary for energies above the barrier, $\varepsilon > 0$. Similarly it may be assumed that the turning points b and c for the real problem can also be followed to complex conjugate points in the complex plane as the energy rises above the potential maximum. Hence the turning point correspondence equation (2.71) can be used to define an effective tunnelling parameter ε by means of the identity

$$\int_b^c k(x)\mathrm{d}(x) = \int_{-\xi_0}^{\xi_0} \kappa(\xi)\mathrm{d}\xi = \int_{-\xi_0}^{\xi_0} (\xi^2 - \xi_0^2)^{1/2}\mathrm{d}\xi = -\mathrm{i}\pi\varepsilon, \tag{2.88}$$

in which it is assumed that the branches of $k(x)$ and $\kappa(\xi)$ are chosen consistently. As written, eqn (2.88) implies that $\arg[k(x)] = \arg[\kappa(\xi)] = \pi/2$ for $\varepsilon < 0$ and that ξ_0 and c lie in the upper, and $-\xi_0$ and b in the lower, half complex plane for $\varepsilon > 0$.

It is evident from eqn (2.86) that the properties of $\Psi(\xi)$ must depend only on ε, and it is shown in Appendix A that the coefficients in the asymptotic forms

$$\begin{aligned}\Psi(\xi) &\overset{\xi\to\infty}{\approx} \kappa^{-1/2}(\xi)\left[C'\exp\left(\mathrm{i}\int_\gamma^\xi \kappa(\xi)\mathrm{d}\xi\right) + C''\exp\left(-\mathrm{i}\int_\gamma^\xi \kappa(\xi)\mathrm{d}\xi\right)\right]\\ &\overset{\xi\to-\infty}{\approx} \kappa^{-1/2}(\xi)\left[B'\exp\left(\mathrm{i}\int_\beta^\xi \kappa(\xi)\mathrm{d}\xi\right) + B''\exp\left(-\mathrm{i}\int_\beta^\xi \kappa(\xi)\mathrm{d}\xi\right)\right]\end{aligned} \tag{2.89}$$

are related by a matrix of the form

$$\begin{pmatrix} C' \\ C'' \end{pmatrix} = \begin{pmatrix} (1+\kappa^2)^{1/2}\mathrm{e}^{-\mathrm{i}\phi} & -\mathrm{i}\kappa \\ \mathrm{i}\kappa & (1+\kappa^2)^{1/2}\mathrm{e}^{\mathrm{i}\phi} \end{pmatrix} \begin{pmatrix} B' \\ B'' \end{pmatrix} \tag{2.90}$$

where

$$\begin{aligned}\kappa &= \exp(-\pi\varepsilon)\\ \phi &= \arg\Gamma\left(\frac{1}{2} + \mathrm{i}\varepsilon\right) \ln|\varepsilon| + \varepsilon,\end{aligned} \tag{2.91}$$

and the phase reference points are taken as $\beta = -(-2\varepsilon)^{1/2}$, $\gamma = (-2\varepsilon)^{1/2}$ for $\varepsilon < 0$ and $\beta = \gamma = 0$ for $\varepsilon \geq 0$.

The remarkable feature of the uniform approximation is that eqns (2.62)–(2.66) impose precisely the same asymptotic structure on the general wavefunction

$$\begin{aligned}\psi(x) &\overset{x\to\infty}{\approx} k^{-1/2}(x)\left[C'\exp\left(\mathrm{i}\int_c^x k(x)\,\mathrm{d}x\right) + C''\exp\left(-\mathrm{i}\int_c^x k(x)\,\mathrm{d}x\right)\right]\\ &\overset{x\to-\infty}{\approx} k^{-1/2}(x)\left[B'\exp\left(\mathrm{i}\int_b^x k(x)\,\mathrm{d}x\right) + B''\exp\left(-\mathrm{i}\int_b^x k(x)\,\mathrm{d}x\right)\right]\end{aligned} \tag{2.92}$$

with the coefficients again related by eqn (2.90) and with the phase reference points at energies above the barrier maximum taken as

$$b = c = \frac{1}{2}(x_+ + x_-), \tag{2.93}$$

where $x_\pm$ are the complex points at which $k^2(x) = 0$.

Equations (2.90)–(2.92) will play an important role in the treatment of tunnelling problems in Chapter 3. It should also be noted that other tunnelling connection formulae, analogous to (2.90) but based on comparison with non-quadratic barriers, are available in the literature (Dickinson 1970).

Behaviour at a potential singularity

A number of applications give rise to singularities at the origin, such that

$$k^2 \overset{x\to 0}{\approx} k_0^2 \alpha x^{-n}, \tag{2.94}$$

in which case the error term $\gamma(x)$, given by eqn (2.69), also becomes singular, unless $\kappa(\xi)$ also has a built-in singularity at ξ_0, which is forced to correspond with $x = 0$ by extension of (2.66). In practice, this may often prove a severe constraint. As an alternative to forcing this correspondence, one may consider replacing $k^2(x)$ throughout by a modified function $k^2_{\text{mod}}(x)$ designed to eliminate the singular part of $\gamma(x)$. Suppose in fact that

$$k^2_{\text{mod}}(x) = k^2(x) - \beta x^{-n}, \tag{2.95}$$

and that contributions to $\gamma(x)$ from $\kappa(\xi)$ as $x \to 0$ are neglected. It is readily verified that

$$\gamma(x) \simeq -k^{1/2}_{\text{mod}}(\mathrm{d}^2 k^{-1/2}_{\text{mod}}/\mathrm{d}x^2) \simeq [n(n-4)/16]x^{-2}. \tag{2.96}$$

Hence the error term $\gamma(x)$ will be insignificant compared with $k^2(x)$ for $n > 2$, and no modification is required. On the other hand, if $n = 2$, $\gamma(x)$ is of the same magnitude as $k^2(x)$, and the following modification (Langer 1937)

$$k^2_{\text{mod}}(x) = k^2(x) - x^{-2}/4 \tag{2.97}$$

ensures that $k^2_{\text{mod}}(x) + \gamma(x) = k^2(x)$, so that ψ_{app} defined in terms of $k_{\text{mod}}(x)$ gives an exact solution to (2.68) at the singularity. Corrections of this type have an obvious relevance to the radial behaviour of rotating systems where they have the effect of converting the centrifugal term $l(l+1)/r^2$ into $\left(l+\frac{1}{2}\right)^2/r^2$. The same modification is also suggested for $n = 1$ because it is $k_{\text{mod}}(x)$ (not $k(x)$) which appears in the formula for $\gamma(x)$. Langer (1937) reached the same conclusion by using a transformation $x = \mathrm{e}^{\xi}$ which removes the singularity to $\xi = -\infty$ (see problem 2.4).

2.4 Higher-order phase integral approximations

Direct extension of the JWKB series to higher terms in Section 2.1 becomes quite complicated, and hence the following related symmetric phase integral series due to Froman (1966) will be adopted. The symmetry lies in the appearance of a single function $q(x)$ in both the exponential and pre-exponential parts of the approximation:

$$\Psi(x) \simeq q^{-1/2}(x) \exp\left(\frac{\mathrm{i}}{\hbar}\int q(x)\mathrm{d}x\right). \tag{2.98}$$

This form ensures a constant current density, via eqn (2.16), and also eliminates terms in odd powers of $\hbar$ in the equation for $q(x)$, because one finds, on using (2.98) to substitute for $\psi(x)$ in (2.1), that $q(x)$ must satisfy an equation in $\hbar^2$, namely

$$p^2(x) - q^2(x) + \hbar^2 q^{1/2}(x)\frac{\mathrm{d}^2}{\mathrm{d}x^2}[q^{-1/2}(x)] = 0. \tag{2.99}$$

(Notice the similarity of structure between the final term and the error $\gamma(x)$ in eqn (2.69).)

The solution to (2.99) is found by setting

$$q(x) = p(x)Y(x) \tag{2.100}$$

and defining the new variable

$$\xi = \int p(x)\mathrm{d}x. \tag{2.101}$$

The function $Y(x)$ is then found after some manipulation to satisfy

$$1 - Y^2 + \hbar^2\varepsilon_0 + \hbar^2[Y^{1/2}(\mathrm{d}^2 \ln Y^{-1/2}/\mathrm{d}\xi^2)] = 0 \tag{2.102}$$

where

$$\varepsilon_0 = p^{-3/2}(x)\frac{\mathrm{d}^2}{\mathrm{d}x^2}[p^{-1/2}(x)]. \tag{2.103}$$

Finally, if Y is expanded in even powers of $\hbar$,

$$Y = Y_0 + \hbar^2 Y_2 + \hbar^4 Y_4 + \ldots, \tag{2.104}$$

one finds, on collecting terms in the first three powers of $\hbar^2$, that

$$\begin{aligned} 1 - Y_0 &= 0 \\ \varepsilon - 2Y_2Y_0^2 &= 0 \\ (Y_2^2 + 2Y_0Y_4) + \frac{Y_0}{2}\frac{\mathrm{d}^2 Y_2}{\mathrm{d}x^2} &= 0 \end{aligned} \tag{2.105}$$

from which

$$Y_0 = 1, \quad Y_2 = \frac{\varepsilon_0}{2}, \quad Y_4 = -\frac{\varepsilon_0^2}{8} - \frac{1}{8}\frac{\mathrm{d}^2\varepsilon_0}{\mathrm{d}\xi^2}. \tag{2.106}$$

Terms up to $\hbar^{12}$ in this series are given by Campbell (1972) and details of a related higher-order series are discussed by Bender et al. (1977).

Froman and Froman (1974) show that eqn (2.106) also holds if the actual momentum $p(x)$ in eqns (2.100) and (2.101) is replaced by a modified function $p_{\mathrm{mod}}(x)$ provided that $\varepsilon_0(x)$ is replaced by

$$\varepsilon_o(x) = \frac{p^2(x) - p_{\mathrm{mod}}^2(x)}{p_{\mathrm{mod}}^2(x)} + p_{\mathrm{mod}}^{-3/2}(x)\frac{\mathrm{d}^2}{\mathrm{d}x^2}p_{\mathrm{mod}}^{-1/2}(x), \tag{2.107}$$

a device which may be used to minimize the higher-order correction terms in a manner similar to that employed at the end of Section 2.3.

An immediate consequence of eqn (2.106) is that $q(x)$, given by eqn (2.100), diverges at the classical turning point, because $\varepsilon_0(x)$ upon which all subsequent terms depend itself diverges. To see this explicitly note, on expressing $p(x)$ in terms of the potential $V(x)$, that

$$\varepsilon_0(x) = \frac{1}{16}\frac{1}{2m}[4V''(E-V)^{-2} + V'^2(E-V)^{-3}], \tag{2.108}$$

which means that the integral in eqn (2.98) cannot be taken from the turning point. Instead one must make the substitution (Froman 1966)

$$\int_a^x q(x)\mathrm{d}x \Rightarrow \frac{1}{2}\int_C q(x)\mathrm{d}x \tag{2.109}$$

where C is the complex contour shown in Fig. 2.9.

Similarly the quantization condition (2.58) goes over to

$$\frac{1}{2\hbar}\int_\Lambda q(x)\mathrm{d}x = \left(v + \frac{1}{2}\right)\pi, \tag{2.110}$$

where the contour Λ now encircles the two turning points, a and b, which are assumed to be joined by a branch cut (Froman 1966). Examples of the use of eqn (2.110), with and without the modification expressed by eqn (2.107), are given by Froman (1966), Froman and Froman (1974, 1981), and in a chemical context by Dunham (1932), Simons et al. (1973), Thakker (1975) and LeRoy (1980). The latter makes use of an extended quantization condition derived from (2.110) up to terms in Y_0 and Y_2 (after integration by parts to remove the term in $(E-V)^{-3}$ in (2.108)), namely

$$\begin{aligned}\left(v + \frac{1}{2}\right)\pi \simeq \frac{(2m)^{1/2}}{\hbar}&\int_a^b [E - V(x)]^{1/2}\mathrm{d}x \\ &+ \frac{1}{96}\frac{\hbar}{(2m)^{1/2}}\int_\Lambda V''(x)[E-V(x)]^{-3/2}\mathrm{d}x,\end{aligned} \tag{2.111}$$

Elements of the Dunham (1932) analysis based on (2.111) are explored in problem 3.2, and methods for numerical evaluation of the integrals required by eqns (2.109) and (2.111) are given by Barwell et al. (1979) and Luppi and Pajunen (1982, 1983, 1984).

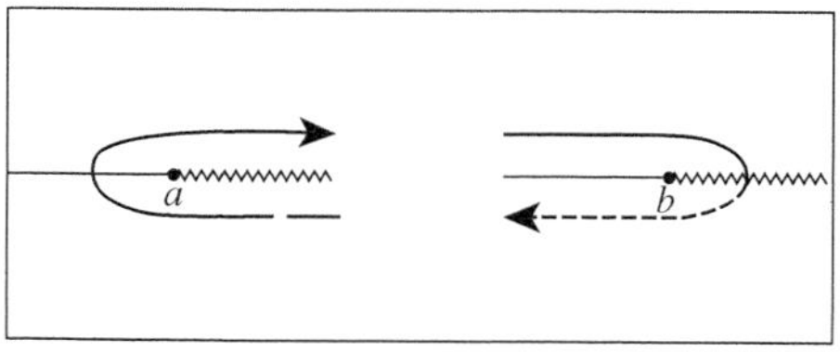

Fig. 2.9 Complex contours around left-hand and right-hand classical turning points a and b respectively. The zig-zag lines are branch cuts, the crossing of which leads to change by π in the argument of $k(x)$.

An instructive example of the use of this higher-order approximation is provided by the scaled quartic oscillator, defined by the Schrödinger equation

$$\left(\frac{\mathrm{d}^2}{\mathrm{d}x^2} + \varepsilon - x^4\right)\psi = 0, \tag{2.112}$$

so that

$$p(x) = (\varepsilon - x^4)^{1/2}. \tag{2.113}$$

The necessary integrals may be deduced by manipulating the identity (Abramowitz and Stegun 1965)

$$\begin{aligned} I_n(\varepsilon) &= \int_\Lambda x^n(\varepsilon - x^4)^{1/2}\mathrm{d}x = \varepsilon^{(n+3)/4}\int_0^1 u^{(n-3)/4}(1-u)^{1/2}\mathrm{d}u \\ &= \varepsilon^{(n+3)/4}\Gamma\left(\frac{n+1}{4}\right)\Gamma\left(\frac{3}{2}\right)\Big/\Gamma\left(\frac{n+7}{4}\right), \end{aligned} \tag{2.114}$$

where $u = x^4/\varepsilon$.

The first integral in eqn (2.111) is simply $I_0(\varepsilon)$ and the second is expressible in terms of $\mathrm{d}^2I_2/\mathrm{d}\varepsilon^2$. Using these results it turns out that

$$\left(v + \frac{1}{2}\right)\pi = a\varepsilon^{3/4} - b\varepsilon^{-3/4} \tag{2.115}$$

where

$$\begin{aligned} a &= \frac{1}{2}\Gamma\left(\frac{1}{4}\right)\Gamma\left(\frac{3}{2}\right)\Big/\Gamma\left(\frac{7}{4}\right) = 1.74804 \\ b &= \frac{1}{2}\left(\frac{5}{16}\right)\Gamma\left(\frac{3}{4}\right)\Gamma\left(\frac{3}{2}\right)\Big/\Gamma\left(\frac{9}{4}\right) = 0.14977. \end{aligned} \tag{2.116}$$

Equation (2.115) can be rearranged as a quadratic equation in $\varepsilon^{3/4}$ with the physically relevant solution

$$\varepsilon^{3/4} = (\pi/2a)\left\{\left(v + \frac{1}{2}\right) + \left[\left(v + \frac{1}{2}\right)^2 + (3\pi)^{-1}\right]^{1/2}\right\} \tag{2.117}$$

obtained by use of the identities (Abramowitz and Stegun 1965)

$$\begin{aligned} \Gamma(1+v) &= v\Gamma(v) \\ 4ab &= \left(\frac{4}{3}\right)\left[\Gamma\left(\frac{3}{2}\right)\right]^2 = \frac{\pi}{3}. \end{aligned} \tag{2.118}$$

Table 2.1 gives a comparison between the first-order, $\varepsilon^{3/4} = (v + \frac{1}{2})\pi/a$, third-order (from eqn (2.117)) and numerically exact eigenvalues for this problem. It is obvious

Table 2.1 First-order, third-order and exact eigenvalues for the scaled quartic oscillator.

v	First order	Third order	Exact
0	0.86715	0.98076	1.06036
1	3.75192	3.81033	3.79967
2	7.41399	7.45580	7.45570
3	11.61151	11.64498	11.64475
4	16.23362	16.26102	16.26183
5	21.21365	21.23840	21.23838
10	50.24015	50.25626	50.25626

that the first-order quantization is seriously deficient for $v = 0$ and then rapidly improves in accuracy for higher v values, as expected because the ratio of the terms in eqn (2.115) is of order $\varepsilon^{3/2}$. Inclusion of the third-order term, $b\varepsilon^{-3/4}$, is, however, seen to yield a significant improvement, even at $v = 10$.

Detailed numerical investigation by Bender et al. (1977) shows, however, that the best approximation for $v = 0$ is actually obtained by the relatively poor third-order estimate in Table 2.1, owing to the asymptotic nature of the series. Hence something is seriously amiss. The problem, which has been analysed in considerable detail by Balian et al. (1978), Vöros (1983) and Lundborg and Froman (1988), is that the quartic potential gives rise to complex zeros of $p(x)$ at $\pm \mathrm{i}\varepsilon^{1/4}$ as well as those at $\pm\varepsilon^{1/4}$ and that all four must be taken into account at low energies. More detail is given in Section A.4 and an improved quantization procedure that takes these additional 'turning points' into account is discussed in Section 3.3.

2.5 Problems

2.1. A solution to the time-dependent Schrödinger equation,

$$\left(-\frac{\hbar^2}{2m}\frac{\partial^2}{\partial x^2} + V(x,t)\right)\Psi = \mathrm{i}\hbar\frac{\partial \Psi}{\partial t},$$

is sought in the form $\Psi = \exp(\mathrm{i}S/\hbar)$, with $S = S_0 + \hbar S_1 + \ldots$. Show that

$$S_0 = \int (p\dot{x} - H)\mathrm{d}t,$$

where

$$H = \frac{1}{2m}p^2(x,t) + V(x,t)$$

and the integral is taken along a classical trajectory $[x(t), y(t)]$ such that $(\mathrm{d}p/\mathrm{d}t) = -(\partial H/\partial x)_t$ and $(dx/dt) = (\partial H/\partial p)_x$.

2.2. (i) Show that the Legendre equation

$$\frac{1}{\sin\theta}\frac{\mathrm{d}}{\mathrm{d}\theta}\left[\sin\theta\frac{\mathrm{d}P_l}{\mathrm{d}\theta}\right] + l(l+1)P_l = 0$$

reduces under the substitution $P_l(\cos\theta) = \sin^{-1/2}\theta\chi(\theta)$ to

$$(\mathrm{d}^2\chi/\mathrm{d}\theta^2) + \left[(l+1/2)^2 + 1/(2\sin\theta)^2\right]\chi = 0.$$

(ii) Verify that the JWKB criterion (2.19) applies for $(l+1/2)\sin\theta \gg 1$, and that in this case $\chi \simeq A\cos[(l+\frac{1}{2})\theta + \alpha]$.

(iii) Employ the identities

$$P_l(z) = (-1)^l P_l(-z), \qquad \int_0^\pi P_l^2(\cos\theta)\sin\theta\mathrm{d}\theta = \left(l+\frac{1}{2}\right)^{-1}$$

to deduce that

$$P_l(\cos\theta) \simeq \left(\frac{2}{\pi(l+\frac{1}{2})\sin\theta}\right)^{1/2} \cos\left[\left(l+\frac{1}{2}\right)\theta - \pi/4\right].$$

Hint: evaluate the integral over the interval $[0:\pi]$ and note that $(l+1/2)\pi) \gg 1$.

2.3. Derive the following stationary phase approximation to the Bessel function, valid for $z \gg n$,

$$J_n(z) = \frac{1\,\mathrm{Re}}{\pi}\int_0^\pi \exp\left[i(z\sin\theta - n\theta)\right]\mathrm{d}\theta \simeq \left(\frac{2}{\pi z}\right)\cos\left(z - \frac{n\pi}{2} - \frac{\pi}{4}\right).$$

2.4. Derive the following expression for the semiclassical phase shift for scattering at energy E under the potential $V(R) = A\mathrm{e}^{-\alpha R}$:

$$\eta = \frac{(2mE)^{1/2}}{\alpha\hbar}\left[\ln\left(\frac{4A}{E}\right) - 2\right] + \frac{\pi}{4}.$$

2.5. Prove that the Langer substitutions, $R = \mathrm{e}^x$ and $\Psi = \mathrm{e}^{x/2}\varphi$, transform the radial equation

$$\frac{\mathrm{d}^2\psi}{\mathrm{d}R^2} + \left(k^2 - U(R) - \frac{l(l+1)}{R^2}\right)\psi = 0$$

to the form

$$\frac{\mathrm{d}^2\varphi}{\mathrm{d}x^2} + \mathrm{e}^{2x}\left(k^2 - U(\mathrm{e}^x) - \left(l+\frac{1}{2}\right)^2\mathrm{e}^{-2x}\right)\varphi = 0,$$

with the singularity at $R = 0$ removed to $x \to -\infty$. Hence deduce the following 'Langer-corrected' approximation to ψ:

$$\psi(R) \simeq k_l^{-1/2}(R)\sin\left(\int_{a_l}^R k_l(R)\mathrm{d}R + \frac{\pi}{4}\right),$$

where

$$k_l^2(R) = k^2 - U(R) - (l+1/2)^2/R^2.$$

2.6. The spherical Bessel function $j_l(kR)$ may be expressed as $j_l(kR) = R^{-1}\psi_l(R)$, where $\psi_l(R)$ satisfies the equation in problem 2.5 with $U(R) = 0$. Verify that the Langer-modified asymptotic forms

$$\begin{aligned}\psi_l(R) &\overset{R \gg a_l}{\simeq} A k_l^{-1/2}(R) \sin\left(\int_{a_l}^{R} k_l(R)\mathrm{d}R + \frac{\pi}{4}\right) \\ &\overset{R \ll a_l}{\simeq} A' |k_l(R)|^{1/2} \exp\left(-\int_{R}^{a_l} |k_l(R)|\mathrm{d}R\right),\end{aligned}$$

with $a_l = (l + \frac{1}{2})/k$, lead to the correct long-range and short-range behaviour

$$\begin{aligned}j_l(kR) &\overset{R\to\infty}{\simeq} (C/kR)\sin(kR - l\pi/2) \\ &\overset{R\to 0}{\simeq} C'(kR)^l.\end{aligned}$$

2.7. The Legendre and Bessel functions, $P_l(\cos\theta)$ and $J_0(z)$, are equal to unity at $\theta = 0$ and $z = 0$ respectively, and satisfy the equations

$$\frac{\mathrm{d}^2 P}{\mathrm{d}\theta^2} + \cot\theta \frac{\mathrm{d}P}{\mathrm{d}\theta} + l(l+1)P = 0$$

and

$$\frac{\mathrm{d}^2 J}{\mathrm{d}z^2} + \frac{1}{z}\frac{\mathrm{d}J}{\mathrm{d}z} + J = 0.$$

Deduce by means of the approximation $P_l(\cos\theta) \simeq A(\theta)J[z(\theta)]$ that

$$P_l(\cos\theta) \simeq (\theta/\sin\theta)^{1/2} J_0\left[\left(l + \frac{1}{2}\right)\theta\right]$$

for $\theta < \pi/2$ provided that $(l + \frac{1}{2})^2 \gg (\pi^2 - 4)/4\pi^2$. Suggest a similar approximation to $P_l(\cos\theta)$ that is valid as $\theta \to \pi$.

3
Quantization

Questions of quantization turn on the proper treatment of normally isolated regions where the JWKB description, expressed by eqn (2.13), breaks down. The most obvious of such special regions are those around the classical turning points, at each of which the analysis in Sections 2.2 and 2.3 yields an associated so-called Maslov (1972) phase correction of $\pi/4$, which is discussed by Delos (1986) and Littlejohn and Robbins (1987), while Appendix A and parts of Section 8.2 bear on its choice in particular situations. Other special regions around potential singularities are most conveniently handled by introducing a Langer (1937) correction term to the physical potential function (see Section 2.3). It is shown in Section 3.1 that systems with no other complications may be quantized by the Bohr–Sommerfeld rule

$$\oint p(q)\mathrm{d}q = (n+\delta)h, \quad n = 0, 1, 2, 3 \ldots \tag{3.1}$$

where n is the quantum number and δ is a Maslov index, which takes the values $\frac{1}{2}$, 0, $\frac{1}{2}$ for vibrational, planar rotation and orbital motion respectively.

The necessary systematic theory to take account of further complications due to tunnelling and curve crossing by means of semiclassical connection formulae is outlined in Section 3.2 and applied to a variety of bound and quasi-bound systems in Sections 3.3 and 3.4. In particular, double minimum problems are quantized by eqn (3.72), while eqns (3.108) and (3.111) give the energies and level widths of tunnelling (or orbiting) resonances. Equations (3.127) and (3.130) or (3.131) and (3.133) give the analogous results for predissociation by curve crossing.

The theory which follows is expounded at the first-order level, which is adequate for most practical applications, but the resulting equations are readily extendable to higher order if necessary (Froman 1970; Froman et al. 1972; Paulsson et al. 1983; Luppi and Pajunen 1983).

3.1 Bohr–Sommerfeld quantization

We consider first the quantization of vibrational and rotational motion in situations where the only non-semiclassical regions lie close to isolated classical turning points and potential singularities.

Vibration

The case of vibrational motion is covered by eqns (2.57) and (2.58), with the quantized energies, E_v, determined by the condition

Semiclassical Mechanics with Molecular Applications. Second Edition. M. S. Child

$$\oint p(x)\mathrm{d}x = \left(v + \frac{1}{2}\right) h, \tag{3.2}$$

or more explicitly,

$$\frac{1}{\hbar}\int_a^b [2m(E_v - V(x)]^{1/2}\mathrm{d}x = \left(v + \frac{1}{2}\right)\pi, \tag{3.3}$$

where a and b denote the turning points. Consequently the Maslov index in (3.1) takes the value $\delta = \frac{1}{2}$ for vibration. Numerical evaluation of the integral is preferably done by means of modified Gaussian quadrature

$$\begin{aligned}\int_a^b [E - V(x)]^{1/2}\mathrm{d}x &= \int_a^b f(x)[(b-x)(x-a)]^{1/2}\mathrm{d}x \\ &\simeq (b-a)^2 \sum_{i=1}^{N} f(x_i)w_i \end{aligned} \tag{3.4}$$

where

$$f(x) = \{[E - V(x)]/[(b-x)(x-a)]\}^{1/2}, \tag{3.5}$$

and the integration points, x_i, and weights, w_i, are listed by Abramowitz and Stegun (1965). The advantage of this transformation is that $f(x)$ and its derivatives are finite even close to the turning points a and b, so that the number of integration points, N, is much smaller than that required by Simpson's rule or trapezoidal quadrature, for example.

There are also a number of practically important cases in which the quantization integral can be evaluated in closed form, as seen for the harmonic oscillator in eqns (2.59) and (2.60). The Morse oscillator, with potential function

$$V(q) = D(1 - \mathrm{e}^{-aq})^2, \tag{3.6}$$

offers a more challenging example. The necessary integral is conveniently evaluated by means of the substitution

$$1 - \mathrm{e}^{-aq} = \sqrt{\varepsilon}\cos u \tag{3.7}$$

where $\varepsilon = (E/D)$, which results in the standard form (see §858.536 of Dwight 1961, for $m = 0$ and 2 and $u = \pi - x$)

$$\begin{aligned}\frac{1}{\hbar}\int_{q_a}^{q_b} \{2m[E - V(q)]\}^{1/2}\mathrm{d}q &= \frac{(2mD)^{1/2}}{a\hbar}\int_0^\pi \frac{\varepsilon \sin^2 u}{[1 - \sqrt{\varepsilon}\cos u]}\mathrm{d}u \\ &= \frac{\pi(2mD)^{1/2}}{a\hbar}\left[1 - \sqrt{1-\varepsilon}\right]. \end{aligned} \tag{3.8}$$

It is then readily shown by comparison with the right-hand side of eqn (3.3) that

$$E_v = \left(v + \frac{1}{2}\right)\omega - \left(v + \frac{1}{2}\right)^2 \omega x \tag{3.9}$$

where

$$\omega = a\hbar(2D/m)^{1/2} \quad \text{and} \quad \omega x = a^2\hbar^2/2m.$$

This is again the exact result (Landau and Lifshitz 1965) and the same is found to be true for radial quantization of the Langer-modified hydrogen atom (see Section 4.4).

Another aspect of vibrational quantization may be understood by interpreting the vibrational–rotational energy levels of a diatomic molecule as vibrational eigenvalues supported by the centrifugally corrected potential function

$$V_J(R) = V(R) + J(J+1)\hbar^2/2\mu R^2, \tag{3.10}$$

where R is the bond length and μ is the reduced mass quantum number of the two atoms:

$$\mu = m_1 m_2/(m_1 + m_2).$$

Thus eqn (3.3) becomes

$$\frac{(2\mu)^{1/2}}{\hbar}\int_{a_{vJ}}^{b_{vJ}} \left(E_{vJ} - V(R) - \frac{J(J+1)\hbar^2}{2\mu R^2}\right)^{1/2} \mathrm{d}R = \left(v + \frac{1}{2}\right)\pi. \tag{3.11}$$

Figure 3.1 shows the pattern of resulting energy levels, E_{vJ}, against the effective potential energy curves $V_J(R)$.

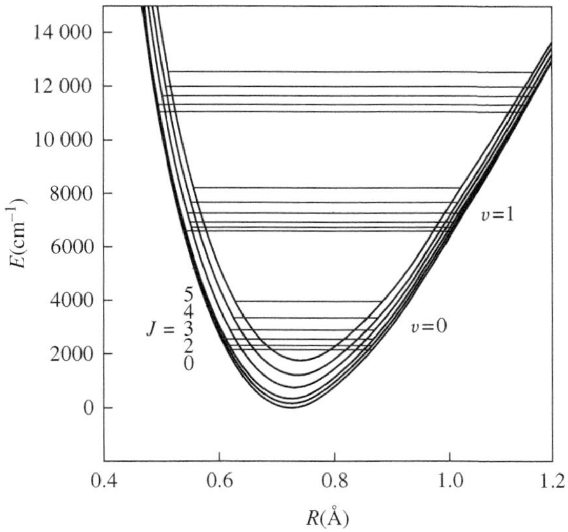

Fig. 3.1 Effective potential curves and vibrational–rotational energy levels for the $X^1\Sigma_g^+$ state of H_2, as reported by LeRoy and Schwartz (1987).

Equation (3.11) is of considerable importance in the context of diatomic spectroscopy, first because it will be seen in Section 6.1 to provide the basis for the RKR inversion procedure, now routinely used to deduce the potential function $V(R)$ from experimental data. Secondly it is evident from the way in which the quantum numbers appear in eqn (3.11) that the energy levels, E_{vJ}, must be expressible as a combined function of $\left(v + \frac{1}{2}\right)$ and $J(J+1)$. Thus Dunham (1932), Simons et al. (1973) and Thakker (1975) have used higher-order variants of eqn (3.11) to relate the coefficients Y_{ij}, in the following Dunham expansion, to successive terms in different types of expansion for the potential $V(R)$ about its equilibrium value R_e (see problem 3.4):

$$E_{vJ} = \sum_{i,\ j} Y_{ij} \left(v + \frac{1}{2}\right)^i [J(J+1)]^j. \tag{3.12}$$

Finally it may be noted from the reduced mass dependence in (3.11) that the dominant first-order contribution to the Y_{ij} in (3.12) must scale on isotopic substitution as

$$Y_{ij}(\mu_1)/Y_{ij}(\mu_2) = (\mu_1/\mu_2)^{-(i+2j)/2}. \tag{3.13}$$

In other words the vibrational term values for different isotopic species must lie on a smooth curve when plotted as a function of the mass-reduced quantum number $\left(v + \frac{1}{2}\right)/\sqrt{\mu}$ (Stwalley 1975). Figure 3.2 illustrates this behaviour for the ground electronic state of the hydrogen molecule. Notice, however, that second-order corrections and non-adiabatic effects, which are certainly perceptible for H_2 (Kolos and Wolniewicz 1968), both disturb this picture, but not on the scale of Figure 3.2.

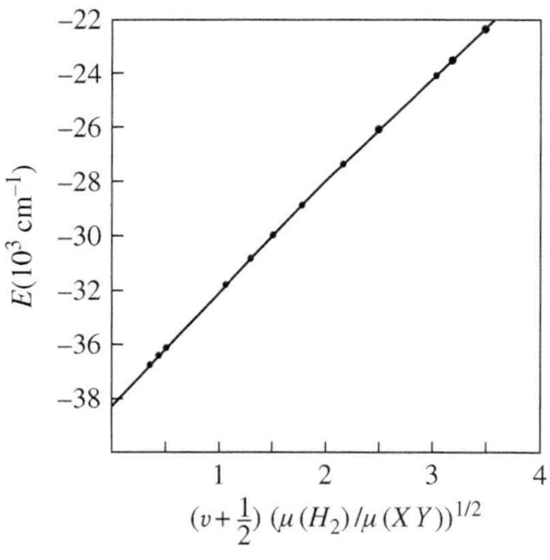

Fig. 3.2 Calculated vibrational energies as a function of mass-reduced quantum numbers for the $X^1\Sigma_g^+$ state of H_2. Points are designated as ○ for H_2, ◇ for HD and □ for D_2. Data from Wolniewicz (1966).

Rotation in a plane

In the case of rotation in a plane with moment of inertia I, the appropriate JWKB wavefunction takes the form

$$\psi(\phi) = Ap^{-1/2}(\phi)\exp\left(\pm\frac{\mathrm{i}}{\hbar}\int_0^{\phi} p(\phi')\mathrm{d}\phi'\right), \tag{3.14}$$

where

$$p(\phi) = [2I(E - V(\phi))]^{1/2} \tag{3.15}$$

provided that the energy lies well above any barriers to rotation in the cyclic potential $V(\phi)$. Hence the periodic boundary condition

$$\psi(\phi) = \psi(\phi + 2\pi) \tag{3.16}$$

requires that

$$\frac{1}{\hbar}\int_0^{2\pi} p(\phi')\mathrm{d}\phi' = 2m\pi, \quad m = 0, \pm 1, \pm 2, \ldots, \tag{3.17}$$

which means by comparison with eqn (3.1) that the Maslov (1972) parameter takes the value $\delta = 0$ for planar rotation.

Note, however, that the case $m = 0$ in eqn (3.17) cannot be reconciled with the requirement that the energy lies well above any barriers to rotation; eqn (3.17) therefore applies with $m = 0$ only if the potential function is independent of ϕ. It will also be seen in Section 3.3 that the presence of a barrier to rotation progressively lifts the double degeneracy of the $m = \pm|m|$ levels as the energy approaches the barrier maximum either from above or below. One other point to note is that the energy levels for different isotopic species must scale smoothly with $m/I^{1/2}$ because of the way in which I contributes to $p(\phi)$.

Orbital motion

The term orbital motion refers to motion in the θ variable in a spherical polar (r, θ, ϕ) system in situations where the angular momentum, p_ϕ, about the z axis is conserved:

$$p_\phi = m\hbar. \tag{3.18}$$

The classical Hamiltonian for motion under a potential $V(\theta)$ then takes the form (Goldstein 1980)

$$H = \frac{1}{2I}\left(p_\theta^2 + p_\phi^2/\sin^2\theta\right) + V(\theta) = E \tag{3.19}$$

showing, at least for $p_\phi \neq 0$, that the term $p_\phi^2/\sin^2\theta$ must give rise to turning points for the θ motion. One is therefore led, by analogy with the vibrational case, to quantize according to the formula

$$\oint p(\theta)\mathrm{d}\theta = \left(n + \frac{1}{2}\right)\hbar \tag{3.20}$$

with a Maslov index of $\delta = \frac{1}{2}$.

Before turning to a number of complications it is convenient to assess the consequences of eqns (3.18)–(3.20) in the case of free orbital motion with $V(\theta) = 0$. The notation is simplified by writing

$$E = \beta^2 \hbar^2 / 2I, \tag{3.21}$$

with β to be determined by the equation

$$\int_{\theta_0}^{\pi-\theta_0} \left(\beta^2 - \frac{m^2}{\sin^2\theta}\right)^{1/2} \mathrm{d}\theta = \left(n + \frac{1}{2}\right)\pi, \tag{3.22}$$

and θ_0 defined by the equation

$$\sin\theta_0 = \sin(\pi - \theta_0) = m/\beta. \tag{3.23}$$

The substitution

$$\sin(x/2) = [\beta/(\beta^2 - m^2)^{1/2}] \sin\theta \tag{3.24}$$

may be verified to cast the integral into a tabulated form, parameterized by the quantity

$$a = (\beta^2 - m^2)/(\beta^2 + m^2), \tag{3.25}$$

namely (Gradsteyn and Ryyzhik 1980)

$$\begin{aligned}\left(n + \frac{1}{2}\right)\pi &= a\beta \int_0^{\pi} \left(\frac{1 + \cos x}{1 + a\cos x}\right) \mathrm{d}x \\ &= \pi\beta\{1 + [(1-a)/(1+a)]^{1/2}\} \\ &= (\beta - |m|)\pi. \end{aligned} \tag{3.26}$$

It follows by means of the substitution $l = |m| + n$ that

$$\beta = l + \frac{1}{2}. \tag{3.27}$$

Consequently the expression for the energy, implied by (3.21),

$$E = \left(l + \frac{1}{2}\right)^2 \hbar^2/2I \tag{3.28}$$

contains a factor $\left(l + \frac{1}{2}\right)^2$ instead of the correct $l(l+1)$.

The difference is obviously small for large l values but it is interesting to trace the origin of the discrepancy by reference to the structure of the quantum mechanical equivalent of (3.19), which scales after factoring out the term $\hbar^2/2I$ to the form

$$\left[\frac{1}{\sin\theta}\frac{\mathrm{d}}{\mathrm{d}\theta}\left[\sin\theta\frac{\mathrm{d}}{\mathrm{d}\theta}\right] + L^2 - \frac{m^2}{\sin^2\theta}\right]\Psi = 0 \tag{3.29}$$

where

$$L^2 = 2EI/\hbar^2. \tag{3.30}$$

Two features of eqn (3.29) must be taken into account. In the first place the 'normal' Schrödinger form from which the JWKB approximation was derived contains only a second-derivative term. Hence it is appropriate to substitute

$$\Psi(\theta) = \sin^{-1/2}\theta\psi(\theta) \tag{3.31}$$

in order to cast (3.29) into the standard form

$$\left(\frac{\mathrm{d}}{\mathrm{d}\theta^2} + L^2 + \frac{1}{4} - \frac{m^2}{\sin^2\theta} + \frac{1}{4\sin^2\theta}\right)\psi = 0. \tag{3.32}$$

Secondly, according to the discussion at the end of Section 2.3 the presence of the $\sin^{-2}\theta$ singularities at $\theta = 0$ and $\theta = \pi$ requires a Langer (1937) or Froman (1966) type of modification to the effective potential that would exactly remove the final term in eqn (3.32). The upshot is that eqn (3.22) is indeed the correct quantization condition provided that β^2 is interpreted not as the square of the total angular momentum, L^2, but as

$$\beta^2 = L^2 + \frac{1}{4}. \tag{3.33}$$

Seen in another way, this means that the quantum mechanical angular momentum differs from its semiclassical counterpart by the term $\frac{1}{4}$ in eqn (3.33), with the result that the semiclassical quantization equation (3.27) does in fact lead to the correct quantum mechanical value,

$$L^2 = \beta^2 - \frac{1}{4} = \left(l + \frac{1}{2}\right)^2 - \frac{1}{4} = l(l+1). \tag{3.34}$$

This argument also implies that the correct quantization condition for perturbed orbital motion should be expressed in the form

$$\frac{(2I)^{1/2}}{\hbar}\int_{\theta_a}^{\theta_b}\left(E + \frac{\hbar^2}{8I} - V(\theta) - \frac{m^2\hbar^2}{2I\sin^2\theta}\right)^{1/2}\mathrm{d}\theta = \left(n + \frac{1}{2}\right)\pi \tag{3.35}$$

where the additional term $\hbar^2/8I$ is equivalent to the term $\frac{1}{4}$ in eqn (3.32).

Steep-walled potentials

As a final point concerning simple quantization it may be noted that the motion of a particle in a box of length a is quantized according to the formula

Table 3.1 Semiclassical eigenvalues for the steep-walled potential $V(x) = \cosh(10x)/\cosh(20)$, in various orders with $\delta = \frac{1}{2}$, from Froman and Froman (1978).

v	First order	Third order	Fifth order	Exact
1	1.4	1.7856	1.93	1.7682
3	6.7	7.03	7.006	7.059
5	15.3	15.61	15.627	15.632
7	27.0	27.315	27.326	27.329
9	41.7	42.004	42.0107	42.0124
11	59.2	59.576	59.5804	59.5812
13	79.6	79.954	79.9570	79.9574
15	102.8	103.076	103.0783	103.0785

$$\oint p \, dx = 2\int_0^a (2mE)^{1/2} dx = a(8mE)^{1/2} = nh \tag{3.36}$$

giving rise to the familiar energy levels

$$E_n = n^2\hbar^2/8ma^2 \tag{3.37}$$

but with n limited to the values $n = 1, 2, 3, \ldots$, corresponding to a Maslov index $\delta = 1$ in eqn (3.1). One is therefore led to wonder how steep the walls of a potential must be to justify this choice.

The point has been addressed in some detail by Froman and Froman (1978) who show (see Table 3.1) that $\delta = \frac{1}{2}$ works very well (albeit with significant third-order corrections for $v \leqslant 5$) for the apparently very-steep-walled potential

$$V(x) = \cosh(10x)/\cosh(20) \tag{3.38}$$

shown in Fig. 3.3.

The same authors also examine the exact quantization conditions for the model V-shaped potential

$$\begin{aligned} V(x) &= -\beta(x+a)\hbar^2/2m \quad & x \leqslant -a \\ &= 0 & -a \leqslant x \leqslant a \\ &= \beta(x-a)\hbar^2/2m & x \geqslant a. \end{aligned} \tag{3.39}$$

Their conclusion is that the proper Maslov index depends on the value of the parameter

$$\gamma = k\beta^{-1/3}, \tag{3.40}$$

where $k^2 = 2mE/\hbar^2$. The value $\delta = 1$ is found to apply when $\gamma \ll 1$ while $\delta = \frac{1}{2}$ applies when $\gamma > 1$, showing again that $\delta = \frac{1}{2}$ becomes increasingly valid as the energy increases.

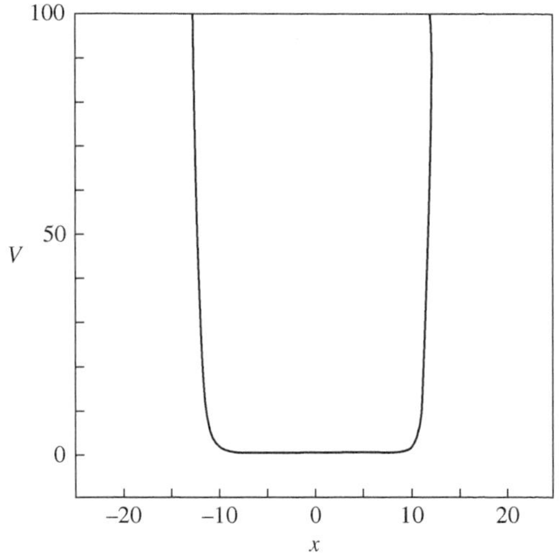

Fig. 3.3 The steep-walled potential given by (3.38).

3.2 Semiclassical connection formulae

The assumption of the previous section is that the motion to be quantized is restricted to a single semiclassical region over which any breakdown of the JWKB approximation is limited to the immediate vicinity of at most two turning points. More complicated situations, involving tunnelling from one JWKB region to another or localized non-adiabatic transitions from one electronic state to another, are conveniently handled by means of standard connection formulae designed to link different JWKB regions. The underlying ideas are due to Froman and Froman (1967), Connor (1968) and Bandrauk and Child (1970); the specific formulation adopted below follows Child (1974a).

Put in mathematical terms, the problem is to follow changes in the amplitudes of the various JWKB terms in passing from one semiclassical region to another. According to eqn (2.13) the generic reference form may be written

$$\psi(R) = k^{-1/2}(R)\left[P'\exp\left(\mathrm{i}\int_p^R k(R)\,\mathrm{d}R\right) + P''\exp\left(-\mathrm{i}\int_p^R k(R)\,\mathrm{d}R\right)\right] \quad (3.41)$$

where

$$k(r) = \{2m[E - V(R)]\}^{1/2}/\hbar, \quad (3.42)$$

and the first point to note is that the phases of the incoming and outgoing coefficients, P'' and P' respectively, depend on the choice of the lower integration limit, p, or phase reference point. For example, $\psi(R)$ could also be expressed as

$$\psi(r) = k^{-1/2}(R)\left[Q'\exp\left(\mathrm{i}\int_q^R k(R)\,\mathrm{d}R\right) + Q''\exp\left(-\mathrm{i}\int_q^R k(R)\,\mathrm{d}R\right)\right] \quad (3.43)$$

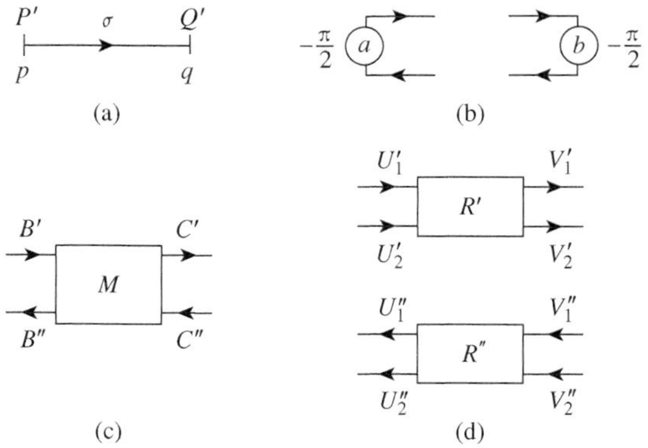

Fig. 3.4 Elementary connection formulae for (a) a change of phase reference point; (b) classical turning points; (c) barrier maxima; and (d) curve crossings. (Taken from Child (1974a). Copyright 1974, with permission from Elsevier.)

where (provided p and q lie in the same semiclassical segment)

$$\begin{pmatrix} Q' \\ Q'' \end{pmatrix} = \begin{pmatrix} e^{i\sigma} & 0 \\ 0 & e^{-i\sigma} \end{pmatrix} \begin{pmatrix} P' \\ P'' \end{pmatrix} \tag{3.44}$$

and

$$\sigma = \int_p^q k(R)\,dR. \tag{3.45}$$

Equation (3.44), which is the simplest type of connection formula, is represented diagrammatically in Fig. 3.4(a).

The second type of connection, already encountered in Section 2.2, relates the incoming and outgoing amplitudes at a classical turning point. Recall from eqns (2.51) and (2.52) that in the region $a \ll R \ll b$ (see Fig. 3.4(b)),

$$\psi(R) \simeq Ak^{-1/2}(R)\sin\left(\int_a^R k(R)\,dR + \frac{\pi}{4}\right)$$

and

$$\psi(R) \simeq Dk^{-1/2}(R)\sin\left(\int_R^b k(R)\,dR + \frac{\pi}{4}\right). \tag{3.46}$$

Hence

$$\begin{pmatrix} A' \\ A'' \end{pmatrix} = A\begin{pmatrix} e^{-i\pi/4} \\ e^{i\pi/4} \end{pmatrix}, \quad \begin{pmatrix} D' \\ D'' \end{pmatrix} = D\begin{pmatrix} e^{i\pi/4} \\ e^{-i\pi/4} \end{pmatrix}. \tag{3.47}$$

The third type of connection to be considered is that between the coefficients (B', B'') and (C', C'') on either side of a potential barrier, as illustrated diagrammatically in Fig. 3.4(c). As shown in Appendix A, the proper connection—due in this context to Connor (1968)—is

$$\begin{pmatrix} C' \\ C'' \end{pmatrix} = \begin{pmatrix} (1+\kappa^2)^{1/2}\mathrm{e}^{-\mathrm{i}\phi} & -\mathrm{i}\kappa \\ \mathrm{i}\kappa & (1+\kappa^2)^{1/2}\mathrm{e}^{\mathrm{i}\phi} \end{pmatrix} \begin{pmatrix} B' \\ B'' \end{pmatrix}, \tag{3.48}$$

where

$$\begin{aligned} \kappa &= \exp(-\pi\varepsilon) \\ \phi &= \arg\Gamma\left(\frac{1}{2} + \mathrm{i}\varepsilon\right) - \varepsilon \ln|\varepsilon| + \varepsilon \overset{|\varepsilon|>1}{\approx} \frac{1}{24|\varepsilon|} + \frac{7}{2880|\varepsilon|\nu^3} + \ldots \end{aligned} \tag{3.49}$$

and ε is the tunnelling parameter

$$\begin{aligned} \varepsilon &= -\frac{1}{\pi}\int_b^c |k(R)|\mathrm{d}R \quad \text{for } E < V_{\max} \\ &= \frac{\mathrm{i}}{\pi}\int_{R_-}^{R_+} k(R)\mathrm{d}R \quad \text{for } E > V_{\max}. \end{aligned} \tag{3.50}$$

Here the phase reference points are taken as the classical turning points for $E < V_{\max}$ and as the real part

$$R_0 = (R_+ + R_-)/2 \tag{3.51}$$

of the complex roots of $k^2(R)$, namely $R_\pm$, for $E > V_{\max}$. Note that in the quadratic approximation

$$V(R) = V_{\max} - \frac{1}{2}m\omega^{*2}(R - R_{\max})^2, \tag{3.52}$$

equations (3.50) may both be shown to reduce to

$$\varepsilon = (E - V_{\max})/\hbar\omega^*, \tag{3.53}$$

which is a convenient approximation at energies close to the barrier maximum.

The significance of the parameter κ is readily appreciated by assuming unit incident flux from the left of the barrier (consistent with $B' = 1$ and $C'' = 0$) and solving eqn (3.48) for the reflected and transmitted amplitudes, B'' and C' respectively. The solution is found to be

$$\begin{aligned} R &= |B''|^2 = \kappa^2/(1+\kappa^2) = 1/[1 + \exp(2\pi\varepsilon)], \\ T &= |C'|^2 = 1/(1+\kappa^2) = 1/[1 + \exp(-2\pi\varepsilon)], \end{aligned} \tag{3.54}$$

which shows that flux is conserved, $R + T = 1$, and that R and T vary smoothly with ε in this uniform approximation. The commonly assumed cruder approximation

$$T \simeq \exp\left[-2\int_b^c |k(R)|\mathrm{d}R\right] = \exp(2\pi\varepsilon), \tag{3.55}$$

which rises to unity at the barrier maximum, is seen to apply only when $T \ll 1$.

The second parameter, ϕ, in eqn (3.49) will later be seen to play a limited but significant role in the theory. At present it is sufficient to note from the values in

Table 3.2 Numerical values of the phase corrections.

x	$\phi(x)$	$\chi(x)$
0.1	0.137	−0.512
0.2	0.150	−0.376
0.3	0.135	−0.287
0.4	0.115	−0.226
0.5	0.096	−0.183
0.6	0.080	−0.152
0.7	0.068	−0.129
0.8	0.058	−0.111
0.9	0.051	−0.098
1.0	0.045	−0.090

Table 3.2 that ϕ is typically small compared with π, but that $\mathrm{d}\phi/\mathrm{d}\varepsilon$ diverges logarithmically at the barrier maximum, $\varepsilon = 0$, a divergence which gives ϕ its special significance (see the discussion after eqn (3.73)). The numerical evaluation of ϕ is quite awkward but Senn (1987) has found that the most convenient procedure is to evaluate $\Gamma\left(N + \frac{1}{2} + \mathrm{i}\varepsilon\right)$ for large N by Stirling's approximation (Abramowitz and Stegun 1965) and then to use the recursion formula $\Gamma(z) = z^{-1}\Gamma(1+z)$ to obtain $\Gamma\left(\frac{1}{2} + \mathrm{i}\varepsilon\right)$.

The final case to be considered (Fig. 3.4(d)) concerns a crossing between the potential curves associated with two so-called diabatic electronic states, which are coupled by some interaction term $V_{12}(R)$ to yield adiabatic terms of the form

$$V_{\pm}(R) = \frac{1}{2}[V_1(R) + V_2(R)] \pm \frac{1}{2}\left\{[V_1(R) - V_2(R)]^2 + 4V_{12}^2(R)\right\}^{1/2}. \tag{3.56}$$

As seen in Appendix A, the connection formulae are most conveniently expressed in terms of changes in the radial factors $\psi_{\pm}(R)$ belonging to the adiabatic states, namely

$$\begin{aligned}
\psi_{\pm}(R) &\overset{R \ll R_x}{\approx} k_{\pm}^{-1/2}(R)\left[P'_{\pm}\exp\left(\mathrm{i}\int_{R_x}^{R} k_{\pm}(R)\mathrm{d}R\right) + P''_{\pm}\exp\left(-\mathrm{i}\int_{R_x}^{R} k_{\pm}(R)\mathrm{d}R\right)\right] \\
&\overset{R \gg R_x}{\approx} k_{\pm}^{-1/2}(R)\left[Q'_{\pm}\exp\left(\mathrm{i}\int_{R_x}^{R} k_{\pm}(R)\mathrm{d}R\right) + Q''_{\pm}\exp\left(-\mathrm{i}\int_{R_x}^{R} k_{\pm}(R)\mathrm{d}R\right)\right]
\end{aligned} \tag{3.57}$$

where R_x denotes the effective curve-crossing point, defined by eqn (3.64) below, and $k_{\pm}(R)$ are given by

$$k_{\pm}(R) = \{2m[E - V_{\pm}(R)]\}^{1/2}/\hbar. \tag{3.58}$$

General connections between the coefficients in eqn (3.57) are given by Nakamura (2007) but present attention will be restricted to energies above the crossing point, where the following formulae (Stückelberg 1932), taken from Appendix A, apply:

$$\begin{pmatrix} Q'_+ \\ Q'_- \end{pmatrix} = \begin{pmatrix} (1-\lambda^2)^{1/2}\mathrm{e}^{\mathrm{i}\chi}, & -\lambda \\ \lambda & (1-\lambda^2)^{1/2}\mathrm{e}^{-\mathrm{i}\chi} \end{pmatrix} \begin{pmatrix} P'_+ \\ P'_- \end{pmatrix}$$

$$\begin{pmatrix} P''_+ \\ P''_- \end{pmatrix} = \begin{pmatrix} (1-\lambda^2)^{1/2}\mathrm{e}^{\mathrm{i}\chi}, & \lambda \\ -\lambda & (1-\lambda^2)^{1/2}\mathrm{e}^{-\mathrm{i}\chi} \end{pmatrix} \begin{pmatrix} Q''_+ \\ Q''_- \end{pmatrix}, \tag{3.59}$$

where

$$\lambda = \mathrm{e}^{-\pi\nu}, \tag{3.60}$$

$$\nu = -\frac{\mathrm{i}}{2\pi}\left(\int_{R_-}^{R_+} [k_-(R) - k_+(R)]\mathrm{d}R\right) \tag{3.61}$$

and

$$\chi = \arg\Gamma(\mathrm{i}\nu) - \nu\ln\nu + \nu + \pi/4 \overset{\nu \gtrsim 1}{\approx} \frac{1}{12\nu} - \frac{1}{360\nu^3} - \frac{1}{1260\nu^5} - \dots. \tag{3.62}$$

Here $R_\pm$ are the complex conjugate points at which the two adiabatic terms, $V_\pm(R)$, intersect, in other words at which

$$[V_1(R) - V_2(R)]^2 + 4V_{12}^2(R) = 0, \tag{3.63}$$

and the effective crossing point, R_x, is defined such that

$$\mathrm{Re}\left(\int_{R_x}^{R_\pm} [k_-(R) - k_+(R)]\mathrm{d}R\right) = 0. \tag{3.64}$$

Equation (3.59) shows that the motion becomes adiabatic (P_+ connects with Q_+, etc.) as $\lambda \to 0$, which means, according to eqn (3.60), that v is very large. The physical factors that govern the magnitude of v are not, however, readily apparent from the sophisticated Stückelberg expression in eqn (3.61). It is therefore often convenient to introduce the following approximations (Landau 1932; Zener 1932):

$$V_i(R) \simeq E_x - F_i(R - R_x) \quad i = 1, 2$$
$$V_{12}(R) \simeq \text{constant}$$

and

$$k_+(R) + k_-(R) \simeq 2mv/\hbar = \text{constant}, \tag{3.65}$$

by means of which eqns (3.56), (3.58) and (3.61) may be reduced to

$$\begin{aligned} \nu &= \frac{1}{2\pi}\mathrm{Im}\left[\int_{R_+}^{R_-} \left(\frac{k_-^2(R) - k_+^2(R)}{k_+(R) + k_-(R)}\right) \mathrm{d}R\right] \\ &\simeq \frac{1}{2\pi}\mathrm{Im}\left(\frac{1}{\hbar v}\int_{R_+}^{R_-} \{[V_1(R) - V_2(R)]^2 + 4{V^2}_{12}(R)\}^{1/2}\mathrm{d}R\right) \\ &= \frac{1}{2\pi}\mathrm{Im}\left(\frac{4\mathrm{i}{V^2}_{12}}{\hbar v\Delta F}\int_{-1}^{1} (1 - y^2)^{1/2}\mathrm{d}y\right) = \frac{{V^2}_{12}}{\hbar v\Delta F}. \end{aligned} \tag{3.66}$$

The substitution

$$R = R_x + (2iV_{12}/\Delta F)y \tag{3.67}$$

has been employed in passing from the second to the third line in eqn (3.66), and

$$\Delta F = |F_1 - F_2|. \tag{3.68}$$

Equation (3.66) shows the explicit dependence of the non-adiabatic switching parameter ν on the coupling strength, V_{12}, the classical velocity at the crossing point, v, and the difference in slopes between the two diabatic potentials. The phase term χ is less apparent, but the values in Table 3.2 show that $|\chi| < \pi/4$ and that $|\chi| \to 0$ as $v \to \infty$. Hence, what is in any case a small phase correction dies away to zero in the adiabatic limit.

The following sections illustrate the use of these connection formulae for various applications—the quantization of double-well and restricted rotation problems and the determination of resonance positions and level widths for two types of predissociation.

3.3 Double minimum potentials and inversion doubling

The case of motion in a double minimum potential, as illustrated in Fig. 3.5, will be taken as a first example. Classically there is a discontinuity in the nature of the motions above and below the barrier; the motion is trapped on one side or the other for $E < V_{max}$ but free to pass over it when $E > V_{max}$. The object is to see how quantum mechanical tunnelling in the first case and barrier reflection in the second affects the picture.

The physical situation requires a combination of the barrier transmission eqn (3.48), turning point conditions (3.47) and the change of phase reference point (3.44). The necessary connections, which are shown in Fig. 3.5, combine to yield

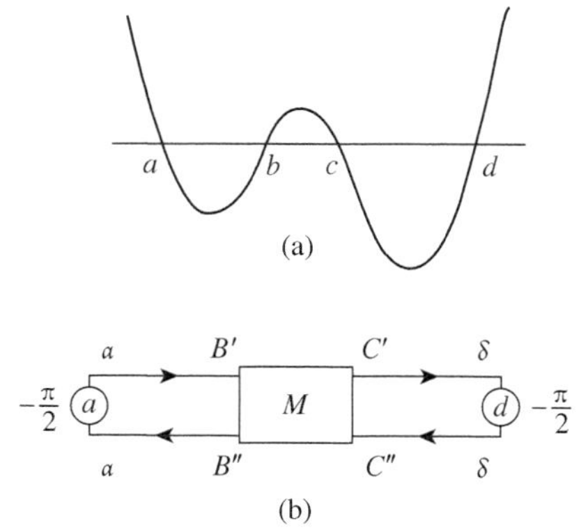

Fig. 3.5 Potential curve and connection diagram for double minimum problems. (Taken from Child (1974a). Copyright 1974, with permission from Elsevier.)

$$\begin{pmatrix} D' \\ D'' \end{pmatrix} = \begin{pmatrix} D\mathrm{e}^{\mathrm{i}\pi/4} \\ D\mathrm{e}^{-\mathrm{i}\pi/4} \end{pmatrix}$$
$$= \begin{pmatrix} \mathrm{e}^{\mathrm{i}\delta} & 0 \\ 0 & \mathrm{e}^{-\mathrm{i}\delta} \end{pmatrix} \begin{pmatrix} (1+\kappa^2)^{1/2}\,\mathrm{e}^{-\mathrm{i}\phi} & -\mathrm{i}\kappa \\ \mathrm{i}\kappa & (1+\kappa^2)^{1/2}\,\mathrm{e}^{\mathrm{i}\phi} \end{pmatrix}$$
$$\times \begin{pmatrix} \mathrm{e}^{\mathrm{i}\alpha} & 0 \\ 0 & \mathrm{e}^{-\mathrm{i}\alpha} \end{pmatrix} \begin{pmatrix} A\mathrm{e}^{-\mathrm{i}\pi/4} \\ A\mathrm{e}^{\mathrm{i}\pi/4} \end{pmatrix} \tag{3.69}$$

where

$$\alpha = \int_a^b k(R)\mathrm{d}R, \quad \delta = \int_c^d k(R)\mathrm{d}R. \tag{3.70}$$

Equation (3.69) rearranges to

$$\begin{pmatrix} D \\ D \end{pmatrix} = \begin{pmatrix} \mathrm{i}(1+\kappa^2)^{1/2}\mathrm{e}^{\mathrm{i}(\alpha+\delta-\phi)} & -\mathrm{i}\kappa\mathrm{e}^{-\mathrm{i}(\alpha-\delta)} \\ \mathrm{i}\kappa\mathrm{e}^{\mathrm{i}(\alpha-\delta)} & -\mathrm{i}(1+\kappa^2)^{1/2}\mathrm{e}^{-\mathrm{i}(\alpha+\delta-\phi)} \end{pmatrix} \begin{pmatrix} A \\ A \end{pmatrix}, \tag{3.71}$$

so that on elimination of the ratio D/A,

$$(1+\kappa^2)^{1/2}\cos(\alpha+\delta-\phi) = -\kappa\cos(\alpha-\delta). \tag{3.72}$$

Equation (3.72) is the required quantization condition, a higher-order version of which is given by Froman et al. (1972). Equations (3.71) may also be solved for the ratio of the amplitudes A and D on the two sides of the barrier which gives information about the nature of the wavefunction; the solution is

$$D/A = (1+\kappa^2)^{1/2}\sin(\alpha+\delta-\phi) - \kappa\sin(\alpha-\delta). \tag{3.73}$$

One important aspect of eqns (3.72) and (3.73), which is seldom emphasized, concerns the role of the phase correction ϕ in cancelling a logarithmic singularity in the energy derivative of $(\alpha+\delta)$ at the barrier maximum. It will be seen below that $\mathrm{d}\alpha/\mathrm{d}E$, for example, determines the inverse semiclassical energy spacing between levels artificially restricted to the left-hand side of the potential barrier, but for reasons explained in Section 4.1 this level spacing is proportional to the classical vibrational frequency, which falls to zero at the barrier maximum. Hence $\mathrm{d}\alpha/\mathrm{d}E$ rises to infinity to make $\alpha+\delta$ discontinuous at $E = V_{\mathrm{max}}$. This divergence is, however, precisely matched by a similar behaviour in $\phi/2$ leading to a smooth energy variation in the phase term $\alpha-\phi/2$ which together with $\delta-\phi/2$ determines the two phase factors in eqns (3.72) and (3.73). To see how this cancellation occurs, suppose that eqn (3.52) applies over some distance range $b' < R < c'$ close to R_{max}. The term α in eqn (3.70) may then be written

$$\alpha = \int_a^{b'} k(R)\,\mathrm{d}R + \frac{(2m)^{1/2}}{\hbar}\int_{b'}^{a}\left[E - V_{\mathrm{max}} + \frac{1}{2}m\omega^{*2}(R-R_{\mathrm{max}})^2\right]^{1/2}\mathrm{d}R, \tag{3.74}$$

which reduces after evaluation of the integral to

$$\alpha = \alpha_0 - \frac{1}{2}\varepsilon \ln |\varepsilon|, \tag{3.75}$$

where the singular behaviour of $\mathrm{d}\alpha/\mathrm{d}E$ is contained in the second term and ε is given by eqn (3.50). It follows by comparison with eqn (3.49) that

$$\alpha - \phi/2 = \alpha_0 - \frac{1}{2}\arg\Gamma(1 + \mathrm{i}\varepsilon) - \frac{1}{2}\varepsilon, \tag{3.76}$$

which no longer contains the logarithmic term. A similar argument also applies to $\delta - \phi/2$. Consequently the two phase terms $\alpha + \delta - \phi$ and $\alpha - \delta$ appearing in (3.72) and (3.73) are smooth functions of E.

The consequences of eqn (3.72) are illustrated in Fig. 3.6 for the symmetric case $\alpha = \delta$. It is evident that the low-energy ($\kappa \gg 1$) roots lie in pairs around the energies at which

$$\cos(2\alpha - \phi) = -1. \tag{3.77}$$

In other words the energy must satisfy the modified Bohr quantization condition

$$\alpha' = \alpha - \phi/2 = \int_a^b k(R)\mathrm{d}R - \phi/2 = \left(v + \frac{1}{2}\right)\pi \tag{3.78}$$

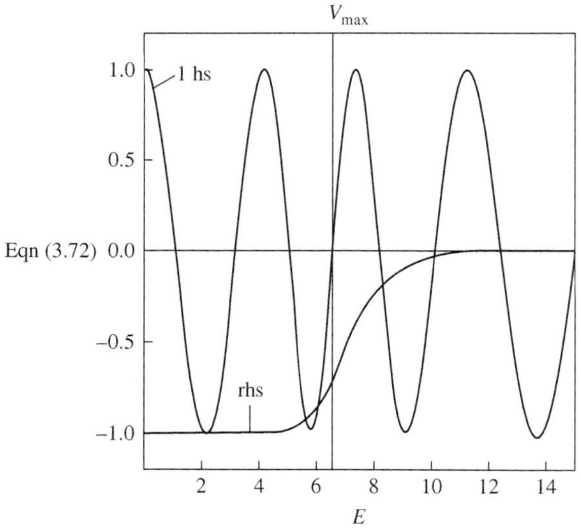

Fig. 3.6 Graphical construction for solution of the double minimum quantization equation (3.72), for the potential $V(x) = (x^2 - \delta^2)^2, \delta = 1.6$ and $m = 1$.

for a bound level in one or other of the isolated wells. At high energies, on the other hand, $\kappa \ll 1$ and eqn (3.72) goes over to

$$\cos(2\alpha - \phi) = 0, \tag{3.79}$$

or

$$2\alpha' = 2\alpha - \phi = \int_a^d k(R)\mathrm{d}R - \phi = \left(v' + \frac{1}{2}\right)\pi, \tag{3.80}$$

which corresponds to Bohr quantization in the full potential at energies above V_{max}.

To proceed further the low-energy ($\kappa \gg 1$) splittings around the roots of (3.78) may be estimated by expanding for the energy dependence of $\alpha'(E)$ in the form

$$\alpha' = \left(v + \frac{1}{2}\right)\pi + \pi(E - E_v)/\hbar\bar{\omega} \tag{3.81}$$

where $\hbar\bar{\omega}$ may be interpreted as the local vibrational energy spacing between the levels in one of the isolated wells (because eqn (3.78) shows that α' increases by π for each energy increment $\hbar\bar{\omega}$). The following approximations

$$\begin{aligned} \cos\alpha' &= \cos\delta' \simeq -(-1)^v(E - E_v)\pi/\hbar\bar{\omega} \\ \sin\alpha' &= \sin\delta' \simeq (-1)^v \end{aligned} \tag{3.82}$$

reduce eqn (3.72) to the form

$$(E - E_v)^2(\pi/\hbar\bar{\omega})^2 \simeq 1 - \frac{\kappa}{(1+\kappa^2)^{1/2}} \simeq \frac{1}{4\kappa^2}. \tag{3.83}$$

The solution may therefore be written, in the light of (3.49) and (3.50), as

$$E = E_v \pm \frac{\hbar\bar{\omega}}{2\pi}\exp\left(-\int_b^c |k(R)|\mathrm{d}R\right), \tag{3.84}$$

provided of course that $\kappa \gg 1$. If not, the solution must be obtained by a similar expansion about the roots of (3.80) if $\kappa \ll 1$, or in the general case by graphical or numerical solution of (3.72).

At this point it is instructive to note that eqn (3.72) actually provides an improved approximation to the quantized levels of the quartic oscillator of eqn (2.112) with E used in place of ε to avoid confusion with the tunnelling parameter. Despite the absence of an explicit barrier the momentum $p(x)$ in (2.113) has two real zeros (turning points) at $x = \pm E^{1/4}$ and two complex ones at $x = \pm \mathrm{i}E^{1/4}$ which is precisely analogous to the double minimum problem at energies above the barrier, but with the added refinement that $\alpha = \pi\varepsilon = aE^{3/4}$ where the parameter a is given by (2.115). It is left to the reader (problem 3.5) to confirm that (3.72) is far superior to (2.115) as a quantization condition.

It is also of interest to examine the nature of the underlying wave-functions as measured by the amplitude ratio D/A, given by eqn (3.73). It is easily verified in any symmetric case that the roots of (3.72) imply the following identity:

$$\sin(\alpha + \delta - \phi) = 2 \sin \alpha' \cos \alpha' = \mp(1 + \kappa^2)^{-1/2} \tag{3.85}$$

where the signs correspond to those in (3.84). Hence, according to (3.73),

$$D/A = \mp 1, \tag{3.86}$$

thereby embodying the known alternation in symmetric and antisymmetric character of successive eigenfunctions.

It is left as an exercise for the reader to analyse the implications of eqns (3.72) and (3.73) in the non-symmetric case, which is relevant, for example, to tunnelling perturbations in the E, F state spectra of the hydrogen molecule.

3.4 Restricted rotation

The barriers to rotation about some axis in a molecule (often a C–C bond) may take various forms, all of which may be handled along the following lines. The chosen example is taken to have three-fold symmetry in order to illustrate how symmetry-determined degeneracies are accommodated.

A typical potential energy function is depicted in Fig. 3.7, together with its associated connection diagram. By contrast with Fig. 3.5, which requires complete reflection at the outermost turning points, the cyclic nature of Fig. 3.7 imposes the condition that the amplitude in any semiclassical segment must remain unchanged on propagation

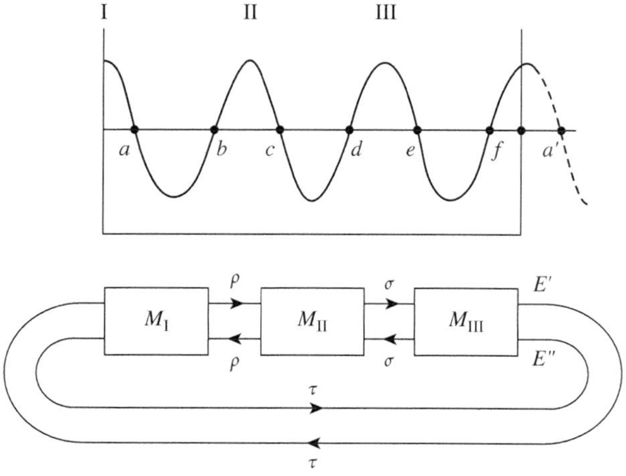

Fig. 3.7 Potential curve and connection diagram for restricted rotation by a three-fold barrier. (Taken from Child (1974a). Copyright 1974, with permission from Elsevier.)

around the cycle. Expressed in mathematical terms this means that if $\mathbf{M}$ denotes the barrier penetration matrix in eqn (3.48), and

$$\mathbf{A} = \begin{pmatrix} A' \\ A'' \end{pmatrix} \quad \text{and} \quad \mathbf{K} = \begin{pmatrix} \mathrm{e}^{\mathrm{i}\sigma} & 0 \\ 0 & \mathrm{e}^{-\mathrm{i}\sigma} \end{pmatrix}, \tag{3.87}$$

where

$$\sigma = \int_a^b k(\theta)\mathrm{d}\theta = \int_c^d k(\theta)\mathrm{d}\theta = \int_e^f k(\theta)\mathrm{d}\theta, \tag{3.88}$$

then the connections in Fig. 3.7 require that

$$\mathbf{A} = \mathbf{M}\,\mathbf{K}\,\mathbf{M}\,\mathbf{K}\,\mathbf{M}\,\mathbf{K}\,\mathbf{A}. \tag{3.89}$$

This means that the matrix $(\mathbf{M}\,\mathbf{K})^3$ must have an eigenvalue equal to unity, or that an eigenvalue of $\mathbf{M}\,\mathbf{K}$ must equal one of the cube roots of unity,

$$\omega_r = \exp(2\pi r\mathrm{i}/3), \quad r = 0, \ \pm 1. \tag{3.90}$$

The quantization condition therefore takes the form

$$\det(\mathbf{M}\,\mathbf{K} - \omega_r\mathbf{I}) = \\ \det\begin{pmatrix} (1+\kappa^2)^{1/2}\mathrm{e}^{\mathrm{i}\sigma-\mathrm{i}\phi} - \omega_r & -\mathrm{i}\kappa\mathrm{e}^{-\mathrm{i}\sigma} \\ \mathrm{i}\kappa\mathrm{e}^{\mathrm{i}\sigma} & (1+\kappa^2)^{1/2}\mathrm{e}^{-\mathrm{i}\sigma+\mathrm{i}\phi} - \omega_r \end{pmatrix} = 0 \tag{3.91}$$

which rearranges to

$$\cos(\sigma - \phi) = \frac{1}{2}(1+\kappa^2)^{-1/2}(\omega_r + \omega_r^{-1}) = (1+\kappa^2)^{-1/2}\cos(2\pi r/3). \tag{3.92}$$

Since $\cos(2\pi r/3) = -1/2$ for both $r = \pm 1$ these two families of solutions have the characteristic double degeneracy associated with functions belonging to the E representations of the C_3 point groups (Cotton 1971). Similarly the $r = 0$ solutions span the A_1 and A_2 representations.

The graphical solution of eqn (3.92) in Fig. 3.8, which brings out this symmetry separation, again assumes that the trigonometric argument $\sigma - \phi$ varies smoothly through the barrier. The justification is similar to that employed for eqn (3.72). In the previous case the variable α acquired a logarithmic term from the left-hand side of the central barrier which was cancelled by the term $-\phi/2$. However the quantity σ now includes two logarithmic terms, because it is enclosed by two barriers, and the correction term is $-\phi$.

Analytical approximations to the eigenvalues are readily obtained by the methods employed in eqns (3.81)–(3.84). For example, at low energies such that $\kappa \gg 1$ the

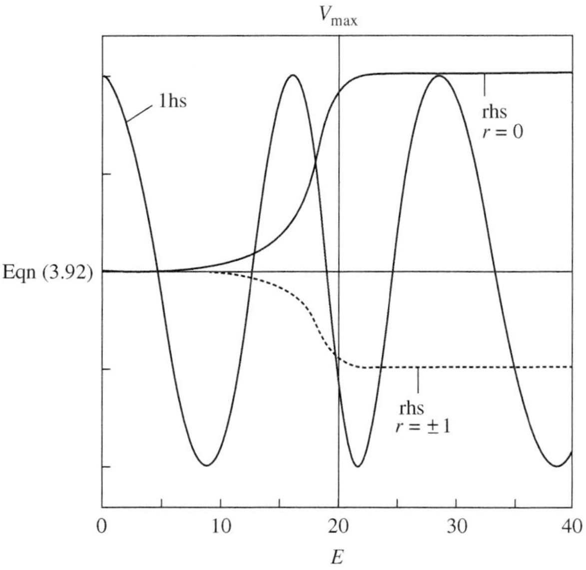

Fig. 3.8 Graphical construction for solution of the three-fold restricted rotation quantization equation (3.92). Roots on the solid and dashed lines correspond to $r = 0$ (symmetry A) and $r = \pm 1$ (symmetry E) respectively.

right-hand side of (3.92) is very small and the solutions cluster in energy around the 'isolated well' eigenvalues, E_v, given by

$$\sigma' = \sigma - \phi = \int_a^b k(\theta)\mathrm{d}\theta - \phi = \left(v + \frac{1}{2}\right)\pi, \tag{3.93}$$

and the splitting shown in Fig. 3.8 may he estimated from the expansion

$$\sigma' = \left(v + \frac{1}{2}\right)\pi + (E - E_v)\pi/\hbar\bar{\omega}, \tag{3.94}$$

where $\hbar\bar{\omega} = \mathrm{d}E_v/\mathrm{d}v$. One readily finds that

$$E_{vr} = E_v - (-1)^v(\hbar\bar{\omega}/\pi)(1 + \kappa^2)^{-1/2}\cos(2\pi r/3). \tag{3.95}$$

Hence the overall splitting at a given value of v is approximately

$$\Delta E \simeq (3\hbar\bar{\omega}/2\pi)(1 + \kappa^2)^{-1/2} \simeq (3\hbar\bar{\omega}/2\pi)\exp\left(-\int_a^b |k(\theta)|\mathrm{d}\theta\right). \tag{3.96}$$

At energies above the barrier a different approximation is valid because $\kappa \simeq 0$ and eqn (3.92) is conveniently rearranged to read

$$\cos(\sigma - \phi) - \cos(2\pi r/3) = -[1 - (1 + \kappa^2)^{-1/2}]\cos(2\pi r/3) \simeq 0. \tag{3.97}$$

It follows that the eigenvalues now lie close to the 'smooth rotation' levels, E_m, given by (compare eqn (3.17))

$$3(\sigma - \phi) = \int_0^{2\pi} k(\theta)\mathrm{d}\theta - 3\phi = 2m\pi, \tag{3.98}$$

where $m = 3n + r$. It is easy to show by means of an expansion similar to that in (3.94) that the A-type levels (with $m = 3n$) show a small splitting given by

$$E = E_m \pm (6\hbar\bar{\omega}'/\pi)[1 - (1 + \kappa^2)^{-1/2}], \tag{3.99}$$

which may be attributed to partial barrier reflection. The remaining, E-type, levels are of course intrinsically doubly degenerate but they are alternately perturbed to high and low energies according to the formula

$$E = E_m \mp (3^{1/2}\hbar\bar{\omega}'/\pi)[1 - (1 + \kappa^2)^{-1/2}], \tag{3.100}$$

for $m = 3n \pm 1$ respectively. The quantity $\hbar\bar{\omega}'$ appearing in eqns (3.99)–(3.100) is of course the local energy splitting of the roots of eqn (3.93). All the above features may be recognized in Fig. 3.8.

3.5 Shape resonances or tunnelling predissociation

The situation illustrated in Fig. 3.9 leads to a more general form of quantization in which one seeks not a sharp energy level, but a combination of the resonance energy, E_v, and its level width, Γ_v, appropriate to the lifetime of the state. The following analysis yields expressions for E_v and Γ_v, in a way that is intended to bring out the interrelation between three different ways in which such resonances may be detected—by sudden changes in the scattering phase shift (Child 1974b), by the breadth of spectroscopic absorption lines, or by the weakness or absence of emission lines indicative of a lifetime short compared with that for fluorescence (Herzberg 1950).

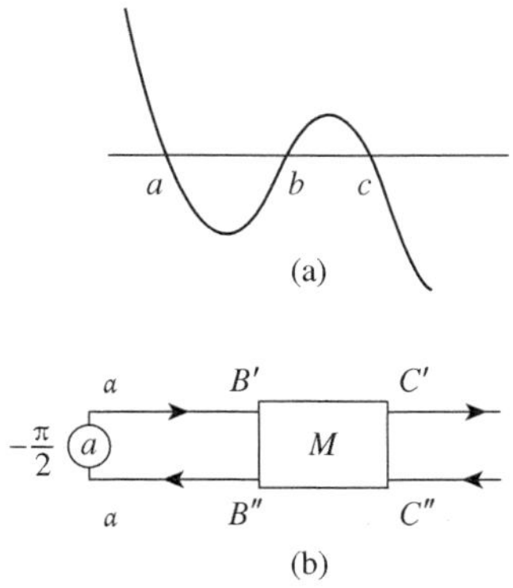

Fig. 3.9 Potential curve and connection diagram for shape resonances. (Taken from Child (1974a). Copyright 1974, with permission from Elsevier.)

The theory is based on the connections implied by Fig. 3.9, which translate, according to eqns (3.44), (3.47) and (3.48), to the identity

$$\begin{aligned}\begin{pmatrix} C' \\ C'' \end{pmatrix} &= \begin{pmatrix} (1+\kappa^2)^{1/2}\mathrm{e}^{-\mathrm{i}\phi} & -\mathrm{i}\kappa \\ \mathrm{i}\kappa & (1+\kappa^2)^{1/2}\mathrm{e}^{\mathrm{i}\phi} \end{pmatrix} \begin{pmatrix} \mathrm{e}^{\mathrm{i}\alpha} & 0 \\ 0 & \mathrm{e}^{-\mathrm{i}\alpha} \end{pmatrix} \begin{pmatrix} A\mathrm{e}^{-\mathrm{i}\pi/4} \\ A\mathrm{e}^{\mathrm{i}\pi/4} \end{pmatrix} \\ &= A\begin{pmatrix} (1+\kappa^2)^{1/2}\exp(\mathrm{i}\alpha - \mathrm{i}\phi - \mathrm{i}\pi/4) + \kappa\exp(-\mathrm{i}\alpha - \mathrm{i}\pi/4) \\ (1+\kappa^2)^{1/2}\exp(-\mathrm{i}\alpha + \mathrm{i}\phi + \mathrm{i}\pi/4) + \kappa\exp(\mathrm{i}\alpha + \mathrm{i}\pi/4) \end{pmatrix} \\ &= AX^{-1}\begin{pmatrix} \mathrm{e}^{\mathrm{i}\gamma} \\ \mathrm{e}^{-\mathrm{i}\gamma} \end{pmatrix}, \end{aligned} \tag{3.101}$$

where

$$X = [1 + 2\kappa^2 + 2\kappa(1+\kappa^2)^{1/2}\cos 2\alpha']^{-1/2} \tag{3.102}$$

$$\gamma = \arctan\left[\left(\frac{(1+\kappa^2)^{1/2} - \kappa}{(1+\kappa^2)^{1/2} + \kappa}\right)\tan\alpha'\right] - \frac{\phi}{2} - \frac{\pi}{4} \tag{3.103}$$

and

$$\alpha' = \alpha - \phi/2 = \int_a^b k(R)\,\mathrm{d}R - \frac{\phi}{2}. \tag{3.104}$$

Equation (3.101) implies that a wavefunction with internal form

$$\psi(R) \overset{a<R<b}{\approx} Xk^{-1/2}(R)\sin\left(\int_a^b k(R)\,\mathrm{d}R + \frac{\pi}{4}\right), \tag{3.105}$$

(compare eqn (3.46) with $A = X$) connects with the external form

$$\psi(R) \overset{R\gtrsim c}{\approx} k^{-1/2}\sin(kR + \eta), \tag{3.106}$$

where k denotes the asymptotic value $k(\infty)$ and the phase term η may be divided into a background contribution, η_0, and a resonant part η_{res}:

$$\begin{aligned}\eta &= \eta_0 + \eta_{\mathrm{res}}, \\ \eta_0 &= \lim_{R\to\infty}\left(\int_c^R k(R)\,\mathrm{d}R - kR - \frac{\phi}{2} + \frac{\pi}{4}\right),\end{aligned}$$

and

$$\eta_{\mathrm{res}} = \arctan\left[\left(\frac{(1+\kappa^2)^{1/2} - \kappa}{(1+\kappa^2)^{1/2} + \kappa}\right)\tan\alpha'\right]. \tag{3.107}$$

To see the physical significance of these equations note first that the energy dependence of the quantity X determines the absorption profile because, as will be seen in Section 5.1, the correct normalization for these intrinsically continuum states requires an energy-independent external normalization coefficient (set equal to unity in eqn (3.106)). Secondly the internal amplitude X and the resonant part of the phase shift, $\eta_{\rm res}$, show linked resonant characteristics at energies sufficiently far below the barrier that $\kappa \gg 1$. For example, according to eqn (3.102), X is typically of order κ^{-1} except at energies such that cos $2\alpha' \simeq -1$; hence at the resonance point itself

$$\alpha' = \int_a^b k(R)\, \mathrm{d}R - \frac{\phi}{2} = \left(v + \frac{1}{2}\right)\pi, \tag{3.108}$$

and X rises to a value of order κ. Similarly $\eta_{\rm res}$ is typically small except when the energy passes through a solution of eqn (3.108) when the change in tan α' causes $\eta_{\rm res}$ to increase abruptly by π. The pattern of these related changes is illustrated in Fig. 3.10.

An explicit expression for the energy width of these resonant features is readily available by expanding about the roots of eqn (3.108) (compare eqns (3.80), (3.81) and (3.94)). Thus in the limit $\kappa \gg 1$ eqns (3.102) and (3.103) imply that

$$X^2(E) \simeq \frac{\hbar\bar{\omega}}{2\pi} \frac{\Gamma_v}{(E - E_v)^2 + \Gamma_v^2/4} \tag{3.109}$$

while

$$\eta_{\rm res} \simeq \arctan[\Gamma_v/2(E_v - E)], \tag{3.110}$$

where in both cases the width term, Γ_v, is given by

$$\Gamma_v \simeq \frac{\hbar\bar{\omega}}{2\pi\kappa^2} = \frac{\hbar\bar{\omega}}{2\pi} \exp\left(-2\int_b^c |k(R)|\mathrm{d}R\right). \tag{3.111}$$

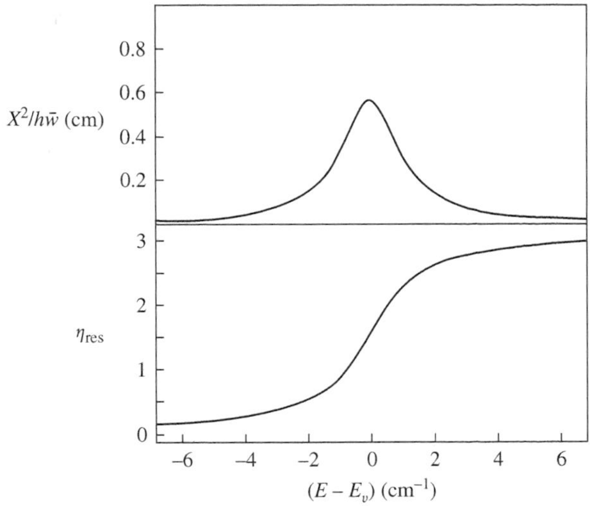

Fig. 3.10 Related changes in the internal amplitude $X^2(E)$ and the resonant part of the phase shift $\eta_{\rm res}(E)$.

It is also easily verified that the integrated absorption intensity from a given resonance depends only on the local level spacing $\hbar\bar{\omega}$, because according to equation (3.109)

$$\int_{-\infty}^{\infty} X^2(E)\mathrm{d}E = \hbar\bar{\omega}, \tag{3.112}$$

a result which neatly fits into the semiclassical normalization scheme for bound and continuum wavefunctions discussed in Section 5.1.

Equation (3.111) also has a natural physical interpretation in terms of the frequency of predissociation events, as measured by the inverse predissociation lifetime

$$\tau_v^{-1} = \frac{\Gamma_v}{\hbar} = \left(\frac{\bar{\omega}}{2\pi}\right) \exp\left(-2\int_b^c |k(R)|\mathrm{d}R\right). \tag{3.113}$$

The factor $\bar{\omega}/2\pi$ gives the classical frequency with which the system encounters the outer turning point (see Section 4.1 for the connection between level spacing $\hbar\bar{\omega}$ and classical frequency), and the exponential factor gives the barrier transmission probability on each such encounter (see eqn (3.55)).

For a final view of this tunnelling predissociation process, relevant to the lifetime of the quasi-bound state, one can impose so-called Siegert (1939) outgoing boundary conditions, $C'' = 0$, on eqn (3.101), which will be seen to result in a complex energy eigenvalue,

$$E = E_v - \mathrm{i}\Gamma_v/2, \tag{3.114}$$

appropriate to the time decay of the quantum mechanical state:

$$|\Psi(R,t)|^2 = |\psi(R)\exp(-\mathrm{i}Et/\hbar)|^2 = |\psi(R)|^2\exp(-\Gamma_v t/\hbar). \tag{3.115}$$

To see this explicitly note that the condition $C'' = 0$ requires that

$$[(1+\kappa^2)^{1/2} + \kappa]\cos\alpha' - \mathrm{i}[(1+\kappa^2)^{1/2} - \kappa]\sin\alpha' = 0, \tag{3.116}$$

in which the first term normally dominates the second. Consequently the roots of eqn (3.116) again lie close to the real energy levels, E_v, at which cos α' vanishes, in other words at which eqn (3.108) is satisfied. It is then a simple matter to expand for α' about $E = E_v$, to obtain

$$\begin{aligned} 2(E - E_v) = -\mathrm{i}\Gamma_v &= \frac{-2\mathrm{i}\hbar\bar{\omega}}{\pi}\left(\frac{(1+\kappa^2)^{1/2} - \kappa}{(1+\kappa^2)^{1/2} + \kappa}\right) \\ &\simeq \frac{-\mathrm{i}\hbar\bar{\omega}}{2\pi}\exp\left(-2\int_b^c |k(R)|\mathrm{d}R\right), \end{aligned} \tag{3.117}$$

in exact agreement with eqn (3.111). This shows that the complex energy formalism gives, at least in the sharp resonance limit, $\kappa \gg 1$, precisely the same prescriptions for both E_v and Γ_v as those required to account for the resonance behaviour of the internal amplitude, $X(E)$, and the phase shift, $\eta(E)$.

As a variant of this complex energy formalism, complex angular momenta are frequently used in nuclear physics to characterize families of Regge poles (Alfaro and Regge 1965) which are analogous to the present resonances. Sukumar and Bardsley (1975), Delos and Carlson (1975) and Connor et al. (1976, 1979, 1980) have used similar semiclassical methods to find these poles at a level well beyond the above linear expansion approximations. Connor (1990a) gives an excellent review, and aspects of the results are discussed in relation to elastic atom–atom scattering in Section 9.2.

Before leaving this section it should be noted that the precise equivalence between the resonance positions and widths associated with the three different prescriptions—internal amplitude, phase shift and Siegert eigenvalue—applies only in the sharp resonance limit, $\kappa \gg 1$. For example, at higher energies the amplitude criterion (3.109) yields the following linewidth formula,

$$\Gamma_v = \frac{\hbar\bar{\omega}}{\pi}\left(\frac{(1+\kappa^2)^{1/2}-\kappa}{[\kappa^2(1+\kappa^2)]^{1/4}}\right), \tag{3.118}$$

which differs from the form implied by η_{res} and by the Siegert quantization eqn (3.116), namely

$$\Gamma_v = \frac{2\hbar\bar{\omega}}{\pi}\left(\frac{(1+\kappa^2)^{1/2}-\kappa}{(1+\kappa^2)^{1/2}+\kappa}\right). \tag{3.119}$$

These differences are not artefacts of the semiclassical approximation because numerical results by LeRoy and Bernstein (1971a) show that different prescriptions may give different values for the resonance parameters, for levels close to the barrier maximum.

3.6 Predissociation by curve crossing

The case of predissociation of a bound state by interaction with a repulsive one is pictured in Fig. 3.11 and modelled by the accompanying connection diagram. Again one can analyse the situation according to the behaviour of the internal amplitude coefficients associated with one or other electronic state (Child 1980), according to the behaviour of the scattering phase shift as measured by the relative phases of Q'_- and Q''_- (Bandrauk and Child 1970), or by Siegert quantization. Only the latter approach will be adopted because it leads most readily to the physically relevant results.

The first step is to notice that the left-hand side of the connection diagram requires, according to (3.44), (3.47) and (3.59), that

$$\begin{pmatrix} Q'_+ \\ Q'_- \end{pmatrix} = \begin{pmatrix} (1-\lambda^2)^{1/2}e^{i\chi} & -\lambda \\ \lambda & (1-\lambda^2)^{1/2}e^{-i\chi} \end{pmatrix} \begin{pmatrix} e^{2i\alpha_+ - i\pi/2} & 0 \\ 0 & e^{2i\alpha_- - i\pi/2} \end{pmatrix}$$

$$\times \begin{pmatrix} (1-\lambda^2)^{1/2}e^{i\chi} & \lambda \\ -\lambda & (1-\lambda^2)^{1/2}e^{-i\chi} \end{pmatrix} \begin{pmatrix} Q''_+ \\ Q''_- \end{pmatrix}, \tag{3.120}$$

where λ and χ are given by eqns (3.60)–(3.62) and

$$\alpha_\pm = \int_{a_\pm}^{R_X} k_\pm(R)\mathrm{d}R. \tag{3.121}$$

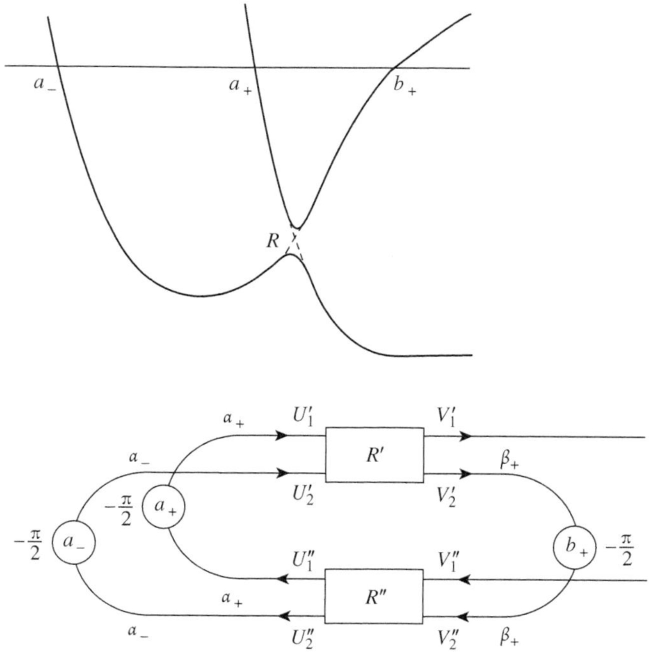

Fig. 3.11 Potential curves and connection diagram for predissociation by curve crossing. (Taken from Child (1974a). Copyright 1974, with permission from Elsevier.)

Secondly, the presence of the turning point, b_+, in Fig. 3.11 requires that

$$Q'_+ = Q''_+ \exp(-2i\beta_+ + i\pi/2). \tag{3.122}$$

Finally, imposition of the Siegert state boundary condition $Q''_- = 0$ in eqn (3.120) yields a second relation between Q'_+ and Q''_+, namely

$$Q'_+ = -i[(1-\lambda^2)\exp(2i\alpha_+ + 2i\chi) + \lambda^2 \exp(2i\alpha_-)]Q''_+. \tag{3.123}$$

Consistency between eqns (3.122) and (3.123) therefore imposes the following complex quantization condition:

$$1 + \exp(2i\theta'_2) = -u[1 + \exp(2i\theta'_+)], \tag{3.124}$$

where θ'_+ and θ'_2 may be regarded as modified adiabatic and diabatic phase integrals respectively,

$$\theta'_+ = \alpha_+ + \beta_+ + \chi = \int_{a_+}^{b_+} k_+(R)\mathrm{d}R + \chi,$$

$$\theta'_2 = \alpha_- + \beta_+ = \int_{a_-}^{R_\chi} k_-(R)\mathrm{d}R + \int_{R_\chi}^{b_+} k_+(R)\mathrm{d}R, \tag{3.125}$$

and u is a curve-crossing transition parameter

$$u = \lambda^{-2} - 1 = \mathrm{e}^{2\pi\nu} - 1, \tag{3.126}$$

with ν and χ given by eqns (3.61) (or (3.66)) and (3.62) respectively. Note that $u \ll 1$ and $u \gg 1$ correspond to the near diabatic ($\nu \to 0$) and near adiabatic ($\nu \to \infty$) limits respectively.

The consequences of eqn (3.124) are now readily assessed. Suppose, for example, that $u \ll 1$; the roots of (3.124) then lie close to energies E_2 at which

$$\theta'_2 = \left(v_2 + \frac{1}{2}\right)\pi \tag{3.127}$$

so that $\exp(2\mathrm{i}\theta'_2) \simeq -1$. This corresponds to Bohr quantization in the modified diabatic potential $V_2(R)$, and it follows by the now familiar expansion procedure,

$$\theta'_2 = \left(v_2 + \frac{1}{2}\right)\pi + (E - E_2)\pi/\hbar\bar{\omega}_2, \tag{3.128}$$

that the complex energy eigenvalue

$$E = E_2 + \Delta_2 - \mathrm{i}\Gamma_2/2 \tag{3.129}$$

with level shift, Δ_2, and width, Γ_2, given by

$$\Delta_2 \simeq \frac{\hbar\bar{\omega}_2}{\pi} u \sin\theta_+ \cos\theta_+, \quad \Gamma_2 \simeq \frac{2\hbar\bar{\omega}_2}{\pi} u \cos^2\theta_+. \tag{3.130}$$

The predicted fluctuations in the linewidth are well attested experimentally (Lefebvre-Brion and Field 1985) and evidence of similar fluctuations in the level shift has also been reported (Schmidt et al. 1988). It is also well known that the linewidth fluctuations provide valuable information about the shape of the repulsive potential $V_1(R)$. A systematic semiclassical inversion is described in Section 6.2.

Equations (3.130) apply in the near diabatic limit, but it is evident from the structure of eqn (3.124) that eqns (3.125)–(3.130) have near adiabatic counterparts when $u \gg 1$. In this case an expansion about the roots, E_+, of the adiabatic quantization equation

$$\theta'_+ = \int_{a_+}^{b_+} k_+(R)\mathrm{d}R + \chi = \left(v_+ + \frac{1}{2}\right)\pi \tag{3.131}$$

Table 3.3 Semiclassical and numerical level widths, Γ_J, and positions, E_J, for resonances with $(v_2, v_+) = (26, 3)$ for the model of Child and Lefebvre (1978). Numerical values were derived from a Lorentzian fit to the energy derivative of the phase shift.

	Γ_J (cm^{-1})		E_J (cm^{-1})	
J	Semiclassical	Numerical	Semiclassical	Numerical
26	5.146	5.302	13 158.1	13 156.9
27	3.078	3.184	13 179.6	13 178.5
28	1.510	1.622	13 201.7	13 200.8
28	1.510	1.622	13 201.7	13 200.8
29	0.466	0.533	13 224.6	13 223.8
30	0.016	0.029	13 248.3	13 247.6
31	0.205	0.172	13 272.8	13 272.0
32	1.087	1.020	13 298.0	13 297.2
33	2.732	2.631	13 323.9	13 323.3
34	5.172	5.024	13 350.5	13 350.1
35	8.518	8.211	13 379.9	13 377.9

leads to the following formulae for the adiabatic level shift and width:

$$\Delta_+ \simeq \frac{\hbar\bar{\omega}_+}{\pi} u^{-1} \sin\theta_2' \cos\theta_2', \quad \Gamma_+ \simeq \frac{2\hbar\bar{\omega}_+}{\pi} u^{-1} \cos^2\theta_2'. \tag{3.132}$$

These limiting formulae were first obtained by Bandrauk and Child (1970) but it was subsequently realized (Child 1976a) that analytical expressions for Δ and Γ are also available for any coupling strengths, u, in the neighbourhood of certain arbitrarily sharp levels. The existence of such sharp resonances is readily understood by reference to eqn (3.124), which is clearly satisfied if ever both eqns (3.128) and (3.131) are simultaneously satisfied, regardless of the value of u. The physical condition is therefore that there should be an accidental coincidence between one notionally diabatic and one notionally adiabatic level. Given this sharp resonance condition it is quite straightforward to analyse the behaviour of neighbouring resonances. All that is necessary is to expand the exponentials to terms quadratic in $(E - E_2)$ and $(E - E_+)$ and to rearrange the resulting equation. The expressions for the resulting resonance positions and widths turn out to be

$$\begin{aligned} E &= (E_2 + xE_+)/(1 + x) \\ \Gamma &= 2\pi x(1 + \gamma x)(E_2 - E_+)^2/[\hbar\bar{\omega}_2(1 + x)^3], \end{aligned} \tag{3.133}$$

where

$$x = u\hbar\bar{\omega}_2/\hbar\bar{\omega}_+, \quad \gamma = \hbar\bar{\omega}_2/\hbar\bar{\omega}_+. \tag{3.134}$$

The entries in Table 3.3, due to Child and Lefebvre (1978), demonstrate the remarkable accuracy of this intermediate coupling approximation, which was first used by Child (1976a) to analyse the fragmentary visible absorption spectrum of IBr (Selin 1962).

3.7 Problems

3.1. Use the substitutions $\sinh(ax) = \sqrt{(1-\varepsilon)/\varepsilon}\sin\theta$, where $\varepsilon = -E/D$, to show that the Bohr quantization condition for the potential $V(x) = -D/\cosh^2(\alpha x)$, yields

$$E_v \simeq -\frac{\alpha^2\hbar^2}{2\mu}\left[\frac{(2\mu D)^{1/2}}{\alpha\hbar} - \left(v+\frac{1}{2}\right)\right]^2.$$

Discuss the origin of discrepancies from the exact result (Landau and Lifshitz 1965)

$$E_v = -\frac{\alpha^2\hbar^2}{2\mu}\left[\frac{(8\mu D + \alpha^2\hbar^2)^{1/2}}{2\alpha\hbar} - \left(v+\frac{1}{2}\right)^2\right].$$

3.2. Dunham (1932) uses the notation $\xi = (r/r_e) - 1$, $F = E/hc$, $U = V/hc$, $B_e = h/8\pi^2\mu c r_e^2$ and $U' = \mathrm{d}U/\mathrm{d}\xi$, together with the expansions

$$U' = A_1U^{1/2} + A_2U + A_3U^{3/2} + \ldots$$
$$(U')^{-1} = A_1^{-1}U^{-1/2}[1 + B_1U^{1/2} + B_2U + \ldots],$$

to show that the second-order quantization condition (2.111) may be written as

$$\begin{aligned}2\pi B_e^{1/2}\left(v+\frac{1}{2}\right) &= \oint (F-U)^{1/2}\mathrm{d}\xi - \frac{1}{32}B_e\oint U'^2(F-U)^{-5/2}\mathrm{d}\xi\\ &= \oint (F-U)^{1/2}(U')^{-1}\,\mathrm{d}U - \frac{1}{32}B_e\oint U'(F-U)^{-5/2}\mathrm{d}U,\\ &= (2\pi/A_1)\left[F + \frac{1}{4}B_2F^2 + \ldots\right] - \frac{\pi B_e}{8}[A_3 + 5A_5F/2 + \ldots].\end{aligned}$$

(i) Show by inverting the quantization condition that to order $(v+\frac{1}{2})^2$

$$F \simeq \frac{1}{16}B_eA_1A_3 + A_1B_e^{1/2}\left(v+\frac{1}{2}\right) - \frac{1}{4}B_eA_1^2B_2\left(v+\frac{1}{2}\right)^2 + \ldots,$$

on the assumptions that $A_3 \ll (v+1/2)B_e^{1/2}$ and $B_eA_1A_5 \ll 1$.

(ii) Assuming the power series expansion

$$U(\xi) = a_0\xi^2[1 + a_1\xi + a_2\xi^2 + \ldots],$$

with $a_0 = \omega_e^2/4B_e$, show by comparing coefficients of powers of ξ that the quantities A_ν and B_ν are given by

$$A_1 = 2a_0^{1/2},\quad A_2 = 2a_1,\quad A_3 = a_0^{-1/2}(3a_2 - 7a_1^2/4),$$
$$B_1 = -(A_2/A_1),\quad B_2 = -[(A_3/A_1) - (A_2/A_1)^2].$$

(iii) Show by retaining only leading terms in B_e/ω_e that

$$\begin{aligned} U_J(\xi) &= U(\xi) + J(J+1)B_e(1+\xi)^{-2} \\ &\simeq J(J+1)B_e + a_0[1+3(1+a_1)\delta](\xi-\delta)^2 \end{aligned}$$

where $\delta = J(J+1)B_e/a_0$.

(iv) Hence obtain the following leading contributions to coefficients in the Dunham expansion:

$$F_{vJ} = \sum_{ij} Y_{ij}\left(v+\frac{1}{2}\right)^i [J(J+1)]^j$$

$$\begin{aligned} &Y_{00} \simeq (B_e/8)(3a_2 - 7a_1^2/4), \quad Y_{10} \simeq \omega_e \\ &Y_{20} \simeq (3B_e/2)(a_2 - 5a_1^2/4), \quad Y_{01} \simeq B_e \\ &Y_{11} \simeq (6B_e^2/\omega_e)(1+a_1). \end{aligned}$$

(v) By means of the substitutions $Y_{20} \simeq -\omega_e x_e$ and $Y_{11} = -\alpha_e$, deduce that

$$Y_{00} \simeq \frac{1}{4}B_e + \frac{1}{12}(\alpha_e\omega_e/B_e) + \frac{1}{144}(\alpha_e^2\omega_e^2/B_e^3) - \frac{1}{4}\omega_e x_e.$$

See Simons et al. (1973) and Thakker (1975) for similar results, derived from alternative forms of potential expansion.

3.3. The quartic oscillator, for which $k^2(x) = (E - x^4)$, gives rise to two real ($\pm x_E$) and two imaginary ($\pm ix_E$) turning points, where $x_E = E^{1/4}$. Verify, by comparison with (2.115), that

$$\int_{-x_E}^{x_E} k(x)\mathrm{d}x = -\mathrm{i}\int_{-\mathrm{i}x_E}^{\mathrm{i}x_E} k(x)\mathrm{d}x = a\alpha^3,$$

where $a = \frac{1}{2}\Gamma(\frac{1}{4})\Gamma(\frac{3}{2})/\Gamma(\frac{7}{4}) \simeq 1.74804$. Hence, by using the four turning point quantization equation (3.72), find the five lowest eigenvalues. Compare with the entries in Table 2.1.

3.4. Demonstrate, for large κ, that the roots of (3.72) are equal to the eigenvalues of the local perturbation matrix

$$H = \begin{pmatrix} E_a & V_{ad} \\ V_{ad} & E_d \end{pmatrix},$$

where E_a is the energy at which $\alpha - \phi/2 = (v_a + \frac{1}{2})\pi$, and similarly for E_d, δ and v_d; also

$$V_{ad}^2 \simeq \frac{\hbar\bar{\omega}_a\hbar\bar{\omega}_d}{\pi^2[(1+\kappa^2)^{1/2}+\kappa]^2} \simeq \left[\frac{\hbar\omega}{2\pi}\exp\left(-\int_b^c |k(x)|\mathrm{d}x\right)\right]^2,$$

with $\hbar\omega = (\hbar\bar{\omega}_a\hbar\bar{\omega}_d)^{1/2}$. Confirm also that the eigenvector components are in the ratio $(-1)^{v_a+v_d}[D(\hbar\bar{\omega}_d)^{-1/2}]/[A(\hbar\bar{\omega}_a)^{-1/2}]$ with A and D determined by (3.73). [Hint: retain only the dominant term in the expansion for (D/A).]

3.5. Use the connection formulae to derive the following quantization equation for motion governed by a cyclic two-fold barrier:

$$\cos(\sigma - \phi) = \pm(1+\kappa^2)^{-1/2},$$

where σ, ϕ and κ have the meanings in (3.49) and (3.88). Hence estimate the five lowest eigenvalues of the Mathieu equation

$$(\mathrm{d}^2\psi/\mathrm{d}x^2) + (\varepsilon - 2q\cos 2x)\psi = 0,$$

for $q = 5$. The tabulated values are -5.800, -5.790, 1.858, 2.099 and 7.449.

3.6. Define

$$P'_+ = A_+ \exp(\mathrm{i}\alpha_+ - \mathrm{i}\pi/4), \quad P'_- = A_- \exp(\mathrm{i}\alpha_- - \mathrm{i}\pi/4),$$
$$Q'_+ = B_+ \exp(-\mathrm{i}\beta_+ + \mathrm{i}\pi/4), \quad Q'_- = \mathrm{e}^{\mathrm{i}\gamma},$$

where $A_\pm$ and B_+ are real, and use eqn (3.59) to prove that

$$A_-^2 = \frac{u(u+1)\cos^2\theta'_+}{[\cos^2\theta'_2 + 2u\cos(\theta'_+ - \theta'_2)\cos\theta'_+\cos\theta'_2 + u^2\cos^2\theta'_+]},$$

where u, θ'_+ and θ'_2 are given by (3.125) and (3.126).

Deduce, for $u \ll 1$ and energies $E \simeq E_2$ such that $\theta'_2(E) \simeq (v_2 + \frac{1}{2})\pi$, that the internal channel 2 amplitude has the following Lorentzian absorption profile:

$$A_-^2(E) \simeq \frac{\hbar\bar{\omega}_2}{2\pi}\frac{\Gamma}{(E - E_2 - \Delta)^2 + \Gamma^2/4},$$

where Γ and Δ are given by (3.130).

4
Angle–action variables

The Bohr–Sommerfeld quantization rule,

$$\frac{1}{2\pi}\oint p\,\mathrm{d}q = (n+\delta)\hbar, \tag{4.1}$$

arose in Section 3.1 from the boundary conditions on the JWKB wave-function. Here it is set in the context of classical angle–action theory from which it evolved (Bohr 1913). The idea is that any bound periodic motion can be simply described in special action–angle variables, termed here (I, α), such that the action, defined in Cartesian variables by the equation

$$I = \frac{1}{2\pi}\oint p\,\mathrm{d}q, \tag{4.2}$$

is a constant of the motion and the conjugate angle, α, varies linearly with time. Some authors use q (Goldstein 1980; Miller 1976) or θ (Percival and Richards 1982) in place of the present α, while others (Born 1960) replace (I, α) by $J = 2\pi I$ and $w = \alpha/2\pi$. Here a Greek letter is preferred for an angle but θ would lead to confusion with the normal polar angle.

The origin of eqn (4.2) is illustrated in Section 4.1 for the simple case of a linear oscillator and a general angle–action formulation of semiclassical mechanics is outlined. Subsequent sections deal with a variety of important separable systems such as degenerate harmonic oscillators, angular momentum, the hydrogen atom and symmetric and asymmetric tops. Care has been taken to define the angles in a manner consistent with quantum mechanical phase conventions, and the main results are listed for convenient reference in Appendix E. The additional problems posed by non-separable systems are treated in Chapter 7. The underlying source in both cases is Born's classic text (Born 1960).

The chapter concludes with an introduction to the phenomenon of 'quantum monodromy', which relates to possible topological impediments to the construction of a global angle–action system. The physical implications have cast interesting light on quantum mechanical level structures associated with barriers on molecular potential energy surfaces.

4.1 The linear oscillator

The motions of a linear oscillator, governed by the Hamiltonian

$$H = \frac{1}{2m}p^2 + V(q), \tag{4.3}$$

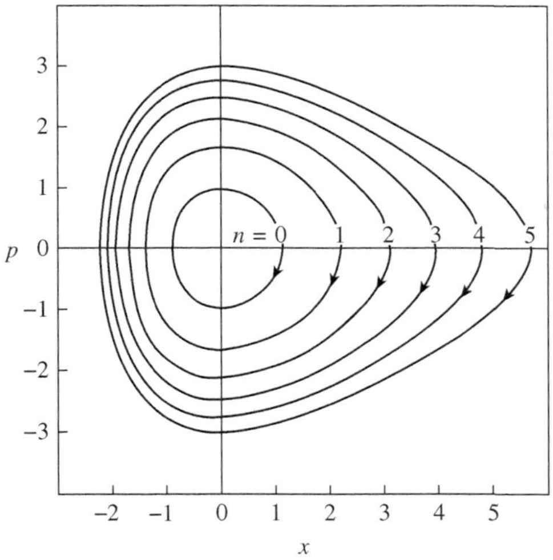

Fig. 4.1 Phase space orbits of the Morse oscillator (see Appendix E), with parameters $D = 3.75$, $a = (2D)^{-1/2}$. The actions are given by $I = n + \frac{1}{2}$ and the corresponding energies are $E_n = n + \frac{1}{2} - (n + \frac{1}{2})^2/4D$.

are used to introduce the concepts. The first point is that the family of periodic orbits, illustrated in Fig. 4.1, may be interpreted as energy contours, $E = H(p, q)$, in the (p, q) plane. Hence, since the energy and the area enclosed, I, by a given orbit (see (4.2)) are both constants, one must be functionally dependent on the other, say

$$E = \tilde{H}(I), \tag{4.4}$$

where the function $\tilde{H}(I)$ henceforth plays the part of a transformed Hamiltonian. On the assumption that Hamilton's equations still apply, in other words that the (I, α) system is canonical (Percival and Richards 1982), it is then a simple matter to confirm that I is indeed a constant of the motion (this is of course already evident from (4.4)) and that α varies linearly with time. The formal equations are

$$\mathrm{d}I/\mathrm{d}t = -(\partial \tilde{H}/\partial \alpha) = 0$$

and

$$\mathrm{d}\alpha/\mathrm{d}t = \partial \tilde{H}/\partial I = \omega(I) = \text{constant}. \tag{4.5}$$

It also follows in turn from the Bohr correspondence principle

$$I = (n + \delta)\hbar, \tag{4.6}$$

implied by (4.1) and (4.2), that the classical frequency $\omega(I)$ in (4.5) is directly related to the quantum mechanical level spacing by the equation

$$\partial \tilde{H}/\partial n = \mathrm{d}E/\mathrm{d}n = \hbar\bar{\omega} \tag{4.7}$$

where $\bar{\omega}$ is interpreted as the local frequency, $\bar{\omega} = \omega[(n+\delta)\hbar]$, at the quantum number of interest.

To put the theory in a more mathematical context it is necessary to recall some aspects of classical canonical transformation theory (see Appendix C and Percival and Richards (1982)). The essential feature is that a transformation from $(p,\ q)$ to a new system $(P,\ Q)$ is automatically canonical if it is 'generated' by partial differentiation of some function $F_1(q,Q)$, $F_2(q,P)$, $F_3(p,Q)$, or $F_4(p,P)$ dependent on one initial and one final variable. In the present context a generator of type $F_2(q,I)$ is most convenient for the $(p,q) \to (I,\alpha)$ transformation and the conjugate variables are generated by the equations

$$p(q,I) = \partial F_2/\partial q, \qquad \alpha(q,I) = \partial F_2/\partial I. \tag{4.8}$$

Similar equations apply, apart from certain sign differences, for the other generators, F_i (Percival and Richards 1982).

So far the function $F_2(q,I)$ is arbitrary, but to be useful the chosen form must ensure that p given by (4.8) is consistent with (4.3) and (4.4), namely

$$H = \frac{1}{2m}p^2 + V(q) = \tilde{H}(I). \tag{4.9}$$

To do this $F_2(q,I)$, which is normally denoted as $S(q,\ I)$ in this context, is chosen to satisfy the Hamilton–Jacobi equation

$$\frac{1}{2m}\left(\frac{\partial S}{\partial q}\right)^2 + V(q) = \tilde{H}(I). \tag{4.10}$$

The solution is the JWKB phase integral, which will be taken initially in the indefinite form

$$S(q,I) = \int^q \{2m[\tilde{H}(I) - V(q)]\}^{1/2}\mathrm{d}q; \tag{4.11}$$

the second of eqns (4.8) then yields the following expression for the angle variable:

$$\alpha(q,I) = \left(\frac{\partial S}{\partial I}\right) = \left(\frac{m}{2}\right)^{1/2}\left(\frac{\partial \tilde{H}}{\partial I}\right)\int^q [\tilde{H}(I) - V(q)]^{-1/2}\,\mathrm{d}q. \tag{4.12}$$

It is readily verified by introducing the classical velocity

$$v(q) = (2/m)^{1/2}[\tilde{H}(I) - V(q)]^{1/2} \tag{4.13}$$

that the above definition indeed ensures that α increases linearly with time, at a frequency $\omega(I)$, because in the light of (4.5), (4.12) and (4.13),

$$\alpha = \omega(I)\int^q \frac{\mathrm{d}q}{v(q)} = \omega(I)t + \text{constant}, \tag{4.14}$$

where t is the time to pass from some fixed point to q.

The simple harmonic oscillator, with Hamiltonian

$$H = \frac{1}{2m}p^2 + \frac{1}{2}m\omega^2 q^2 = E, \tag{4.15}$$

may be used to show how the theory works in practice. Equation (4.2) for the action yields

$$I = \frac{1}{2\pi}\oint p\ \mathrm{d}q = \frac{m\omega}{\pi}\int_{-q_0}^{q_0} (q_0^2 - q^2)^{1/2}\mathrm{d}q, \tag{4.16}$$

where $\pm q_0$ denote the turning points, with

$$q_0 = (2E/m)^{1/2}/\omega, \tag{4.17}$$

and the momentum has been taken as negative on the lower branch of the orbit in Fig. 4.1. Hence, on evaluating the integral,

$$I = m\omega q_0^2/2 = E/\omega \tag{4.18}$$

or

$$E = \tilde{H}(I) = I\omega. \tag{4.19}$$

In other words the frequency $\partial\tilde{H}/\partial I$ is independent of I or E, which is indeed the hallmark of harmonic motion.

The second step in the transformation is to use eqn (4.12) to relate the angle α to the Cartesian variables $(p,\ q)$. This requires a choice of fixed integration limit to determine the origin of α, which in turn imposes a phase convention (see below); the convenient choice for later applications is

$$\begin{aligned}\alpha &= \left(\frac{m\omega^2}{2}\right)^{1/2}\int_q^{q_0}\left(I\omega - \frac{1}{2}m\omega^2 q^2\right)^{-1/2}\mathrm{d}q \\ &= \int_q^{q_0}\frac{\mathrm{d}q}{(q_0^2 - q^2)^{1/2}} = \arccos(q/q_0).\end{aligned} \tag{4.20}$$

Consequently

$$q = (2I/m\omega)^{1/2}\cos\alpha, \qquad p = -(2Im\omega)^{1/2}\sin\alpha, \tag{4.21}$$

where the sign of p has been chosen to be consistent with eqn (4.8) and with $\mathrm{d}q/\mathrm{d}t = \partial H/\partial p = p/m$. Notice that eqn (4.20) strictly applies only for positive momentum $p(q)$; to follow the negative momentum branch of the orbit the sign is changed and a different branch of the inverse cosine is adopted to ensure continuity in α. Equations (4.21) define the Cartesian $(q,\ p)$ to angle–action (I, α) transformation.

Notice that eqns (4.1) (with $\delta = \frac{1}{2}$ for vibrational motion), (4.2) and (4.19) lead to the correct equation for the harmonic oscillator energy levels

$$E_n = \left(n + \frac{1}{2}\right) \hbar\omega \tag{4.22}$$

as already demonstrated by eqn (2.60). Similar arguments apply for other systems with the functional form for $\tilde{H}(I)$ differing from system to system; for example, the Morse oscillator yields an additional term, quadratic in $(n + \frac{1}{2})$, while in the case of the hydrogen atom the transformed Hamiltonian, H, varies as I^{-2} (see Section 4.4). In each case, however, the action I is strictly quantized in units of $\hbar$, but the pattern of energy levels varies from system to system according to the functional form of the classical angle–action Hamiltonian.

Turning to the connection with quantum mechanics, there is unfortunately no rigorous unitary transformation from Cartesian to angle–action variables (Carruthers and Nieto 1968; Leaf 1969; Augustin and Rabitz 1979), but one can nevertheless develop a coherent semiclassical theory by associating the action, in classical mechanics, with a semiclassical operator

$$\hat{I} = -\mathrm{i}\hbar(\partial/\partial\alpha) + \delta\hbar, \tag{4.23}$$

which conveniently includes the Maslov (1972) term δ from eqn (4.1) without disturbing the proper commutation relation

$$[\hat{I}, \alpha] = -\mathrm{i}\hbar. \tag{4.24}$$

Equation (4.23) has the remarkable feature that the periodic eigenfunctions of $\hat{I}$ are identical for all systems:

$$\phi_n(\alpha) = (2\pi)^{-1/2} \exp(\mathrm{i}n\alpha), \tag{4.25}$$

with n taking the values $0, 1, 2, \ldots$ for vibrational or orbital motion and $0, \pm1, \pm2, \ldots$ for rotational motion. One then sees by application of (4.23) that

$$\hat{I}\phi_n(\alpha) = (n + \delta)\hbar\, \phi_n(\alpha), \tag{4.26}$$

as required for consistency with eqns (4.1) and (4.2). Moreover, since $\hat{I}$ commutes with the obvious quantum mechanical analogue of $\tilde{H}(I)$, namely

$$\hat{H} = \tilde{H}(\hat{I}), \tag{4.27}$$

it follows that the eigenfunctions of $\hat{I}$ are also eigenfunctions of $\hat{H}$, with eigenvalues

$$E_n = \tilde{H}[(n + \delta)\hbar]. \tag{4.28}$$

One should also bear in mind that precise knowledge of the action variable implies complete uncertainty in the classical angle.

Equation (4.25) will play an important role in tracing the Heisenberg correspondence between classical Fourier components and quantum mechanical matrix elements (see Section 5.2), in a phase convention such that the wavefunction is positive for coordinates corresponding to $\alpha \simeq 0$ (thus eqns (4.20)–(4.21) are consistent with the assumption that all $\psi_n(q)$ are positive at their outer classical turning points). The burden of any calculation rests on expressing the appropriate operator in terms of angle–action variables. The necessary transformations are given for various important separable systems in the following sections and collected in Appendix E. Other aspects of the theory of transformations in classical and semiclassical mechanics are discussed in Appendix C.

4.2 The degenerate harmonic oscillator

The degenerate harmonic oscillator, with Hamiltonian

$$H = \frac{1}{2\mu}(p_x^2 + p_y^2) + \frac{1}{2}\mu\omega^2(x^2 + y^2), \tag{4.29}$$

is of particular interest as an introduction to the concept of classical degeneracy and also as a vehicle for comparing the treatment of vibrational angular momentum in classical and quantum mechanics (note that μ has been introduced to denote the mass in eqn (4.29), to avoid confusion with the angular momentum quantum number).

To see the concept of classical degeneracy in its simplest form note that the separability of eqn (4.29) allows a transformation to angle–action variables (I_x, α_x) and (I_y, α_y), in terms of which, by analogy with eqns (4.16)–(4.22),

$$H(I_x, I_y) = (I_x + I_y)\omega = (n_x + n_y + 1)\hbar \tag{4.30}$$

and

$$x = (2I_x/\mu\omega)^{1/2}\cos\alpha_x, \quad y = (2I_y/\mu\omega)^{1/2}\cos\alpha_y. \tag{4.31}$$

The classical degeneracy lies in the fact that the transformed Hamiltonian $\tilde{H}(I_x, I_y)$ depends on a linear combination of the individual actions I_x and I_y. Consequently the dynamics may be simplified by performing a further transformation to the new variables

$$\begin{aligned} I &= I_x + I_y, \quad \alpha_I = (\alpha_x + a_y)/2, \\ J &= I_x - I_y, \quad \alpha_J = (\alpha_x - \alpha_y)/2, \end{aligned} \tag{4.32}$$

where the factors of 1/2 ensure the unit Jacobian required of a canonical transformation (Percival and Richards 1982). One then sees that $\tilde{H}$ depends only on I,

$$\tilde{H} = I\omega = (n + 1)\hbar, \tag{4.33}$$

which means that I, J and α_J are all constants of the motion, while α_I varies linearly with time

$$\alpha = \omega t + \text{constant}. \tag{4.34}$$

The physical implication is that I determines the energy of the system while J and α_J determine the nature of the various degenerate states. This secondary role becomes apparent by combining eqns (4.31) and (4.32) in the form

$$x = [(I + J)/\mu\omega]^{1/2} \cos(\alpha_I + \alpha_J), \quad y = [(I - J)/\mu\omega]^{1/2} \cos(\alpha_I - \alpha_J), \tag{4.35}$$

because it may be verified by eliminating α_I, which carries the time dependence, that eqns (4.35) describe motion around an ellipse (as shown in Fig. 4.2) whose principal axes have lengths

$$r_{\pm}^2 = (\mu\omega)^{-1}[I \pm (I^2 \cos^2 2\alpha_J + J^2 \sin^2 2\alpha_J)^{1/2}], \tag{4.36}$$

with the major axis inclined at an angle ϕ_0 to the x axis, where

$$\tan 2\phi_0 = (I^2 - J^2)^{1/2} \cos 2\alpha_J / J. \tag{4.37}$$

In other words J and α_J jointly determine the fixed orientation and eccentricity, $\varepsilon = (1 - r_-^2/r_+^2)^{1/2}$, of the ellipse. In classical mechanics both J and α_J are continuous variables, but in quantum mechanics $J/\hbar$ is limited by eqns (4.32) to the values $n, n-2, \ldots, -n$, where n is the quantum number in eqn (4.33).

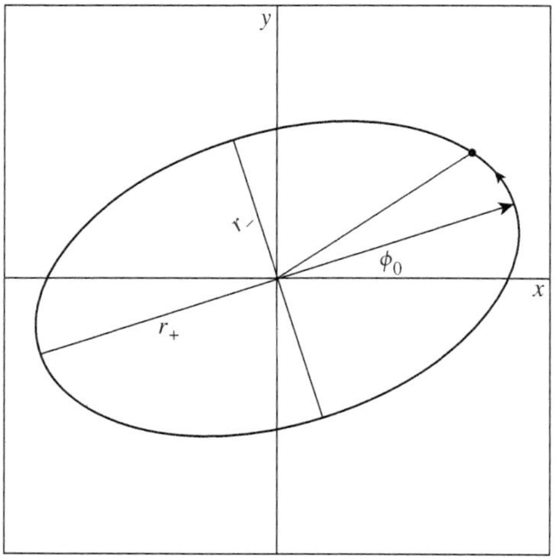

Fig. 4.2 A closed elliptical orbit of the degenerate harmonic oscillator. The quantities $r_{\pm}$ and ϕ_0 are given by eqns (4.36) and (4.37) in the Cartesian formulation and by (4.53) and (4.54) in the polar representation. The angle α_I increases in the direction of the arrow if $I_\phi > 0$.

An alternative formulation, which highlights the vibrational angular momentum, may be obtained by working in polar coordinates, with the Hamiltonian

$$H = \frac{1}{2\mu}(p_r^2 + p_\phi^2/r^2) + \frac{1}{2}\mu\omega^2 r^2. \tag{4.38}$$

Determination of the angle–action variables is, however, more complicated. The first step is to separate the Hamilton–Jacobi equation,

$$H = \frac{1}{2\mu}\left[\left(\frac{\partial S}{\partial r}\right)^2 + \frac{1}{r^2}\left(\frac{\partial S}{\partial \phi}\right)^2\right] + \frac{1}{2}\mu\omega^2 r^2 = E, \tag{4.39}$$

in the form

$$p_\phi = \partial S/\partial\phi = M \tag{4.40}$$

$$p_r = \partial S/\partial r = \pm(2\mu E - M^2/r^2 - \mu^2\omega^2 r^2)^{1/2}, \tag{4.41}$$

where the sign of p_r depends on whether r is on the increasing or decreasing branch of its closed orbit. The angular and radial actions, I_ϕ and I_r respectively, are then expressed in terms of E and the separation constant M by the equations

$$\begin{aligned} I_\phi &= \frac{1}{2\pi}\oint M\mathrm{d}\phi = M \\ I_r &= \frac{1}{2\pi}\oint p_r\,\mathrm{d}_r = \frac{1}{\pi}\int_a^b (2\mu E - M^2/r^2 - \mu^2\omega^2 r^2)^{1/2}\mathrm{d}r. \end{aligned} \tag{4.42}$$

The radial integral may be cast by the substitution $z = r^2$ into the standard form (Gradsteyn and Ryyzhik 1980)

$$I_r = \frac{1}{2\pi}\int_{a^{1/2}}^{b^{1/2}} (A + Bz + Cz^2)^{1/2} z^{-1}\mathrm{d}z = \frac{1}{4}[B(-C)^{-1/2} - 2(-A)^{1/2}], \tag{4.43}$$

where a and b are the turning point values of r and

$$A = -M^2,\ B = 2\mu E,\ C = -\mu^2\omega^2. \tag{4.44}$$

Consequently, on making these substitutions in eqn (4.42),

$$E = \tilde{H}(I_r, I_\phi) = (2I_r + |I_\phi|)\omega. \tag{4.45}$$

Note that the radial frequency $\omega_r = \partial H/\partial I_r = 2\omega$ while $\omega_\phi = \omega$ because the radial motion executes two cycles through $r_\pm$ during one cycle of the ellipse in Fig. 4.2. Again the transformed Hamiltonian depends on a linear combination of I_r and I_ϕ, which plays the role of the total action

$$I = 2I_r + |I_\phi|, \tag{4.46}$$

while I_ϕ, which is the vibrational angular momentum, is taken as the second action variable.

The picture is completed by deriving expressions for the conjugate angle variables, α_I and α_ϕ, from the equations analogous to (4.11), namely

$$\alpha_I = \left(\frac{\partial S}{\partial I}\right) = \mu\omega \int_r^b \frac{\mathrm{d}r}{(2\mu\omega I - \mu^2\omega^2 r^2 - I_\phi^2/r^2)^{1/2}}$$
$$= \frac{1}{2}\arccos\left(\frac{\mu\omega r^2 - I}{(I^2 - I_\phi^2)^{1/2}}\right) \tag{4.47}$$

and

$$\alpha_\phi = \left(\frac{\partial S}{\partial I_\phi}\right) = \phi - I_\phi \int_b^r \frac{\mathrm{d}r}{r(2\mu\omega I - \mu^2\omega^2 r^2 - I_\phi^2/r^2)^{1/2}}$$
$$= \phi - \frac{1}{2}\left(\frac{I_\phi}{|I_\phi|}\right)\arccos\left(\frac{I\mu\omega r^2 - I_\phi^2}{\mu\omega r^2(I^2 - I_\phi^2)^{1/2}}\right). \tag{4.48}$$

Note that the outer turning point, $b = I + (I^2 - I_\phi^2)^{1/2}$, has been taken as the fixed integration limit. The radial integrals were evaluated by the substitution employed for (4.42), and the sign of the square root and branches of the inverse cosines were chosen to be consistent with continuous motion round the cycle.

Put in a more convenient form, eqn (4.47) implies that

$$\mu\omega r^2 = I + (I^2 - I_\phi^2)^{1/2}\cos 2\alpha_I \tag{4.49}$$

while eqns (4.48) and (4.49) rearrange to

$$\phi = \alpha_\phi + \arctan\{I_\phi \tan\alpha_I/[I + (I^2 - I_\phi^2)^{1/2}]\}. \tag{4.50}$$

These two equations determine the time dependence of r and ϕ; I, I_ϕ and α_ϕ are constants because $\tilde{H}$ is independent of α_I, α_ϕ and I_ϕ, while α_I varies as

$$\alpha_I = \omega t + \text{constant} \tag{4.51}$$

because $\tilde{H} = I\omega$ and $\dot{\alpha}_I = \partial\tilde{H}/\partial I = \omega$. Elimination of α_I between (4.49) and (4.50) again results in the equation of an ellipse,

$$\frac{\mu\omega r^2\cos^2(\phi - \alpha_\phi)}{I + (I^2 - I_\phi^2)^{1/2}} + \frac{\mu\omega r^2\sin^2(\phi - \alpha_\phi)}{I - (I^2 - I_\phi^2)^{1/2}} = 1. \tag{4.52}$$

Its principal axes have lengths given by

$$\mu\omega r_\pm^2 = I \pm (I^2 - I_\phi^2)^{1/2}, \tag{4.53}$$

and the angle ϕ_0 that fixes the orientation in Fig. 4.2 is now simply

$$\phi_0 = \alpha_\phi. \tag{4.54}$$

This is a much cleaner specification than that given by eqns (4.36) and (4.37). Useful variants of (4.47)–(4.49) are collected in Appendix E.2.

Equations (4.42), (4.45) and (4.46) also provide immediate contact with the quantum mechanical picture (Pauling and Wilson 1935). Given that the motions in ϕ and r are rotational and vibrational in character respectively, the proper quantization equations are

$$I_\phi = m\hbar \quad \text{and} \quad I_r = \left(v + \frac{1}{2}\right)\hbar. \tag{4.55}$$

Hence, according to eqn (4.45) with $n = 2v + |m|$,

$$E = (2v + |m| + 1)\hbar\omega = (n+1)\hbar\omega, \tag{4.56}$$

subject to the familiar restrictions $m = n, n-2, \ldots, -n$, because $v \geq 0$ and states with $\pm m$ values must be degenerate since (4.38) depends only on I_ϕ^2.

It should also be noted that although the classical motion is restricted to a fixed ellipse, the angle variables are completely uncertain in any quantum mechanical state with defined n and m. Hence the classical equivalent of such a state is the complete family of all fixed ellipses with the same total action and the same eccentricity. Consequently the calculation of any matrix element (see Section 5.2) requires integration (via the angle variables) over cycles of all members of this family.

4.3 Angular momentum

The angle–action theory of angular momentum has similarities with the polar description of the degenerate harmonic oscillator. The square of the total angular momentum plays the role of an effective Hamiltonian,

$$J^2 = p_\theta^2 + p_\phi^2/\sin^2\theta, \tag{4.57}$$

and the corresponding Hamilton–Jacobi equation

$$J^2 = \left(\frac{\partial S}{\partial \theta}\right)^2 + \frac{1}{\sin^2\theta}\left(\frac{\partial S}{\partial \theta}\right)^2 \tag{4.58}$$

separates in the form

$$\begin{aligned} p_\phi &= \left(\frac{\partial S}{\partial \phi}\right) = M \\ p_\theta &= \left(\frac{\partial S}{\partial \theta}\right) = \left(J^2 - \frac{M^2}{\sin^2\theta}\right)^{1/2}. \end{aligned} \tag{4.59}$$

The action equations therefore yield

$$I_\phi = \frac{1}{2\pi}\oint p_\phi \mathrm{d}\phi = M \tag{4.60}$$

and

$$I_\theta = \frac{1}{2\pi}\oint p_\theta \, \mathrm{d}\theta = \frac{1}{\pi}\int_{\theta_0}^{\pi-\theta_0}\left(J^2 - \frac{M^2}{\sin^2\theta}\right)^{1/2}\mathrm{d}\theta$$
$$= J - |M|. \tag{4.61}$$

Here $\theta_0 = \arcsin(M/J)$ and the integral was evaluated by use of eqns (3.21)–(3.26). Furthermore, following this previous discussion, the properly quantized values of I_ϕ and I_θ are

$$I_\phi = m\hbar \quad \text{and} \quad I_\theta = \left(v + \frac{1}{2}\right)\hbar, \tag{4.62}$$

giving rise by virtue of (4.61) to

$$J = \left(v + |m| + \frac{1}{2}\right)\hbar = \left(j + \frac{1}{2}\right)\hbar. \tag{4.63}$$

Equations (4.62)–(4.63) take no account of any quantum mechanical complications, but the discussion of orbital motion in Section 3.1 shows that the quantum mechanical equivalent of (4.57), obtained after removing terms in $\partial/\partial\theta$ from the Schrödinger equation, differs from it by a term $\hbar^2/4$;

$$\hat{J}^2_{\text{quantum}} = J^2_{\text{classical}}(\hat{p}_\theta, \hat{p}_\phi) - \hbar^2/4. \tag{4.64}$$

Hence, according to eqn (4.63), the proper quantized values are

$$J^2_{\text{quantum}} = j(j+1)\hbar^2. \tag{4.65}$$

However, this *ad hoc* correction has no part in a coherent semiclassical development and J will henceforth be ascribed the value $\left(j + \frac{1}{2}\right)\hbar$ as specified by eqn (4.63).

Returning to the classical picture, it is convenient to adopt J and M as action variables and to seek the forms of their angular conjugates, α_J and α_M. Of these α_J varies linearly with time under the Hamiltonian

$$H = BJ^2 \tag{4.66}$$

at frequency

$$\omega(J) = \partial H/\partial J = 2BJ \tag{4.67}$$

and α_M is a constant of the motion. The convenient phase convention sets $\alpha_J = 0$ and $\alpha_M = \phi + \pi$ at $\theta = \theta_0$; the analogues of (4.12) or (4.47)–(4.48) then become

$$\alpha_J = \left(\frac{\partial S}{\partial J}\right) = \int_{\theta_0}^{\theta}\left(\frac{\partial p_\theta}{\partial J}\right)\mathrm{d}\theta = \int_{\theta_0}^{\theta}\frac{J\sin\theta \mathrm{d}\theta}{(J^2\sin^2\theta - M^2)^{1/2}}$$
$$= \arccos\left(\frac{J\cos\theta}{(J^2 - M^2)^{1/2}}\right), \tag{4.68}$$

and

$$
\begin{aligned}
\alpha_M = \left(\frac{\partial S}{\partial M}\right) &= \int_{-\pi}^{\phi}\left(\frac{\partial p_\phi}{\partial M}\right)\mathrm{d}\phi + \int_{\theta_0}^{\theta}\left(\frac{\partial p_\theta}{\partial M}\right)\mathrm{d}\theta \\
&= \phi + \pi - \int_{\theta_0}^{\theta}\frac{M\ \mathrm{d}\theta}{\sin\theta(J^2\sin^2\theta - M^2)^{1/2}} \\
&= \phi + \pi - \arctan\left(\frac{(J^2\sin^2\theta - M^2)^{1/2}}{M\cos\theta}\right). \qquad (4.69)
\end{aligned}
$$

The integrals in (4.68) and (4.69) are conveniently cast into standard forms (Gradsteyn and Ryyzhik 1980) by one or other of the substitutions

$$
\cos\theta = (M/J)\sin\xi = [1-(M/J)^2]^{1/2}\sin\zeta, \qquad (4.70)
$$

with the branches of the inverse trigonometric functions being chosen so as to increase around the cyclic orbit.

It will prove useful for the evaluation of matrix elements in Section 5.2 to rearrange eqns (4.57), (4.59), (4.68) and (4.69) to the forms

$$
\cos\theta = (1 - M^2/J^2)^{1/2}\cos\alpha_J \qquad (4.71)
$$

$$
\sin\theta = (J^2\sin^2\alpha_J + M^2\cos^2\alpha_J)^{1/2}/J \qquad (4.72)
$$

$$
\phi = \alpha_M + \arctan[(J/M)\tan\alpha_J] - \pi \qquad (4.73)
$$

$$
p_\theta = \frac{J(J^2 - M^2)^{1/2}\sin\alpha_J}{(J^2\sin^2\alpha_J + M^2\cos^2\alpha_J)^{1/2}} \qquad (4.74)
$$

and

$$
p_\phi = M. \qquad (4.75)
$$

It may be verified from eqn (4.71)–(4.73) that the motion traces out the following path:

$$
\hat{r} = (\sin\theta\cos\phi, \sin\theta\sin\phi, \cos\theta) = -\mathbf{x}'\sin\alpha_J + \mathbf{y}'\cos\alpha_J, \qquad (4.76)
$$

where

$$
\begin{aligned}
\mathbf{x}' &= (-\sin\alpha_M, \cos\alpha_M, 0) \\
\mathbf{y}' &= (-\cos\beta\cos\alpha_M, -\cos\beta\sin\alpha_M, \sin\beta) \qquad (4.77) \\
\mathbf{z}' &= (\sin\beta\cos\alpha_M, \sin\beta\sin\alpha_M, \cos\beta),
\end{aligned}
$$

and

$$
\sin\beta = (1 - M^2/J^2)^{1/2} \quad \text{or} \quad \cos\beta = M/J. \qquad (4.78)
$$

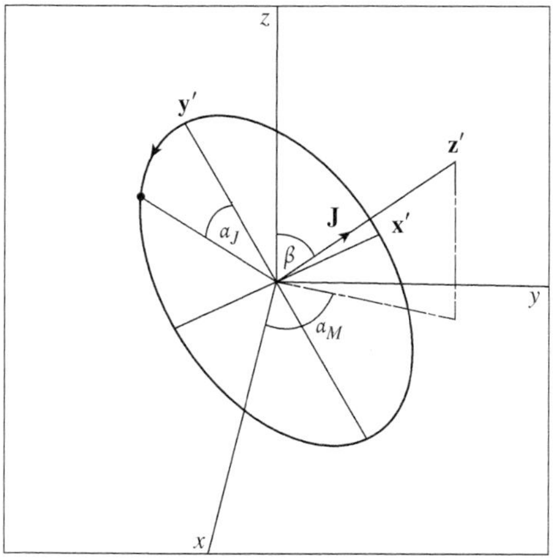

Fig. 4.3 The geometrical significance of the angular momentum action–angle variables; $\cos\beta = M/J$.

Equations (4.77) and (4.78) correspond to rotation about a fixed angular momentum vector, aligned parallel to $\mathbf{z}'$ with spherical polar angles β and α_M. Fig. 4.3 provides a geometrical interpretation.

It should be noted that the angles β and α_M which specify the orientation of $\mathbf{z}'$ are fixed for any classical orbit, but that α_M is completely uncertain in quantum mechanics, an uncertainty which is commonly visualized as precession of $\mathbf{z}'$ about $\mathbf{z}$ (Zare 1988). Consequences of the uncertainties in α_J and α_M arise in the theory of angular momentum matrix elements in Section 5.3 and in the semiclassical account of $3j$ and $6j$ symbols in Appendix C.4.

4.4 The hydrogen atom

No discussion of angle–action variables would be complete without reference to the Bohr–Sommerfeld theory of the hydrogen atom. The following exposition, which is largely taken from Born (1960), is also of interest in extending the degeneracy to three classical degrees of freedom.

The Hamiltonian

$$H = \frac{1}{2\mu}(p_r^2 + p_\theta^2/r^2 + p_\phi^2/r^2 \sin^2\theta) - \frac{e^2}{r} \tag{4.79}$$

gives rise to the Hamilton–Jacobi equation

$$H = \frac{1}{2\mu}\left[\left(\frac{\partial S}{\partial r}\right)^2 + \frac{1}{r^2}\left(\frac{\partial S}{\partial \theta}\right)^2 + \frac{1}{r^2\sin^2\theta}\left(\frac{\partial S}{\partial \phi}\right)^2\right] - \frac{e^2}{r} = E \tag{4.80}$$

which separates into three parts:

$$p_\phi = \left(\frac{\partial S}{\partial \phi}\right) = M \tag{4.81}$$

$$p_\theta^2 + \frac{M^2}{\sin^2\theta} = \left(\frac{\partial S}{\partial \theta}\right)^2 + \frac{M^2}{\sin^2\theta} = L^2 \tag{4.82}$$

$$p_r^2 + \frac{L^2}{r^2} = \left(\frac{\partial S}{\partial r}\right)^2 + \frac{L^2}{r^2} = 2\mu\left(E + \frac{e^2}{r}\right). \tag{4.83}$$

The next step is to relate the separation constants M and L and the energy E to the actions I_ϕ, I_θ and I_r; the necessary integrals have already appeared in eqns (4.43) and (4.61):

$$I_\phi = \frac{1}{2\pi}\oint p_\phi \, \mathrm{d}\phi = M \tag{4.84}$$

$$\begin{aligned} I_\theta &= \frac{1}{2\pi}\oint p_\theta \, \mathrm{d}\theta = \frac{1}{\pi}\int_{\theta_0}^{\pi-\theta_0}\left(L^2 - \frac{M^2}{\sin^2\theta}\right)^{1/2} \mathrm{d}\theta \\ &= L - |M| \end{aligned} \tag{4.85}$$

$$\begin{aligned} I_r &= \frac{1}{2\pi}\oint p_r \, \mathrm{d}r = \frac{1}{\pi}\int_a^b \left(2\mu E + \frac{2\mu e^2}{r} - \frac{L^2}{r^2}\right)^{1/2} \mathrm{d}r \\ &= (-2\mu E)^{-1/2}\mu e^2 - L. \end{aligned} \tag{4.86}$$

It is evident from (4.84)–(4.86) that E depends on $L + I_r$ while L itself is the sum of I_θ and I_ϕ; hence the natural procedure is to define three new actions

$$\begin{aligned} N &= I_r + I_\theta + |I_\phi| \\ L &= I_\theta + |I_\phi| \\ M &= I_\phi, \end{aligned} \tag{4.87}$$

of which the transformed Hamiltonian $\tilde{H}$ depends only on N:

$$E = \tilde{H}(N) = -\frac{\mu e^4}{2N^2}. \tag{4.88}$$

Consequently the three actions N, L and M and the two conjugate angles α_L and α_M are all constants of the motion, while α_N increases linearly with time according to the equation

$$\alpha_N = \omega(N)t + \text{constant} \tag{4.89}$$

where

$$\omega(N) = \partial\tilde{H}/\partial N = \mu e^4/N^3. \tag{4.90}$$

One also finds, by adopting the quantization rules appropriate to rotational motion in ϕ and vibrational or librational motion in r and θ, that (see Section 4.1)

$$\begin{aligned} I_\phi &= m\hbar \\ I_\theta &= \left(v_\theta + \frac{1}{2}\right)\hbar \\ I_r &= \left(v_r + \frac{1}{2}\right)\hbar. \end{aligned} \tag{4.91}$$

Consequently

$$\begin{aligned} N &= (v_r + v_\theta + |m| + 1)\hbar = n\hbar \\ L &= \left(v_\theta + |m| + \frac{1}{2}\right)\hbar = \left(l + \frac{1}{2}\right)\hbar \\ M &= m\hbar. \end{aligned} \tag{4.92}$$

Finally, on substituting in eqn (4.88), the energy levels take the exact quantum mechanical values:

$$E_n = -\frac{\mu e^4}{2n^2\hbar^2}. \tag{4.93}$$

The discussion is completed by relating the three angle variables α_N, α_L and α_M to the polar coordinates r, θ and ϕ. The phase convention for α_L and α_M is the same as for α_J and α_M in Section 4.3, and the zero of α_N is taken when r is at its outer turning point:

$$\alpha_N = \left(\frac{\partial S}{\partial N}\right) = \int_r^b \left(\frac{\partial p_r}{\partial N}\right) \mathrm{d}r = \left(\frac{\mu^2 e^4}{N^3}\right) \int_r^b \frac{r\mathrm{d}r}{R(r)}, \tag{4.94}$$

$$\begin{aligned} \alpha_L &= \left(\frac{\partial S}{\partial L}\right) = \int_r^b \left(\frac{\partial p_r}{\partial L}\right) \mathrm{d}r + \int_{\theta_0}^{\theta} \left(\frac{\partial p_\theta}{\partial L}\right) \mathrm{d}\theta \\ &= -L \int_r^b \frac{\mathrm{d}r}{rR(r)} + L \int_{\theta_0}^{\theta} \frac{\sin\theta \; \mathrm{d}\theta}{(L^2 \sin^2\theta - M^2)^{1/2}}, \end{aligned} \tag{4.95}$$

and

$$\begin{aligned} \alpha_M &= \left(\frac{\partial S}{\partial M}\right) \int_{\theta_0}^{\theta} \left(\frac{\partial p_\theta}{\partial M}\right) \mathrm{d}\theta + \int_{-\pi}^{\phi} \left(\frac{\partial p_\theta}{\partial M}\right) \mathrm{d}\phi \\ &= \phi + \pi - M \int_{\theta_0}^{\theta} \frac{\mathrm{d}\theta}{\sin\theta (L^2 \sin^2\theta - M^2)^{1/2}}, \end{aligned} \tag{4.96}$$

where

$$R(r) = (2\mu E r^2 + 2\mu e^2 r - L^2)^{1/2}, \tag{4.97}$$

in which the signs of the square root terms are taken according to the directions of the motions in θ and r.

The integrals with respect to θ are given by (4.66) and (4.67) and those with respect to r are conveniently reduced to standard forms by the substitution

$$r = \rho(1 + \varepsilon \cos u) \tag{4.98}$$

where

$$\varepsilon = (N^2 - L^2)^{1/2}/N. \tag{4.99}$$

Thus, with the help of (4.98),

$$\alpha_N = \int_0^u (1 + \varepsilon \cos u)\mathrm{d}u = u + \varepsilon \sin u \tag{4.100}$$

$$\alpha_L = -\psi + \arccos\left(\frac{L\cos\theta}{(L^2 - M^2)^{1/2}}\right) \tag{4.101}$$

$$\alpha_M = \phi + \pi - \arctan\left(\frac{(L^2\sin^2\theta - M^2)^{1/2}}{M\cos\theta}\right), \tag{4.102}$$

where in eqn (4.101)

$$\psi = (1 - \varepsilon^2)^{1/2} \int \frac{\mathrm{d}u}{1 + \varepsilon\cos u} = 2\ \arctan\left[\left(\frac{1-\varepsilon}{1+\varepsilon}\right)^{1/2} \tan\left(\frac{u}{2}\right)\right]. \tag{4.103}$$

The following equations, obtained by rearrangement of (4.97)–(4.103), show how the polar variables (r, θ, ϕ) depend on $(u,\ \alpha_L,\ \alpha_M)$, u being a more convenient variable than α_N in most applications:

$$r = \rho(1 + \varepsilon\cos u) \tag{4.104}$$

$$\cos\theta = \sin\beta\cos(\alpha_L + \psi) \tag{4.105}$$

$$\sin\theta\exp(\pm \mathrm{i}\phi) = -[\cos\beta\cos(\alpha_L + \psi) \pm \mathrm{i}\sin(\alpha_L + \psi)]\exp(\pm\mathrm{i}\alpha_M), \tag{4.106}$$

with the following additional identities:

$$\begin{aligned} \cos\psi &= \frac{\cos u + \varepsilon}{1 + \varepsilon\cos u} \\ \sin\psi &= \frac{(1-\varepsilon^2)^{1/2}\sin u}{1 + \varepsilon\cos u} \\ \rho &= N^2/\mu e^2 \\ \cos\beta &= M/L. \end{aligned} \tag{4.107}$$

To see the physical significance of eqns (4.103)–(4.107), notice first that r is related to ψ by the standard polar representation of an ellipse with the origin at its focus

$$r = \frac{\rho(1-\varepsilon^2)}{1-\varepsilon\cos\psi}. \tag{4.108}$$

Secondly, the radius vector $\mathbf{r}$ may be verified to trace out such an ellipse in the $\mathbf{x}'$, $\mathbf{y}'$ plane according to equation

$$\mathbf{r} = -r\sin(\psi+\alpha_L)\mathbf{x}' + r\cos(\psi+\alpha_L)\mathbf{y}', \tag{4.109}$$

where the body-fixed axes $(\mathbf{x}', \mathbf{y}', \mathbf{z}')$ are aligned in such a way that

$$\begin{aligned} \mathbf{x}' &= (-\sin\alpha_M, \cos\alpha_M, 0) \\ \mathbf{y}' &= (-\cos\beta\cos\alpha_M, -\cos\beta\sin\alpha_M, \sin\beta) \\ \mathbf{z}' &= (\sin\beta\cos\alpha_M, \sin\beta\sin\alpha_M, \cos\beta). \end{aligned} \tag{4.110}$$

Figure 4.4 illustrates the orientation of the ellipse in the $(\mathbf{x}', \mathbf{y}')$ plane together with a construction for the eccentric anomaly u, while Fig. 4.5 shows the overall orientation with respect to space-fixed axes.

The angle–action variables, or the equivalent quantum numbers, therefore play the following roles in determining the periodic elliptical motion. The total action N scales the overall size of the ellipse via the parameter

$$\rho = N^2/\mu e^2 = n^2 a_0, \tag{4.111}$$

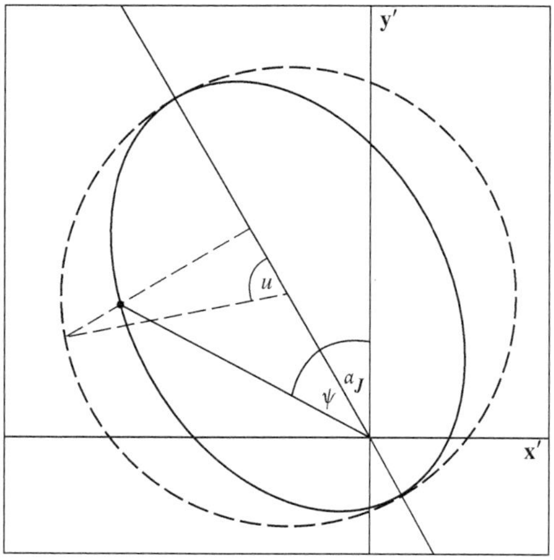

Fig. 4.4 Angle–action description of an elliptical hydrogenic orbit in the $(\mathbf{x}', \mathbf{y}')$ plane; u is the eccentric anomaly. The dashed circle, with radius ρ, is centred on the midpoint of the ellipse.

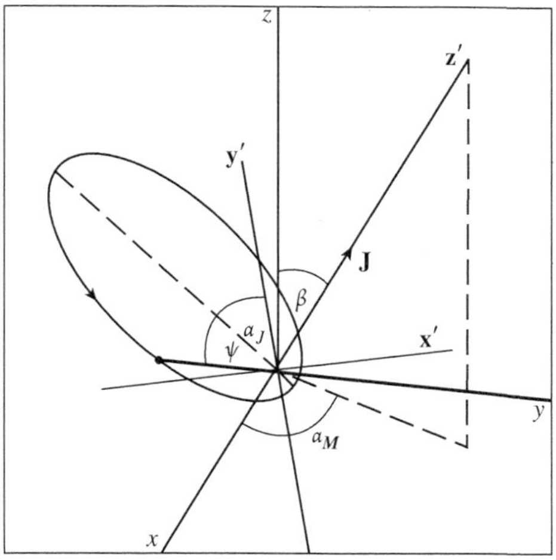

Fig. 4.5 Geometrical significance of the hydrogenic angle–action variables; $\cos\beta = M/L$. α_N increases in the direction of the arrow if $J > 0$.

where $a_0 = \hbar^2/\mu e^2$ is the Bohr radius. The ratio L/N determines the ellipticity

$$\varepsilon = (N^2 - L^2)^{1/2}/N = \frac{1}{n}\left[n^2 - \left(l + \frac{1}{2}\right)^2\right]^{1/2}, \tag{4.112}$$

and the orientation with respect to the z axis is given by

$$\cos\beta = M/L = 2m/(2l+1), \tag{4.113}$$

while α_M gives the angle ϕ at which the (x', y') and (x, y) planes intersect. Finally α_L fixes the orientation of the ellipse in the (x', y') plane and α_N determines the time evolution of u and ψ, according to eqns (4.89), (4.100) and (4.107).

This is the most ambitious example to be tackled here, but readers should note that Born (1960) reworks the theory in cylindrical polar coordinates to take account of the Stark effect arising from the imposition of a uniform electric field (details are set as an exercise in problem 4.4). Aspects of the H_2^+ problem are also outlined in Born (1960) and later discussed by Strand and Reinhardt (1979) and by Pajunen (1981). Finally, Leopold and Percival (1980) have used a hydrogenic angle–action model to treat the ground state of the helium atom.

4.5 Symmetric and asymmetric tops

The theory of rigid symmetric tops involves a relatively trivial extension of the angular momentum theory in Section 4.3, but the treatment of asymmetric and non-rigid tops raises several points of interest. For example, in the asymmetric case the motion bifurcates into a-type or c-type precessional motion, reminiscent of the near K_a and near K_c labels adopted in conventional spectroscopy (Herzberg 1945), and the familiar asymmetry splittings are found to have a tunnelling interpretation similar to that associated with restricted rotation in Section 3.4. The same ideas may be generalized to understand the fine rotational structure seen in the high angular momentum states of non-rigid symmetric and spherical tops (Harter and Patterson 1984). The account that follows is largely taken from papers by Augustin and Miller (1974) and Harter (1986, 1988).

It is convenient to start the discussion by extending the angular momentum theory of Section 4.3 to include the full Euler angle system (ϕ, θ, ψ) illustrated in Fig. 4.6. The following equations for the body-fixed components of J, taken from Augustin and Miller (1974), provide a starting point:

$$\begin{aligned} J_x &= p_\phi \frac{\sin\psi}{\sin\theta} + p_\theta \cos\psi - p_\psi \frac{\cos\theta \sin\psi}{\sin\theta} \\ J_y &= p_\phi \frac{\sin\psi}{\sin\theta} + p_\theta \sin\psi - p_\psi \frac{\cos\theta \cos\psi}{\sin\theta} \\ J_z &= p_\psi . \end{aligned} \tag{4.114}$$

From these it follows that

$$J^2 = p_\theta^2 + \frac{1}{\sin^2\theta}(p_\phi^2 + p_\psi^2 - 2p_\phi p_\psi \cos\theta), \tag{4.115}$$

and one finds, by generalization of the arguments in Section 4.3 using J^2 as the effective Hamiltonian, that a transformation to angle–action variables $(J, K, M, \alpha_J, \alpha_K, \alpha_M)$ may be generated by the equations

$$\begin{aligned} &K = p_\psi, \quad M = p_\phi \\ \alpha_J &= \int_{\theta_0}^{\theta} \left(\frac{\partial p_\theta}{\partial J}\right) \mathrm{d}\theta = J \int_{\theta_0}^{\theta} \frac{\sin\theta \, \mathrm{d}\theta}{(J^2 \sin^2\theta + 2MK\cos\theta - M^2 - K^2)^{1/2}} \\ &= \arccos\left(\frac{J^2\cos\theta - MK}{[(J^2 - K^2)(J^2 - M^2)]^{1/2}}\right) \end{aligned} \tag{4.116}$$

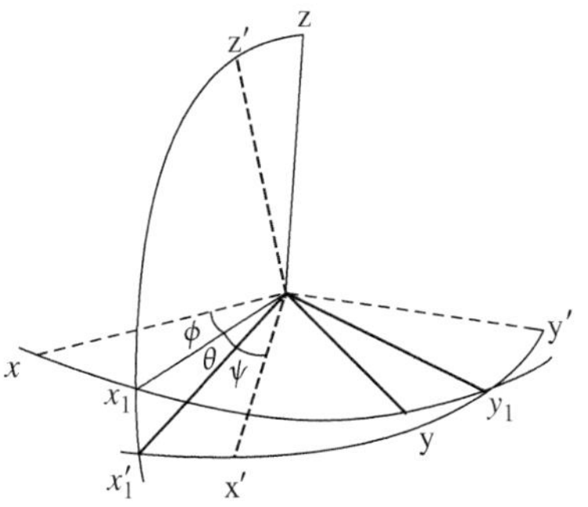

Fig. 4.6 The Euler angles (ϕ, θ, ψ) in the convention of Brink and Satchler (1968). By permission of Oxford University Press.

$$\begin{aligned}\alpha_K &= \int_{-\pi}^{\psi}\left(\frac{\partial p_\psi}{\partial K}\right)\mathrm{d}\psi + \int_{\theta_0}^{\theta}\left(\frac{\partial p_\theta}{\partial K}\right)\mathrm{d}\theta \\ &= \psi + \pi - \int_{\theta_0}^{\theta}\frac{(K - M\cos\theta)\ \mathrm{d}\theta}{\sin\theta(J^2\sin^2\theta + 2MK\cos\theta - M^2 - K^2)^{1/2}} \\ &= \psi + \pi - \arccos\left(\frac{K\cos\theta - M}{\sin\theta(J^2 - K^2)^{1/2}}\right),\end{aligned} \tag{4.117}$$

and, finally, by the symmetry between K and M in (4.116),

$$\alpha_M = \phi + \pi - \arccos\left(\frac{M\cos\theta - K}{\sin\theta(J^2 - M^2)^{1/2}}\right). \tag{4.118}$$

Note that Augustin and Miller (1974) would replace α_K by $\alpha_K + \pi/2$ and similarly for α_M, but that these changes simply correspond to a different choice of integration limit.

The immediate purpose of equations (4.116)–(4.118) is to express the body-fixed components in eqn (4.114) in terms of the angle:action variables; the result takes the following remarkably simple form:

$$\begin{aligned}J_x &= (J^2 - K^2)^{1/2}\cos\alpha_K \\ J_y &= -(J^2 - K^2)^{1/2}\sin\alpha_K \\ J_z &= K,\end{aligned} \tag{4.119}$$

while the space-fixed components (J_X, J_Y, J_Z) are given by similar equations, except that K is replaced by M and α_K by $-\alpha_M$ throughout. This sign difference is associated with sign differences in the space- and body-fixed quantum mechanical commutators (Van Vleck 1951) and equivalent Poisson brackets, $[\hat{f}, \hat{g}] \to -i\hbar\{f, g\}$ (Landau and Lifshitz 1965), where

$$\{f, g\} = \sum_v [(\partial f/\partial p_v)(\partial g/\partial q_v) - (\partial f/\partial q_v)(\partial g/\partial p_v)]. \tag{4.120}$$

The difference lies in the signs of the angles α_K and α_M, and different choices for the origins of these angles are equally acceptable (e.g. Augustin and Miller 1974; Gray and Davis 1989) with respect to the Poisson bracket; the choices implied by eqn (4.118) will be seen, according to eqn (5.52), to conform with the normal angular momentum phase conventions.

Equations (4.120) are the starting point for any rotational top problem. For example, in the rigid asymmetric case the body-fixed Hamiltonian,

$$H = (AJ_x^2 + BJ_y^2 + CJ_z^2)/\hbar^2, \tag{4.121}$$

transforms to

$$\hbar^2 H = (J^2 - K^2)(A\cos^2\alpha_K + B\sin^2\alpha_K) + CK^2, \tag{4.122}$$

where by convention $A > B > C$. Since α_K appears in eqn (4.122) the body-fixed projection K is no longer a constant of the motion although both J and M are still conserved. The problem therefore reduces to one of motion in the (K, α_K) system, except in the oblate limit $(A = B)$ when K is conserved. Similarly, an interchange of the x and z axes in (4.121) would lead to

$$\hbar^2 H = AK'^2 + (J^2 - K'^2)(B\sin^2\alpha_{K'} + C\cos^2\alpha_{K'}), \tag{4.123}$$

so that K' is a constant of the motion in the prolate top limit $(B = C)$. Equations (4.122) and (4.123) therefore take the form of $K_c = K$ and $K_a = K'$ representations of the Hamiltonian respectively, with the two sets of variables related by

$$K' = (J^2 - K^2)^{1/2}\cos\alpha_K$$

and

$$\alpha_{K'} = \arctan\left(\frac{\sin\alpha_K (J^2 - K^2)^{1/2}}{K}\right). \tag{4.124}$$

Harter and Patterson (1984) have devised an elegant geometrical picture that unifies these two descriptions (see also Harter 1988); they set

$$K = J\cos\beta, \tag{4.125}$$

and visualize β and α_K as spherical polar angles which specify the orientation of the principal axes **a**, **b** and **c** with respect to the fixed total angular momentum vector.

Figure 4.7 shows what is termed the rotational surface representation of the resulting Hamiltonian,

$$\hbar^2 H = J^2(A\sin^2\beta\cos^2\alpha + B\sin^2\beta\sin^2\alpha + C\cos^2\beta), \tag{4.126}$$

where the subscript K has been dropped for notational convenience. The picture is drawn in such a way that the energy is represented by distance from the origin to the surface, so that the contour lines correspond to constant energy trajectories. The notable features are that stable precessional orbits occur around the maximal and minimal inertial axes, **a** and **c**, but that precession around **b** is unstable (note that H has a saddle point with respect to α and β at $\alpha = \beta = \pi/2$). Secondly, the relative areas occupied by a-type and c-type orbits depend on the asymmetry parameter (Herzberg 1945)

$$\kappa = (A + C - 2B)/(A - C), \tag{4.127}$$

which takes the values $\kappa = \pm 1$ in the prolate and oblate limits respectively; for example, one may verify that the separatrix with energy BJ^2 between a-type and c-type orbits cuts the ac plane $(\alpha = 0)$ at the angle $\beta = \frac{1}{2}\arccos\kappa$.

Figure 4.7 is also helpful for semiclassical quantization purposes. The symmetry of the diagram with respect to $\pm K_v$ rotations about the positive and negative ends of the **a** or **c** axes ensures an exact double degeneracy in the classical limit, which is

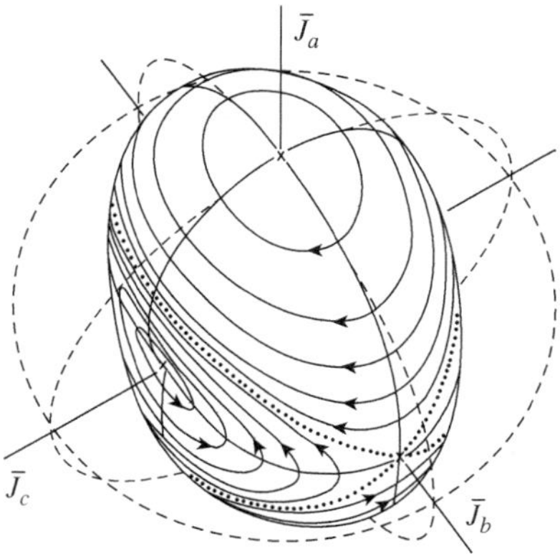

Fig. 4.7 A rotational energy surface representation of the motions of an asymmetric top. (Reprinted with permission from Harter and Patterson (1984). Copyright 1984, AIP Publishing LLC.)

closely approximated for the high K_c and high K_a states of most asymmetric tops. However, one sees from the familiar correlation plot of reduced energies

$$e_n(\kappa) = \left[E_n(\kappa) - \frac{1}{2}(A+C)J(J+1)\right] \Big/ (A-C), \tag{4.128}$$

shown in Fig. 4.8, that these close classical degeneracies are progressively removed on approaching the oblique dashed line that follows the reduced energy of the separatrix, $E = BJ(J+1)$, as a function of the asymmetry parameter κ. The semiclassical interpretation is that these asymmetry splittings arise from tunnelling from one equivalent classical orbit to the other.

Colwell and Handy (1978) and Pajunen (1985) perform the quantization procedure along the lines of the barrier penetration problems discussed in Sections 3.3 and 3.4. Harter and Patterson (1984) adopt an equivalent but superficially different approach, which generalizes more readily to the spherical top situations discussed below. Both sets of authors use the primitive quantization condition

$$A_n = \int_0^{2\pi} K_\nu(E_n, \alpha_\nu)\mathrm{d}\alpha_\nu = 2\pi n, \quad n = J, J-1, \ldots \tag{4.129}$$

to determine the mean energies, E_n, of the doublets, where the function $K_v(E_n, \alpha_v)$ is determined by (4.122) or (4.123) for $\nu = c$ or a as the case may be. Harter and Patterson (1984) then recognized that the tunnelling splittings in, for example, eqns (3.84) and (3.99) may be attributed to local Hamiltonian interaction matrix element of the form (see problem 3.5)

$$H_{ij} = g_t S, \tag{4.130}$$

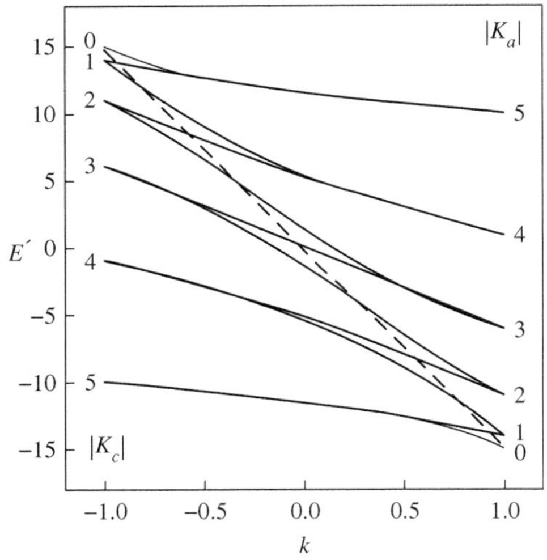

Fig. 4.8 Reduced asymmetric top energy levels for $J = 5$ as functions of the asymmetry parameter κ. The dashed line follows the energy of the unstable point on the **b** axis in Fig. 4.7.

where $g_{\rm t}$ is the number of tunnelling paths from one equivalent classical orbit to another and S depends on the tunnelling integral δ according to the equation

$$S = \mathrm{e}^{-\delta_\nu}/(\partial A_n/\partial E_n), \tag{4.131}$$

this integral

$$\delta_\nu = \mathrm{i}\int_{C_v} K_\nu(E_n, \alpha_\nu)\mathrm{d}\alpha_\nu \tag{4.132}$$

being obtained for c-type orbits by setting $\alpha_c = \pi/2 + \mathrm{i}\alpha'_c$ in (4.126), so that

$$K_c = \left(\frac{A - (E\hbar^2/J^2) - (A-B)/\cosh^2\alpha'_c}{A - C - (A-B)\cosh^2\alpha'_c}\right). \tag{4.133}$$

The practical procedure, in the case of a rigid asymmetric top, is therefore to determine E_n by eqn (4.129) and then to obtain the splittings by diagonalization of the matrix

$$H = \begin{pmatrix} E_n & 2S_n \\ 2S_n & E_n \end{pmatrix}, \tag{4.134}$$

where the factor 2 arises from the existence of two tunnelling paths—one through each end of the **b** axis in Fig. 4.7. The results given in Table 4.1 show excellent agreement with the quantum mechanical eigenvalues for the moderate to high J states of a 'most asymmetric top' with $\kappa = 0$.

Harter and Patterson (1984) have extended these ideas to the high J rotational spectra of spherical tops, with remarkable effect. The most striking example is SF_6, for which one may show by group theoretical arguments (Hecht 1960) that the first symmetry-allowed non-rigid (centrifugal stretching) contributions to the Hamiltonian give rise to

$$\hbar^2 H = BJ^2 + DJ^4 + 4t_{044}[J_x^4 + J_y^4 + J_z^4 - 3(J_x^2 J_y^2 + J_y^2 J_z^2 + J_z^2 J_x^2)]. \qquad (4.135)$$

The J dependence is more complicated than in eqn (4.121), but it is still relatively simple to substitute for the J_i in terms of the spherical polar angles (β, α) by use of (4.120) and (4.124). Hence the resulting rotational surface Hamiltonian (Harter 1986),

$$\hbar^2 H = BJ^2 + DJ^4 + \frac{1}{2} J^4 t_{044}(35\cos^4\beta - 30\cos^2\beta + 3 + 5\sin^4\beta\cos 4\alpha), \qquad (4.136)$$

may again be projected on to the surface of a sphere, as shown in Fig. 4.9, where the bulges and dimples indicate high and low energies respectively.

The high-energy orbits are seen to process around the six four-fold symmetry axes, giving rise to six-fold near degeneracies. Similarly, the sets of eight equivalent low-energy orbits around the three-fold symmetry axes give rise to eight-fold classical degeneracies, which are conveniently labelled by angular momentum components K_4 and K_3, respectively, bearing in mind that the octahedral symmetry imposes limits $K_r = 0, 1 \ldots \mathrm{mod}(r-1)$ on the distinct values. This means, following Harter and Patterson (1977, 1984), that there are four types of splitting patterns for the six-fold multiplets, designated $0_4, 1_4, 2_4$, and 3_4. Their O_h symmetry labels and energy level formulae are given in Table 4.2, in which the symbols S, T, U, ..., etc. denote tunnelling

Table 4.1 Asymmetric top eigenvalues for rotational constants $A = 0.2$, $B = 0.4$, $C = 0.6\ \mathrm{cm}^{-1}$, from Harter and Patterson (1984). E_n are the mean eigenvalues and S_n the tunnelling splittings.

n	E_n(QM) (cm^{-1})	E_n(SC) (cm^{-1})	S_n(QM) (cm^{-1})	S_n(SC) (cm^{-1})
$J = 10$				
10	63.181	63.176	2.16(−7)	1.96(−7)
9	57.851	57.842	1.93(−5)	1.87(−5)
8	53.147	53.133	6.98(−4)	6.91(−4)
7	49.113	49.084	1.27(−2)	1.29(−2)
6	45.833	45.781	1.11(−1)	1.20(−1)
$J = 20$				
20	246.352	246.347	—	1.25(−14)
19	235.360	235.354	2.84(−12)	2.66(−12)
18	224.977	224.970	2.55(−10)	2.49(−10)
17	215.214	215.205	1.42(−8)	1.40(−8)
16	206.083	206.071	5.33(−7)	5.29(−7)
15	197.630	197.586	1.41(−5)	1.41(−5)
14	189.810	189.777	2.71(−4)	2.72(−4)
13	182.721	182.684	3.76(−3)	3.80(−3)
12	176.448	176.382	3.64(−2)	3.75(−2)
11	171.132	171.029	2.18(−1)	2.38(−1)

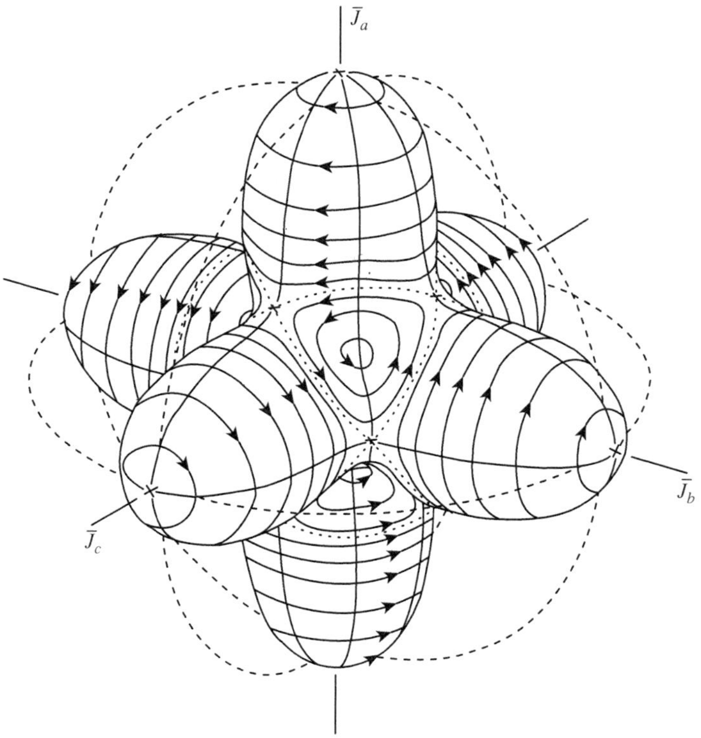

Fig. 4.9 A rotational energy surface representation of the motions of a semi-rigid octahedral top. Note the implication of six-fold and eight-fold classical degeneracies associated with motions around the four-fold and three-fold axes, respectively; the tunnelling splitting patterns that remove them are given in Table 4.2. (Reprinted with permission from Harter and Patterson (1984). Copyright 1984, AIP Publishing LLC.)

Table 4.2 Octahedral superfine clusters. Symmetries and splitting patterns. From Harter and Patterson (1977).

	0_4	1_4	2_4	3_4
A1	$E_n + 4S_n + T_n$		$E_n + 2S_n + T_n$	
A2			$E_n - 4S_n + T_n$	
E	$E_n - 2S_n + T_n$			
T1	$E_n - T_n$	$E_n - 2S_n$		$E_n - 2S_n$
T2		$E_n + 2S_n$	$E_n - T_n$	$E_n + 2S_n$
	0_3	1_3	2_3	
A1	$E_n + 3S_n + 3T_n + U_n$			
A2	$E_n - 3S_n + 3T_n - U_n$			
E		$E_n + 3T_n$	$E_n + 3T_n$	
T1	$E_n + S_n - T_n - U_n$	$E_n - 2S_n - T_n$	$E_n - 2S_n - T_n$	
T2	$E_n - S_n - T_n + U_n$	$E_n + 2S_n - T_n$	$E_n + 2S_n - T_n$	

elements between nearest axes, next to nearest axes, ..., etc. Each multiplet is six-fold degenerate with a mean energy E_n. Similar results are also given for the eight-fold clusters centred around the three-fold axes, which consequently fall into three types. Analysis of the spectrum of SF_6 (Bordé and Bordé 1982) confirms the qualitative validity of this picture in every detail. Moreover, the results in Table 4.3 show that the semiclassical estimates for E_n and S_n for a whole range of clusters are in excellent agreement with the quantum mechanical values (E_n values are measured with respect to $BJ(J+1)$). One sees, for $J = 30$, that the anisotropy in the centrifugal distortion, represented by the bulges in the diagram, gives rise to energy splittings of the order

Table 4.3 SF_6 high rotational eigenvalues (Harter and Patterson 1984). E_n and S_n have the same meanings as in Table 4.1; n_4 and n_3 refer to quantization around the four-fold and three-fold axes respectively.

	E_n(QM) (cm^{-1})	E_n(SC) (cm^{-1})	S_n(QM) (cm^{-1})	S_n(SC) (cm^{-1})
n_4		$J = 30$		
30	5.31×10^{-4}	5.29×10^{-4}	2.63(−11)	2.50(−11)
29	3.54	3.53	9.55(−10)	9.57(−10)
28	2.04	2.03	1.53(−8)	1.56(−8)
27	0.80	0.79	1.42(−7)	1.47(−7)
26	−0.20	−0.22	8.32(−7)	8.65(−7)
25	−0.97	−1.00	2.34(−6)	3.20(−6)
n_3				
30	-3.57×10^{-4}	-3.54×10^{-4}	1.75(−7)	1.68(−7)
29	−2.48	−2.44	2.02(−6)	2.15(−6)
28	−1.68	−1.66	6.59(−6)	7.40(−6)
n_4		$J = 88$		
88	4.206×10^{-2}	4.205×10^{-2}	1.73(−23)	1.62(−23)
87	3.728	3.727	2.17(−21)	2.13(−21)
86	3.275	3.273	1.31(−19)	1.30(−19)
85	2.845	2.844	5.12(−19)	5.11(−18)
84	2.440	2.438	1.44(−16)	1.44(−16)
83	2.056	2.055	3.12(−15)	3.13(−15)
82	1.695	1.694	5.41(−14)	5.44(−14)
81	1.356	1.355	7.71(−13)	7.77(−13)
80	1.038	1.037	9.19(−12)	9.28(−12)
79	0.741	0.739	9.30(−11)	9.40(−11)
78	0.463	0.462	8.04(−10)	8.15(−10)
77	0.205	0.204	5.99(−9)	6.09(−9)
76	−0.033	−0.034	3.86(−8)	3.93(−8)
75	−0.252	−0.254	2.15(−7)	2.20(−7)
74	−0.452	−0.454	1.04(−6)	1.06(−6)
73	−0.633	−0.636	4.23(−6)	4.42(−6)
72	−0.794	−0.798	1.44(−5)	1.56(−5)
n_3				
88	$-2.806 \times {}^{-2}$	-2.804×10^{-2}	1.24(−11)	1.17(−11)
87	−2.493	−2.491	6.06(−10)	5.97(−10)
86	−2.204	−2.201	1.37(−8)	1.37(−8)
85	−1.938	−1.935	1.90(−7)	1.94(−7)
84	−1.697	−1.693	1.79(−6)	1.82(−6)
83	−1.482	−1.477	1.17(−5)	1.22(−5)
82	−1.297	−1.290	5.07(−5)	5.63(−5)

of 10^{-4} cm^{-1}, while the possibility of tunnelling from one equivalent orbit to another causes further splittings which range in magnitude from 10^{-5} cm^{-1}, for orbits close to a separatrix, to 10^{-10} cm^{-1} for those at the extreme ends of the four-fold axes. The extreme superfine splitting of less than 2×10^{-21} cm^{-1} represents less than one tumble of the SF_6 molecule in 50 000 years, but splittings of this order are of course beyond any possibility of measurement. Robbins et al. (1989, 1990) give an alternative analysis of these splitting patterns, by which the tunnelling integrals are derived from complex trajectories and the quantization is achieved by a powerful generalization of the connection formula method, designed to take into account multiple tunnelling paths from one symmetry-related orbit to another.

4.6 Quantum monodromy

The previous sections have tacitly assumed the existence of classical actions for any Hamiltonian system, but recent developments in the mathematical literature have cast new light on the situation by focussing on global aspects of the angle–action picture (Cushman and Bates 1997). In particular, 'quantum monodromy' has been coined to highlight the quantum mechanical consequences of an isolated fixed point of the classical Hamiltonian which acts as a topological impediment to the global construction of a system of action variables, even in separable systems. The topic has excited considerable interest in the applied mathematics literature (Cushman and Bates 1997) as well as providing insight into the quantum mechanical eigenvalue structure in the vicinity of a potential barrier (Child et al. 1999; Jacobson and Child 2001; Child 2007).

The simplest illustration is for a scaled 'champagne bottle' or 'Mexican hat' Hamiltonian

$$H = \frac{1}{2}\left[p_r^2 + \frac{p_\phi^2}{r^2}\right] - \frac{1}{2}r^2 + \beta r^4, \tag{4.137}$$

in which the units have been scaled such that $\hbar = m = 1$. Thus the motion is constrained by a ring of equivalent fixed points, with $p_r = p_\phi = 0$, $r_e = (1/4\beta)^{1/2}$ and $H_{\min} = -\varepsilon_0 = -(1/16\beta)$. There are also 'relative equilibria' with $p_r = (\partial H/\partial r)_{p_\phi} = 0$ for arbitrarily small fixed p_ϕ. Finally and most importantly there is an 'isolated' fixed point at $r = p_r = p_\phi = H = 0$, with no neighbouring points, because the centrifugal term has a negative divergent second derivative for arbitrarily small p_ϕ. In addition the parameter β determines the number of bound states with $p_\phi = 0$ below the barrier at $H = 0$, by the formula $v + 1/2 = (8\varepsilon_0/3\pi)$ (Child 1998).

Hamiltonians of this type may be taken to model the bending vibrations of a 'quasi-linear' molecule, with an equilibrium bent geometry corresponding to $r_e = (4\beta)^{-1/2}$. The small amplitude motions around r_e when $H < 0$, model the molecular bending vibration, while the angular motion in the ϕ variable corresponds to rotation about the principal inertial axis, with angular momentum $p_\phi = k$, and quantum number $\ell = k$. The reason for this apparently redundant notation will become apparent later. A change of character occurs as the energy increases above the barrier, at which the reduced moment of inertia, r^2, vanishes. The bending–rotational motions of the previously bent molecule go over to the degenerate vibrational bending modes of a linear molecule—now with vibrational rather than rotational angular momentum k.

This situation has been well recognized in the molecular physics literature since the work of Dixon (1964) and Johns (1967), who brought out the local character of the quantum numbers, and the associated classical actions, by employing what appears below as a radial quantum number, v, for states with energies below the barrier and a linear molecule quantum number $n = 2v + |\ell|$ for those above. The factor of two arises because there are two cycles of the radial motion for each cycle of ϕ. The spectroscopic notation is $|v, \ell\rangle \equiv |v_b, K_a\rangle$ for the 'bent molecule' states and $|n, \ell\rangle \equiv |v_l, \ell\rangle$ for the 'linear molecule' ones (Johns 1967). To explain the rationale for this distinction, Fig 4.10(a) gives an overall view of the eigenvalues, ε, of the reduced Hamiltonian, H, for the parameter value $\beta = 0.00625$, roughly appropriate for the H_2O molecule. The radial quantum number, $v = 0, 1, \ldots$, is the ranking number for the eigenvalues in a given column, and the families with a given value are seen to lie on smooth roughly parabolic curves provided that $\varepsilon(v, \ell) < 0$ for $\ell = k = 0$. The second derivative $\left(\partial^2\varepsilon/\partial k^2\right)_v$, which is equal to twice the rotational constant, A_v, increases with increasing v in response to the diminishing moment of inertia. However the situation changes abruptly at higher energies, because it will be shown below that the function $\varepsilon(v, \ell)$ has a singularity at $\varepsilon = k = \ell = 0$, such that $(\partial\varepsilon/\partial\ell)_v$ and the higher derivatives with respect to ℓ are undefined. On the other hand the dashed curves in Fig 4.10(a), which join alternate eigenvalues with common linear molecule quantum numbers $n = 2v + |\ell|$, cross the line $k = 0$ smoothly at positive energies, but have undefined derivatives $(\partial\varepsilon/\partial\ell)_n$ when $k = 0$ and $\varepsilon < 0$. The upshot is that any

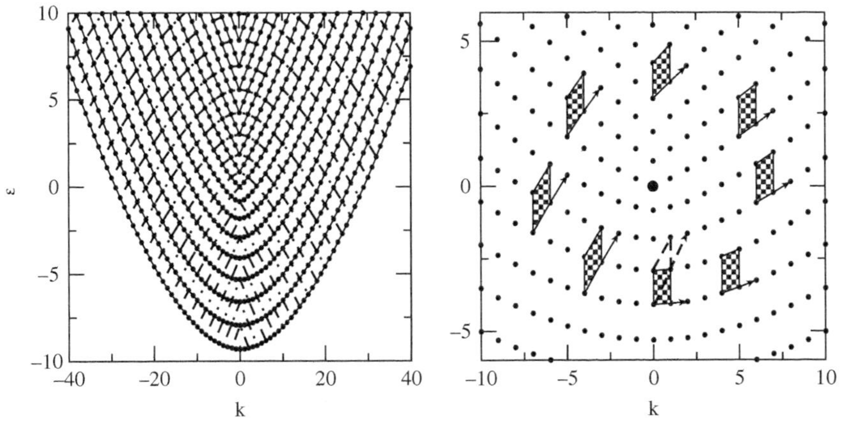

Fig. 4.10 (a) Eigenvalues (points) of the champagne bottle Hamiltonian for $\beta = 0.00625$. Solid lines join points with common values of the vibrational quantum number, v, and dotted ones those with a common value of $n = 2v + |k|$. (b) An expanded view around the fixed point $(0, 0)$, showing quantum monodromy as a primitive cell of the quantum lattice is taken 'once round' the isolated fixed point at $\epsilon = k = 0$.

particular eigenstate can be ascribed either of the two 'local' quantum numbers, v or n, because both solid and dashed curves can be drawn through any particular point in the diagram. The smoothness of the solid curves at $k = 0$ and $\varepsilon < 0$ makes v more convenient for interpretive purposes at low energies, while n is a more convenient quantum number for the high energy states.

Before attempting to analyse the situation in mathematical detail, it is useful to examine Fig 4.10(b) in relation to the possible construction of a 'global' system of classical actions or quantum numbers, such that the coordinates (k, ε) of every point in the physically observable quantum lattice can be expressed as single-valued functions of the quantum numbers (v, ℓ). The diagram traces out changes in the shape of a unit cell as it is taken around the origin according to the rule that the side lengths with respect to the lower left vertex correspond to quantum number changes $\delta\ell = \delta v = 1$. The smooth rotation of the arrowed projection of the base of each cell confirms the smoothness of this graphical construction. The failure of the final (dashed) cell to coincide with its original counterpart is clear evidence of the impossibility of constructing a global quantum number system, even for this separable system. Circuits which avoid the origin present no such problem, but the difficulty in encircling the origin always occurs, regardless of whether (v, ℓ), (n, ℓ) or any other combination of quantum numbers is used to define the cells and regardless of the chosen initial and final locations. The term 'monodromy' is chosen as a Greek-based term associated with paths taken 'once round' the origin.

To put the situation into a mathematical context, the quantum numbers are defined as $\ell = k$ and $v\,(\varepsilon, k)$, as given by the Bohr–Sommerfeld quantization condition

$$\begin{aligned}[v\,(\varepsilon, k) + 1/2]\,\pi &= \int_{r_1}^{r_2} \sqrt{2\varepsilon - \frac{k^2}{r^2} + r^2 - 2\beta r^4}\mathrm{d}r \\ &= \frac{1}{2}\int_b^a z^{-1}\sqrt{2\varepsilon z - k^2 + z^2 - 2\beta z^3}\mathrm{d}z, \\ &= \frac{1}{\sqrt{2\beta}}\int_b^a \frac{\left[\varepsilon z - k^2 + \beta z^3\right]\mathrm{d}z}{z\sqrt{(a-z)(z-b)(z-c)}},\end{aligned} \tag{4.138}$$

in which $z = r^2$, and the roots of the cubic polynomial are written as $a > b > c$. Note that a term $\varepsilon z + z^2 - 3\beta z^3$ in the numerator, which is equal to half the derivative of the cubic polynomial, has been integrated out to reach the final expression. The final form is written to suggest evaluation in terms of elliptic integrals (Gradsteyn and Ryyzhik 1980), although we shall limit attention to behaviour arising from coalescence between the inner turning point, b, and the origin as $k \to 0$ for small ε values. The roots may then be approximated as

$$a \simeq (2\beta)^{-1}, \qquad b = \sqrt{\varepsilon^2 + k^2} - \varepsilon, \qquad c = -\sqrt{\varepsilon^2 + k^2} - \varepsilon. \tag{4.139}$$

Moreover, since $a \gg b, c$ the terms involving $\varepsilon z - k^2$ in the numerator may be evaluated by approximating $(a - z) \simeq a$. It follows by use of standard integrals (Gradsteyn and Ryyzhik 1980) that

$$\left[v(\varepsilon,k)+\frac{1}{2}\right]\pi = v_0(\varepsilon,k) - \frac{\varepsilon}{2}\ln\sqrt{\varepsilon^2+k^2} - \frac{k}{2}\arctan(\varepsilon/k) - \frac{|k|\pi}{2}$$
$$= v_0(\varepsilon,k) - \frac{1}{2}\mathrm{Im}\left[(k+\mathrm{i}\varepsilon)\ln(k+\mathrm{i}\varepsilon)\right] - \frac{|k|\pi}{2}, \tag{4.140}$$

where $v_0(\varepsilon,k)$ is an analytic function, which vanishes as $k \to 0$.

It will be seen below that the logarithmic singularity in eqn (4.140) is responsible for the essential features of Fig 4.10, although it appears at first sight that its influence would be compensated by a semiclassical phase correction (Child 1998; compare eqn (3.49)):

$$\eta(\varepsilon,k) = -\eta(-\varepsilon,k) = \frac{\varepsilon}{4}\ln\left(\frac{\varepsilon^2+k^2}{4}\right) - \frac{\varepsilon}{2} + \frac{k}{2}\arctan\left(\frac{\varepsilon}{k}\right) - \arg\Gamma\left(\frac{|k|+1}{2}+\mathrm{i}\frac{\varepsilon}{2}\right) \tag{4.141}$$

in the modified quantization condition

$$[v(\varepsilon,k)+1/2]\,\pi = \int_{r_1}^{r_2}\sqrt{2\varepsilon - \frac{k^2}{r^2} + r^2 - 2\beta r^4}\mathrm{d}r + \eta(\varepsilon,k). \tag{4.142}$$

However, the influence of $\eta(\varepsilon,k)$ is limited to small values of k and ε, because the cancelled terms are largely restored by the asymptotic behaviour of the term $\arg\Gamma\left[(|k|+1)/2+\mathrm{i}\varepsilon/2\right]$. Numerical investigation shows that the largest value $\eta(\varepsilon,0) = 0.15$ occurs at $\varepsilon = 0.365$ and that $|\eta(\varepsilon,0)| < 0.002$ for $|k| > 2$. By analogy with Section 3.3, the most significant feature is that a singularity in $(\partial\eta/\partial\varepsilon)_k$ at $k = \varepsilon = 0$ cancels a similar singularity in $(\partial\eta/\partial\varepsilon)_k$ to allow a finite interval between the quantum eigenvalues in the vicinity of the barrier. Thus the pattern of eigenvalues is largely dictated by the properties of $v(\varepsilon,k)$ alone, apart from minor quantum corrections close to the origin.

In particular the limiting values of the partial derivatives $(\partial v/\partial k)_\varepsilon$ and $(\partial v/\partial\varepsilon)_k$, as $k \to 0_\pm$, are given by

$$\left(\frac{\partial v}{\partial k}\right)_\varepsilon \simeq -\frac{1}{2\pi}\left[\arctan\left(\frac{\varepsilon}{k}\right) + \frac{\pi|k|}{2k}\right] \to \begin{cases} 0, & \varepsilon < 0, \quad k \to 0_\pm \\ -1/2, & \varepsilon > 0, \quad k \to 0_+ \\ +1/2, & \varepsilon > 0, \quad k \to 0_- \end{cases}$$
$$\left(\frac{\partial v}{\partial \varepsilon}\right)_k \simeq \frac{1}{2\pi}\left[\left(\frac{\partial v_0}{\partial\varepsilon}\right)_k - \ln\sqrt{\varepsilon^2+k^2}\right] \tag{4.143}$$

on the assumption that $\arctan(\varepsilon/k)$ is taken as a principal value. Taken together with the partial derivative identity $(\partial v/\partial k)_\varepsilon(\partial k/\partial\varepsilon)_v(\partial\varepsilon/\partial v)_k = -1$, these results explain the eigenvalue pattern associated with the v labelling around the origin, because $(\partial\varepsilon/\partial v)_k$ depends on the energy interval between successive eigenvalues, which is always positive. Moreover the converse behaviour associated with the n labelling is readily explained by the identity $(\partial n/\partial k)_\varepsilon = 2(\partial v/\partial k)_\varepsilon + k/|k|$.

Finally, eqn (4.143) may be used both to explain and to quantify the monodromy in Fig 4.10(b). Note first that displacements $(\delta\ell,\delta v)$ in the quantum numbers are related to those in the observables $(\delta\varepsilon,\delta k)$ by the tangent relations

$$\begin{pmatrix} \delta\ell \\ \delta v \end{pmatrix} = \begin{pmatrix} D_{\ell k} & D_{\ell\varepsilon} \\ D_{vk} & D_{v\varepsilon} \end{pmatrix} \begin{pmatrix} \delta k \\ \delta\varepsilon \end{pmatrix} = \begin{pmatrix} 1 & 0 \\ a & b \end{pmatrix} \begin{pmatrix} \delta k \\ \delta\varepsilon \end{pmatrix}, \tag{4.144}$$

where the elements of the tangent matrix D are the partial derivatives, $D_{vk} = a = (\partial v/\partial k)_\varepsilon$ and $D_{v\varepsilon} = b = (\partial v/\partial \varepsilon)_k$, evaluated at the origin (lower left vertex) of any unit cell. Thus the changes in the shapes of the unit cells in Fig 4.10(b) reflect changes in the partial derivatives, according to the inverse of eqn (4.144). For present purposes it is, however, more convenient to quantify changes in the quantum number increment vector $(\delta\ell, \delta v)^T$ by assuming a circuit that brings $(\delta k, \delta\varepsilon)^T$ back to its original location. The following argument is due to Bates (1991), who applied eqn (4.143) and (4.144) in the separate regions $k > 0$ and $k < 0$ and used the continuity of $(\delta k, \delta\varepsilon)^T$ to pass from one half plane to the other. The relevant tangent matrices are denoted as $D_{\pm,\pm}$, where the first index indicates $k \to 0_\pm$ and the second is the sign of ε. The individual elements in (4.144) are therefore given by $a_{+,+} = -a_{-,+} = -1/2$, $a_{+,-} = -a_{-,-} = 0$, $b_{+,+} = b_{-,+}$ and $b_{+,-} = b_{-,-}$. It then follows, for example, that $(\delta k, \delta\varepsilon)^T$ at $\varepsilon > 0$ is related to the initial values of $(\delta\ell, \delta v)^T$ on a circuit that starts at $k \to 0_+$ with $\varepsilon < 0$, by the equation

$$\begin{pmatrix} \delta k \\ \delta\varepsilon \end{pmatrix}_+ = (D_{-,+})^{-1} \begin{pmatrix} \delta\ell_i \\ \delta v_i \end{pmatrix}, \tag{4.145}$$

because the quantum number increment vector $(\delta\ell, \delta v)^T$ is preserved within the half plane. Similarly the final values $(\delta\ell_f, \delta v_f)^T$ satisfy

$$\begin{pmatrix} \delta\ell_f \\ \delta v_f \end{pmatrix} = (D_{-,-}) \begin{pmatrix} \delta k \\ \delta\varepsilon \end{pmatrix}_- = (D_{-,-})\,(D_{-,+})^{-1}\,(D_{+,+}) \begin{pmatrix} \delta k \\ \delta\varepsilon \end{pmatrix}_+. \tag{4.146}$$

Taken together, these two equations imply that

$$\begin{pmatrix} \delta\ell_f \\ \delta v_f \end{pmatrix} = M \begin{pmatrix} \delta\ell_i \\ \delta v_i \end{pmatrix}, \tag{4.147}$$

where the monodromy matrix is given by

$$M = (D_{-,-})\,(D_{-,+})^{-1}\,(D_{+,+})\,(D_{-,+})^{-1} = \begin{pmatrix} 1 & 0 \\ 1 & 1 \end{pmatrix}. \tag{4.148}$$

It is illuminating to confirm that the same result may be obtained by allowing the arctangent in (4.143) to increase smoothly around the circuit, so that $a_f = a_i + 1$, because the junction lies in the zone $\varepsilon < 0$. Thus

$$M = D_f D_i^{-1} = \begin{pmatrix} 1 & 0 \\ a+1 & b \end{pmatrix} \begin{pmatrix} 1 & 0 \\ a & b \end{pmatrix}^{-1} = \begin{pmatrix} 1 & 0 \\ 1 & 1 \end{pmatrix}. \tag{4.149}$$

Seen in this light, the monodromy is associated with passing from one Riemann sheet of $v(k, \varepsilon)$ to the next, because it is readily verified that after N circuits

$$M^N = \begin{pmatrix} 1 & 0 \\ N & 1 \end{pmatrix}. \tag{4.150}$$

Although the model above is restricted to two degrees of freedom, Fig. 4.10(b) provides very useful insight into the energy level structures of real molecules, which are normally represented by power series in the quantum numbers. For example, Child et al. (1999) show that the highly excited bending progressions of H_2O (in the quantum number v_2) reproduce the pattern in the diagram for a variety of stretching quantum number combinations (v_1, v_3). The angular momentum is taken as $k = K_a$ in conventional spectroscopic notation (Herzberg 1945), with the plotted energies taken as for the $J = K_a$ rotational level for the relevant vibrational state. The asymmetry splittings in Fig. 4.8 can be ignored on the scale of the diagram. In addition the local curvature of the solid lines in the monodromy plot may be interpreted in terms of a local, K_a-dependent principal rotational constant $\tilde{A}_v(K_a)$, including contributions from the centrifugal correction terms in a normal spectroscopic expansion. Winnewisser et al. (2006) have taken this idea further by showing, for a variety of quasi-linear species, that the local end-over-end rotational constant $\tilde{B}_v(K_a)$ varies smoothly with K_a at fixed v_b for vibrational states below the barrier but has a discontinuity in $(\partial\tilde{B}/\partial K_a)_v$ for those above the barrier. This observation may be understood on the basis that $\tilde{B}_v(K_a) = (\partial E/\partial P)_v$, where $P = J(J+1)$. Thus, by analogy with the argument after eqn (4.143),

$$\left(\frac{\partial\tilde{B}}{\partial K_a}\right)_v = \left(\frac{\partial^2 E}{\partial P \partial K_a}\right)_v = -\left(\frac{\partial^2 E}{\partial P \partial v}\right)_{K_a} \Big/ \left(\frac{\partial v}{\partial K_a}\right)_E, \tag{4.151}$$

in which the final factor is the analogue of $(\partial v/\partial k)_\varepsilon$ in eqn (4.143), which controls the monodromy.

Another important model is the spherical pendulum, which undergoes a transition from swinging to rotational motion as the energy increases, again with an isolated fixed point at zero angular momentum and with energy equal to the swinging limit (Cushman and Bates 1997). Jacobson and Child (2001) have used the associated quantum monodromy to predict very significant changes in various spectroscopic parameters of HCP as the energy increases towards the barrier. Applications to a variety of other systems have been reviewed by Sadovskii and Zhilinskii (2006) and Child (2007).

4.7 Problems

4.1. Given the Morse oscillator Hamiltonian

$$H = \frac{1}{2m}p^2 + D[1 - \exp(-aq)]^2,$$

use the substitution $1 - \exp(-aq) = \sqrt{\varepsilon}\cos u$, where $\varepsilon = H(I)/D$ in eqn (4.12), to show that q and p are related to proper angle–action variables (α, I) by the equations

$$\exp(-aq) = [1 - \varepsilon]/\{1 + \sqrt{\varepsilon}\cos\alpha\}$$
$$p = -\sqrt{2mD}\sqrt{\varepsilon(1-\varepsilon)}\sin\alpha/[1 + \sqrt{\varepsilon}\cos\alpha].$$

Note from problem 3.1 that $H(I) = D\{1 - [1 - aI/\sqrt{2mD}]^2\}$ where $I = (v + 1/2)\hbar$.

4.2. Employ parabolic coordinates, such that $x = \frac{1}{2}(\xi^2 - \eta^2)$ and $y = \xi\eta$, for analysis of the uncoupled (2:1) resonance Hamiltonian (Noid et al. 1979)

$$H = \frac{1}{2}[p_x^2 + p_y^2 + \omega^2(4x^2 + y^2)].$$

The classical type 3 generator (see Appendix C.1)

$$F_3(p_x, p_y, \xi, \eta) = -\frac{1}{2}(\xi^2 - \eta^2)p_x - \sqrt{\xi\eta}\; p_y$$

such that $x = -(\partial F_3/\partial p_x)$ and $y = -(\partial F_3/\partial p_y)$ may be used to perform the necessary canonical transformation.

(i) Show, by means of the identities $p_\xi = -(\partial F_3/\partial \xi)$ and $p_\eta = -(\partial F_3/\partial \eta)$, that

$$p_x = (\xi p_\xi + \eta p_\eta)/(\xi^2 + \eta^2), \quad p_y = (-\eta p_\xi + \xi p_\eta)/(\xi^2 + \eta^2).$$

(ii) Hence verify that the Hamiltonian transforms to

$$H = \frac{1}{2}[p_\xi^2 + p_\eta^2 + \omega^2(\xi^6 + \eta^6)]/(\xi^2 + \eta^2).$$

(iii) Deduce that the proper actions, obtained by separating the solution of the Hamilton–Jacobi equation, may be expressed as

$$J_\xi = \oint (2\xi^2 E - \omega^2\xi^6 + K)^{1/2} \mathrm{d}\xi$$

$$J_\eta = \oint (2\eta^2 E - \omega^2\eta^6 - K)^{1/2} \mathrm{d}\eta,$$

where E is the energy and K is a separation constant.

4.3. (i) Use the expressions in Appendix E.3(ii) to derive the following angle–action expressions for a scaled degenerate harmonic oscillator with $\mu\omega = 1$.

$$r + \mathrm{i}p_r \pm p_\phi/r = \frac{\sqrt{I \pm L}}{r}\left[\sqrt{I \pm L}\mathrm{e}^{\mathrm{i}\alpha_I} + \sqrt{I \mp L}\mathrm{e}^{-\mathrm{i}\alpha_I}\right]\mathrm{e}^{-\mathrm{i}\alpha_I}$$

and

$$\sqrt{I \pm L}\mathrm{e}^{\mathrm{i}\alpha_I} + \sqrt{I \mp L}\mathrm{e}^{-\mathrm{i}\alpha_I} = \sqrt{2}r\mathrm{e}^{\pm\mathrm{i}\phi\mp\mathrm{i}\alpha_I},$$

where $L = p_\phi = I_\phi$.

(ii) Using the notation $\alpha_\phi = \alpha_L$, deduce that

$$\begin{aligned} a_d &= \frac{1}{2}(r + \mathrm{i}p_r + p_\phi/r)\mathrm{e}^{-\mathrm{i}\phi} = \sqrt{(I+L)/2}\mathrm{e}^{-\mathrm{i}\alpha_I - \mathrm{i}\alpha_L} \\ a_g &= \frac{1}{2}(r + \mathrm{i}p_r - p_\phi/r)\mathrm{e}^{-\mathrm{i}\phi} = \sqrt{(I-L)/2}\mathrm{e}^{-\mathrm{i}\alpha_I + \mathrm{i}\alpha_L} \\ a_d^\dagger = (a_d)^* &= \frac{1}{2}(r - \mathrm{i}p_r + p_\phi/r)\mathrm{e}^{\mathrm{i}\phi} = \sqrt{(I+L)/2}\mathrm{e}^{+\mathrm{i}\alpha_I + \mathrm{i}\alpha_L} \\ a_g^\dagger = (a_g)^* &= \frac{1}{2}(r - \mathrm{i}p_r - p_\phi/r)\mathrm{e}^{\mathrm{i}\phi} = \sqrt{(I+L)/2}\mathrm{e}^{+\mathrm{i}\alpha_I - \mathrm{i}\alpha_L}. \end{aligned}$$

These are the classical analogues of the right-hand and left-hand quantum-mechanical annihilation and creation operators (Cohen-Tannoudji et al. 1977).

4.4. Particles with position vectors $\mathbf{R}$ and $\mathbf{r}$ have angular momenta $\mathbf{j}_1$ and $\mathbf{j}_2$ respectively, which combine to a resultant $\mathbf{J}$. Prove, by reference to Fig. 4.3 and/or eqns (4.77), that if the angles α_1 and α_2 conjugate to the magnitudes of $\mathbf{j}_1$ and $\mathbf{j}_2$ respectively are measured from the common normal to $\mathbf{j}_1$ and $\mathbf{j}_2$, then the geometrical angle χ between $\mathbf{R}$ and $\mathbf{r}$ is given by

$$\cos\chi = \cos\alpha_1 \cos\alpha_2 + [(J^2 - j_1^2 - j_2^2)/2j_1 j_2] \sin\alpha_1 \sin\alpha_2.$$

4.5. Use eqn (4.107) to express α_N in eqn (4.100) as an integral over the physical angle ψ in Fig. 4.4. Deduce that the area of the ellipse subtended by ψ increases linearly with time (Kepler's second law).

4.6. The Hamiltonian for motion of a hydrogen atom in an electric field, $\mathcal{E}$, may be expressed in parabolic coordinates $x = \xi\eta\cos\varphi$, $y = \xi\eta\sin\varphi$ and $z = (\xi^2 - \eta^2)/2$ in the scaled form (Born 1960)

$$H = \frac{1}{2(\xi^2+\eta^2)}[p_\xi^2 + p_\eta^2 + (\xi^{-2} + \eta^{-2})p_\varphi^2 + \mathcal{E}(\xi^4 - \eta^4) - 4].$$

Show, by introducing separation constants (x_1, x_2) such that $x_1 + x_2 = 2$, that the generator, S, to angle–action variables must satisfy

$$p_\xi = \partial S/\partial\xi = (f_1(\xi))^{1/2}, \qquad p_\eta = \partial S/\partial\eta = (f_2(\xi))^{1/2}, \qquad p_\varphi = J_\varphi$$

where

$$\begin{aligned} f_1(\xi) &= 2E\xi^2 + 2x_1 - \xi^{-2}J_\varphi^2 - \mathcal{E}\xi^4 \\ f_2(\eta) &= 2E\eta^2 + 2x_2 - \eta^{-2}J_\varphi^2 + \mathcal{E}\eta^4, \end{aligned}$$

E being the energy.

Obtain the following iterative approximation to the energy, valid to first order in $\mathcal{E}$:

$$E = -\frac{1}{2J^2} - \frac{3}{2}\mathcal{E} J J_e,$$

where $J = J_\xi + J_\eta + J_\varphi$ and $J_e = J_\eta - J_\xi$.

Hint: The required quantization integrals are given to first order in $\mathcal{E}$ by (see Appendix II of Born 1960)

$$J_\xi = \frac{1}{2\pi}\oint p_\xi \mathrm{d}\xi = \frac{1}{2}\left[-J_\varphi + \frac{x_1}{(-2E)^{1/2}} + \frac{1}{4}\mathcal{E}(-2E)^{-3/2}\left(J_\varphi^2 + \frac{3x_1^2}{2E}\right)\right]$$

$$J_\eta = \frac{1}{2\pi}\oint p_\eta \mathrm{d}\eta = \frac{1}{2}\left[-J_\varphi + \frac{x_2}{(-2E)^{1/2}} - \frac{1}{4}\mathcal{E}(-2E)^{-3/2}\left(J_\varphi^2 + \frac{3x_2^2}{2E}\right)\right].$$

5 Matrix elements

The calculation of matrix elements raises several interesting points. It is first shown, in Section 5.1, that the JWKB normalization factors for bound and continuum states differ only by a factor dependent on the density of states. Secondly, Section 5.2 shows how an angle–action representation for the wavefunction leads directly to the Heisenberg correspondence between matrix elements and classical Fourier components. Thirdly, Franck–Condon and curve-crossing matrix elements often show informative interference patterns which are clearly revealed by the stationary phase and uniform approximations outlined in Section 5.3. Finally, the more awkward non-curve-crossing cases examined in Section 5.4 provide a further illustration of the uniform approximation approach.

5.1 Semiclassical normalization

Three types of semiclassical wavefunction have been introduced in previous chapters: the primitive JWKB forms (Sections 2.1–2.2), their uniform variants (Section 2.3) and the angle–action functions (Section 4.1), of which only the last are properly normalized, and these apply only to bound states. The JWKB and uniform expressions, on the other hand, impose no restriction on the type of motion and it will be shown that the normalization factors for bound and continuum motion differ only by a term dependent on the density of states. The derivation given below is presented from a physical standpoint, but the more drastic approximations may be removed by transformation from the angle–action or energy–time pictures along the lines outlined in Appendix C.

It is convenient first to list the types of wavefunctions to be considered. In the case of rotational motion (uninhibited by turning points),

$$\psi^{\mathrm{JWKB}}(\phi) = Ck^{-1/2}(\phi)\exp\left(\mathrm{i}\int_{\phi_0}^{\phi} k(\phi)\mathrm{d}\phi\right), \tag{5.1}$$

where

$$k(\phi) = \{2I[E - V(\phi)]\}^{1/2}/\hbar; \tag{5.2}$$

there are no turning point complications, hence no uniform corrections are required. The JWKB wavefunction for vibrational or translational motion take similar forms,

$$\psi^{\mathrm{JWKB}}(x) = Ak^{-1/2}(x)\sin\left(\int_{a}^{x} k(x)\mathrm{d}x + \frac{\pi}{4}\right), \tag{5.3}$$

Semiclassical Mechanics with Molecular Applications. Second Edition. M. S. Child

where

$$k(x) = \{2m[E - V(x)]\}^{1/2}/\hbar, \tag{5.4}$$

but now some form of uniform correction is required to remove the turning point singularities. The most convenient form for practical applications is the Airy approximation (2.76),

$$\psi^{\text{Airy}}(x) = \pi^{1/2} A \zeta^{1/4}(x) k^{-1/2}(x) Ai(-\zeta(x)), \tag{5.5}$$

with the argument defined by the equation

$$\zeta(x) = \left(\frac{3}{2} \int_a^x k(x) \mathrm{d}x \right)^{2/3}; \tag{5.6}$$

recall from Section 2.4 that $k(x)$ is taken to have argument $-3i\pi/2$ when $x < a$ in order to make $\zeta(x)$ real and negative. The factor $\pi^{1/2}$ is included in eqn (5.5) to ensure equivalence with (5.3) for $x \gg a$ via the asymptotic form (Abramowitz and Stegun 1965)

$$Ai(-\zeta) \overset{\zeta \gg 1}{\sim} \pi^{-1/2} \zeta^{-1/4} \sin \left(\frac{2}{3} \zeta^{3/2} + \pi/4 \right). \tag{5.7}$$

Thus $\psi^{\text{JWKB}}(x)$ and $\psi^{\text{Airy}}(x)$ carry the same normalization constant A.

Normalization of the rotational function is trivial because the necessary integral,

$$\int_0^{2\pi} |\psi(\phi)|^2 \mathrm{d}\phi = C^2 \int_0^{2\pi} k^{-1}(\phi) \mathrm{d}\phi = 1, \tag{5.8}$$

may be expressed in terms of the classical time period T or frequency ω,

$$T = \frac{2\pi}{\omega} = \int_0^{2\pi} \left(\frac{\mathrm{d}\phi}{\mathrm{d}t} \right)^{-1} \mathrm{d}\phi \tag{5.9}$$

by recognizing that $\mathrm{d}\phi/\mathrm{d}t = k(\phi)\hbar/I$. Thus, on combining (5.8) and (5.9),

$$C = (I/T\hbar)^{1/2} = (I\hbar\omega/2\pi\hbar^2)^{1/2}. \tag{5.10}$$

An additional factor of $\hbar$ has been included in both numerator and denominator so that $\hbar\omega$ may be identified with the quantum mechanical level spacing (see Section 4.1).

The vibrational case is more complicated because $k^{-1/2}(x)$ diverges at the turning points, a and b say, and because eqn (5.3) is valid only in the classical range $a < x < b$. To the extent, however, that the lost contributions from the non-classical regions are compensated by terms from the integrable singularities at a and b,

$$\begin{aligned} \int_{-\infty}^{\infty} [\psi^{\text{JWKB}}(x)]^2 \mathrm{d}x &\simeq A_{\text{b}}^2 \int_a^b k^{-1}(x) \sin^2 \left(\int_a^b k(x) \mathrm{d}x + \frac{\pi}{4} \right) \mathrm{d}x \\ &= \frac{A_{\text{b}}^2}{2} \int_a^b k^{-1}(x) \left[1 - \cos \left(2 \int_a^b k(x) \mathrm{d}x + \frac{\pi}{2} \right) \right] \mathrm{d}x, \end{aligned} \tag{5.11}$$

where A_b carries the subscript to indicate bound motion. A further approximation is introduced by neglecting the oscillatory part of the integrand, after which it is a simple matter to relate the integral $\int_a^b k^{-1}(x)\mathrm{d}x$ to the classical time period T and frequency ω in a manner similar to that in eqn (5.9), so that

$$A_b = (2m\hbar\omega/\pi\hbar^2)^{1/2}. \tag{5.12}$$

Despite the approximations involved, eqn (5.12) proves remarkably accurate for x values well separated from the classical turning points and this latter deficiency is removed by replacing ψ^{JWKB} by ψ^{Airy} in the turning point regions. As evidence, Fig. 5.1 shows a comparison between ψ^{Airy} and the exact $v = 2$ and $v = 4$ harmonic oscillator wavefunctions. This success is, however, better explained by stationary phase evaluation of the integral in (C.54).

Turning to the continuum case, the appropriate normalization for spectroscopic purposes is to a delta function of energy, which is equivalent to a density of one state per unit of energy (Child 1974b);

$$\int_{-\infty}^{\infty} \psi_E(x)\psi_{E'}(x)\mathrm{d}x = \delta(E - E'), \tag{5.13}$$

where $\delta(E - E')$ is conveniently represented by the integral (Dirac 1958)

$$\delta(E - E') = \frac{1}{\pi\hbar}\int_0^{\infty} \cos[(E - E')t/\hbar]\mathrm{d}t. \tag{5.14}$$

The equivalent of the bound state normalization therefore involves inserting two JWKB terms, with slightly different energies, in eqn (5.13), restricting the integration

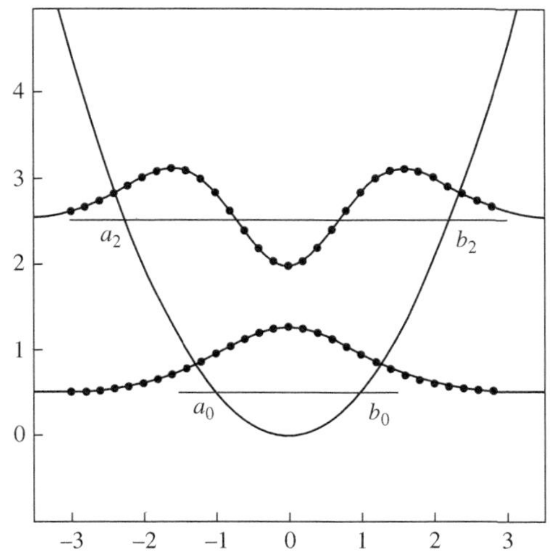

Fig. 5.1 Normalized uniform Airy approximation (points) to the $v = 0$ and $v = 2$ harmonic oscillator wavefunctions (solid curves).

range to $x > a$ and retaining only the slowly varying component of the integrand. The integral in (5.13) then reduces to

$$\int_{-\infty}^{\infty} \psi_E(x)\psi_{E'}(x)\mathrm{d}x \simeq$$
$$\frac{1}{2}A_\mathrm{c}\, A'_\mathrm{c} \int_a^\infty \cos\left(\int_a^x \{k(x) - k'(x)\}\mathrm{d}x/[k(x)k'(x)]^{1/2}\right)\mathrm{d}x. \tag{5.15}$$

The next step is to write

$$\begin{aligned} k(x) - k'(x) &= [k^2(x) - k'^2(x)]/[k(x) + k'(x)] \\ &= (E - E')/\hbar\bar{v}(x), \end{aligned} \tag{5.16}$$

where the potential energy contributions to $k^2(x)$ and $k'^2(x)$ have cancelled and $\bar{v}(x)$ is the mean classical velocity,

$$\bar{v}(x) = [k(x) + k'(x)]\hbar/2m \simeq [k(x)k'(x)]^{1/2}\hbar/m. \tag{5.17}$$

Remember that E and E' are assumed to be arbitrarily close. The object of these manipulations is to convert eqn (5.15) into an integral over time, by means of the substitution $\mathrm{d}t = \mathrm{d}x/\bar{v}(x)$; thus with $t = 0$ at $x = a$,

$$\begin{aligned} \int_{-\infty}^{\infty} \psi_E(x)\psi_{E'}(x)\mathrm{d}x &\simeq [A_\mathrm{c}^2\hbar/2m] \int_0^\infty \cos[(E - E')t/\hbar]\mathrm{d}t \\ &= [A_\mathrm{c}^2\pi\hbar^2/2m]\ \delta(E - E'). \end{aligned} \tag{5.18}$$

It follows by comparison with eqn (5.14) that

$$A_\mathrm{c} = (2m/\pi\hbar^2)^{1/2}. \tag{5.19}$$

In other words the bound state and continuum normalization constants, given by (5.12) and (5.19), differ by a factor $(\hbar\omega)^{1/2}$, where $\hbar\omega$ is the bound state energy level separation. The physical interpretation of this difference is that a bound state effectively concentrates all the continuum density from an energy band of width $\hbar\omega$ in its neighbourhood. Consequently, if some measure of the bound state intensity, say a Franck–Condon factor, is divided by the local level spacing $\hbar\omega$, then the resulting intensity per unit energy will vary smoothly across the dissociation limit. As an example, Fig. 5.2 shows the calculated variation of the X–B' absorption intensity in molecular hydrogen (Allison and Dalgarno 1971). Smith (1971) gives a similar plot of experimental data from the spectrum of the O_2 molecule.

The reader should note in conclusion that the above normalization arguments take little note of the classically forbidden parts of the wavefunction. It is therefore perhaps not surprising that the classically forbidden harmonic oscillator amplitude is actually in error by a factor of order 1.04 for the $n = 0$ state, with smaller errors for higher n values (Uzer and Child 1980). The same authors find more drastic errors in the forbidden region for scattering under an exponential potential with a range comparable to the de Broglie wavelength, but a failure of the semiclassical approximation is not unexpected in these circumstances.

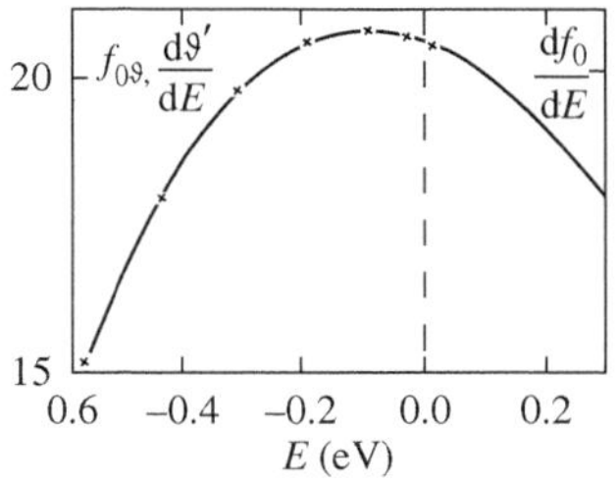

Fig. 5.2 Continuity between $f_{0v'}(dv'/dE)$ and df_0/dE in passing from the discrete to the continuous $X^1\Sigma_g^+ - B'^1\Sigma_u^+$ calculated absorption spectrum of H_2. (Reprinted with permission from Allison and Dalgarno (1971). Copyright 1971, AIP Publishing LLC.)

5.2 Matrix elements and Fourier components: the Heisenberg correspondence

The Heisenberg correspondence between quantum mechanical matrix elements and classical Fourier components played an important part in the old quantum theory. To see its origin, recall that the angle–action wavefunction in eqn (4.25) takes the same form for any periodic system, namely

$$\phi_n(\alpha) = \frac{1}{(2\pi)^{1/2}} \exp(in\alpha) \tag{5.20}$$

in which the angle variable always varies linearly with time and the quantum number n is related to the classical action, I, in the form

$$I = (n + \delta)\hbar, \tag{5.21}$$

where δ is the Maslov parameter. An analogous form may be assumed to apply for the semiclassical action operator. The obvious procedure is to associate the classical function, say $A(I, \alpha)$, with a semiclassical equivalent $A(\hat{I}, \alpha)$ where, according to eqn (4.23),

$$\hat{I} = -i\hbar(\partial/\partial\alpha) + \delta\hbar, \tag{5.22}$$

so that

$$A(\hat{I}, \alpha)\phi_n(\alpha) = A[(n + \delta)\hbar, \alpha]\phi_n(\alpha). \tag{5.23}$$

Unfortunately, however, matrix elements constructed according to eqn (5.23) are not Hermitian, $\langle n|a|n'\rangle \neq \langle n'|A|n\rangle^*$, because in one case $\hat{I}$ brings down n and in the other case n'. Hence it is natural to employ the mean quantum number

$$\bar{n} = \frac{1}{2}(n + n') \tag{5.24}$$

when evaluating the classical function $A(I, \alpha)$. In other words

$$\begin{aligned} \langle n'|A|n\rangle &= \int_0^{2\pi} \phi_{n'}^*(\alpha) A(\hat{I}, \alpha)\phi_n(\alpha)\,\mathrm{d}\alpha \\ &\simeq \frac{1}{2\pi}\int_0^{2\pi} A[(\bar{n}+\delta)\hbar, \alpha]\exp[\mathrm{i}(n-n')\alpha]\,\mathrm{d}\alpha. \end{aligned} \tag{5.25}$$

The Fourier equivalent of (5.25) is then obtained by substituting

$$\alpha = \bar{\omega}t, \tag{5.26}$$

where $\bar{\omega}$ is the frequency of the mean orbit with action $\bar{I} = (\bar{n}+\delta)\hbar$. Thus

$$\langle n'|A|n\rangle \simeq \frac{1}{T}\int_0^T A[(\bar{n}+\delta)\hbar, \bar{\omega}t]\exp[\mathrm{i}(n-n')\bar{\omega}t]\,\mathrm{d}t, \tag{5.27}$$

T being the time period

$$T = 2\pi/\bar{\omega}. \tag{5.28}$$

Equation (5.27) implies in particular that the quantum mechanical mean value is approximated by the classical time average,

$$\bar{A} = \langle n|A|n\rangle \simeq \frac{1}{T}\int_0^T A[(n+\delta)\hbar, \bar{\omega}t]\,\mathrm{d}t. \tag{5.29}$$

Various equivalent forms of eqns (5.27) and (5.29) are available. For example, it is unnecessary to follow the explicit angle–action transformation; the integrals are equally well evaluated in terms of the time dependence of the Cartesian variables around the mean orbit:

$$\langle n'|A|n\rangle \simeq \frac{1}{T}\int_0^T A[p(t), q(t)]\exp[\mathrm{i}\bar{\omega}(n-n')t]\,\mathrm{d}t. \tag{5.30}$$

A superficially quite different form

$$\langle n'|A|n\rangle \simeq \frac{2}{T}\int_a^b \frac{A[x, p(x)]}{\bar{v}(x)}\cos\left(\int_a^x [k(x') - k'(x')]\,\mathrm{d}x'\right)\mathrm{d}x \tag{5.31}$$

may also be obtained directly from the JWKB forms (5.3) and (5.12). Alternatively (5.30) may be recovered by making the following substitutions in (5.31):

$$\bar{v}(x) = \mathrm{d}x/\mathrm{d}t = \bar{k}(x)\hbar/m = \{2[\bar{E} - V(x)]/m\}^{1/2}, \tag{5.32}$$

$$E - \bar{E} \simeq (n - \bar{n})\hbar\bar{\omega}, \tag{5.33}$$

$$k^2(x) - k'^2(x) = 2m(E - E')/\hbar^2, \tag{5.34}$$

$$(n-n')t = \int_a^x [(n-n')\bar{\omega}/\bar{v}]\,\mathrm{d}x = \int_a^x [k(x') - k'(x')]\,\mathrm{d}x', \tag{5.35}$$

and

$$T = 2\int_a^b \frac{\mathrm{d}x}{\bar{v}(x)}. \tag{5.36}$$

Equation (5.31) implies in particular that the mean value is given by

$$\bar{A} = \langle n|A|n\rangle \simeq \int_a^b A[x, p(x)]v^{-1}(x)\mathrm{d}x \bigg/ \int_a^b v^{-1}(x)\mathrm{d}x. \tag{5.37}$$

As a measure of the accuracy of the Fourier approximation (5.25), we first consider matrix elements of the creation operator

$$a^\dagger = \frac{1}{2^{1/2}}(x - \mathrm{i}p) \tag{5.38}$$

between states of the scaled harmonic oscillator with Hamiltonian

$$H = \frac{1}{2}p^2 + \frac{1}{2}x^2 = n + \frac{1}{2}. \tag{5.39}$$

Equations (4.7) and (4.21), with $m = \omega = \hbar = 1$, yield the identity

$$a^\dagger = \left(n + \frac{1}{2}\right)^{1/2} \mathrm{e}^{\mathrm{i}\alpha}. \tag{5.40}$$

Hence, according to eqn (5.27),

$$\begin{aligned}\langle n'|a^\dagger|n\rangle &= \left(\bar{n} + \frac{1}{2}\right)^{1/2} \frac{1}{2\pi}\int_0^{2\pi} \exp[\mathrm{i}(n - n' + 1)\alpha]\,\mathrm{d}\alpha \\ &= \left(\bar{n} + \frac{1}{2}\right)^{1/2} \delta_{n',n+1} = (n+1)^{1/2}\delta_{n',n+1},\end{aligned} \tag{5.41}$$

because the selection rule $n' = n + 1$ requires that $\bar{n} = n + \frac{1}{2}$. This result is exact, with regard both to the selection rule and to the magnitude and sign of the matrix element, while the sign follows from the choice of the outer turning point as the phase reference in eqn (4.20). Repetition of the argument with $a^\dagger$ replaced by $(a^\dagger)^v$ yields, however,

$$\langle n'|(a^\dagger)^v|n\rangle \simeq \left[n + \frac{v}{2} + \frac{1}{2}\right]^{v/2} \delta_{n',n+v}, \tag{5.42}$$

whereas the exact result may be shown to be

$$\begin{aligned}\langle n+v|(a^\dagger)^v|n\rangle &= [(n+1)(n+2)\dots(n+v)]^{1/2} \\ &= [\Gamma(n+v+1)/\Gamma(n+1)]^{1/2} \\ &\overset{n\gg 1}{\approx} \left(n + \frac{v}{2} + \frac{1}{2}\right)^{v/2}\left(1 - \frac{(v-1)v(v+1)}{24(n + v/2 + \frac{1}{2})^2}\right),\end{aligned} \tag{5.43}$$

after using Abramowitz and Stegun (1965) to approximate the ratio of gamma functions. Equation (5.42) carries the correct selection rule but there is now an error in magnitude which increases with increasing v, a failure which can be understood on the basis that the mean orbit is less and less able to approximate both initial and final states as the quantum number change increases.

Another test of the approximation may be illustrated by considering the mean value of $\mathrm{e}^{-\gamma x}$ in a scaled harmonic oscillator state. To vary the approach, eqn (5.37) will be taken as the starting point although the result follows equally directly from (5.30). The necessary integrals

$$\langle n|\mathrm{e}^{-\gamma x}|n\rangle = \int_{-x_n}^{x_n} \mathrm{e}^{-\gamma x}(x_n^2 - x^2)^{-1/2}\mathrm{d}x \Big/ \int_{-x_n}^{x_n} (x_n^2 - x^2)^{-1/2}\mathrm{d}x, \tag{5.44}$$

where $x_n = (2n+1)^{1/2}$, are readily cast by means of the substitution

$$x = x_n \cos\theta \tag{5.45}$$

into the form

$$\begin{aligned}\langle n|\mathrm{e}^{-\gamma x}|n\rangle &\simeq \frac{1}{\pi}\int_0^{\pi} \exp(-\gamma x_n \cos\theta)\mathrm{d}\theta \\ &= I_0[(2n+1)^{1/2}\gamma],\end{aligned} \tag{5.46}$$

where $I_0(z)$ is the modified Bessel function (Abramowitz and Stegun 1965). Figure 5.3 compares the predictions of eqn (5.46) with the exact expression (Pechukas and Light 1966)

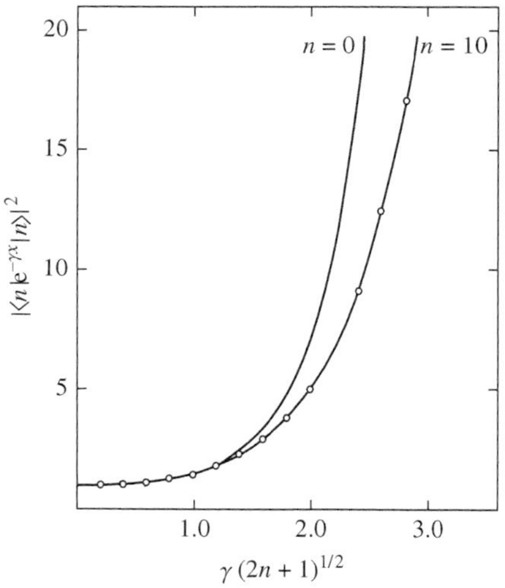

Fig. 5.3 The functional dependence of $|\langle n|\mathrm{e}^{-\gamma x}|n\rangle|^2$ on $(2n+1)^{1/2}\gamma$ for $n = 0$ and $n = 10$, compared with the semiclassical values (points) which apply for all n. (Reprinted from Uzer and Child (1980) by permission of Taylor and Francis.)

$$\langle n|e^{-\gamma x}|n\rangle = \exp(\gamma^2/4)L_n(-\gamma^2/2), \tag{5.47}$$

where $L_n(z)$ denotes the Laguerre polynomial. Clearly the range of validity of the semiclassical approximation increases, for given γ, as the quantum number increases, and decreases for given n as γ increases. The explanation for both trends is that the dominant contribution to the quantum mechanical integral, for large enough γ, eventually comes from outside the classical region $-x_n < x < x_n$ and hence cannot be taken into account by the classical integral, but that this problem is delayed to larger γ the larger the value of x_n. Uzer and Child (1980), who give a detailed analysis, also find, not surprisingly, that the off-diagonal matrix elements are even less reliable for large γ, because an additional mean orbit approximation is required.

The above examples show the strengths and weaknesses of the Heisenberg correspondence principle in a one-dimensional context. Two other examples will be taken to show how it works in higher-dimensional situations. The first concerns matrix elements of the shift operators between angular relevant eigenstates. The procedure is to substitute classical equivalents for the relevant operators in the quantum mechanical expressions (Brink and Satchler 1968),

$$\hat{J}_x \pm i\hat{J}_y = -e^{\pm i\phi}(\cot\theta\hat{p}_\phi \mp i\hat{p}_\theta), \tag{5.48}$$

and then to use eqns (4.71)–(4.75) to replace the polar variables by their angle–action equivalents. The resulting operators are found to take the remarkably simple forms,

$$J_x \pm iJ_y = (J^2 - M^2)^{1/2}\exp(\pm i\alpha_M). \tag{5.49}$$

Hence on making the semiclassical substitutions

$$J = \left(j + \frac{1}{2}\right)\hbar \quad \text{and} \quad M = m\hbar, \tag{5.50}$$

and employing the angle–action wavefunctions

$$\psi_{jm}(\alpha_j, \alpha_m) = \frac{1}{2\pi}\exp[i(j\alpha_j + m\alpha_m)], \tag{5.51}$$

the matrix element reduces to

$$\begin{aligned}\langle j', m'|J_x \pm iJ_y|j, m\rangle &= \frac{[(\bar{j} + \frac{1}{2})^2 - \bar{m}^2]^{1/2}}{4\pi^2}\int_0^{2\pi}\int_0^{2\pi} e^{i[(j-j')\alpha_j + (m-m'\pm 1)\alpha_m]}\,d\alpha_j d\alpha_m \\ &= [j(j+1) - m(m\pm 1)]^{1/2}\delta_{j',j}\delta_{m',m\pm 1},\end{aligned} \tag{5.52}$$

where the final line follows because $\bar{m} = m \pm \frac{1}{2}$ for $m' = m \pm 1$. Equation (5.52) is of course the exact quantum mechanical result.

The second example is the matrix element $\langle n, l+1, m|r\cos\theta|n, l, m\rangle$, which determines the magnitude of the Stark effect in the spectrum of atomic hydrogen. Changes in the principle quantum number have been excluded to avoid undue complexity in the resulting expression. The appropriate angle–action wavefunctions are

$$\psi_{nlm}(\alpha_n, \alpha_l, \alpha_m) = (2\pi)^{-3/2} \exp[\mathrm{i}(n\alpha_n) + l\alpha_l + m\alpha_m)], \tag{5.53}$$

and the operator may be expressed, with the help of eqns (4.90) and (4.103)–(4.107), as

$$r\cos\theta = d[(\cos u + \varepsilon)\cos\alpha_l + (1-\varepsilon^2)^{1/2}\sin u\,\sin\alpha_l], \tag{5.54}$$

where

$$\begin{aligned} d &= \left[n^2\hbar^2/\mu e^2\left(\bar{l}+1/2\right)\right]\left[\left(\bar{l}+1/2\right)^2 - m^2\right]^{1/2}, \\ \varepsilon &= \left[1 - \left(\bar{l}+1/2\right)^2/n^2\right]^{1/2} \end{aligned} \tag{5.55}$$

and the variable u is related to the angle α_n by eqn (4.100), namely

$$\alpha_n = u + \varepsilon\sin u. \tag{5.56}$$

The terms $\cos\alpha_l$ and $\sin\alpha_l$ in eqn (5.54) impose the known selection rule $\Delta l = \pm 1$ on the matrix element and the integral with respect to α_n is most conveniently performed in terms of the variable u. Hence, after integrating over α_l and α_M,

$$\begin{aligned} \langle n, l+1, m|r\cos\theta|n, l, m\rangle &= \frac{d}{4\pi}\int_0^{2\pi}[(\cos u + \varepsilon) + \mathrm{i}(1-\varepsilon^2)^{1/2}\sin u] \times (1 + \varepsilon\cos u)\,du \\ &= \frac{3}{4}\varepsilon d. \end{aligned} \tag{5.57}$$

It follows, in the light of (5.55) with $\bar{l} = l + \frac{1}{2}$, that

$$\langle n, l+1, m|r\cos\theta|n, l, m\rangle = \frac{3n\hbar}{4\mu e^2(l+1)}\left\{[(l+1)^2 - m^2][n^2 - (l+1)^2]\right\}^{1/2} \tag{5.58}$$

The exact result may be shown to be (Condon and Shortley 1957)

$$\langle n, l+1, m|r\cos\theta|n, l, m\rangle = \frac{3n\hbar}{2\mu e^2}\left(\frac{[(l+1)^2 - m^2][n^2 - (l+1)^2]}{4(l+1)^2 - 1}\right)^{1/2}. \tag{5.59}$$

Again one sees that the Heisenberg correspondence principle is remarkably accurate when the changes in the quantum numbers are small. Readers are referred to reviews by Dickinson (1980) and Dickinson and Richards (1982) for further examples of this correspondence principle approach.

5.3 Franck–Condon and curve-crossing matrix elements

The Fourier approximation outlined in the previous section is limited to situations in which the initial and final motions are well approximated by some common mean classical orbit. At the opposite extreme Franck–Condon transitions, accompanying a change of electronic state, may give rise to drastic changes in the nature of the associated orbit. The following stationary phase and uniform approximations then become applicable. One finds, on inspecting the relevant orbits, that the same theory also applies to curve-crossing situations. Furthermore, the bound or continuum character of the initial and final motions is reflected only in the multiplicative normalization factor (see Section 5.1).

Several relevant situations are illustrated in Fig. 5.4, namely (a) Franck–Condon transitions, (b) predissociation and (c) non-adiabatic transitions. The common feature is that there is some point R_x (or perhaps several points in more complicated situations) at which the classical momenta on the two orbits are equal and hence at which a classical Franck–Condon (momentum and position conserving) transition can take place (Miller 1974). The relevance of such points is most easily understood by considering a primitive semiclassical evaluation of the matrix element. The wavefunctions are taken as suitably normalized (see Section 5.1) JWKB terms

$$\psi_i(R) \simeq A_i k_i^{-1/2}(R) \sin\left(\int_{a_i}^{R} k_i(R)\mathrm{d}R + \frac{\pi}{4}\right), \tag{5.60}$$

and the integral,

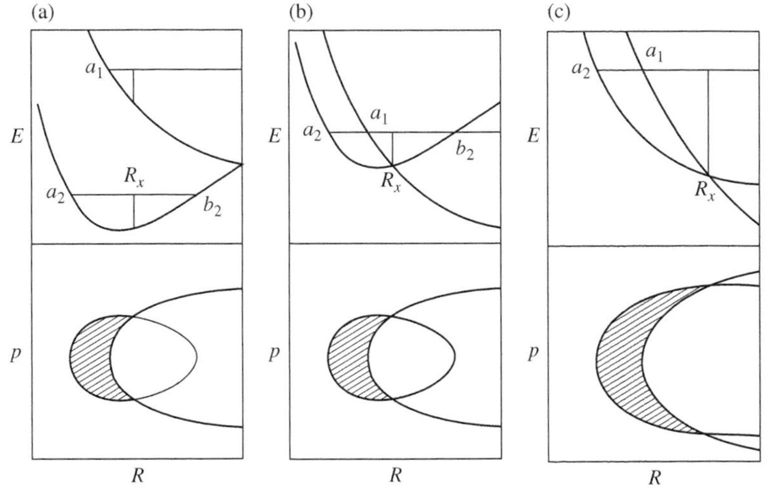

Fig. 5.4 Potential curves and phase orbits for (a) Franck–Condon transitions, (b) predissociation and (c) non-adiabatic transitions.

$$\langle \psi_1 | M_{12} | \psi_2 \rangle = \int_0^\infty \psi_1(R) M_{12}(R) \psi_2(R)\, \mathrm{d}R, \tag{5.61}$$

is evaluated by the stationary phase approximation. As usual

$$k_i(R) = \{2m[E_i - V_i(R)]\}^{1/2}/\hbar \tag{5.62}$$

and a_i denotes the classical turning point. The procedure is first to decompose the integrand into a sum of rapidly and slowly varying terms and to retain only the second, which alone can give rise to a point of stationary phase; thus

$$\langle \psi_1 | M_{12} | \psi_2 \rangle \simeq \frac{A_1 A_2}{4} \int_0^\infty g(R)\{\exp[\mathrm{i}f(R)] + \exp[-\mathrm{i}f(R)]\}\, \mathrm{d}R, \tag{5.63}$$

where

$$f(R) = \int_{a_2}^{R} k_2(R')\mathrm{d}R' - \int_{a_1}^{R} k_1(R')\mathrm{d}R' \tag{5.64}$$

and

$$g(R) = k_1^{-1/2}(R) k_2^{-1/2}(R) M_{12}(R). \tag{5.65}$$

The next step is to argue, as in Section 2.2, that the dominant contribution to the integral comes from the point (or points) R_x at which the phase term $f(R)$ is stationary with respect to R. The resulting stationarity condition

$$\mathrm{d}f/\mathrm{d}R = k_2(R) - k_1(R) = 0 \tag{5.66}$$

clearly corresponds to the orbit-crossing point R_x in Fig. 5.4. The approximation is completed by performing the quadratic expansion about R_x,

$$f(R) \simeq f(R_x) + \frac{1}{2} f''(R_x)(R - R_x)^2, \tag{5.67}$$

and noting according to eqns (5.62) and (5.66) that

$$\begin{aligned} f''(R_x) &= [(\mathrm{d}k_2/\mathrm{d}R) - (\mathrm{d}k_1/\mathrm{d}R)]_{R_x} \\ &= -(F_1 - F_2)/\hbar v_x, \end{aligned} \tag{5.68}$$

where F_i denote the slopes of the potentials $V_i(R)$ and v_x is the classical velocity at the crossing point,

$$F_i = -(\mathrm{d}V_i/\mathrm{d}R)_{R_x} \tag{5.69}$$

$$v_x = k_1(R_x)\hbar/m = k_2(R_x)\hbar/m. \tag{5.70}$$

It is convenient to assume, without loss of generality, that $F_1 > F_2$, a choice which ensures that $f(R_x)$ corresponds to the (positive) shaded area in Fig. 5.4, and that $f''(R_x) < 0$. Otherwise one simply reverses the signs in eqns (5.64) and (5.69).

It is then readily verified, by formally extending the integration range to $-\infty < R - R_x < \infty$ (on the basis that extending a rapidly oscillatory range cannot affect the integral), and making use of the standard integrals

$$\int_{-\infty}^{\infty} \exp\left[\pm \mathrm{i}\gamma^2 x^2/2\right] \mathrm{d}x = \frac{(2\pi)^{1/2}}{\gamma} \exp(\pm \mathrm{i}\pi/4), \tag{5.71}$$

that

$$\langle \psi_1 | M_{12} \psi_2 | \rangle \simeq \frac{1}{2} A_1 A_2 g(R_x) [2\pi / f''(R_x)]^{1/2} \cos[f(R_x) - \pi/4]. \tag{5.72}$$

Alternatively, on back-substitution from eqns (5.12), (5.19), (5.64), (5.65) and (5.68)–(5.70),

$$\begin{aligned} \langle \psi_1 | M_{12} | \psi_2 \rangle \simeq & \left[\frac{2[\hbar\bar{\omega}_1 \hbar\bar{\omega}_2] M_{12}^2(R_x)}{\pi \hbar \nu_x \Delta F} \right]^{1/2} \\ & \times \sin\left(\int_{a_2}^{R_x} k_2(R) \mathrm{d}R - \int_{a_1}^{R_x} k_1(R) \mathrm{d}R + \frac{\pi}{4} \right), \end{aligned} \tag{5.73}$$

where the local vibrational energy spacings, $\hbar\bar{\omega}_i$, are placed in square brackets to indicate that they are to be omitted if the relevant state, ψ_i, is a continuum one. For example, the golden rule predissociation formula yields

$$\begin{aligned} \Gamma &= 2\pi |\langle \psi_1 | V_{12} | \psi_2 |^2 \\ &\simeq \left(\frac{4\hbar\bar{\omega}_2 V_{12}^2(R_x)}{\hbar \nu_x \Delta F} \right) \sin^2\left(\int_{a_2}^{R_x} k_2(R_x) \mathrm{d}R - \int_{a_1}^{R_x} k_1(R) \mathrm{d}R + \frac{\pi}{4} \right), \end{aligned} \tag{5.74}$$

where the term in $\hbar\bar{\omega}_1$ is omitted because the potential $V_1(R)$ is repulsive (see Fig. 5.4(b)). It may be verified that eqn (5.74) goes over exactly to the weak interaction ($u \ll 1, \chi \simeq -\pi/4$) limit of eqn (3.130). Thus the present semiclassical perturbation formula agrees with that obtained by the connection formula method.

As derived, eqn (5.73) assumes a single transition point R_x but the result is easily generalized to include a sum over R_x when more than one crossing occurs. The physical condition for such a multiple crossing is conveniently obtained by rewriting eqn (5.66) in the form

$$E_1 - E_2 = V_1(R) - V_2(R), \tag{5.75}$$

which clearly has multiple roots only if the difference function $V_1(R) - V_2(R)$ passes through a turning point. The existence of such multiple roots naturally yields a more complicated interference pattern, as exemplified by Fig. 5.5 which shows the calculated Franck–Condon intensity for transitions from the $v = 0$ state of a scaled harmonic

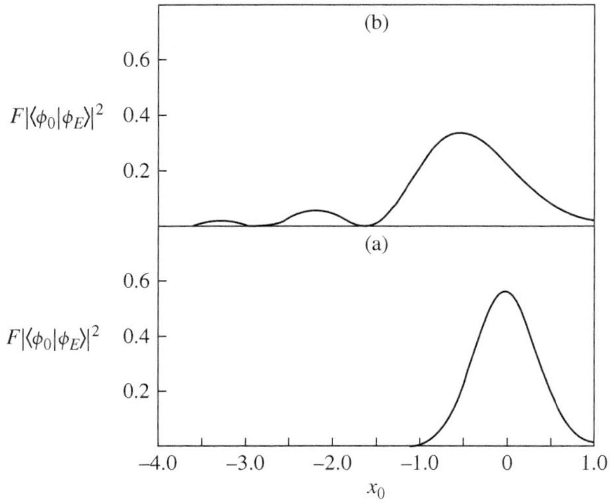

Fig. 5.5 Overlap between the $v = 0$ scaled harmonic oscillator wavefunction and the wavefunction for a linear potential, plotted as a function of the displacement, x_0, between the harmonic oscillator equilibrium and the continuum classical turning point, (a) for the slope parameter $F = 10$ and (b) for $F = 0.2$.

oscillator to two different linear continuum potentials. The scaled slope $F = 10.0$ relevant to Fig. 5.5(a) allows only a single root close to the classical range of the oscillator and one sees a typical reflection pattern, whereas the much smaller slope, $F = 0.2$, employed for Fig. 5.5(b) leads to the onset of supernumerary oscillations. A famous experimental example, seen in the B $\leftarrow$ E emission spectrum of I_2, was first recognized by Tellinghuisen (1985).

While the derivation of eqn (5.73) is relatively straightforward, the result suffers from the spurious singularities inherent in any primitive semiclassical approximation. Two types may be recognized in eqn (5.73). Either v_x may vanish owing to a coalescence between the crossing radius, R_x, and the turning points, a_1 and a_2 in Fig. 5.4, or ΔF may vanish in the double root case if the energy difference, $E_1 - E_2$, coincides with a stationary value of $V_1(R) - V_2(R)$. The latter situation has been briefly discussed by Hunt and Child (1978), but the former is of more general interest. Miller (1968) has suggested the following uniform Airy approximation to remedy the deficiency and an alternative uniform harmonic approximation due to Hunt and Child (1978) is introduced later in the section.

The uniform Airy approximation takes the form

$$\langle\psi_1|M_{12}|\psi_2\rangle = 2\pi^{1/2}P_x^{1/2}\eta^{1/4}Ai(-\eta), \tag{5.76}$$

where

$$P_x = [\hbar\bar{\omega}_1\hbar\bar{\omega}_2]M_{12}^2(R_x)/2\pi\hbar v_x\Delta F \tag{5.77}$$

and

$$\eta = \left[\frac{3}{2}\left(\int_{a_2}^{R_x} k_2(R)\mathrm{d}R - \int_{a_1}^{R_x} k_1(R)\mathrm{d}R\right)\right]^{2/3} \quad \text{for } R_x > a_i$$
$$\eta = \left[\frac{3}{2}\left(\int_{R_x}^{a_2} |k_2(R)|\mathrm{d}r - \int_{R_x}^{a_1} |k_1(R)|\mathrm{d}r\right)\right]^{2/3} \quad \text{for } R_x < a_i. \tag{5.78}$$

Again the term $\hbar\bar{\omega}_i$ is omitted if motion in the ith channel is unbound. The strength of (5.76) is, as usual, that the divergence of $v_x^{-1/2}$ in the term $P_x^{1/2}$ is cancelled by the disappearance of $\eta^{1/4}$. Hence $\langle\psi_1|M_{12}|\psi_2\rangle$ remains finite. Complications due to the existence of outer turning points, at which v_x might also vanish, are discussed at the end of the section, and modifications to take into account rapid variation in the coupling term $M_{12}(R)$ have been given by Child (1972).

First it is illuminating to see how Miller's result (5.76), which was quoted without proof, can be derived in a completely uniform manner. The argument, due to Child (1975) and Connor (1981), starts by replacing the above JWKB functions by their uniform Airy counterparts

$$\psi_i(R) \simeq \pi^{1/2} A_i \xi_i^{1/4}(R) k_i^{-1/2}(R) Ai[-\xi_i(R)], \tag{5.79}$$

where

$$\xi_i(R) = \left(\frac{3}{2}\int_{a_i}^{R} k_i(R)\mathrm{d}R\right)^{2/3} \tag{5.80}$$

and

$$Ai(-\xi) = \frac{1}{2\pi}\int_{-\infty}^{\infty} \exp(\mathrm{i}u^3/3 - \mathrm{i}\xi u)\,\mathrm{d}u. \tag{5.81}$$

The resulting integral therefore takes the form

$$\langle\psi_1|M_{12}|\psi_2\rangle = \frac{A_1 A_2}{4\pi}\int_0^{\infty}\int_{-\infty}^{\infty}\int_{-\infty}^{\infty} G(R)\exp[\mathrm{i}F(u,v,R)]\,\mathrm{d}u\,\mathrm{d}v\,\mathrm{d}R, \tag{5.82}$$

with

$$F(u,v,R) = \frac{u^3}{3} + \frac{v^3}{3} - \xi_1(R)u - \xi_2(R)v \tag{5.83}$$

and

$$G(R) = M_{12}(R)[\xi_1(R)\xi_2(R)]^{1/4}[k_1(R)k_2(R)]^{-1/2}. \tag{5.84}$$

Now Appendix B, which concerns the evaluation of multiple integrals with oscillatory integrands, shows that the functional form of the appropriate uniform approximation depends on the number and on the types of stationary phase points. In the present case, it is readily verified by use of the identity, derived from (5.80),

$$d\xi_i/dR = k_i(R)\xi_i^{-1/2}(R) \tag{5.85}$$

that the phase term $F(u, v, R)$ has two stationary points, (u_a, v_a, R_x) and (u_b, v_b, R_x), specified by

$$k_1(R_x) = k_2(R_x) \tag{5.86}$$

and

$$(u_a, v_a) = -(u_b, v_b) = [-\xi_1^{1/2}(R_x), \xi_2^{1/2}(R_x)]. \tag{5.87}$$

Equation (5.86) is obviously identical with the earlier primitive stationary phase condition, while eqn (5.87) ensures that

$$\begin{aligned} F(u_b, v_b, R_x) &= -F(u_a, v_a, R_x) \\ &= \int_{a_2}^{R_x} k_2(R)dR - \int_{a_1}^{R_x} k_1(R)dR. \end{aligned} \tag{5.88}$$

Given these two transition points, the proper uniform approximation, according to the discussion in Section B.1, is expressed in terms of the Airy function, regardless of the dimensionality of the parent integral (see eqns (B.14) and (B.22)):

$$\begin{aligned} \langle\psi_1|M_{12}|\psi_2\rangle = \frac{A_1A_2}{4\pi}\pi^{1/2}\exp[i(F_a+F_b)]\Big[\left(P_a^{1/2}+P_b^{1/2}\right)\eta^{1/4}Ai(-\eta) \\ -i\left(P_a^{1/2}-P_b^{1/2}\right)\eta^{-1/4}Ai'(-\eta)\Big], \end{aligned} \tag{5.89}$$

where

$$\eta = \left[\frac{3}{4}(F_b - F_a)\right]^{2/3} \tag{5.90}$$

and, for a three-dimensional parent integral,

$$P_\nu = \pm(2\pi i)^3 G_\nu^2/\det \parallel \partial^2 F/\partial u \partial v \parallel . \tag{5.91}$$

Here the quantity $\det \parallel \partial^2 F/\partial u\partial v \parallel$ is the Hessian determinant of the second derivative matrix and the upper and lower signs are taken for $\nu = b$ and a respectively, on the assumption that $F_b > F_a$. The result simplifies in the present case because the phase sum $F_a + F_b$ is identically zero. Furthermore the two probability factors, P_a and P_b, turn out to be identical;

$$\begin{aligned} P_a = P_b &= (2\pi)^3 G(R_x)^2\left[4\xi_1^{1/2}\xi_2^{1/2}(\partial k_1/\partial R - \partial k_2/\partial R)\right]^{-1} \\ &= 2\pi^3(\hbar^2/m)^2 M_{12}^2(R_x)/\hbar v_x \Delta F. \end{aligned} \tag{5.92}$$

It follows that

$$\langle\psi_1|M_{12}|\psi_2\rangle \simeq \frac{A_1A_2}{4\pi}\pi^{1/2}\left(P_a^{1/2}+P_b^{1/2}\right)\eta^{1/4}Ai(-\eta)$$
$$= \{2[\hbar\bar{\omega}_1\hbar\bar{\omega}_2]M_{12}(R_x)^2/\hbar v_x\Delta F\}^{1/2}\eta^{1/4}Ai(-\eta), \tag{5.93}$$

with the argument of the Airy function given by

$$\eta = \left[\frac{3}{4}(F_b - F_a)\right]^{2/3} = \left[\frac{3}{2}\left(\int_{a_2}^{R_x} k_2(R)\mathrm{d}R - \int_{a_1}^{R_x} k_1(R)\mathrm{d}R\right)\right]^{2/3}. \tag{5.94}$$

It is assumed as usual that $k_i(R) = |k_i(R)|\exp(-3i\pi/2)$ in the forbidden case, so that eqns (5.93) and (5.94) coincide exactly with (5.76) and (5.78).

As it stands eqn (5.93) properly takes into account only a possible coalescence between the crossing point, R_x, and the left-hand turning points, a_1 and a_2. The situation in which R_x lies simultaneously close to a left-hand turning point on one potential and a right-hand one on the other, as might occur in the situation depicted in Fig. 5.6, is, however, readily handled by replacing η in eqn (5.93) by the alternative parameter

$$\eta' = \left[\frac{3}{2}\left(\int_{a_1}^{R_x} k_1(R)\mathrm{d}R + \int_{R_x}^{b_2} k_2(R)\mathrm{d}R\right)\right]^{2/3}. \tag{5.95}$$

$\eta'^{1/4}$ then vanishes in just the right way to cancel the divergence of $v_x^{-1/2}$, when $a_1 = b_2 = R_x$.

To conclude this section it is convenient to mention an alternative uniform approximation designed to emphasize the Condon reflection implications of Fig. 5.6. It takes the following form:

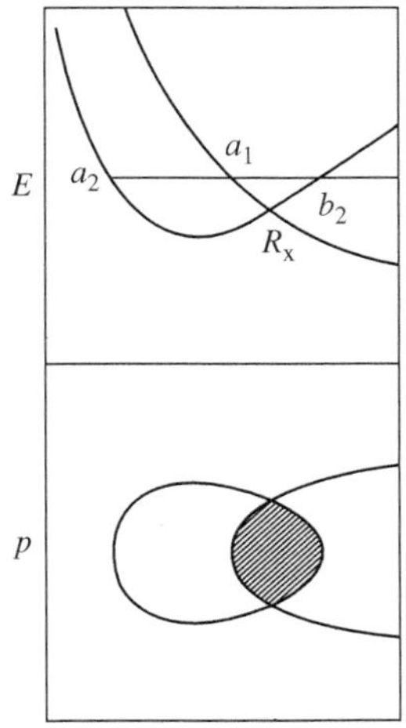

Fig. 5.6 Potential curves and phase orbits for an outer curve crossing.

$$\langle\psi_1|M_{12}|\psi_2\rangle = (2\pi)^{1/2}P_x^{1/2}(2v+1-\xi^2)^{1/4}\phi_v(\xi), \tag{5.96}$$

which corrects an expression given by Hunt and Child (1978) by including a missing factor of $(\pi/2)^{1/2}$; it may be seen as a special case of (B.105) with $A = 0$ and $P_a = P_b$. Here P_x is again defined by eqn (5.77), $\phi_v(\xi)$ is the vth harmonic oscillator wavefunction,

$$\phi_v(\xi) = (2^v v!\pi^{1/2})^{-1/2}H_v(\xi)\exp(-\xi^2/2), \tag{5.97}$$

and ξ is determined by the transcendental equation

$$\begin{aligned}\int_{a_2}^{R_x} k_2(R)\mathrm{d}R - \int_{a_1}^{R_x} k_1(R)\mathrm{d}R &= \int_{-\xi_v}^{\xi} \left(\xi_v^2-\xi^2\right)^{1/2}\mathrm{d}\xi \\ &= \frac{1}{2}\xi\left(\xi_v^2-\xi^2\right)^{1/2} + \left(v+\frac{1}{2}\right)[\arcsin(\xi/\xi_v)+\pi/2] \quad -\xi_v<\xi<\xi_v \\ &= -\frac{\mathrm{i}}{2}\xi\left(\xi^2-\xi_v\right)^{1/2} - \mathrm{i}\left(v+\frac{1}{2}\right)\operatorname{arcsinh}(\xi/\xi_v) \quad \xi<-\xi_v \\ &= \frac{\mathrm{i}}{2}\xi\left(\xi^2-\xi_v^2\right)^{1/2} + \mathrm{i}\left(v+\frac{1}{2}\right)[\operatorname{arcsinh}(\xi/\xi_v)-\mathrm{i}\pi] \quad \xi>\xi_v,\end{aligned} \tag{5.98}$$

where $\xi_v = (2v+1)^{1/2}$.

The choice between this uniform harmonic approximation and the previous Airy form is not one of accuracy because both are seen from Fig. 5.7 to be remarkably

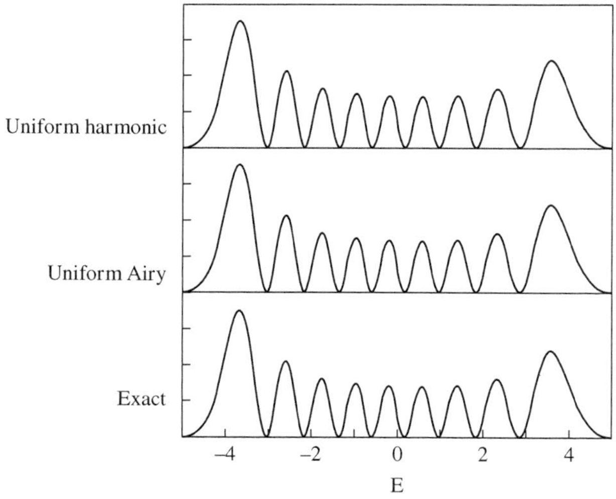

Fig. 5.7 Comparison between exact, uniform Airy and uniform harmonic representations of the bound free Franck–Condon factor between the $v = 0$ state of a scaled harmonic oscillator and the continuum wavefunction for an exponential potential.

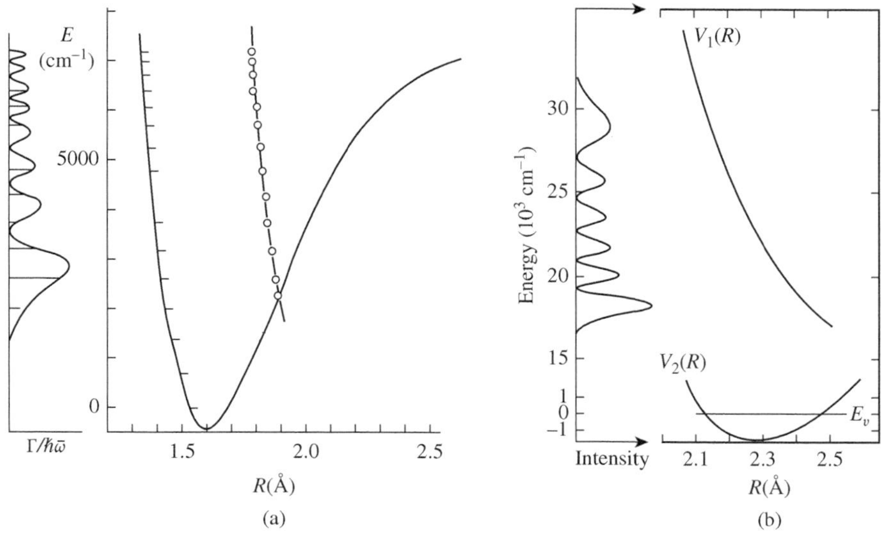

Fig. 5.8 Interference structure associated with (a) predissociation and (b) bound free transitions, indicating uniform Airy and uniform harmonic functional forms respectively.

accurate; far more so, for example, than the crude Condon reflection principle by which the continuum wavefunction is approximated by a delta function at its turning point (Herzberg 1950). From a conceptual point of view, however, the Airy approximation gives a convenient picture of the functional variation of a sequence of matrix elements involving interaction with a fixed continuum, as seen in Fig. 5.8(a), because the phase area mapping that underlies the approximation is unbounded. Conversely the harmonic approximation is more appropriate in situations where a variable continuum interacts with a fixed bound state (see Fig. 5.8(b)). Methods for analysis of these two types of behaviour are outlined in Chapter 6.

5.4 Matrix elements for non-curve-crossing situations

The above treatment of Franck–Condon and curve-crossing matrix elements put little emphasis on the nature of the interaction function $M_{12}(R)$, because the transition was assumed to be localized close to the roots of (5.75). In non-curve-crossing situations, however, the functional variation of $M_{12}(R)$ plays a more central role. Two types of approximation then become applicable, one being a variant of the Fourier method outlined in Section 5.2 and the other being derived by generalization of the uniform approach employed in Section 5.3.

The generalized Fourier approach simply replaces eqn (5.30) by

$$\begin{aligned}\langle\psi_1|M_{12}|\psi_2\rangle &= \frac{1}{\pi\hbar}\int_0^\infty M_{12}[R(t)]\cos\left(\frac{1}{\hbar}\int_0^{t'}[W_1(t')-W_2(t')]\mathrm{d}t'\right)\mathrm{d}t,\\ W_i(t) &= E_i - V_i[R(t)],\end{aligned}\tag{5.99}$$

where the time dependence of the trajectory $R(t)$ is obtained by integration of the equation

$$\mathrm{d}R/\mathrm{d}t = \bar{v}(R) = \{2[\bar{E} - \bar{V}(R)]/m\}^{1/2}, \tag{5.100}$$

applicable to motion under some mean potential $\bar{V}(R)$ at a mean energy $\bar{E}$. Thus the difference term $W_1(t) - W_2(t)$ makes allowance for different potential functions $V_i(R)$ for the two wavefunctions $\psi_i(R)$.

The second type of approximation is of the stationary phase type, but with the form of the interaction function explicitly taken into account. It is assumed, for illustrative purposes, that

$$M_{12}(R) = M_{12}^0 \exp[-\alpha(R)], \tag{5.101}$$

and that the wavefunctions are again approximated by the uniform Airy expressions in eqns (5.79)–(5.80). The subsequent argument is similar in outline to that at the end of Section 5.3 except that eqn (5.82) is replaced by

$$\langle\psi_1|M_{12}|\psi_2\rangle = \frac{A_1 A_2}{4\pi} \int_0^\infty \int_{-\infty}^\infty \int_{-\infty}^\infty G'(R) \exp[\mathrm{i}F'(u, v, R)]\, \mathrm{d}u\, \mathrm{d}v\, \mathrm{d}R, \tag{5.102}$$

where

$$F'(u, v, R) = \frac{u^3}{3} + \frac{v^3}{3} - u\xi_1(R) - v\xi_2(R) + \mathrm{i}\alpha(R) \tag{5.103}$$

and

$$G'(u, v, R) = M_{12}^0[\xi_1(R)\xi_2(R)]^{1/4}[k_1(R)k_2(R)]^{-1/2}, \tag{5.104}$$

with the functions $\xi_i(R)$ again given by (5.80).

As in Section 5.3 the nature of the approximation depends on the number of solutions to the stationary phase equations, which yield

$$(u, v) = [\pm\xi_1^{1/2}(R), \pm\xi_2^{1/2}(R)] \tag{5.105}$$

and

$$\pm k_2(R) \pm k_1(R) = \mathrm{i}\alpha'(R). \tag{5.106}$$

Alternatively, eqn (5.106) may be written

$$\kappa_2(R) \pm \kappa_1(R) = \pm\alpha'(R), \tag{5.107}$$

where

$$\kappa_i(R) = -\mathrm{i}k_i(R) = \{2m[V_i(R) - E_i]\}^{1/2}/\hbar. \tag{5.108}$$

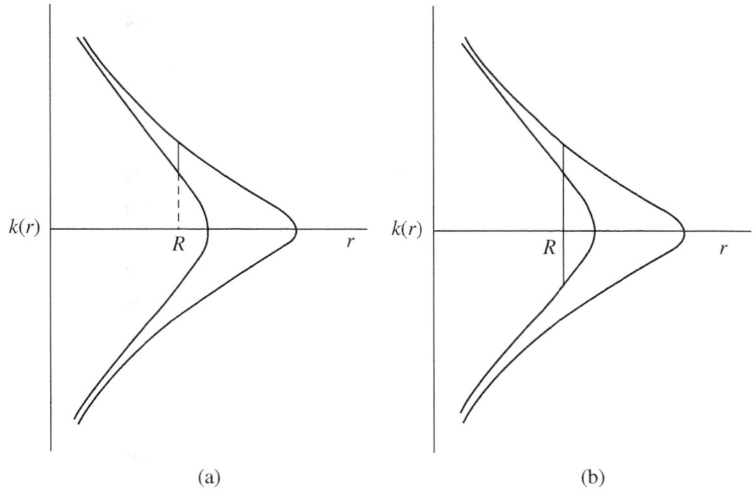

Fig. 5.9 Construction for the transition points given by eqns (5.110) and (5.115) respectively. (Reprinted from Uzer and Child (1980) by permission of Taylor and Francis.)

Note that eqns (5.107) and (5.108) have real solutions provided the transition point, R_x, is classically forbidden in both channels, that is $R_x < a_i$ for $i = 1, 2$.

The nature of possible solutions of eqn (5.107) depends on the relative magnitudes of $\kappa_1(R)$ and $\kappa_2(R)$ and on the sign of $\alpha'(R)$. It is assumed for illustrative purposes that $\kappa_2(R) - \kappa_1(R) > 0$ and that $\alpha'(R) > 0$, in which case the negative sign on the right-hand side of (5.107) is irrelevant. One or other of the two types of solution illustrated in Fig. 5.9 may occur but in each case eqn (5.107) admits only a single root, which means, according to the discussion in Appendix B, that the following simple exponential approximation applies:

$$\langle\psi_1|M_{12}|\psi_2\rangle = \frac{A_1 A_2}{4\pi}\frac{(2\pi\mathrm{i})^{3/2}G'(R)\exp[\mathrm{i}F'(u,v,R)]}{(\det||\partial^2F'/\partial u\partial v||)^{1/2}}. \tag{5.109}$$

It is evident from Fig. 5.9 that the choice of cases depends on the relative magnitudes of $\kappa_2(a_1)$ and $\alpha'(a_1)$. Case (a) applies when $\kappa_2(a_1) > \alpha'(a_1)$ and the transition point is given by

$$\kappa_2(R_x) - \kappa_1(R_x) = \alpha'(R_x). \tag{5.110}$$

This sign choice in (5.107) is consistent with

$$u(R_x) = -\xi_1^{1/2}(R_x), \quad v(R_x) = \xi_2^{1/2}(R_x), \tag{5.111}$$

$$\begin{aligned} F'(u,v,R_x) &= \frac{2}{3}\xi_1^{3/2}(R_x) - \frac{2}{3}\xi_2^{3/2}(R_x) + \mathrm{i}\alpha(R_x) \\ &= \mathrm{i}\left(\int_{R_x}^{a_2}\kappa_2(R)\mathrm{d}R - \int_{R_x}^{a_1}\kappa_1(R)\mathrm{d}R + \alpha(R_x)\right), \end{aligned} \tag{5.112}$$

and

$$\det||\partial^2 F'/\partial u\partial v|| = -4uv(u\xi_1'' + v\xi_2'' - \alpha'') - 2u{\xi_1'}^2 - 2v{\xi_2'}^2$$
$$= 4\mathrm{i}\xi_1^{1/2}\xi_2^{1/2}[\kappa_2'(R_x) - \kappa_1'(R_x) - \alpha''(R_x)]. \tag{5.113}$$

Thus, on collecting the various terms, the matrix element becomes

$$\langle\psi_1|M_{12}|\psi_2\rangle = [\hbar\bar{\omega}_1\hbar\bar{\omega}_2]^{1/2} m M_{12}^0 \exp\left(\int_{R_x}^{a_2}\kappa_2(R)\mathrm{d}R - \int_{R_x}^{a_1}\kappa_1(R)\mathrm{d}R - \alpha(R_x)\right) \Big/ \hbar^2\{2\pi\kappa_1(R_x)\kappa_2(R_x) \times [\kappa_2'(R_x) - \kappa_1'(R_x) - \alpha''(R_x)]\}^{1/2}. \tag{5.114}$$

Case (b) on the other hand applies when $\kappa_2(a_1) < \alpha'(a_1)$, in which case eqns (5.110)–(5.114) are replaced by

$$\kappa_2(R_x) + \kappa_1(R_x) = \alpha'(R_x), \tag{5.115}$$

$$u(R_x) = \xi_1^{1/2}(R_x), \quad v(R_x) = \xi_2^{1/2}(R_x), \tag{5.116}$$

$$F'(u, v, R_x) = -\frac{2}{3}\xi_1^{3/2}(R_x) - \frac{2}{3}\xi_2^{3/2}(R_x) + \mathrm{i}\alpha(R_x)$$
$$= \mathrm{i}\left(\int_{R_x}^{a_2}\kappa_2(R)\mathrm{d}R + \int_{R_x}^{a_1}\kappa_1(R)\mathrm{d}R + \alpha(R_x)\right), \tag{5.117}$$

$$\det||\partial^2 F'/\partial u\partial v|| = -4\mathrm{i}\xi_1^{1/2}\xi_2^{1/2}[\kappa_2'(R_x) + \kappa_1'(R_x) - \alpha''(R_x)], \tag{5.118}$$

and

$$\langle\psi_1|M_{12}|\psi_2\rangle = [\hbar\bar{\omega}_1\hbar\bar{\omega}_2]^{1/2} m M_{12}^0 \exp\left(-\int_{R_x}^{a_2}\kappa_2(R)\mathrm{d}R - \int_{R_x}^{a_1}\kappa_1(R)\mathrm{d}R - \alpha(R_x)\right) \Big/ \hbar^2\{-2\pi\kappa_1(R_x)\kappa_2(R_x) \times [\kappa_2'(R_x) + \kappa_1'(R_x) - \alpha''(R_x)]\}^{1/2}. \tag{5.119}$$

Equation (5.110), which takes the form of a Landau approximation (Landau and Lifshitz 1965), was first given in the chemical literature by Pack and Dahler (1969) and Bieniek (1987); the need for case (b) (i.e. eqn (5.119)) was first recognized by Uzer and Child (1980), who are responsible for the present derivation and who also give a number of tests of the validity of (5.99), (5.110) and (5.119) in realistic contexts. Both types of approximation work well, for example, when applied to the Jackson and Mott (1932) model for vibrational energy transfer.

5.5 Problems

5.1. Assuming that $p = m(\mathrm{d}q/\mathrm{d}t)$, consistent with a Hamiltonian of the form $H = (p^2/2m) + V(q)$, deduce from (5.25) that

$$\langle n'|p|n\rangle = -\mathrm{i}(n-n')m\omega(\bar{n})\langle n'|q|n\rangle.$$

Derive the corresponding quantum mechanical result by considering matrix elements of the commutator $[\hat{H}, \hat{q}]$.

5.2. Use the angle–action forms in Appendix E.2, and the result in problem 5.1, to derive the following Heisenberg matrix elements for the Morse oscillator, with $x_e = 0$. Note that the choice of angle origin, $\alpha = 0$, implies a phase convention such that the wavefunctions are positive at the outer turning point.

$$\langle n'|\mathrm{e}^{-ax}|n\rangle = \sqrt{1-\varepsilon}\, f(\Delta n, \varepsilon)$$

$$\langle n'|p|n\rangle = -\mathrm{i}\sqrt{2mD}(1-\varepsilon) f(\Delta n, \varepsilon)$$

$$\begin{aligned}\langle n'|x|n\rangle &= -\frac{\sqrt{1-\varepsilon}}{a\Delta n} f(\Delta n, \varepsilon) \quad \text{for} \quad n \neq n' \\ &= \frac{1}{a}\ln\left(\frac{1+\sqrt{1-\varepsilon}}{2(1-\varepsilon)}\right) \quad \text{for} \quad n = n',\end{aligned}$$

where $\Delta n = |n - n'|$ and (see § 858.536 of Dwight 1961)

$$f(\Delta n, \varepsilon) = \frac{\sqrt{1-\varepsilon}}{\pi}\int_0^\pi \frac{\cos \Delta n\alpha}{1+\sqrt{\varepsilon}\cos\alpha}\mathrm{d}\alpha = (-1)^{\Delta n}\left(\frac{1-\sqrt{1-\varepsilon}}{\sqrt{\varepsilon}}\right)^{\Delta n},$$

in which ε is evaluated at the average energy of the relevant states.

The same scaling rules may be verified to apply to the exact quantum mechanical matrix elements (Sage and Child 1989).

Hints:

$$\frac{1}{\pi}\int_0^\pi \frac{\cos m\alpha \mathrm{d}\alpha}{1+\sqrt{\varepsilon}\cos\alpha} = \frac{f(m,\varepsilon)}{\sqrt{1-\varepsilon}} \quad \text{for} \quad m = 0, 1, 2,$$

$$\frac{1}{\pi}\int_0^\pi \ln(1+\sqrt{\varepsilon}\cos\alpha)\mathrm{d}\alpha = \ln\left(\left[1+\sqrt{1-\varepsilon}\right]/2\right).$$

5.3. Use the Heisenberg correspondence principle to approximate the following matrix elements:

$$\langle j'm'|\cos\theta|jm\rangle = \sqrt{\frac{(j+1)^2-m^2}{(2j+1)(2j+3)}}\delta_{j'j+1}\delta_{m'm} + \sqrt{\frac{j^2-m^2}{(2j-1)(2j+1)}}\delta_{j'j-1}\delta_{m'm}$$

and

$$\langle j'm'|\sin\theta e^{i\phi}|jm\rangle = -\sqrt{\frac{(j+m+1)(j+m+2)}{(2j+1)(2j+3)}}\delta_{j'j+1}\delta_{m'm+1} + \sqrt{\frac{(j-m)(j-m-1)}{(2j-1)(2j+1)}}\delta_{j'j-1}\delta_{m'm+1}.$$

The necessary angle–action expressions are collected in Appendix E.4.

[As in problem 5.2, the sign of the matrix element depends on the choice of zero for the angle variables.]

5.4. Consider the overlap integral between a scaled harmonic oscillator ground state ψ_0 and a linear continuum wavefunction ψ_1, consistent with the following equations, with $\alpha > 0$:

$$\left(\frac{1}{2}\frac{d^2}{dx^2} + \frac{1}{2}(1-x^2)\right)\psi_0 = 0, \quad \left(\frac{1}{2}\frac{d^2}{dx^2} + \frac{\alpha}{2}(x-x_0)\right)\psi_1 = 0.$$

(i) Deduce from (5.66) or (5.75) that the transition points are real provided that $\alpha^2 + 4(1+\alpha x_0) > 0$, that at most one is classically accessible if $\alpha > 2$, and that two are classically accessible if $\alpha < 2$ and $(\alpha^2+4) > -4\alpha x_0 > 0$.

(ii) Given the functional forms

$$\psi_0(x) = \pi^{-1/4}\exp(-x^2/2)$$
$$\psi_1(x) = Ai[-\alpha(x-x_0)] = \frac{1}{2\pi}\int_{-\infty}^{\infty}\exp\left(i\frac{u^3}{3} - i\alpha u(x-x_0)\right)du,$$

show that

$$S_{01} = \int_{-\infty}^{\infty}\psi_0(x)\psi_1(x)dx = (4\pi)^{1/4}\exp\left(\frac{\alpha^6}{12} + \frac{\alpha^3 x_0}{2}\right)Ai\left(\frac{\alpha^4}{4} + \alpha x_0\right).$$

Hence, by use of the asymptotic forms (2.77), verify that the 'Condon reflection approximation', $S_{01} \simeq \alpha^{-1}\psi(x_0)$, is valid for $\alpha^4 \gg 4\alpha x_0 \gg 4$.

(iii) Plot S_{01} as a function of x_0 for $\alpha = 5$ and $\alpha = 0.5$, noting the change in behaviour in relation to the inequalities in (i).

6

Semiclassical inversion methods

The semiclassical inversion of experimental data exploits the dependence of quantization conditions, expectation values and various interference effects on phase integrals that contain explicit reference to the underlying potential energy function. Such phase integrals are amenable to an elegant Abelian transformation that allows one to extract the classical turning points at any chosen energy, which trace out the required potential function.

The applications below cover inversion of energy level positions, expectation values, predissociation linewidths, spectroscopic intensities and elastic and inelastic scattering amplitudes. An account is also given of the long-range theory of LeRoy and Bernstein (1970, 1971b) which can lead to very precise knowledge of the dissociation energy.

6.1 The RKR method

The classic RKR method (Rydberg 1931; Klein 1932; Rees 1946) determines the potential energy function, $V(R)$, by manipulating the Bohr quantization condition,

$$\left[v(E, J) + \frac{1}{2}\right]\pi = \frac{(2\mu)^{1/2}}{\hbar}\int_{a(E,\,J)}^{b(E,\,J)}\left(E - V(R) - J(J+1)\frac{\hbar^2}{2mR^2}\right)^{1/2} \mathrm{d}R, \quad (6.1)$$

which determines the vibrational–rotational energy levels, $E(\nu, J)$, and hence also, by differentiation, the generalized rotational constant,

$$B(v, J) = (\partial E/\partial J)_v/(2J+1). \quad (6.2)$$

In many applications, the necessary data are available at $J = 0$, $E(v, 0)$ being the vibrational term value and $B(v, 0)$ the normal rotational constant (Herzberg 1950). The following theory is, however, presented as a method for extracting the centrifugally corrected potential:

$$V_J(R) = V(R) + J(J+1)\frac{\hbar^2}{2mR^2}, \quad (6.3)$$

at an arbitrary J value. Of course once $V_J(R)$ is known, $V(R)$ itself is readily determined by subtracting the centrifugal term. It is assumed throughout that $v(E, J)$ is a continuous function of the continuous variables E and J, although of course only discrete values of v and J are physically available. Additional information on the form of $v(E, J)$ may also be obtained by combining the mass-reduced quantum numbers from isotopic variants of the species in question (Stwalley 1973, 1975, 1978; see also Section 3.1).

The first step in the RKR procedure is to define the function

$$A(U, J) = \pi \int_{E_{\min}}^{U} \frac{(\nu + \frac{1}{2})\mathrm{d}E}{(U - E)^{1/2}} = 2\pi \int_{-1/2}^{v(U,\, J)} [U - E(v, J)]^{1/2} \mathrm{d}v, \tag{6.4}$$

where the second integral is related to the first by partial integration and $v(U, J)$ is to be interpreted as the vibrational quantum number at energy U and angular momentum J. It follows by substituting from eqn (6.1) that

$$A(U, J) = \frac{(2\mu)^{1/2}}{\hbar} \int_{E_{\min}}^{U} \left[\int_{a(E,\, J)}^{b(E,\, J)} \left(\frac{E - V_J(R)}{E - U} \right)^{1/2} \mathrm{d}R \right] \mathrm{d}E. \tag{6.5}$$

It at this point that the Abelian transformation is performed. The trick is to recognize that the integration domain is taken over the shaded area in Fig. 6.1, so that on reversing the order of integration and using the second of the following identities,

$$\int_p^q \left(\frac{q - x}{x - p} \right)^{1/2} \mathrm{d}x = \int_p^q \left(\frac{x - p}{q - x} \right)^{1/2} \mathrm{d}x = \frac{1}{2}\pi(q - p), \tag{6.6}$$

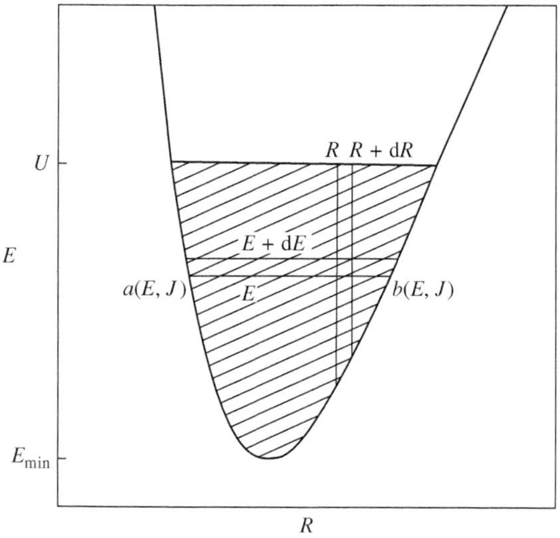

Fig. 6.1 The integration domain employed in eqn (6.5).

eqn (6.5) transforms to

$$
\begin{aligned}
A(U,J) &= \frac{(2\mu)^{1/2}}{\hbar}\int_{a(U,\,J)}^{b(U,\,J)}\left[\int_{V_J(R)}^{U}\left(\frac{E-V_J(R)}{U-E}\right)^{1/2}\mathrm{d}E\right]\mathrm{d}R \\
&= \frac{\pi(2\mu)^{1/2}}{2\hbar}\int_{a(U,\,J)}^{b(U,\,J)}\left(U-V(R)-J(J+1)\frac{\hbar^2}{2\mu R^2}\right)\mathrm{d}R. \qquad (6.7)
\end{aligned}
$$

Taken in conjunction with eqn (6.4), this means that

$$
\begin{aligned}
&\int_{a(U,\,J)}^{b(U,\,J)}\left(U-V(R)-J(J+1)\frac{\hbar^2}{2\mu R^2}\right)\mathrm{d}R \\
&\qquad = \frac{4\hbar}{(2\mu)^{1/2}}\int_{-1/2}^{v(V,\,J)}[U-E(v,J)]^{1/2}\mathrm{d}v, \qquad (6.8)
\end{aligned}
$$

where the left-hand integral depends on the desired potential function, $V(R)$, and the right one on the function $E(v,J)$, which is determined by the experimental energy levels. Finally, differentiation of eqn (6.8) by U and $J(J+1)$ in turn leads, when combined with (6.2), to the two working RKR equations

$$
f(U,J) = \int_{a(U,\,J)}^{b(U,\,J)}\mathrm{d}R = b(U,J)-a(U,\,J) = \frac{2\hbar}{(2\mu)^{1/2}}\int_{-1/2}^{v(U,\,J)}\frac{\mathrm{d}v}{[U-E(v,J)]^{1/2}}, \qquad (6.9a)
$$

and

$$
g(U,J) = \int_{a(U,\,J)}^{b(U,\,J)}\frac{\mathrm{d}R}{R^2} = \frac{1}{a(U,J)}-\frac{1}{b(U,J)} = \frac{2(2\mu)^{1/2}}{\hbar}\int_{-1/2}^{v(U,\,J)}\frac{B(v,J)\mathrm{d}v}{[U-E(v,J)]^{1/2}}. \qquad (6.9b)
$$

Given the values of the two integrals, $f(U,J)$ and $g(U,J)$, it is a simple matter to determine the turning points, $a(U,J)$ and $b(U,J)$, in the form

$$
\begin{aligned}
a(U,J) &= \frac{1}{2}\left\{[f^2(U,J)+4f(U,J)/g(U,J)]^{1/2}-f(U,J)\right\}, \\
b(U,J) &= \frac{1}{2}\left\{[f^2(U,J)+4f(U,J)/g(U,J)]^{1/2}+f(U,J)\right\},
\end{aligned} \qquad (6.10)
$$

and it is the U dependence of these turning points that traces out the form of the centrifugally corrected potential, $V_J(R)$, at the J value in question.

The important conclusions are that the turning point separation, $b(U,J)-a(U,J)$, at any given energy U is determined by the vibrational energy function $E(v,J)$ from $v=-1/2$ to $v(U,J)$, while the difference between the reciprocal function $[a(U,J)]^{-1}$ and $[b(U,J)]^{-1}$ requires the additional knowledge of the rotational constant function $B(v,J)$. Practical algorithms for implementation of the method are discussed by Tellinghuisen (1972) and Dickinson (1972); in particular it is important that the quadrature scheme should accommodate the singularity at the upper integration limit,

Table 6.1 The RKR potential for the $X^1\Sigma^+$ state of CO. δE and δB are discrepancies (obs–calc) between experimental values and those derived from the RKR potential.

v	$E(v,0)+Y_{00}$ (cm^{-1})	$B(v,0)$ (cm^{-1})	$a(v,0)$ (Å)	$b(v,0)$ (Å)	δE (cm^{-1})	δB (10^{-4} cm^{-1})
0	1081.7778	1.922 512 5	1.083 325	1.178 790	−0.0009	+0.009
1	3225.0502	1.905 007 4	1.053 438	1.219 632	−0.0013	+0.014
2	5341.8409	1.887 502 7	1.034 238	1.249 908	+0.0055	+0.033
3	7432.2173	1.869 998 3	1.019 412	1.275 935	+0.0227	+0.041
4	9496.2479	1.852 494 4	1.007 131	1.299 542	+0.2229	−0.022
5	11 534.0025	1.834 990 8	1.996 566	1.321 565	−0.0076	−0.029
6	13 545.5522	1.817 487 6	0.987 252	1.342 468	−0.0040	+0.068
7	15 530.9695	1.799 984 7	0.978 902	1.362 542	+0.0022	−0.023
8	17 490.3285	1.782 482 1	0.971 325	1.381 982	−0.0203	+0.026
9	19 423.7047	1.764 980 0	0.964 381	1.400 929	−0.0071	+0.034
10	21 331.1753	1.747 478 2	0.957 917	1.419 487	−0.0258	+0.001
11	23 212.8188	1.729 976 7	0.952 018	1.437 736	−0.0197	+0.054
12	25 068.7152	1.712 475 7	0.946 461	1.455 741	−0.0330	+0.009
13	26 898.9457	1.694 975 0	0.941 251	1.473 553	−0.0301	+0.051
14	28 703.5928	1.677 476 0	0.936 348	1.491 215	−0.0415	+0.029
15	30 482.7399	1.659 974 6	0.931 721	1.508 763	−0.0391	+0.033
16	32 236.4713	1.642 475 0	0.927 342	1.526 230	−0.0472	+0.045
17	33 964.8717	1.624 975 7	0.923 188	1.543 641	−0.0458	+0.006
18	35 668.0262	1.607 476 8	0.919 239	1.561 021	−0.0466	+0.030
19	37 346.0199	1.589 978 3	0.915 478	1.578 391	−0.0458	−0.021
20	38 998.9374	1.572 480 1	0.911 889	1.595 771	−0.0377	−0.037
21	40 626.8630	1.554 982 3	0.908 461	1.613 178	−0.0315	−0.063
22	42 229.8794	1.537 484 8	0.905 181	1.630 629	−0.0176	−0.136
23	43 808.0682	1.519 987 7	0.902 039	1.648 140	+0.0014	−0.171
24	45 361.5087	1.502 491 0	0.899 026	1.665 724	+0.0222	−0.237
25	46 890.2779	1.484 994 6	0.896 133	1.683 396	+0.0513	−0.322
26	48 394.4497	1.467 498 6	0.893 353	1.701 169	+0.0842	−0.365
27	49 874.0946	1.450 003 0	0.890 679	1.719 057	+0.1181	−0.414
28	51 329.2788	1.432 507 7	0.888 104	1.737 072	+0.1556	−0.439

where $U = E(v, J)$. It is also useful to include a small correction (see problems 3.2 and 3.3),

$$Y_{00} = \frac{B_e}{4} + \frac{\alpha_e\omega_e}{12B_e} + \frac{(\alpha_e\omega_e)^2}{144B_e^3} - \frac{\omega_e x_e}{4}, \tag{6.11}$$

to each energy level; it is the leading term in the Dunham (1932) expansion (obtained by higher-order quantization) and implies that the lower integration limits in (6.9a) should be replaced by $v_{\min}$, the value at which the modified energy $E(v,0) + Y_{00}$ vanishes. Finally, of course, a consistent set of units must be employed, so that $\hbar^2/2\mu$ would be replaced by $h/8\pi^2\mu c$ if energies are replaced by their equivalent wavenumber units.

As to the accuracy of the inversion, the only approximation lies in use of the Bohr quantization condition. Table 6.1, which is taken from Mantz et al. (1971), lists the input energy levels and rotational constants for the $X^1\Sigma^+$ state of CO, together with the resulting turning points $a(v, 0)$ and $b(v, 0)$ and the discrepancies, δE and δB, between the spectroscopic values and those calculated from the Schrödinger equation based on the RKR potential function. It is evident that the energy levels are reproduced to within 0.2 cm^{-1} and the rotational constants to within 0.0001 cm^{-1}. To remedy even such small defects LeRoy (1980) has suggested an iterative scheme for the incorporation of third-order quantization corrections.

It is pertinent at this point also to mention two alternative RKR equations (Miller 1971) which are similar to those employed in the inversion of elastic scattering data (see Section 6.4). They exploit the similarity in the way in which the energy E and the quantity $J(J+1)$ enter the quantization equation, a similarity which may be emphasized by introducing the notations $K = J(J+1)$ and

$$K(R) = \frac{2\mu R^2}{\hbar^2}[E - V(R)]. \tag{6.12}$$

K (or $J(J+1)$) in eqn (6.1) is then treated as functionally dependent on E and v. A pictorial summary relevant to quantization at a given energy is shown in Fig. 6.2. The first shows that the area used to determine $v(E, K)$ decreases as $K \to K_{\max}$ where $v(E, K_{\max}) = -1/2$. Similarly the turning points $a(E, K)$ and $b(E, K)$ of $V_K(R)$ in Fig. 6.2(a) appear as turning points of $K(R)$ in Fig. 6.2(b), with a coalescence at $K' = K_{\max}$.

The additional (Miller 1971) relations are obtained by defining the following function $B(U, K)$ which plays the part of $A(U, J)$ in the earlier discussion:

$$B(U, K) = \pi \int_K^{K_{\max}} \frac{\left[v(E, K) + \frac{1}{2}\right] \mathrm{d}K'}{(K' - K)^{1/2}} = 2\pi \int_{-1/2}^{v(E,\, K)} [K(E, v) - K]^{1/2} \mathrm{d}v, \tag{6.13}$$

where the two integrals are related by integration by parts. The subsequent manipulations are also similar to those employed above. Thus on combining eqns (6.1) and (6.13) and reversing the order of integration over the domain depicted in Fig. 6.2(b),

$$\begin{aligned}
B(U, K) &= \frac{(2\mu)^{1/2}}{\hbar} \int_K^{K_{\max}} \left[\int_{a(U,\, K')}^{b(U,\, K')} \left(\frac{E - V(R) - K'/R^2}{K' - K} \right)^{1/2} \mathrm{d}R \right] \mathrm{d}K' \\
&= \frac{(2\mu)^{1/2}}{\hbar} \int_K^{K_{\max}} \left[\int_{a(U,\, K')}^{b(U,\, K')} \frac{1}{R} \left(\frac{K(R) - K'}{K' - K} \right)^{1/2} \mathrm{d}R \right] \mathrm{d}K' \\
&= \frac{(2\mu)^{1/2}}{\hbar} \int_{a(U,\, K)}^{b(U,\, K)} \left[\frac{1}{R} \int_K^{K(R)} \left(\frac{K(R) - K'}{K' - K} \right)^{1/2} \mathrm{d}K' \right] \mathrm{d}R \\
&= \frac{\pi}{2} \frac{(2\mu)^{1/2}}{\hbar} \int_{a(U,\, K)}^{b(U,\, K)} \frac{1}{R} [K(R) - K] \mathrm{d}R \\
&= \frac{\pi}{2} \frac{(2\mu)^{1/2}}{\hbar} \int_{a(U,\, K)}^{b(U,\, K)} R \left(E - V(R) - \frac{K\hbar^2}{2\mu R^2} \right) \mathrm{d}R.
\end{aligned} \tag{6.14}$$

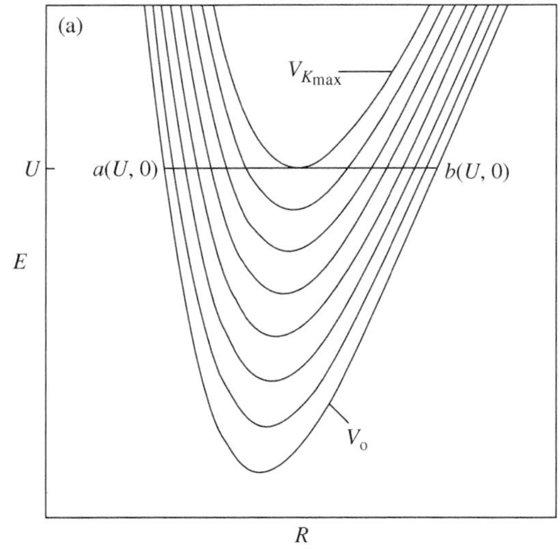

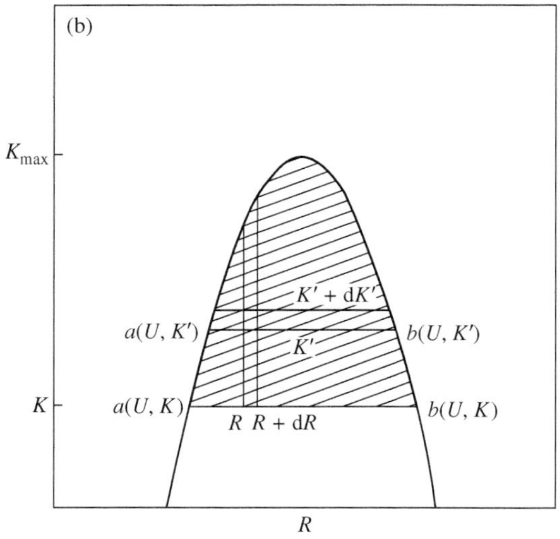

Fig. 6.2 (a) The K dependence of relevant potential energy curves. (b) The integration domain employed in eqn (6.14).

It remains by comparing derivatives of (6.13) and (6.14) with respect to E and K to conclude that

$$\frac{1}{2}[b^2(E,K) - a^2(E,K)] = \frac{\hbar^2}{\mu}\int_{-1/2}^{v(E,\,K)} \left(\frac{\partial K}{\partial E}\right) \frac{dv}{[K(E,v) - K]^{1/2}}$$
$$= \frac{\hbar^2}{\mu}\int_{-1/2}^{v(E,\,K)} \frac{1}{B(v,K)} \frac{dv}{[K(E,v) - K]^{1/2}} \tag{6.15}$$

and

$$\ln[b(E,K)/a(E,K)] = 2\int_{-1/2}^{v(E,K)} \frac{dv}{[K(E,v)-K]^{1/2}}. \tag{6.16}$$

Equations (6.15) and (6.16) are less convenient for common applications than (6.9a) and (6.9b) because data on the (E, v) dependence of K (or $J(J+1)$) are less readily available than those on the (v, J) dependence of E. However, eqn (6.16) is similar to the Firsov (1953) and Vollmer (1969) working equations for Vollmer inversion of elastic scattering data, which is discussed in Section 6.4. In addition Stwalley (1973) has shown how the above equations can be manipulated to determine the moments

$$M_s = \int_{a(U,\,K)}^{b(U,\,K)} R^s M(R)\mathrm{d}R$$

of a molecular property $M(R)$ from knowledge of its state-specific mean values $\langle M\rangle_v^K$ together with the term values, $E(v, K)$, and rotational constants, $B(v, K)$.

Other more complicated energy level patterns are also amenable to RKR-related inversion schemes. For example, the quantization equation for the symmetric double-well problem depends on the energy dependence of the phase integrals $\alpha(E)$ and $\varepsilon(E)$, taken over the potential well and barrier regions respectively; according to eqn (3.72)

$$\cos[2\alpha(E) - \phi(E)] = -[1+\kappa^2(E)]^{-1/2}, \tag{6.17}$$

where

$$\kappa(E) = \exp[-\pi\varepsilon(E)] \tag{6.18}$$

and $\phi(E)$ is a phase correction determined by $\varepsilon(E)$. Hence, if one can find smooth functions $\alpha(E)$ and $\varepsilon(E)$ that satisfy (6.17) at the known eigenvalues, then the generalizations of (6.9a) with α or ε in place of $\left(v+\frac{1}{2}\right)\pi$ will determine the widths of the well and the barrier as functions of E respectively. Pajunen and Child (1980) have developed a practical scheme based on this idea.

It may also be noted that Child and Nesbitt (1988) have adapted the first (vibrational) RKR equation (6.9a) to determine the classical turning points for a given vibrational state at different J values, and hence to deduce the underlying potential function from purely rotational data.

6.2 Inversion of predissociation linewidth and intensity data

The interference patterns associated with predissociation linewidths and with the intensities of bound to continuum spectroscopic transitions (see Figs. 6.3 and 6.4) contain information about the energy or quantum number dependence of relevant phase integrals, from which information about the underlying potential functions may be extracted by variants of the Abelian transformation technique.

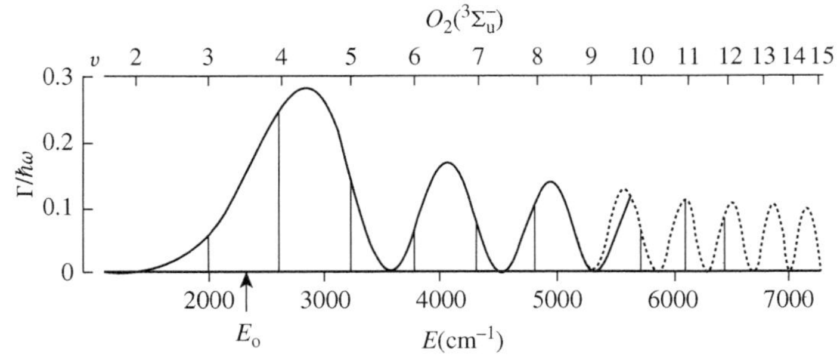

Fig. 6.3 Construction for recovery of the phase function $\phi(E)$ for inversion of predissociation linewidth data. (Reproduced from Child (1974c) by permission of the Royal Society of Chemistry.)

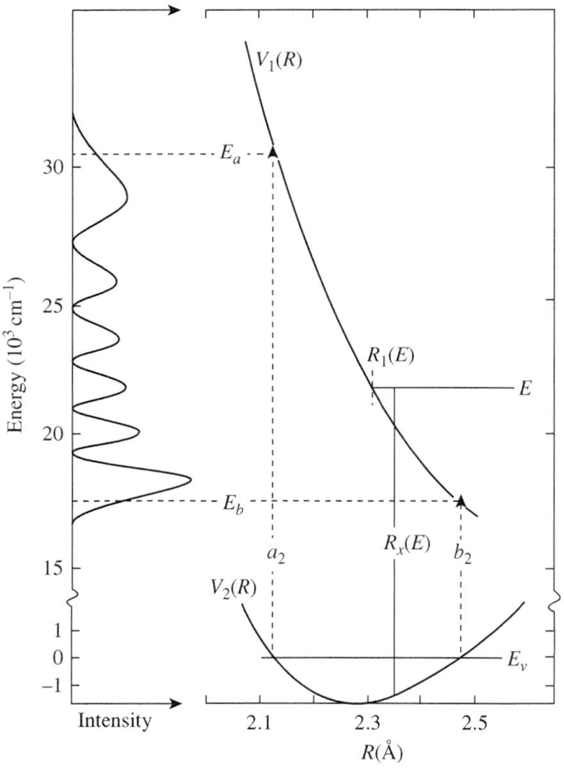

Fig. 6.4 Potential curves and notation employed in inversion of continuum emission data. (Reprinted with permission from Child et al. (1983). Copyright 1983, AIP Publishing LLC.)

The predissociation case illustrated in Fig. 6.3 will be taken as the first example, using the following combination of eqns (5.74) and (5.76) to represent the linewidth in the form

$$\Gamma(v) = \frac{4\hbar\bar{\omega}_2 V_{12}^2(R_x)}{\hbar v_x \Delta F}\eta^{1/2}(v)Ai^2[-\eta(v)], \tag{6.19}$$

where

$$\eta(v) = \frac{2}{3}[\varphi(v)]^{3/2}, \tag{6.20}$$

$$\begin{aligned}\varphi(v) &= \frac{(2\mu)^{1/2}}{\hbar}\int_{a_+}^{b_+}[E(v) - V_+(R)]^{1/2}\mathrm{d}R \qquad E > E_x \\ &= \mathrm{e}^{-3\mathrm{i}\pi/2}\frac{(2\pi)^{1/2}}{\hbar}\int_{b_-}^{c_-}[V_-(R) - E(v)]^{1/2}\mathrm{d}R \quad E < E_x,\end{aligned} \tag{6.21}$$

in which $V_\pm(R)$ are defined in the notation of Section 5.3 by the equations

$$\begin{aligned}V_+(R) &= V_1(R) \quad R < R_x \\ &= V_2(R) \quad R > R_x \\ V_-(R) &= V_2(R) \quad R < R_x \\ &= V_1(R) \quad R > R_x.\end{aligned} \tag{6.22}$$

The inversion procedure depends on the similarity between (6.22) and (6.1), with $\varphi(v)$ taking the role of $(v + \frac{1}{2})\pi$. Consequently if the linewidth pattern can be analysed to determine the energy dependence of φ, eqn (6.9a) may be generalized to read

$$b_+(U) - a_+(U) = b_2(U) - a_1(U) = \frac{2}{\pi}\frac{(2\mu)^{1/2}}{\hbar}\int_0^{\varphi(U)}\frac{\mathrm{d}\varphi}{[U - E(\varphi)]^{1/2}}, \tag{6.23}$$

where $E(\varphi)$ is the inverse function to $\varphi(E)$, which is itself shorthand for $\varphi[v(E)]$. The practical value of this result is that the turning points $b_2(U)$ of the bound potential $V_2(R)$ may be assumed to be available by the normal RKR method; hence eqn (6.23) determines the repulsive potential $V_1(R)$ via the turning points $a_1(U)$.

The same ideas also apply to situations in which the potential $V_1(R)$ cuts the repulsive branch of $V_2(R)$ or runs roughly parallel to it (Child 1980). The only changes are that $\varphi(v)$ in eqn (6.20) is replaced for $E > E_x$ by

$$\varphi'(v) = \frac{(2\mu)^{1/2}}{\hbar}\left(\int_{a_2}^{R_x}[E(v) - V_2(R)]^{1/2}\mathrm{d}R - \int_{a_1}^{R_x}[E(v) - V_1(R)]^{1/2}\mathrm{d}R\right)$$

or for $E < E_x$ by

$$\begin{aligned}\varphi'(v) = \mathrm{e}^{-3\mathrm{i}\pi/2}\frac{(2\pi)^{1/2}}{\hbar}\bigg(&\int_{R_x}^{a_2}[V_2(R) - E(v)]^{1/2}\mathrm{d}R \\ &- \int_{R_x}^{a_1}[V_1(R) - E(v)]^{1/2}\mathrm{d}R\bigg),\end{aligned} \tag{6.24}$$

and that eqn (6.23) is replaced by

$$a_1(U) - a_2(U) = \frac{2}{\pi}\frac{(2\pi)^{1/2}}{\hbar}\int_0^{\varphi'(U)} \frac{\mathrm{d}\varphi'}{[U - E(\varphi')]^{1/2}}. \tag{6.25}$$

The only awkward step in the procedure concerns extraction of the function $\varphi(E)$ (or $\varphi'(E)$) which is achieved by expressing eqn (6.19) in the form

$$\Gamma(v) = C\hbar\bar{\omega}_2(E - E_x)^{-1/2}\eta^{1/2}(v)Ai^2[-\eta(v)], \tag{6.26}$$

and assuming a linear divergence between the potential curves close to their intersection. Equation (6.21) (or (6.24)) then yields

$$\varphi(E) = \frac{2}{3}[(E - E_x)/E^*]^{3/2}, \tag{6.27}$$

where

$$E^* = [\hbar|F_1F_2|/(2\mu)^{1/2}(F_1 - F_2)]^{2/3}, \tag{6.28}$$

F_i being used to denote $-(\partial V_i/\partial R)$ at the crossing point R_x. Hence, close to this point

$$\Gamma/\hbar\bar{\omega}_2 \simeq C(E^*)^{-1/2}Ai^2[-(E - E_x)/E^*]. \tag{6.29}$$

Secondly, the known asymptotic behaviour of $Ai(-\eta)$ (eqn (2.77)) means that at energies well above E_x

$$\Gamma(E - E_x)^{1/2}/\hbar\bar{\omega}_2 \overset{\eta\gg 1}{\simeq} \pi C\sin^2[\varphi(E) + \pi/4]. \tag{6.30}$$

The problem is to estimate the quantities C, E_x and E^*, of which the first may be estimated from the largest value of the composite function on the left of eqn (6.30), given a preliminary estimate for E_x. This value of C is then used in eqn (6.29) to refine the estimate of E_x and to determine E^* from the magnitudes of the linewidths close to and below the crossing point. Finally, only E_x and C are retained and the best smooth function of $\varphi(E)$ against $(E - E_x)^{3/2}$ is fit to the linewidth data (Child 1974a).

A similar approach may be used to analyse fluctuations in bound to continuum absorption or emission spectra (Child et al.; LeRoy et al. 1983; 1988) such as those reported by Breford and Engleke (1978). In this case the uniform harmonic approximation (eqns (5.96)–(5.98)) is more convenient than the Airy approximation because it builds a connection between the number of intensity maxima and the nodal structure of the relevant vibrational wavefunction. The necessary equations may be expressed, in the notation of Fig. 6.4, as

$$\begin{aligned} M_{vE} &= \int_0^{\infty} \psi_{2v}(R)M(R)\psi_{1E}(R)\,\mathrm{d}R \\ &= \frac{1}{2}M(R_x)[\hbar\bar{\omega}_2/\hbar u_x\Delta F_x]^{1/2}[2v + 1 - \xi^2(E)]^{1/4}\phi[\xi(E)], \end{aligned} \tag{6.31}$$

where $\phi_v(\xi)$ is the vth harmonic oscillator wavefunction

$$\phi_v(\xi) = (2^v v! \pi^{1/2})^{-1/2} H_v(\xi) \exp(-\xi^2/2) \tag{6.32}$$

with ξ determined over the classically accessible region by the equation

$$\begin{aligned}\Delta(E) &= \frac{(2\mu)^{1/2}}{\hbar}\left(\int_{a_1(E)}^{R_x(E)} [E - V_1(R)]^{1/2} \mathrm{d}R + \int_{R_x(E)}^{b_2} [E_v - V_2(R)]^{1/2} \mathrm{d}R\right) \\ &= -\frac{1}{2}\xi(2v + 1 - \xi^2)^{1/2} + \left(v + \frac{1}{2}\right) \arccos\, [\xi/(2v+1)]^{1/2}.\end{aligned} \tag{6.33}$$

Here $R_x(E)$ is the Franck–Condon transition point at which the kinetic energies (and hence also the momenta) in the two channels are equal;

$$E - V_1(R_x) = E_v - V_2(R). \tag{6.34}$$

Note that eqn (6.33) differs from (5.98) by subtraction of $(v + \frac{1}{2})\pi$ from both sides followed by an overall sign reversal. The present forms are adopted to conform with the notation of Child et al. (1983). LeRoy et al. (1988) also use similar notation except that the subscripts 1 and 2 are exchanged throughout.

It must be emphasized that eqn (6.31) carries no implication that the vibrations supported by $V_2(R)$ are harmonic—only that the integral $M_{vE}(E)$ has the same qualitative structure as a function of E as does the wavefunction $\psi_{2v}(R)$. The approximation is also extremely accurate in a quantitative sense, as may be seen in Fig. 5.7.

The inversion proceeds, as illustrated in Fig. 6.5, by matching ξ values for the known maxima and zeros of $\phi_v^2(\xi)$ with the energies at which the spectroscopic intensity has maxima and minima; the resulting function $\xi(E)$ is central to all subsequent manipulations. Note in particular that

$$\frac{\mathrm{d}\Delta}{\mathrm{d}E} = -(2v + 1 - \xi^2)^{1/2} \frac{\mathrm{d}\xi}{\mathrm{d}E} = \frac{1}{2} \frac{(2\mu)^{1/2}}{\hbar} \int_{a_1(E)}^{R_x(E)} \frac{\mathrm{d}R}{[E - V_1(R)]^{1/2}}, \tag{6.35}$$

where terms in $(\mathrm{d}R_x/\mathrm{d}E)$ arising from differentiation of (6.33) have cancelled out by virtue of (6.34). Two different procedures are now followed according to whether $V_1(R)$ is assumed to be known, with $V_2(R)$ to be determined (LeRoy et al. 1988) or vice versa (Child et al. 1983).

In the former case eqn (6.35) is in effect an equation for $R_x(E)$, because the left-hand side is experimentally determined by the function $\xi(E)$ and the integral is readily computed for any test value of $R_x(E)$. The required function $V_2(R)$ is therefore readily determined by rearranging eqn (6.34) to read

$$V_2(R_x) = V_1(R_x) + E_{2v} - E(R_x), \tag{6.36}$$

where $E(R_x)$ is the inverse function to $R_x(E)$.

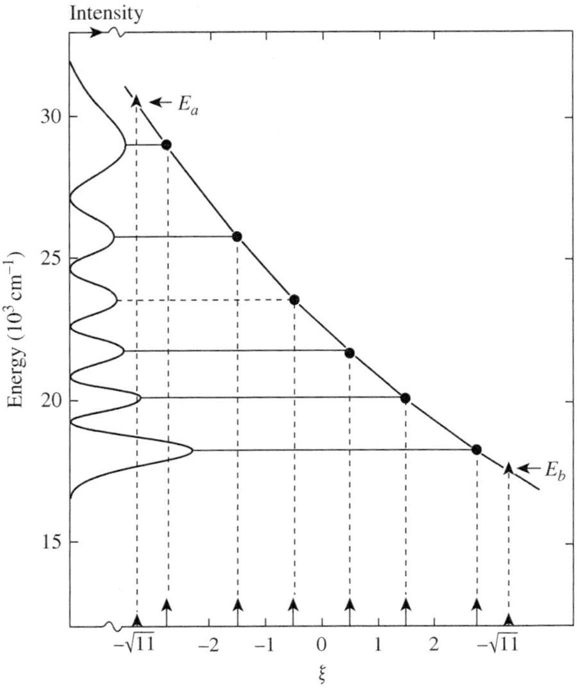

Fig. 6.5 Construction of the function $\xi(E)$ employed in eqn (6.35). (Reprinted with permission from Child et al. (1983). Copyright 1983, AIP Publishing LLC.)

The case in which $V_1(R)$ is unknown is less convenient, but Child et al. (1983) were able to obtain the expression

$$a_1(E) = \frac{1}{2}[b_2 + R_x(E)] - \frac{2\hbar}{\pi(2\mu)^{1/2}} \int_{E_b}^{E} \frac{d\Delta}{dE'} \frac{dE'}{(E - E')^{1/2}}$$
$$- \frac{1}{\pi} \int_{R_x(E)}^{b_2} \arcsin\left(1 - \frac{2[E_\nu - V_2(R)]}{[E - V_1(R)]}\right) dR. \tag{6.37}$$

The first integral is readily evaluated from the experimentally determined energy dependence of $\xi(E)$; but the second integral is more awkward because it depends both on the unknown function $V_1(R)$ and on $R_x(E)$ which cannot be determined independently of $V_1(R)$. Fortunately, however, it is readily verified that $R_x(E)$ lies between b_2 and $a_1(E)$ for any $E > E_b$ (see Fig. 6.4). Hence, once an initial segment of $V_1(R)$ is known on some interval $[a_1^0, b_2]$, eqn (6.34) may be used to extend the range of knowledge to smaller values of $a_1(E)$, with the required value of $R_x(E)$ being given by (6.34). Child et al. (1983) have designed an iterative scheme to implement this procedure, which is quite stable to initial assumptions about the form of the function $V_1(R)$.

6.3 LeRoy–Bernstein extrapolation to dissociation limits

The Abelian transformation approach is not the only way to exploit the presence of the potential function $V(R)$ in the quantization equation. LeRoy and Bernstein (1970, 1971b) have shown how knowledge of the limiting behaviour

$$V(R) \overset{R\to\infty}{\sim} D - C_n/R^n \tag{6.38}$$

may be exploited to linearize the Birge–Sponer plot used to determine dissociation limits (Herzberg 1950). The same theory also leads to analytical expressions for the near dissociation behaviour of vibrational term separations and rotational constants (LeRoy 1972, 1980).

The argument depends on the form of the derivative of eqn (6.1),

$$\left(\frac{\mathrm{d}\nu}{\mathrm{d}E}\right) = \frac{1}{2\pi}\frac{(2\mu)^{1/2}}{\hbar}\int_{a_\nu}^{b_\nu}\frac{\mathrm{d}R}{[E_\nu - V(R)]^{1/2}}, \tag{6.39}$$

in which the dominant contributions to the integral come from the turning point regions. Moreover, since each such contribution is readily shown to depend inversely on the local potential derivative $(\mathrm{d}V/\mathrm{d}R)$, the inner turning point contribution is roughly independent of energy while that from the outer turning point rapidly increases as the energy approaches the dissociation limit.

The mathematical consequence is that the approximation

$$\left(\frac{\mathrm{d}\nu}{\mathrm{d}E}\right) \hat{=} \frac{1}{2\pi}\frac{(2\mu)^{1/2}}{\hbar}\int_{0}^{b_\nu}[E_\nu - D + C_n/R^n]^{-1/2}\mathrm{d}R \tag{6.40}$$

becomes increasingly valid as E_ν approaches D, where b_ν is given by

$$E_\nu = V(b_\nu) \simeq D - C_n/b_\nu^n. \tag{6.41}$$

The substitution

$$z = (b_\nu/R)^n \tag{6.42}$$

then yields

$$\left(\frac{\mathrm{d}\nu}{\mathrm{d}E}\right) = \frac{1}{2\pi}\frac{(2\mu)^{1/2}}{\hbar}C_n^{1/n}(D - E_\nu)^{-(n+2)/2n}\int_0^1 z^{(2-n)/2n}(1-z)^{-1/2}\,\mathrm{d}z, \tag{6.43}$$

which rearranges, after evaluation of the integral, to (LeRoy and Bernstein 1971b)

$$\left(\frac{\mathrm{d}E}{\mathrm{d}v}\right) = K_n(D - E_\nu)^{(n+2)/2n}, \tag{6.44}$$

where

$$K_n = \frac{2\pi\hbar n}{(2\mu)^{1/2}}C_n^{-1/n}\frac{\Gamma(1+\frac{1}{n})}{\Gamma\left(\frac{1}{2}+\frac{1}{n}\right)\Gamma(\frac{1}{2})}. \tag{6.45}$$

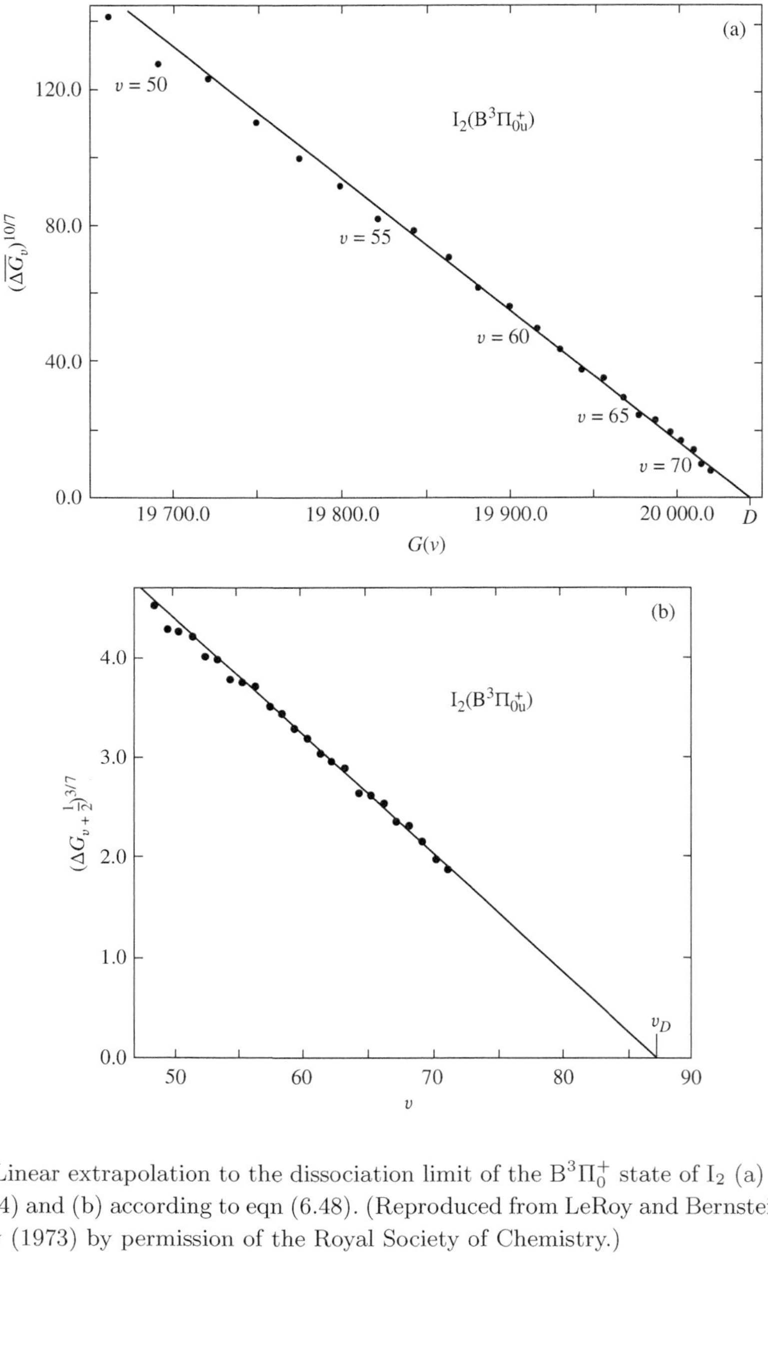

Fig. 6.6 Linear extrapolation to the dissociation limit of the $B^3\Pi_0^+$ state of I_2 (a) according to eqn (6.44) and (b) according to eqn (6.48). (Reproduced from LeRoy and Bernstein (1971b) and LeRoy (1973) by permission of the Royal Society of Chemistry.)

An alternative integrated version of (6.44) is

$$E_\nu = D - X_n(v_D - v)^{2n/(n-2)}, \tag{6.46}$$

where v_D is the non-integral value of v at the dissociation limit and

$$X_n = [(n-2)K_n/2n]^{2n/(n-2)}. \tag{6.47}$$

Finally, eqns (6.44) and (6.46) may be rearranged to yield

$$\left(\frac{\mathrm{d}E}{\mathrm{d}v}\right) = \left(\frac{2n}{n-2}\right) X_n(v_D - v)^{(n+2)/(n-2)}. \tag{6.48}$$

The validity of this approach to extrapolation for dissociation limits is amply demonstrated by the plots in Fig. 6.6 of appropriate powers of $\mathrm{d}E/\mathrm{d}v$ against E_v and v respectively. The experimental data apply to the $\mathrm{B}^3\Pi(0_u^+)$ state of I_2 with $\mathrm{d}E/\mathrm{d}v$ estimated as $\Delta G_{v+\frac{1}{2}} = E_{v+1} - E_v$. For further details of this approach and for extension to the near dissociation behaviour of the various rotational constants $B_v, D_v, H_v, \ldots$, the reader is referred to reviews by LeRoy (1973, 1980) and Stwalley (1978).

6.4 Inversion of elastic scattering data

It is shown in Chapter 9 that the rainbow oscillations in the elastic scattering differential cross-section depend on the l dependence of the classical deflection function

$$\chi(\lambda) = \pi - 2\lambda \int_{a_\lambda}^{\infty} \frac{\mathrm{d}R}{R^2[k^2 - U(R) - \lambda^2/R^2]^{1/2}} \tag{6.49}$$

and of the related semiclassical phase shift,

$$\eta(\lambda) = \int_{a_\lambda}^{\infty} [k^2 - U(R) - \lambda^2/R^2]^{1/2}\,\mathrm{d}R - \int_{a_\lambda^0}^{\infty} (k^2 - \lambda^2 R^2)^{1/2}\,\mathrm{d}R, \tag{6.50}$$

where

$$\lambda = \left(l + \frac{1}{2}\right), \quad k = (2\mu E)^{1/2}\hbar, \quad U(R) = 2\mu V(R)/\hbar^2. \tag{6.51}$$

The discussion in Section 9.3 also reveals a second source of interference information in the oscillatory variation with energy of the integral cross-section for mutual scattering of identical particles, which is governed in turn by the energy dependence of the s-wave phase shift. Both the λ dependence of $\chi(\lambda)$ or $\eta(E, \lambda)$ and the energy dependence of $\eta_0(E)$ will now be shown to be amenable to inversion by the Abelian integral method.

Suppose in the first place that eqn (6.49) is rearranged, by analogy with (6.12)–(6.15), to read

$$\chi(\lambda) = 2\lambda \left(\int_{a^0(K)}^{\infty} \frac{\mathrm{d}R}{R[K^0(R) - K]^{1/2}} - \int_{a(K)}^{\infty} \frac{\mathrm{d}R}{R[K(R) - K]^{1/2}} \right), \tag{6.52}$$

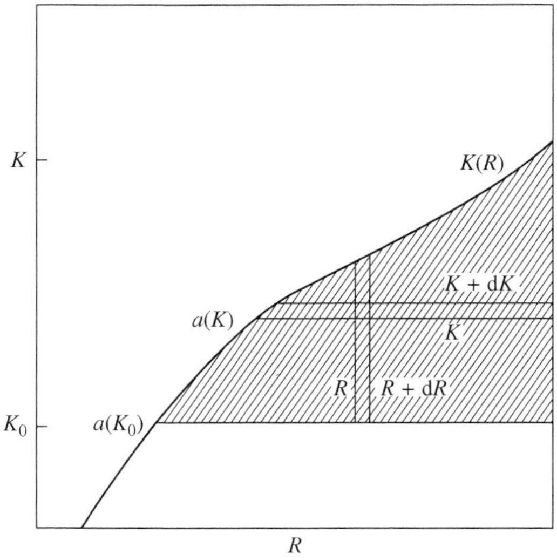

Fig. 6.7 The integration domain employed in eqn (6.54).

where

$$K = \lambda^2, \quad K(R) = R^2[k^2 - U(R)], \quad K^0(R) = R^2k^2, \tag{6.53}$$

and $a(K)$ and $a^0(K)$ are the assumedly single roots of the two square root factors, subject to the limitation that the energy lies above the classical orbiting limit (see Section 9.1). The familiar Abelian transformation process is now initiated by multiplying eqn (6.52) through by $(\lambda^2 - \lambda_0^2)^{-1/2}$ and integrating with respect to λ to obtain

$$\begin{aligned}\int_{\lambda_0}^{\infty} \frac{\chi(\lambda)\mathrm{d}\lambda}{(\lambda^2 - \lambda_0^2)^{1/2}} &= \int_{K_0}^{\infty}\left(\int_{a^0(K)}^{\infty} \frac{\mathrm{d}R}{R\{[K^0(R) - K](K - K_0)\}^{1/2}}\right)\mathrm{d}K \\ &- \int_{K_0}^{\infty}\left(\int_{a(K)}^{\infty} \frac{\mathrm{d}R}{R\{[K(R) - K](K - K_0)\}^{1/2}}\right)\mathrm{d}K.\end{aligned} \tag{6.54}$$

Finally, on reversing the order of integration over the domain shown in Fig. 6.7 and evaluating the resulting inner integrals by use of the identity

$$\int_p^q \frac{\mathrm{d}x}{[(q-x)(x-p)]^{1/2}} = \pi, \tag{6.55}$$

one finds that

$$\int_{\lambda_0}^{\infty} \frac{\chi(\lambda)\mathrm{d}\lambda}{(\lambda^2-\lambda_0^2)^{1/2}} = \int_{a^0(K_0)}^{\infty} \left(\int_{K_0}^{K^0(R)} \frac{\mathrm{d}K}{R\{[K^0(R)-K](K-K_0)\}^{1/2}} \right) \mathrm{d}R$$

$$- \int_{a(K_0)}^{\infty} \left(\int_{K_0}^{K(R)} \frac{\mathrm{d}K}{R\{[K(R)-K](K-K_0)\}^{1/2}} \right) \mathrm{d}R$$

$$= \pi \left(\int_{a^0(K_0)}^{\infty} \frac{\mathrm{d}R}{R} - \int_{a(K_0)}^{\infty} \frac{\mathrm{d}R}{R} \right)$$

$$= \pi \ln[a(K_0)/a^0(K_0)]. \tag{6.56}$$

But $a^0(K_0) = K_0^{1/2}/k$ and $K_0 = \lambda_0^2$; hence

$$ka(K_0) = \lambda_0 \exp\left(\frac{1}{\pi} \int_{\lambda_0}^{\infty} \frac{\chi(\lambda)\mathrm{d}\lambda}{(\lambda^2-\lambda_0^2)^{1/2}} \right), \tag{6.57}$$

which determines the point $R = a(K_0)$ at which $K(R) = K_0$. The corresponding value of the potential energy $V(a)$ is given by

$$V(a) = \hbar^2 U(a)/2\mu = E - \lambda_0^2 \hbar^2/2\mu a^2, \tag{6.58}$$

after rearrangement of eqn (6.53). Equations (6.57) and (6.58) are the basis of the Firsov (1953) inversion procedure, which has been applied with considerable success by Buck (1975).

Variants of the Firsov (1953) equations, due to Vollmer (1969), may be obtained by recognizing that

$$\chi(\lambda) = 2(\partial\eta/\partial\lambda), \tag{6.59}$$

and that

$$\int_{\lambda_0}^{\infty} \frac{\chi(\lambda)\mathrm{d}\lambda}{(\lambda^2-\lambda_0^2)^{1/2}} = -\frac{1}{\lambda}\frac{\mathrm{d}}{\mathrm{d}\lambda_0} \left(\int_{\lambda_0}^{\infty} \chi(\lambda)(\lambda^2-\lambda_0^2)^{1/2}\mathrm{d}\lambda \right). \tag{6.60}$$

Hence, on substituting for $\chi(\lambda)$ in terms of $\eta(\lambda)$ and integrating by parts, eqn (6.57) transforms to

$$ka(K_0) = \lambda_0 \exp\left(\frac{2}{\pi\lambda_0}\frac{\mathrm{d}}{\mathrm{d}\lambda_0} \int_{\lambda_0}^{\infty} \frac{\eta(\lambda)\lambda\,\mathrm{d}\lambda}{(\lambda^2-\lambda_0^2)^{1/2}} \right). \tag{6.61}$$

Again the potential energy at the point $R = a$, given by (6.61), is determined by (6.58).

Turning to the information extracted from symmetry oscillations in the integral cross-section (see Section 9.3), knowledge of the s-wave phase shift, $\eta_0(E)$, alone is insufficient to determine other than purely repulsive potentials; it must usually be supplemented by knowledge of the bound state rotationless energy levels (Miller 1971).

The convenient quantities for the inversion may be combined as a composite variable $\tilde{\eta}(E)$, taken as $\eta_0(E)$ for $E > 0$ and $[v(E) + 1/2]\,\pi$ for $E < 0$, where

$$\eta_0(E) = \frac{(2\mu)^{1/2}}{\hbar}\left(\int_{a(E)}^{\infty}[E - V(R)]^{1/2}\mathrm{d}R - \int_0^{\infty}E^{1/2}\mathrm{d}R\right)$$
$$\left[v(E) + \frac{1}{2}\right]\pi = \frac{(2\mu)^{1/2}}{\hbar}\left(\int_{a(E)}^{b(E)}[E - V(R)]^{1/2}\mathrm{d}R\right). \tag{6.62}$$

It follows by the now familiar procedure that

$$\begin{aligned}&\int_{V_{\min}}^{U}\frac{\mathrm{d}\tilde{\eta}}{\mathrm{d}E}\frac{\mathrm{d}E}{(U - E)^{1/2}}\\ &= \pi\int_{V_{\min}}^{0}\frac{\mathrm{d}v}{\mathrm{d}E}\frac{\mathrm{d}E}{(U - E)^{1/2}} + \int_0^U\frac{\mathrm{d}\eta_0}{\mathrm{d}E}\frac{\mathrm{d}E}{(U - E)^{1/2}}\\ &= \frac{\pi}{2}\frac{(2\mu)^{1/2}}{\hbar}\int_0^{a(U)}\mathrm{d}R = \frac{\pi}{2}\frac{(2\mu)^{1/2}}{\hbar}a(U).\end{aligned} \tag{6.63}$$

The steps involved in eqn (6.63) are to substitute for $\tilde{\eta}(E)$ from eqn (6.62), to reverse the integration order with respect to R and E, and to evaluate the resulting inner integral by use of eqn (6.55). It follows by simple rearrangement that the energy-dependent turning point, which traces out the repulsive branch of $V(R)$ for $U > 0$, is given by

$$a(U) = \frac{2}{\pi}\frac{\hbar}{(2\mu)^{1/2}}\left(\int_{V_{\min}}^{0}\frac{\pi\,\mathrm{d}v}{[U - E(v)]^{1/2}} + \int_0^U\frac{\mathrm{d}\eta_0}{\mathrm{d}E}\frac{\mathrm{d}E}{(U - E)^{1/2}}\right). \tag{6.64}$$

Notice, by comparison with eqn (6.9a), that knowledge of the vibrational energy levels, $E(v)$, alone determines only the turning point separation, $b(U)$– $a(U)$, but that additional information about the energy derivative of $\eta_0(E)$ fixes the absolute position of the repulsive potential wall.

In conclusion it may be noted that Child and Gerber (1979) have extended the above ideas to the inversion of Stückelberg (1932) oscillations in the differential cross-section for inelastic scattering from coupled monotonically repulsive potentials (see Section 9.4). The inversion scheme carries echoes not only of the present section, but also of Section 6.2, because the Stückelberg oscillations are scattering manifestations of the interferences responsible for fluctuations in bound state Franck–Condon factors. It is also pertinent to reference a number of successful non-semiclassical schemes for inversion of scattering data by Gerber and Shapiro (1976), Shapiro and Gerber (1976) and Gerber et al. (1978).

6.5 Problems

6.1. The energy levels for a Morse potential, $V(x) = D(1 - e^{-ax})^2$, may be expressed as

$$E(v) = D\left[1 - \left(1 - a\hbar(v + 1/2)/\sqrt{2\mu D}\right)^2\right].$$

Confirm that the first RKR equation (6.9a) yields the correct turning point separation, at energy U,

$$f(U) = b(U) - a(U) = \frac{1}{a}\ln\left[\frac{\sqrt{D} + \sqrt{U}}{\sqrt{D} - \sqrt{U}}\right].$$

Hint: substitute $z = 1 - a\hbar(v + 1/2)/\sqrt{2\mu D} = \sqrt{(1 - E/D)}$.

6.2. It was shown in Section 4.4 that the principal quantum number for the hydrogen atom may be expressed as a sum of vibrational and angular momentum quantum numbers, $n = v + l + 1$. Show in atomic units, such that $\hbar = m = 1$, that RKR inversion of the energy formula

$$E(v, l) = -\frac{1}{2n^2}$$

yields the Langer-corrected effective potential function

$$V_l(r) = -\frac{1}{r} + \frac{(l + 1/2)^2}{r^2}.$$

Hints: Note that the RKR integrals are taken over v, and that the rotational constant $B_v = (2l+1)^{-1}\left(\partial E/\partial l\right)_v = (2l+1)^{-1}n^{-3}$. Evaluate the second integral by substituting $n = z^{-1}$.

6.3. The long-range form of the ground state potential of $BeAr^+$ is known to be well approximated by $V(R) \simeq D - C_4/R^4$, with $C_4 = 95\,340$ cm^{-1}Å^4 (LeRoy 1972). Given the energy levels in the table below, deduce the dissociation energy and the number of bound energy levels:

v	0	1	2	3	4
E_v (cm^{-1})	0.00	344.98	672.42	982.48	1275.39

Hint: the constant C_4 determines the coefficient K_4 and X_4 in eqns (6.44), (6.48).

7
Non-separable bound motion

The quantization of non-separable bound systems is best appreciated in the light of the possible phase space structures, discussed in Section 7.1. Emphasis is placed on the role of the so-called invariant tori, which are generalizations of the phase orbits of Section 4.1. Section 7.2 describes how the simplest situations may be quantized by an extension of the Bohr–Sommerfeld approach, due to Einstein (1917), Brillouin (1926a, b) and Keller (1958) (EBK). However, resonant situations, in which the frequencies of motion in different degrees of freedom are rationally related, introduce two types of complication. The first is that unstable resonances inevitably lead to bifurcation into two or more coexisting types of motion at a given energy, which means that different procedures are required in different parts of phase space and that non-classical corrections may be important at the boundaries between these different regions. Section 7.3 shows how difficulties of this type may be handled by methods due to Jaffé and Brumer (1980) and Sibert et al. (1982), while Section 7.4 offers an alternative to the above methods, based on a Fourier representation of the torus due originally to Percival and Pomphrey (1973, 1976). The second type of complication, which is more serious, is that interactions between several such resonances may lead to classical chaos; the extent to which such chaotic systems may be quantized by classical perturbation and adiabatic switching techniques is explored in Sections 7.5 and 7.6. Finally, Section 7.7 deals with quantization by use of periodic orbits as suggested by Gutzwiller (1967, 1969, 1970, 1990) and Berry and Tabor (1976, 1977). Reviews of the various approaches have been given by Percival (1977), Handy (1980), Noid et al. (1981) and Ezra et al. (1987).

7.1 Phase space structures

As a preliminary it is convenient to start with a separable two-oscillator system described by the Hamiltonian

$$H = \frac{1}{2m}(p_x^2 + p_y^2) + V_x(x) + V_y(y), \tag{7.1}$$

where it is assumed, for purposes of illustration, that

$$V_x(x) = D(1 - \mathrm{e}^{-ax})^2 \quad \text{and} \quad V_y(y) = \frac{1}{2}y^2. \tag{7.2}$$

Figure 7.1 shows three projections of a typical classical trajectory on to different planes in the four-dimensional phase space (p_x, p_y, x, y). The starting conditions were chosen such that the frequencies of the x and y motions were non-commensurate, in which

Semiclassical Mechanics with Molecular Applications. Second Edition. M. S. Child

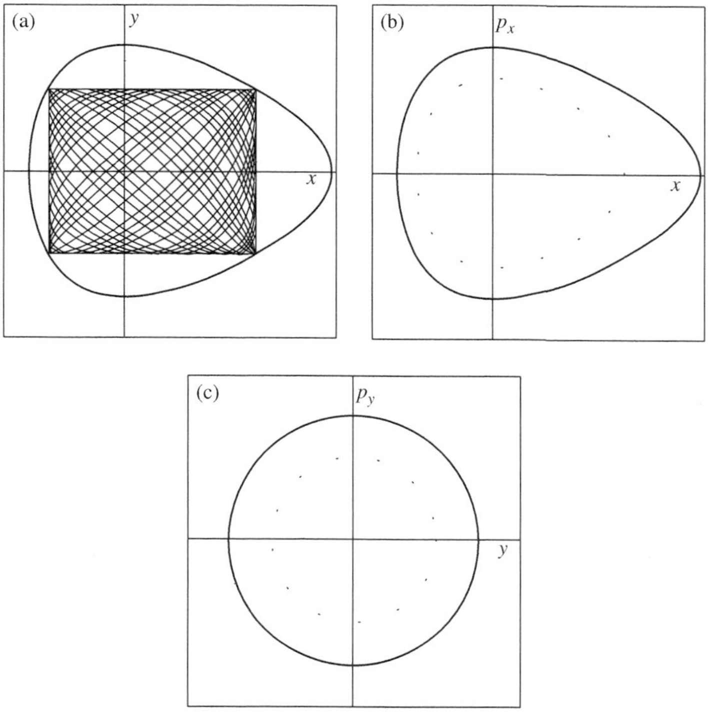

Fig. 7.1 (a) Lissajous figure; Poincaré sections taken at (b) $y = 0$, $p_y > 0$, and (c) $x = 0$, $p_x > 0$ respectively for the Hamiltonian (7.1) with $D = 12.5$ and $a = (2D)^{-1/2}$. The bounding curves are fixed by the total energy.

case the motion is said to be *quasi-periodic*, meaning that the trajectory will never close on itself, but will eventually come arbitrarily close to any point on a particular two-dimensional surface in phase space—two-dimensional because the four variables are subject to two constraints imposed by conservation of the two components of the total energy,

$$E = E_x + E_y, \tag{7.3}$$

or equivalently by conservation of the two associated actions I_x and I_y.

This quasi-periodic character ensures that the Lissajous figure in Fig. 7.1(a) will ultimately fill the region bounded by the rectangular *caustics* $x = x_a$, x_b and $y = y_a$, y_b which determine the limits of the separate x and y motions at energies E_x and E_y respectively. Figure 7.1(b) shows a different projection by means of a so-called Poincaré surface of section, which is constructed by plotting simultaneous (x, p_x) values whenever the trajectory crosses a given line, $y = y_0$ (in the range $y_a < y_0 < y_b$), with a given sign of p_y, say $p_y > 0$. The resulting points provide snapshots of the underlying phase space orbit defined by the equation

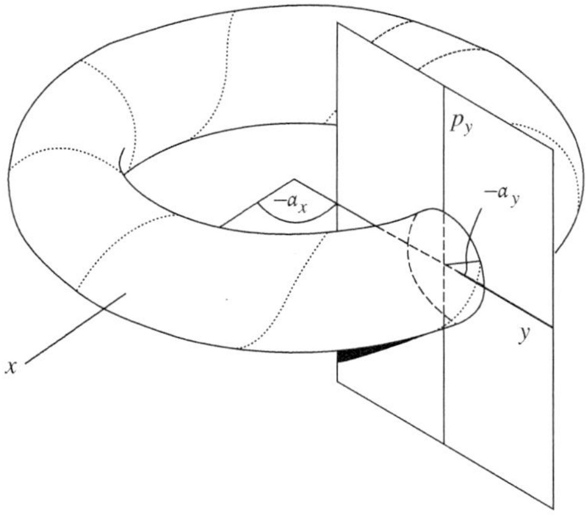

Fig. 7.2 Quasi-periodic motion over the surface of an invariant torus, drawn to illustrate the Poincaré section. The dots representing the trajectory will ultimately cover the surface.

$$E_x = \frac{1}{2m}p_x^2 + V_x(x). \tag{7.4}$$

Such points will ultimately become infinitely dense on this orbit. Analogous considerations apply to the (y, p_y) surface of section taken at $x = x_0$ and $p_x > 0$, which is plotted in Fig. 7.1(c).

An alternative way to summarize these properties is to visualize the motion as lying on the two-dimensional surface of a so-called invariant torus, as shown in Fig. 7.2. The representation is such that the two angle variables, α_x and α_y, designate the phases associated with the two topologically distinct paths around the torus, which correspond to the angles around the orbits in Figs. 7.1(b) and 7.1(c) respectively. Thus the choice of a given sectioning plane, $x = x_0$ with $p_x > 0$, defines a particular angle α_x while the section cut through the torus in Fig. 7.2, corresponding to the (y, p_y) section in Fig. 7.1(c), is one in which angles are measured as α_y. Finally, the Lissajous figure in Fig. 7.1(a) may be seen as a side projection of the spiral motion in Fig. 7.2 after multiple passes around the torus. In other words the toroidal representation encompasses all three projections in Fig. 7.1.

The next important consideration is that any particular division of a given total energy between E_x and E_y has its own particular torus, neatly nested with respect to the others in the separable case, as shown in Fig. 7.3. Furthermore, the central member of this nested set, which acts as a focus for the rest of the family, corresponds to motion along the x axis in Fig. 7.1(a); it would therefore appear in Fig. 7.1(c) as a central point at $y = y_e$, $p_y = 0$, and in Fig. 7.1(b) as a bounding orbit circumscribing all others at the energy in question. Motions of this type are necessarily periodic in time, with a period equal to that for the x motion at the given total energy. Of course the same remarks apply with x and y exchanged throughout, so that there are always two simple periodic orbits with frequency ratios $\omega_x : \omega_y = 1 : 0$ or $0 : 1$. Other rational frequency ratios will also yield periodic behaviour but the generic behaviour will be

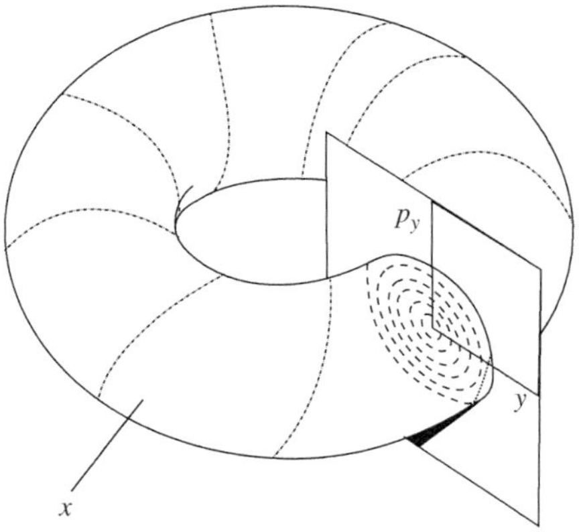

Fig. 7.3 Nested invariant tori and their surfaces of section.

quasi-periodic in the sense described above. This concludes the qualitative picture of the separable case.

The non-separable case is more complicated because the onset of coupling between the two degrees of freedom may have a variety of consequences. The simplest, covered by the Kolmogorov (1954, 1957), Arnol'd (1961, 1963), Moser (1973) theorem, is that the structure of the phase space is qualitatively undisturbed by some weak perturbation. The main changes are that the caustics bounding the Lissajous figures are no longer strictly rectangular and that the shapes of, for example, the (y, p_y) sections depend on the x value at which they are taken. However, as first shown by Einstein (1917), the enclosed area in the surface of section is invariant to the choice of sectioning plane (provided it cuts with the correct topology), a result of vital importance for the EBK quantization methods discussed in Section 7.2.

The next possibility is that the onset of intermode coupling can generate new types of motion, as illustrated in Fig. 7.4 for the local mode Hamiltonian

$$H = \frac{1}{2m}(p_a^2 + p_b^2) + \lambda x_a x_b + D(1 - \mathrm{e}^{-\beta x_a})^2 + D(1 - \mathrm{e}^{-\beta x_b})^2. \qquad (7.5)$$

At sufficiently low energies one finds the expected pattern of nested box-like Lissajous figures (Fig. 7.4(a)) giving rise to a simple pattern of concentric orbits in the surface of section (Fig. 7.4(b)). Above a certain energetic threshold, however, one observes a bifurcation into two distinct trajectory types, as shown in Figs. 7.4(c) and (d), while the surface of section bifurcates into two regions divided by the dashed separatrix shown in Fig. 7.4(e). In the present example the inner and outer regions are occupied by trajectories of the types in Fig. 7.4(c) and Fig. 7.4(d) respectively. The onset of this behaviour is associated with a change in the character of the periodic motion along the x axis in Fig. 7.4(a) and of its image, the fixed point, in Poincaré section at the centre of Fig. 7.4(b); the motion is stable with respect to small perturbations below the critical energy but becomes unstable above, and the fixed point changes character from elliptic to hyperbolic (Percival and Richards 1982). A fuller analysis of

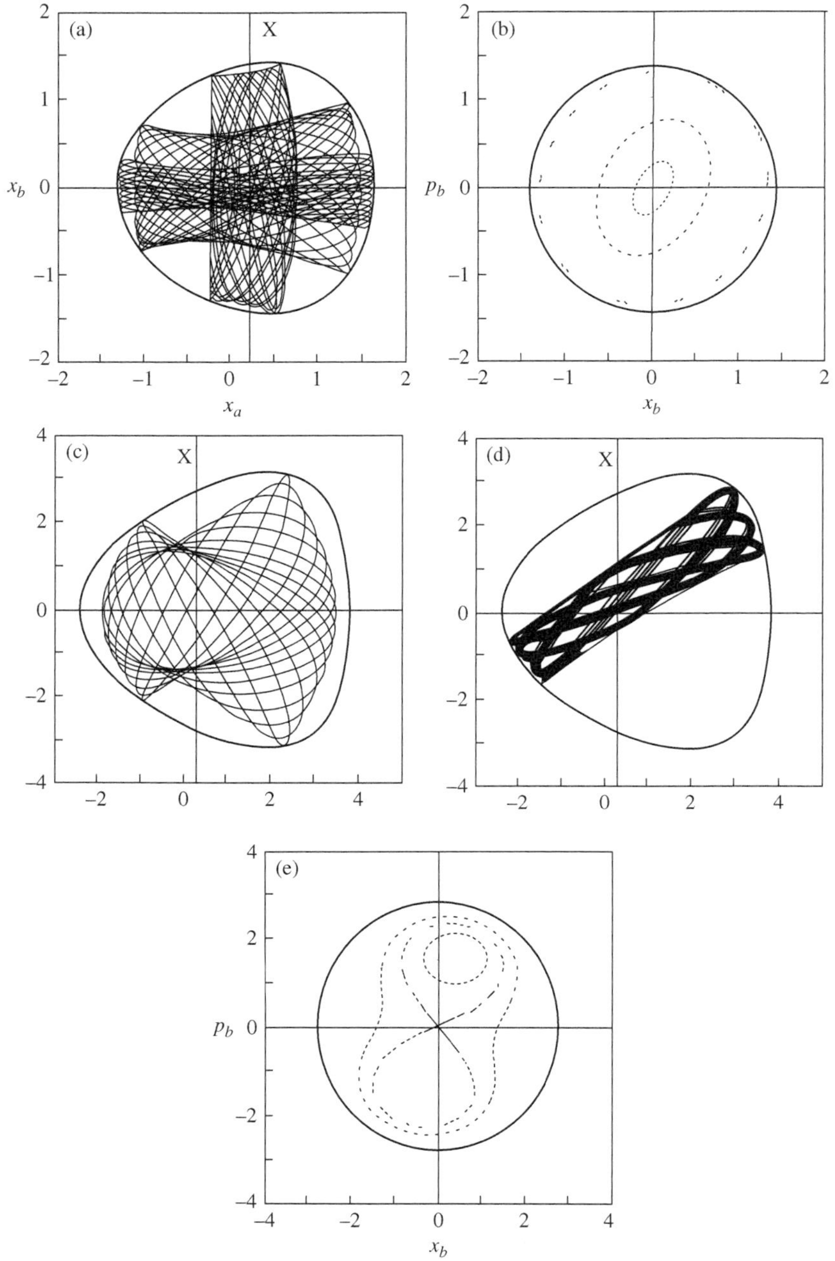

Fig. 7.4 Aspects of motion under the local mode Hamiltonian (7.5): (a) Lissajous figures at $E = 2.0$; (b) Poincaré sections at X in panel (a); (c) a normal mode trajectory at $E = 4.0$; (d) a local model trajectory at $E = 4.0$; (e) Poincaré sections at X in panels (c) and (d), with a figure-of-eight separatrix between them.

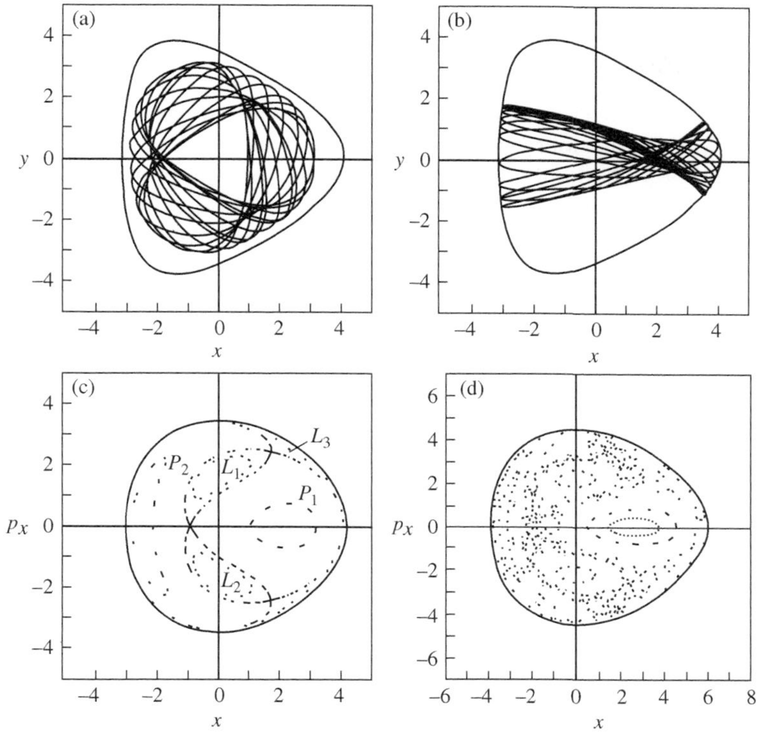

Fig. 7.5 Henon–Heiles Hamiltonian (7.6): (a) a processional trajectory; (b) a librational trajectory; (c) the $y = 0, p_y > 0$ surface of section at $E = 6$, in which P_i indicate processional and L_i librational modes; and (d) evidence of chaos at $E = 10$.

the resulting classical dynamics, taken from the work of Jaffé and Brumer (1980) and Sibert (1982), is given in Section 7.3.

A more complicated example of a similar bifurcation pattern is shown in Fig. 7.5 for the Henon and Heiles (1964) Hamiltonian

$$H = \frac{1}{2}(p_x^2 + p_y^2 + x^2 + y^2) + \lambda(x^2 y - y^3/3). \tag{7.6}$$

The potential function now has reflection symmetry in the lines $y = 0$ and $y = \pm\sqrt{3}x$ which are themselves related by a three-fold rotation axis. Thus there are three symmetrically related librational modes of the type in Fig. 7.5(b), which are marked as L_1, L_2 and L_3 in Fig. 7.5(c). There are also two precessional modes of the type in Fig. 7.5(a)—one moving clockwise and the other anticlockwise, whose positions in the surface of section are marked as P_1 and P_2. Consequently the surface of section shows four stable fixed points at the centres of P_1, P_2, L_1 and L_2, while the outer boundary belongs to a fifth stable orbit corresponding to the locus of L_3. Figure 7.5(c) also shows three unstable hyperbolic points on the separatrices between these different types of motion.

The next stage in the phase space structure is to follow the onset of classical chaos as the energy increases further. The diagnostic feature in the surface of section is that trajectories in the neighbourhood of any hyperbolic points no longer intersect the section on closed concentric curves but begin to occupy finite areas (see Fig. 7.5(d)), which spread out as the energy increases, to occupy the whole space. At the same time the caustics of the Lissajous figures in coordinate space become blurred, the power spectrum obtained by Fourier transformation shows broad lines instead of sharp peaks, and neighbouring trajectories diverge exponentially from each other in phase space (Noid et al. 1977, 1981; Percival 1977). These are indications that the underlying invariant tori have disappeared or become so convoluted as to be unrecognizable, although trajectories that are chaotic in the long term are frequently found to hover in the vicinity of one of the last vestigial tori before wandering off to find another (Contopoulos 1971; Reinhardt and Jaffé 1981). Aspects of the onset of this chaotic behaviour are addressed in Appendix D and some of its quantum mechanical implications are beginning to be understood (Casati 1985; MacKay et al. 1984; Berry et al. 1987; Tabor 1989; Gutzwiller 1990). It is beyond the scope of this book to explore these matters in detail, but the reader should note that the perturbation theory and adiabatic switching methods described in Sections 7.5 and 7.6, which were developed to handle quasi-periodic systems, can sometimes yield quite accurate quantum mechanical energy levels, even for classically chaotic systems (Skodje et al. 1985; Swimm and Delos 1979; Jaffé and Reinhardt 1982). The demonstration of periodic orbit 'scars' on the wavefunctions of classically chaotic systems (Heller 1984) is of particular interest. Analyses of the dynamics of magnetically perturbed Rydberg atoms (Robnik 1981; Delos et al. 1984; Wintgen and Friedrich 1986; Delande and Gay 1986, 1987) and of Rydberg molecules (Lombardi et al. 1988) should also be noted.

7.2 Einstein–Brillouin–Keller quantization

The EBK quantization scheme is based on the fact that the action integral along any path around the invariant torus is invariant to deformations of that path, provided that the topology is preserved (Einstein 1917). Hence, by extension of the Bohr quantization rule, the necessary conditions are specified as

$$\oint \mathbf{p} \,.\, \mathrm{d}\mathbf{x} = (n_k + \delta_k)h, \quad n_k = 0, 1, \ldots \tag{7.7}$$

where the subscript k specifies the topology of the path and δ_k is the Maslov index appropriate for that path (see below); the number of k values is equal to the number of degrees of freedom. The problems are to evaluate these line integrals and to find energies and trajectory starting conditions such that eqn (7.7) is simultaneously satisfied for all k.

The conceptually simplest approach is to compute the integrals as areas in appropriate surfaces of section (Noid and Marcus 1975), a method extensively reviewed by Noid et al. (1980). As illustrated in more detail below, the choice of section automatically restricts the quantization path to a particular phase plane (p_k, x_k) so that eqn (7.7) reduces to

$$\oint p_k \, \mathrm{d}x_k = (n_k + \delta_k)h. \tag{7.8}$$

This approach is most easily applied to motion in two degrees of freedom because there is only a remote chance of catching a given trajectory in two or more sectioning planes at the same time; but higher-dimensional versions of the method have been outlined by Noid et al. (1980), Stine and Noid (1983) and Noid and Sumpter (1985).

An alternative due to Sorbie and Handy (1976), which extends more readily to higher-dimensional systems, is to accumulate each of the N partial integrals specified by eqn (7.8) separately, as the trajectory winds round the torus until it 'almost closes', and then to correct for the number of cycles in the various degrees of freedom k. This method is not, however, strictly equivalent to the surface of section approach, and to date only a perturbative justification has been given (Handy 1980). Some care is clearly required in the choice of coordinates, because it is readily verified, for a separable system in (x_1, x_2), that a rotation of axes by an angle θ to (x_1', x_2') would

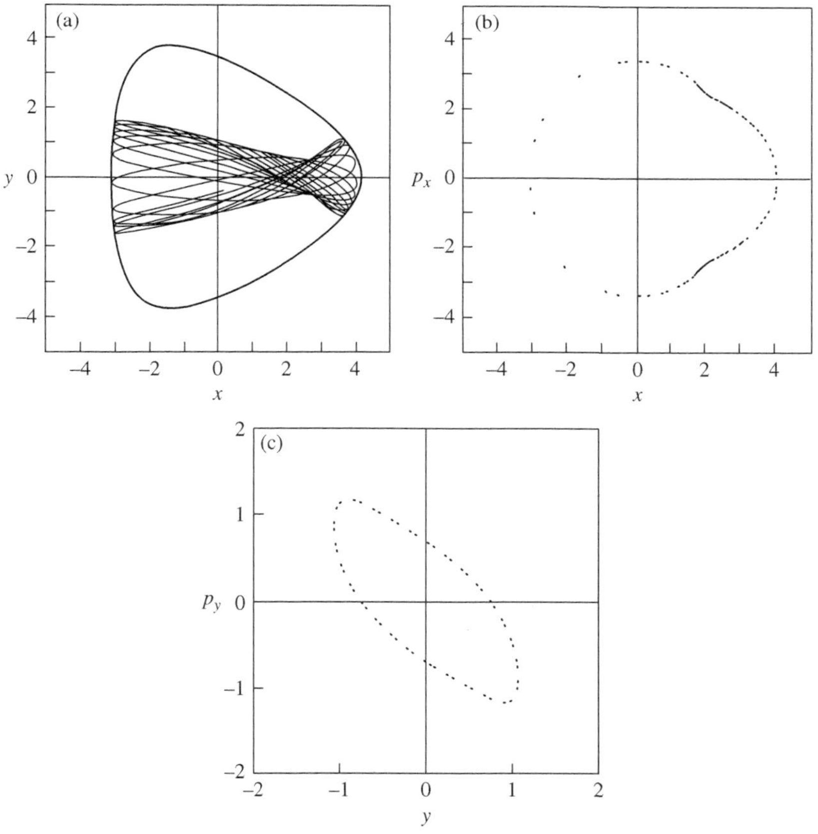

Fig. 7.6 (a) A libration trajectory, with its associated Poincaré sections at (b) $y = 0$ and (c) $x = 0$.

yield $N_1' = N_1 \cos^2\theta + N_2 \sin^2\theta$, where N_k denotes the integral in (7.8). Hence the undoubted success of the method may be due to a choice of axes adapted to the separable part of the Hamiltonian and to the relatively small perturbations encountered in the test applications.

Both methods require a choice of quantization coordinates (x_k, p_k) which ensure topological independence of the integration paths, a requirement which demands prior inspection of the Lissajous figures. For example, as discussed in Section 7.1, the Henon–Heiles Hamiltonian given by eqn (7.6) supports two types of trajectory, termed librational and precessional modes, as shown in Figs. 7.6 and 7.7 respectively. The box-like caustics in Fig. 7.6 show that modes of this type are vibrational in both x and y degrees of freedom; hence both have Maslov index $\delta_k = \frac{1}{2}$. The lines $x = 0$ and $y = 0$ which cut opposite caustics, and otherwise lie inside the caustic box in Fig. 7.6(a), are therefore appropriate sectioning planes. The action integrals themselves,

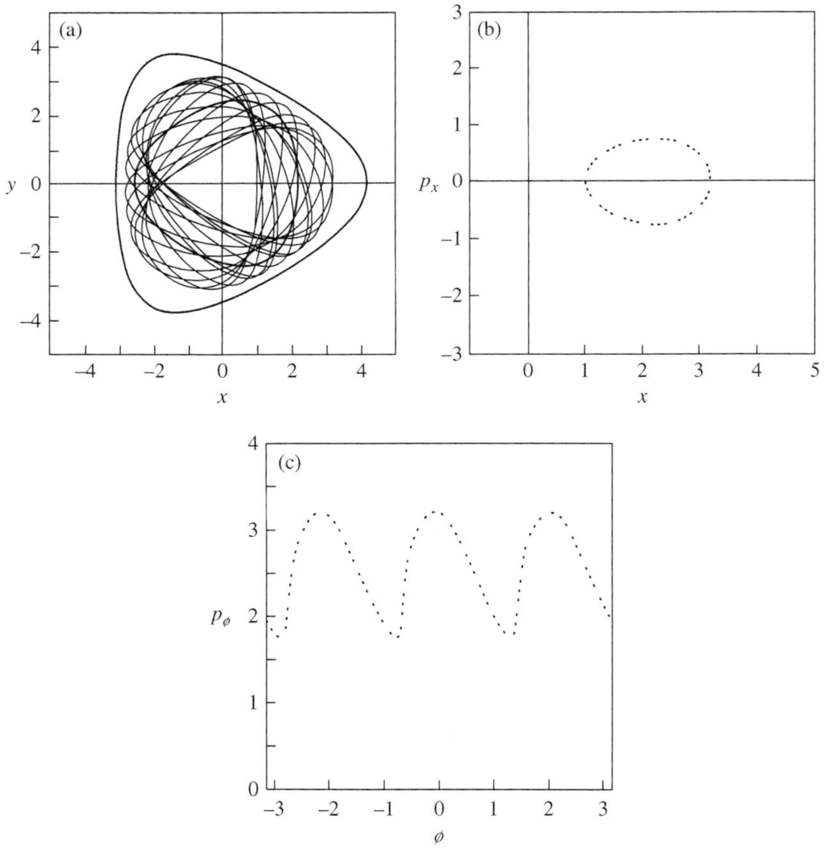

Fig. 7.7 (a) A precessional trajectory, with its associated Poincaré sections at (b) $y = 0$ and (c) a circle within the Lissajous figure in panel (a).

represented by the enclosed areas in Figs. 7.6(b) and (c), are most conveniently evaluated by transforming to angle–action variables

$$\begin{aligned} I_k &= \frac{1}{2}[(x_k - x_k^0)^2 + (p_k - p_k^0)^2], \\ \alpha_k &= -\arctan\left[(p_k - p_k^0)/(x_k - x_k^0)\right], \end{aligned} \tag{7.9}$$

where (x_k^0, p_k^0) is a convenient point in the centre of the section in Fig. 7.6(b) or 7.6(c). The canonical nature of this transformation ensures that

$$\oint p_k \, \mathrm{d}x_k = \int_0^{2\pi} I_k \mathrm{d}\alpha_k, \tag{7.10}$$

but the second of these integrals is more readily evaluated by quadrature than the first, because I_k is relatively insensitive to α_k.

The precessional modes in Fig. 7.7 pose a new problem because the lines $x = 0$ and $y = 0$ are topologically equivalent for motions of this type. Hence it is necessary to take the second section as a line of constant

$$r = (x^2 + y^2)^{1/2}, \tag{7.11}$$

chosen to lie inside the caustic boundaries, and to construct the section in terms of an angle φ and its conjugate momentum p_φ, given by

$$\varphi = \arctan(y/x), \quad p_\varphi = xp_y - yp_x. \tag{7.12}$$

Furthermore the Maslov index is $\delta_k = 0$ for this rotational component of the motion, while quantization in the vibrational section (x, p_x) is handled as before. Another common situation, in which the quantization coordinates must be matched to the shapes of the Lissajous figures, arises when the frequencies are rationally related—as, for example, in the Fermi resonance cases treated by Noid et al. (1979); see also problem 7.1. Complications due to very twisted trajectory forms in two degrees of freedom may also be avoided by supplementing a single surface of section with the following constraint on the total action until the trajectory almost closes:

$$\oint \mathbf{p}.\mathrm{d}\mathbf{q} = (N + l/8)h, \tag{7.13}$$

where l is the number of caustics encountered (Noid and Marcus 1975, 1986). Finally, however the details are handled, it is unnecessary to search for individual quantizing trajectories, because it is a simple matter to extract the energies from a suitable analytical fit to the energy surface $E(n_1, n_2)$ determined by non-integral values of the n_k (Sumpter and Noid 1986).

Table 7.1 shows that a modified version of this EBK quantization scheme gives satisfactory agreement with the quantum mechanical eigenvalues for the Henon–Heiles system. The modification concerns treatment of the librational modes, which yield a false three-fold degeneracy when quantized by the primitive conditions

$$\oint p_x \, \mathrm{d}x = (2n_x + 1)\pi \quad \text{and} \quad \oint p_y \, \mathrm{d}y = (2n_y + 1)\pi.$$

To overcome this, Noid et al. (1977) argue that the (y, p_y) section is topologically equivalent to one-third of the circular (φ, p_φ) section shown in Fig. 7.7(a); hence by

Table 7.1 Comparison between quantum mechanical eigenvalues for the Henon–Heiles potential and semiclassical values obtained by surface of section (Noid and Marcus 1977).

n	l	Symmetry	Quantum†	Semiclassical
0	0	A	0.9986	0.9947
1	±1	E	1.9901	1.9863
2	0	A	2.9562	2.9506
2	±2	E	2.9853	2.9815
3	±1	E	3.9260	3.9233
3	±3	A	3.9849‡	3.9803
4	0	A	4.8702	4.8573
4	±2	E	4.8987	4.8954
4	±4	E	4.9863	4.9821
5	±1	E	5.8170	5.8160
5	±3	A	5.8743‡	5.8713
5	±5	E	5.9913	5.9869

† Calculated for $\lambda^2 = 1/80$

‡ Mean energies of close A_1, A_2 doublets.

Table 7.2 Comparison between quantum mechanical vibrational energies of SO_2 and H_2O and those obtained by the Sorbie–Handy method (Colwell and Handy 1978).

	SO_2		H_2O	
State	Quantum (cm^{-1})	Semiclassical (cm^{-1})	Quantum (cm^{-1})	Semiclassical (cm^{-1})
0,0,0	1529.60	1529.11	4651.98	4645.02
0,1,0	2045.81	2045.29	6249.55	6242.30
0,2,0	2556.21	2555.67	7811.53	7804.34
1,0,0	2685.63	2684.93	8369.29	8358.20
0,0,1	2889.53	889.06	8472.75	8467.42

continuity with the precessional modes (for which $\oint p_\varphi \mathrm{d}\varphi = 2\pi l$) the proper conditions should be taken as

$$\oint p_y \, \mathrm{d}y = \frac{2\pi l}{3} \quad \text{and} \quad \oint p_x \, \mathrm{d}x + \oint p_y \, \mathrm{d}y = 2\pi(n+1).$$

This splits the three-fold degeneracies into A- and E-type components, but does nothing to treat the A_1, A_2 splittings of the A levels for $l = \pm 3, \pm 6, \ldots$, etc. A more soundly based uniform quantization scheme, applicable to this awkward Henon–Heiles case, is outlined in Section 7.5, and a simpler related situation relevant to the local–normal mode dichotomy, illustrated in Fig. 7.4, is discussed in Section 7.3.

To complete this section Table 7.2 gives sample results obtained by the Sorbie–Handy method for systems with more than two coupled oscillators, which are much more difficult to handle by the surface of section procedure (but see Stine and Noid 1983 and Noid and Sumpter 1985).

7.3 Uniform quantization at a resonance

The hyperbolic point on the separatrix between two types of motion was identified in Section 7.1 as the signature of an unstable periodic orbit or Chirikov (1979) resonance, and the properties of such resonances may be exploited to obtain a uniform semiclassical description of the neighbouring motions. The argument is that a resonance constitutes a near classical degeneracy, and that any classically degenerate system is amenable to a canonical transformation, such that the entire time dependence is concentrated in a single 'fast' degree of freedom (see Section 4.2). The resonance therefore constitutes a fixed point in the slow part of the phase space, and motions close to it may be analysed in terms of averages (taken over the fast motion), thereby reducing the dimensionality of the quantization problem.

The following local mode, coupled Morse oscillator system described by the Hamiltonian

$$H = \frac{1}{2\mu}(p_a^2 + p_b^2) + \frac{1}{2\mu_{ab}} p_a p_b + V(q_a) + V(q_b), \tag{7.14}$$

where

$$V(q) = D[1 - \exp(-\beta q)]^2, \tag{7.15}$$

offers a convenient illustration. One starts by transforming to Morse angle–action variables, (α_v, I_v) $v = a, b$ (see Appendix E.2), as described in full by Jaffé and Brumer (1980). The essence of the result is, however, more simply obtained by following Sibert et al. (1982) and approximating the momenta in the coupling term by the harmonic forms (see eqn (4.21))

$$p_v = -(2I_v \mu\omega)^{1/2} \sin\alpha_v \tag{7.16}$$

where

$$\omega = \beta(2D/\mu)^{1/2}; \tag{7.17}$$

all other terms are treated exactly. The result takes the form

$$H = \sum_{v=a,b} (I_v - xI_v^2)\omega + \lambda(I_a I_b)^{1/2}[\cos(\alpha_a - \alpha_b) - \cos(\alpha_a + \alpha_b)], \tag{7.18}$$

where

$$x = \omega/4D, \quad \lambda = \mu\omega/2\mu_{ab}. \tag{7.19}$$

The next step is to replace the individual actions I_v by their sum and difference, I and J respectively, and to define associated conjugate angles α_I and α_J:

$$\begin{aligned} I &= I_a + I_b, \quad \alpha_I = \tfrac{1}{2}(\alpha_a + \alpha_b), \\ J &= I_a - I_b, \quad \alpha_J = \tfrac{1}{2}(\alpha_a - \alpha_b), \end{aligned} \tag{7.20}$$

where the factors of $\frac{1}{2}$ ensure that the transformation is canonical. The effect of these substitutions is to reduce the Hamiltonian to the form

$$H = H^{(0)} - H^{(1)}, \tag{7.21}$$

where

$$\begin{aligned} H^{(0)} &= \left(I - \frac{x}{2}I^2\right)\omega \\ H^{(1)} &= \frac{1}{2}J^2 x\omega - \tfrac{\lambda}{2}(I^2 - J^2)^{1/2}(\cos 2\alpha_J - \cos 2\alpha_I). \end{aligned} \tag{7.22}$$

The cases of interest are ones in which x and λ are small ($x \ll 1$ and $\lambda \ll \omega$). Hence $H^{(1)}$ may be regarded as a perturbation to the zeroth-order motion under $H^{(0)}$, for which I is a constant of the motion and α_I varies with time as

$$\alpha_I = \alpha_I^0 + (1 - xI)\omega t. \tag{7.23}$$

Similarly, since neither J nor α_J appear in $H^{(0)}$ they are also both constants of the zeroth-order motion. In other words, as expected by the general arguments in Section 4.1, all motion is concentrated in the fast angle variable α_I.

Barring complications due to further so-called Chirikov resonances between the resulting α_I and α_J motions, which may lead to chaos (Chirikov 1979), the effect of the perturbation may be taken into account by averaging H over one cycle of the α_I motion, thereby eliminating all reference to α_I in the resulting effective Hamiltonian

$$H_{\text{eff}} = \frac{1}{2\pi}\int_0^{2\pi} H \mathrm{d}\alpha_I = H^{(0)} - H_J, \tag{7.24}$$

where

$$H_J = \frac{x\omega}{2}J^2 - \frac{\lambda}{2}(I^2 - J^2)^{1/2}\cos 2\alpha_J. \tag{7.25}$$

The effect of this averaging is to make I a strict constant of motion under H_{eff}, whose value has only a parametric effect on the motion in J and α_J. The negative sign in eqns (7.21) and (7.24) is introduced to produce a positive coefficient in the 'kinetic-energy-like' term involving J^2 in (7.25).

The most obvious consequence of this procedure is to concentrate attention solely on the slow (J, α_J) degree of freedom. One also sees from Fig. 7.8 that the form of H_J allows the phase space to bifurcate into two regions, just as in Fig. 7.4, depending on the relative values of the anharmonicity and coupling parameters, ωx and λ respectively, and on the magnitude of the total action I.

If, for example, $\omega x = 0$, it is readily verified that H_J has a maximum value of $\lambda I/2$ at $J = 0$, $\alpha_J = \pi/2$ and a minimum value of $-\lambda I/2$ at $J = 0$, $\alpha_J = 0$. These two fixed points correspond to the two normal modes of the coupled harmonic system, and the time dependence of the action difference J along an orbit of constant H_{eff} measures

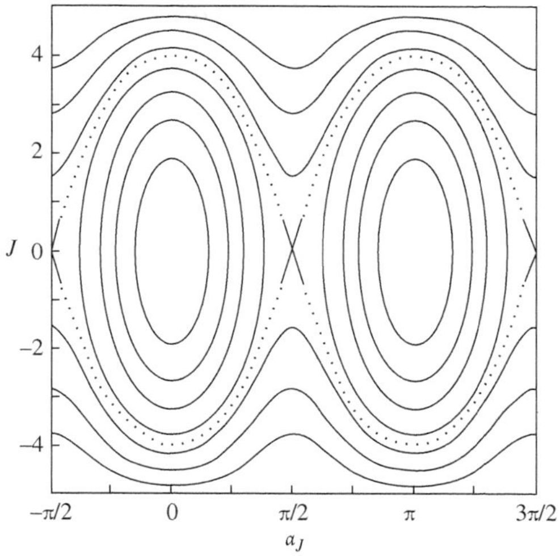

Fig. 7.8 Phase space orbits of the Hamiltonian H_J given by (7.25).

the flow of excitation from oscillator a to b and vice versa. The opposite local mode limit, with $\lambda = 0$, is readily shown to lead to a constant action difference, J.

The intermediate case in Fig. 7.8 merits closer analysis. In the first place the condition that the point $J = 0$, $\alpha_J = \pi/2$ lies at a saddle point may be used to find a condition for coexistence of the two types of motion, namely (since $\partial^2 H_{\text{eff}}/\partial\alpha_J^2 < 0$ at this point)

$$\partial^2 H_{\text{eff}}/\partial J^2 = \omega x - \lambda/2I < 0. \tag{7.26}$$

This sets a lower bound, $I > \lambda/2\omega$, on the onset of local mode behaviour. It is also readily verified that the inner vibrational-like normal modes span the energy range $-\lambda I/2 < H_{\text{eff}} < \lambda I/2$, while the outer rotational-like local modes have energies bounded by $\lambda I/2 < H_{\text{eff}} < (1 + I\omega x/\lambda)\lambda I/2$. Finally, the separatrix, with energy $H_{\text{eff}} = \lambda I/2$, cuts the line $\alpha_J = 0$ at

$$|J| = (\lambda/\omega)[(2I\omega x/\lambda) - 1]^{1/2}, \tag{7.27}$$

which means by reference to Fig. 7.8 that larger $|J|$ values lie wholely outside the (normal mode) resonance zone. Figure 7.9 illustrates the resulting division of the action space (I_a, I_b) into local and normal mode regions, where I_a and I_b are given according to eqn (7.20) by $(I + J)/2$ and $(I - J)/2$ respectively.

Turning to the question of quantization, Sibert et al. (1982) neglect the term J^2 under the square root in eqn (7.25), a device which converts the equation to that of a vertical pendulum. The eigenvalues may therefore be identified with those of the corresponding Schrödinger equation, which may be expressed in the standard Mathieu form (Abramowitz and Stegun 1965)

$$\left(\frac{\mathrm{d}^2}{\mathrm{d}\xi^2} + a - 2q\cos 2\xi\right)\psi = 0, \tag{7.28}$$

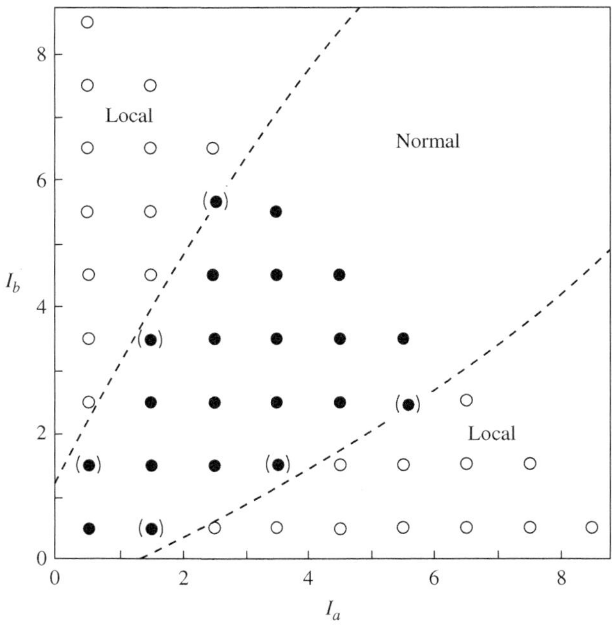

Fig. 7.9 Division of the action space into local and normal regions. The characters of the indicated points are those associated with EBK quantizing trajectories; those in parentheses cannot be quantized at the EBK level. (Taken from Child and Halonen (1984) with permission.)

by means of the substitutions

$$\xi = \alpha - \frac{\pi}{2}, \qquad a = \frac{2E_J}{x\omega}, \qquad q = \frac{\lambda I}{2x\omega}. \tag{7.29}$$

There is, however, a minor complication in that the quantum numbers $n = (I/\hbar) - 1$ and $j = J/\hbar$ in the uncoupled local mode regime ($\lambda = 0$) are restricted according to eqn (7.20) to the alternate values

$$j = \pm n, \ \pm(n-2), \ldots. \tag{7.30}$$

Consequently alternate eigenvalues of the Mathieu equation must be neglected, giving rise to the typical pattern shown, for $n = 5$, in Fig. 7.10.

Another approach which avoids neglect of J under the square root sign is to solve eqn (7.30) for J as a function of α and E (where $E = H_J$) and then to vary E until the following periodic double barrier quantization condition, which is derived in problem 3.6, is satisfied:

$$\cos[\theta(E) - \varphi(E)] = \pm[1 + \kappa^2(E)]^{-1/2}, \tag{7.31}$$

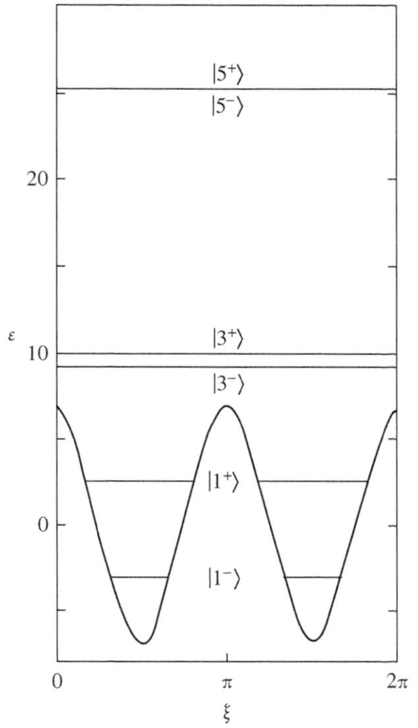

Fig. 7.10 Mathieu quantization for $n = 5$ by use of eqn (7.28). Note that only odd states $|J^{\pm}\rangle$ are physically relevant for odd values of n. (Taken from Child and Halonen (1984) with permission.)

where

$$\begin{aligned}\theta(E) &= \int_{-\xi_a}^{\xi_a} J(\xi, E)\,\mathrm{d}\xi, \quad E < \lambda I/2 \\ &= \int_{-\pi/2}^{\pi/2} J(\xi, E)\,\mathrm{d}\xi, \quad E > \lambda I/2, \\ \kappa(E) &= \exp[-\pi\varepsilon(E)],\end{aligned}$$

$$\begin{aligned}\varepsilon(E) &= -\frac{1}{\pi}\int_{\xi_a}^{2\pi-\xi_a} |J(\xi, E)|\mathrm{d}\xi, \quad E < \lambda I/2 \\ &= \frac{1}{\pi}\int_{\xi_c^*}^{\xi_c} J(\xi, E)\mathrm{d}\xi, \qquad E > \lambda I/2,\end{aligned} \tag{7.32}$$

and $\varphi(E)$ is given in terms of $\varepsilon(E)$ by eqn (3.49). The real turning point ξ_a is shown in Fig. 7.10, and ξ_c and ξ_c^* are the complex roots of $J(\xi, E)$ (with real part $\pi/2$) when $E > \lambda I/2$.

Table 7.3 Quantum mechanical and semiclassical eigenvalues and local mode splittings for the model of Sibert et al. (1982).

	Quantum		Eqn (7.28)		Eqn (7.31)	
$n\ j^{\pm}$	$E(\mathrm{cm}^{-1})$	$\Delta(\mathrm{cm}^{-1})$	$E(\mathrm{cm}^{-1})$	$\Delta(\mathrm{cm}^{-1})$	$E(\mathrm{cm}^{-1})$	$\Delta(\mathrm{cm}^{-1})$
11^{+}	3675.55	53.7	3675.80	53.1	3675.4	43.9
11^{-}	3729.27		3728.91		3719.3	
22^{+}	7220.74	16.9	7221.11	16.2	7224.2	13.5
22^{-}	7237.64		7237.29		7237.7	
20	7419.40		7418.93		7412.0	
33^{+}	10595.25	2.7	10595.58	2.1	10597.0	0.2
33^{-}	10597.97		10597.65		10597.2	
31^{+}	10895.73	97.6	10895.51	97.7	10888.2	97.3
31^{-}	10993.28		10993.19		10985.5	
44^{+}	13794.94	0.3	13794.93	0.1	13795.1	0.0
44^{-}	13795.26		13795.07		13795.1	
42^{+}	14271.82	40.5	14272.37	37.4	14276.8	38.9
42^{-}	14312.30		14309.74		14315.7	
40^{+}	14507.14		14509.16		14500.4	
55^{+}	16824.49	0.4	16824.64	0.0	16824.7	0.0
55^{-}	16824.88		16824.64		16824.7	
53^{+}	17488.37	8.8	17490.36	5.4	17490.3	2.1
53^{-}	17497.19		17495.78		17492.4	
51^{+}	17786.17	131.1	17784.62	133.6	17774.8	136.2
51^{-}	17917.30		17918.18		17911.0	

The results obtained by reduction to the Mathieu equation and by solution of eqn (7.31) are compared with the exact eigenvalues of the quantum mechanical equivalent of (7.14) in Table 7.3. It is evident that the Mathieu approximation gives an excellent representation for all but the very closest local mode doublets. The uniform approximation (7.31) is somewhat less accurate but still represents the relative magnitudes of the splitting patterns and of the mean eigenvalues quite well.

7.4 Fourier representation of the torus

The quasi-periodicity of the motion on an invariant torus implies that any function of the dynamical variables may be represented as a Fourier series in the true but unknown angle variables $\{\alpha_a\}$. The following quantization scheme, due to Percival and Pomphrey (1973, 1976) and later implemented in a different way by Eaker and Schatz (1984), Eaker et al. (1984), and Martens and Ezra (1985, 1987), is based on requiring consistency between Hamilton's equations of motion and the Fourier components in the expansions

$$\begin{aligned}\mathbf{q}(\mathbf{I},\boldsymbol{\alpha}) &= \sum_{\mathbf{k}} q_{\mathbf{k}}(\mathbf{I})\exp(i\mathbf{k}.\boldsymbol{\alpha})\\ \mathbf{p}(\mathbf{I},\boldsymbol{\alpha}) &= \sum_{\mathbf{k}} p_{\mathbf{k}}(\mathbf{I})\exp(i\mathbf{k}.\boldsymbol{\alpha}).\end{aligned} \tag{7.33}$$

In principle $\mathbf{q}$ and $\mathbf{p}$ may be any set of conjugate coordinates and momenta, but the scheme becomes particularly simple if they are chosen as Cartesian variables, scaled to a common mass m, so that

$$\mathbf{p} = m\dot{\mathbf{q}}. \tag{7.34}$$

In terms of the Fourier components this means that

$$\mathbf{p}_{\mathbf{k}} = \mathrm{i}m\mathbf{q}_{\mathbf{k}} \sum_{a'} k_{a'}\omega_{a'}, \tag{7.35}$$

where $\{\omega_{a'}\}$ are the natural frequencies of the quasi-periodic motion,

$$\omega_{a'} = \mathrm{d}\alpha_{a'}/\mathrm{d}t. \tag{7.36}$$

Quantization is again based on the EBK equation (7.7), with the necessary action integrals formally evaluated by the equation

$$I_a = \frac{1}{2\pi}\int_0^{2\pi} \mathbf{p}\cdot(\partial\mathbf{q}/\partial\alpha_a)\mathrm{d}\alpha_a. \tag{7.37}$$

An equivalent equation may be obtained by recognizing that eqn (7.37) applies at any values of the remaining angle variables, $\{\alpha_b\}$ with $b \neq a$, because any choice of $\{\alpha_b\}$ constitutes a Poincaré section. Hence on averaging (7.37) over the α_b and substituting from (7.33),

$$\begin{aligned} I_a &= \frac{1}{(2\pi)^v}\int_0^{2\pi}\int_0^{2\pi}\cdots\int_0^{2\pi} \mathbf{p}\cdot(\partial\mathbf{q}/\partial\alpha_a)\mathrm{d}\alpha_a\mathrm{d}\alpha_b\ldots\mathrm{d}\alpha_v \\ &= \sum_{\mathbf{k}}\sum_{\mathbf{k}'} \mathrm{i}k_a\mathbf{p}_{\mathbf{k}'}\cdot\mathbf{q}_{\mathbf{k}}\delta_{\mathbf{k}',-\mathbf{k}} = \mathrm{i}\sum_{\mathbf{k}} k_a\mathbf{q}_{\mathbf{k}}\cdot\mathbf{p}^*_{\mathbf{k}}, \end{aligned} \tag{7.38}$$

after use of the identity

$$\frac{1}{(2\pi)^v}\int_0^{2\pi}\int_0^{2\pi}\cdots\int_0^{2\pi} \exp[\mathrm{i}(\mathbf{k}+\mathbf{k}')\cdot\boldsymbol{\alpha}]\mathrm{d}\alpha_a\mathrm{d}\alpha_b\ldots\mathrm{d}\alpha_v = \delta_{\mathbf{k}',-\mathbf{k}}. \tag{7.39}$$

Taken together with eqn (7.35), this yields the fundamental working equation

$$I_a = m\sum_i\sum_{\mathbf{k}}\sum_{a'} k_a|q_{i\mathbf{k}}|^2 k_{a'}\omega_{a'}, \tag{7.40}$$

where $|q_{i\mathbf{k}}|$ is the magnitude of the Fourier component of the ith coordinate labelled by $\mathbf{k} = (k_a, k_b, \ldots, k_v)$, and the final sum is taken over the values $a' = a, b, \ldots, v$.

The approach adopted by Percival and Pomphrey (1973, 1976) is to determine the components $q_{i\mathbf{k}}$ by demanding consistency with the second set of Hamiltonian equations (the first are satisfied by (7.34))

$$\dot{p}_i = -(\partial V/\partial q_i) = F_i, \tag{7.41}$$

which means in the light of (7.35) that

$$m \sum_{a'} (k_{a'} \omega_{a'})^2 \mathbf{q_k} = -\mathbf{F_k}, \tag{7.42}$$

where the elements of $\mathbf{F_k}$ are the Fourier components of the forces. The aim is therefore to find a set of Fourier components which satisfy eqn (7.42) and for which the I_a given by eqn (7.40) take the values

$$I_a = (n_a + \delta_a)\hbar. \tag{7.43}$$

The desired quantized energy is then readily obtained by averaging the Hamiltonian $H[\mathbf{p}(\mathbf{I}, \boldsymbol{\alpha}), \mathbf{q}(\mathbf{I}, \boldsymbol{\alpha})]$ over the angles $\{\alpha_a\}$.

The following exactly soluble harmonic model offers a convenient illustration of the practical details:

$$H = \frac{1}{2m}(p_x^2 + p_y^2) + \frac{1}{2}(\lambda_{xx} x^2 + 2\lambda_{xy} xy + \lambda_{yy} y^2). \tag{7.44}$$

The Fourier expansions are written as

$$\begin{aligned} x &= \sum_{k_1} \sum_{k_2} x_{k_1 k_2} \exp[\mathrm{i}(k_1 \alpha_1 + k_2 \alpha_2)] \\ y &= \sum_{k_1} \sum_{k_2} y_{k_1 k_2} \exp[\mathrm{i}(k_1 \alpha_1 + k_2 \alpha_2)]. \end{aligned} \tag{7.45}$$

Taking into account the linearity of the force vector

$$\begin{pmatrix} F_x \\ F_y \end{pmatrix} = - \begin{pmatrix} \lambda_{xx} & \lambda_{xy} \\ \lambda_{xy} & \lambda_{yy} \end{pmatrix} \begin{pmatrix} x \\ y \end{pmatrix}, \tag{7.46}$$

Equation (7.42) yields

$$m(k_a \omega_a + k_b \omega_b)^2 \begin{pmatrix} x_{k_a k_b} \\ y_{k_a k_b} \end{pmatrix} = \begin{pmatrix} \lambda_{xx} & \lambda_{xy} \\ \lambda_{xy} & \lambda_{yy} \end{pmatrix} \begin{pmatrix} x_{k_a k_b} \\ y_{k_a k_b} \end{pmatrix}. \tag{7.47}$$

This requires that the coefficient on the left-hand side must be an eigenvalue of the λ matrix. However, a 2×2 matrix can have only two eigenvalues and two distinct non-zero eigenvectors, where the word distinct is taken to allow simultaneous sign reversal of k_a and k_b because $x_{k_a k_b} = x_{k_{-a} k_{-b}}$ etc. if x is to be real. Each choice of a non-zero eigenvector implies a choice of $(k_a,\, k_b)$; the two most convenient are $(k_a, k_b) = (1, 0)$ and (0,1), in which case $m\omega_a^2$ and $m\omega_b^2$ are the two eigenvalues of the λ matrix. The corresponding eigenvectors may be expressed as

$$\begin{pmatrix} x_{a0} \\ y_{a0} \end{pmatrix} = c_a \begin{pmatrix} \cos\gamma \\ -\sin\gamma \end{pmatrix}, \quad \begin{pmatrix} x_{0b} \\ y_{0b} \end{pmatrix} = c_b \begin{pmatrix} \sin\gamma \\ \cos\gamma \end{pmatrix}, \tag{7.48}$$

where $\tan 2\gamma = 2\lambda_{xy}/(\lambda_{xx} - \lambda_{yy})$ and c_a and c_b must be chosen for consistency with (7.40) and (7.43) as

$$c_v = \left[\left(n_v + \frac{1}{2}\right)\hbar\Big/2m\omega_v\right]^{1/2}, \quad v = a \text{ or } b. \tag{7.49}$$

Finally, the resulting expressions for x, y, p_x and p_y, such as

$$\begin{aligned} x &= x_{a0}\exp(\mathrm{i}\alpha_a) + x_{-a0}\exp(-\mathrm{i}\alpha_a) + x_{0b}\exp(\mathrm{i}\alpha_b) + x_{-0b}\exp(-\mathrm{i}\alpha_b) \\ &= 2(c_a\cos\gamma\cos\alpha_a + c_b\sin\gamma\sin\alpha_b), \end{aligned} \tag{7.50}$$

may be substituted in eqn (7.44) to obtain

$$H = \left(n_a + \frac{1}{2}\right)\hbar\omega_a + (n_b + \frac{1}{2})\hbar\omega_b. \tag{7.51}$$

Of course these results, which are exact, could have been obtained more directly by first transforming the original Hamiltonian to normal coordinates (Wilson et al. 1954), but the same ideas apply to more complicated situations. The natural first step is to use the harmonic potential terms to obtain initial approximations for the fundamental coefficients x_{01}, x_{10}, etc., as described above. Higher coefficients are then derived successively in terms of these, and the whole set iterated for self-consistency. The computational demands are small compared with any trajectory approach and there is in principle no limitation on the number of coupled degrees of freedom. However, as implemented to date, results have been obtained only at the EBK quantization level. It should also be noted that Percival and Pomphrey (1973, 1976) encountered difficulties in near degenerate situations, probably because the convergence of the iteration scheme depends on the choice of the zeroth-order Hamiltonian. For example, one cannot expect to converge to a torus on one side of the separatrix in Fig. 7.4 from a starting point on the other; the correct topology of the torus must be introduced at the outset.

An alternative Fourier-based quantization scheme, introduced by Eaker et al. (1984) and Eaker and Schatz (1984), and perfected by Martens and Ezra (1985, 1987), obtains the frequencies, ω_a, and Fourier components, $q_{i\mathbf{k}}$, from numerically determined classical trajectories. This involves finding the Fourier components of n independent coordinates, q_i, and adjusting the energy and trajectory starting conditions until eqns (7.43) are satisfied for all n actions I_a simultaneously. The key technical feature (Martens and Ezra 1985) is to impose two carefully chosen window functions over the time series to be transformed, in order to ensure that the frequencies, ω_v, and amplitudes, q_{ik_v}, are accurately determined. Applications of this procedure to a variety of resonant and non-resonant systems have been reviewed by Ezra et al. (1987).

7.5 Classical perturbation theory

Applications of classical perturbation theory are conveniently introduced by assuming a Hamiltonian of the angle–action form, with a single perturbation term,

$$H(\mathbf{J}, \boldsymbol{\varphi}) = H_0(\mathbf{J}) + \lambda H_1(\mathbf{J}, \boldsymbol{\varphi}), \tag{7.52}$$

because terms in higher powers of λ cause no serious additional complications. The idea is to perform successive canonical transformations to new conjugate variables, $(\mathbf{I}, \boldsymbol{\alpha})$ say, such that the transformed Hamiltonian

$$K(\mathbf{I}, \boldsymbol{\alpha}) = K_0 + \lambda K_1 + \lambda^2 K_2 + \ldots \tag{7.53}$$

contains only terms (up to a certain power in λ) of an appropriate standard or 'normal' form (Birkhoff 1927). The simplest such standard form is one in which the K_k are all expressible as functions of the final actions $\{I_v\}$, in which case K is readily quantized by the familiar substitutions

$$I_v = (n_v + \delta_v)\hbar. \tag{7.54}$$

However, the discussion in Section 7.3 shows that this primitive quantization may become unreliable in the presence of a resonance between the zero-order frequencies

$$\omega_v^0 = (\partial H_0/\partial J_v). \tag{7.55}$$

In such cases the terms K_k can be reduced to forms that are block diagonal in an appropriately defined total action and the residual motion analysed by the uniform quantization procedures introduced in Section 7.3.

Two approaches have been employed for the necessary canonical transformations. The commonest, followed, for example, by Chapman et al. (1976) in the non-degenerate case and by Birkhoff (1927), Gustavson (1966), Swimm and Delos (1979), Jaffé and Reinhardt (1982) and Warnock and Ruth (1987) in the resonant case, is to employ a type 2 classical generator of the form

$$F_2(\boldsymbol{\varphi}, \mathbf{I}) = \mathbf{I}\cdot\boldsymbol{\varphi} + W(\boldsymbol{\varphi}, \mathbf{I}), \tag{7.56}$$

so that

$$\begin{aligned} \mathbf{J} &= (\partial F_2/\partial \boldsymbol{\varphi}) = \mathbf{I} + (\partial W/\partial \boldsymbol{\varphi}) \\ \boldsymbol{\alpha} &= (\partial F_2/\partial \mathbf{I}) = \boldsymbol{\varphi} + (\partial W/\partial \mathbf{I}) \end{aligned}, \tag{7.57}$$

where the bold symbols imply identities for the corresponding components (J_ν, φ_ν).

Non-degenerate case

It is assumed initially that the system is non-degenerate, in a sense to be defined below, in which case the identity $H(\mathbf{J}, \boldsymbol{\varphi}) = K(\mathbf{I})$ requires that

$$H[\mathbf{I} + (\partial W/\partial \boldsymbol{\varphi}), \boldsymbol{\varphi}] = K[\mathbf{I}, \boldsymbol{\varphi} + (\partial W/\partial \mathbf{I})], \tag{7.58}$$

where $W(\mathbf{I},\varphi)$ is chosen to eliminate terms in φ on the right-hand side. The expansion

$$W = W_0 + \lambda W_1 + \dots \tag{7.59}$$

allows one to set up the following leading terms of a systematic scheme by collecting terms in equivalent powers of λ, with W conventionally taken to be of order λ:

$$K_0(\mathbf{I}) = H_0(\mathbf{I}) \tag{7.60}$$

$$\lambda K_1(\mathbf{I}) = \lambda H_1(\mathbf{I},\varphi) + \sum_\nu \left(\frac{\partial H_0}{\partial J_\nu}\right)\left(\frac{\partial W_0}{\partial \varphi_\nu}\right) \tag{7.61}$$

$$\begin{aligned}\lambda^2 K_2(\mathbf{I}) = {} & \lambda \sum_\nu \left(\frac{\partial H_1}{\partial J_\nu}\right)\left(\frac{\partial W_0}{\partial \varphi_\nu}\right) + \lambda \sum_\nu \left(\frac{\partial H_0}{\partial J_\nu}\right)\left(\frac{\partial W_1}{\partial \varphi_\nu}\right) + \\ & + \frac{1}{2}\sum_{\nu\mu}\left(\frac{\partial^2 H_0}{\partial J_\nu \partial J_\mu}\right)\left(\frac{\partial W_0}{\partial \varphi_\nu}\right)\left(\frac{\partial W_0}{\partial \varphi_\mu}\right).\end{aligned} \tag{7.62}$$

To see how the expansion works in practice, suppose that $H_1(\mathbf{I},\varphi)$ and $W_0(\varphi,\mathbf{I})$ are expanded in the Fourier series

$$H_1(\mathbf{I},\varphi) = c_0(\mathbf{I}) + \sum_{\mathbf{k}} c_{\mathbf{k}}(\mathbf{I}) \exp(\mathrm{i}\mathbf{k}\cdot\varphi) \tag{7.63}$$

$$W_0(\varphi,\mathbf{I}) = \sum_{\mathbf{k}\neq \mathbf{0}} w_{\mathbf{k}}(\mathbf{I}) \exp(\mathrm{i}\mathbf{k}\cdot\varphi)$$

with $c_{-\mathbf{k}}(\mathbf{I}) = c_{\mathbf{k}}(\mathbf{I})$, to ensure that $H_1(\mathbf{I},\varphi)$ is real. The leading term $K_0(\mathbf{I})$ of the transformed Hamiltonian is given by eqn (7.60). Proceeding to eqn (7.61), we see that $K_1(\mathbf{I}) = c_0(\mathbf{I})$, while the coefficients of $W_0(\varphi,\mathbf{I})$ are chosen as $w_{\mathbf{k}}(\mathbf{I}) = \mathrm{i}\lambda c_{\mathbf{k}}(\mathbf{I})/(\mathbf{k}\cdot\boldsymbol{\omega}^0)$ in order to eliminate the angle-dependent terms. Moreover the perturbation expansion requires that the $|w_{\mathbf{k}}(\mathbf{I})|$ should be small compared with unity. The inequality

$$|\mathbf{k}\cdot\boldsymbol{\omega}^0| = |k_1\omega_1^0 + k_2\omega_2^0 + k_3\omega_3^0 + \dots| \gg |c_{\mathbf{k}}(\mathbf{I})|, \tag{7.64}$$

for all positive and negative values of k_ν, is therefore a necessary condition for convergence of the expansion. Knowledge of $W_0(\varphi,\mathbf{I})$ also serves to determine $K_2(\mathbf{I})$, because the presence of both positive and negative Fourier components of $(\partial H_1/\partial J_\nu)$ and $(\partial W_0/\partial \varphi_\nu)$ will combine to produce angle-independent terms on the right of eqn (7.62). One finds, after straightforward manipulations, that (Born 1960)

$$\begin{aligned} K(\mathbf{I}) &= K_0(\mathbf{I}) + \lambda K_1(\mathbf{I}) + \lambda^2 K_2(\mathbf{I}) + \dots \\ &= H_0(\mathbf{I}) + \lambda c_0(\mathbf{I}) - \lambda^2 \sum_{\mathbf{k}}\sum_\nu k_\nu \frac{\partial}{\partial I_\nu}\left[\frac{|c_{\mathbf{k}}(\mathbf{I})|^2}{(\mathbf{k}\cdot\boldsymbol{\omega}^0)}\right],\end{aligned} \tag{7.65}$$

where the sum over $\mathbf{k}$ is restricted to vectors $(k_1, k_2, \dots)$ for which $\mathbf{k}\cdot\boldsymbol{\omega}^0 > 0$. A proper choice for $W_1(\varphi,\mathbf{I})$ then eliminates the angle-dependent terms in eqn (7.62) and allows the computation of $K_3(\mathbf{I})$ and so on.

Degenerate case: the Birkoff–Gustavson normal form

Turning to the more complicated degenerate or resonant situation, we consider the nature of motions under the scaled degenerate harmonic oscillator Hamiltonian

$$H_0 = \frac{1}{2}(p_a^2 + p_b^2 + q_a^2 + q_b^2), \tag{7.66}$$

for which $\omega_a = \omega_b = 1$ and $\mathbf{k}\cdot\boldsymbol{\omega} = 0$ for any term in eqn (7.65) with $k_a = -k_b$. The new feature is that we know from the discussion in Section 4.2 that there are three constants of the motion under H_0: the total action, I, the vibrational angular momentum, written here as I_2, and the angle α_2, conjugate to I_2. Hence problems arising from the divergent denominators $(\mathbf{k}\cdot\boldsymbol{\omega})^{-1}$ may be avoided by transforming any perturbed Hamiltonian to a Birkhoff (1927) or Gustavson (1966) *normal form* which can depend on any function of these three conserved variables. It actually proves more convenient in practice to work in terms of four possible homogeneous quadratic forms, which are given, together with their angle–action equivalents, by

$$\begin{aligned} \pi_0 &= (p_a^2 + p_b^2 + q_a^2 + q_b^2)/2 &&= I \\ \pi_1 &= (p_a^2 - p_b^2 + q_a^2 - q_b^2)/2 &&= \sqrt{I^2 - I_2^2}\cos 2\alpha_2 \\ \pi_2 &= q_a p_b - q_b p_a &&= I_2 \\ \pi_3 &= q_a p_a + q_b p_b &&= \sqrt{I^2 - I_2^2}\sin 2\alpha_2. \end{aligned} \tag{7.67}$$

Only three of them are independent because

$$\pi_0^2 = \pi_1^2 + \pi_2^2 + \pi_3^2. \tag{7.68}$$

Expressed in formal terms, these polynomials define the *null space* of the system, in the sense that they have zero Poisson bracket with H_0:

$$\{H_0, \pi_\gamma\} = \sum_\nu \left[\frac{\partial H_0}{\partial p_\nu}\frac{\partial \pi_\gamma}{\partial q_\nu} - \frac{\partial H_0}{\partial q_\nu}\frac{\partial \pi_\gamma}{\partial p_\nu}\right] = \sum_\nu \left[p_\nu \frac{\partial \pi_\gamma}{\partial q_\nu} - q_\nu \frac{\partial \pi_\gamma}{\partial p_\nu}\right] = 0. \tag{7.69}$$

Any functions purely dependent on these π_γ variables are therefore constants of the motion, which contribute to the transformed Hamiltonian K, while any remaining terms in $H(\mathbf{p},\mathbf{q})$ are treated as *range terms*, to be eliminated by successive canonical transformations. Without going into detail, it should also be understood that the theory is not limited to $1:1$ frequency ratios. Situations in which $\omega_a : \omega_b \simeq n_a : n_b$ may be handled by restricting the null space to $\pi_0 = I_a$ and $\pi_1 = I_b$, where $I_\nu = (p_\nu^2 + q_\nu^2)/2$; and appropriate factors n_ν would multiply the square brackets in eqn (7.69).

Turning to the details for the $1:1$ resonance, it is convenient to follow Swimm and Delos (1979) in considering the sequence of Cartesian transformations, analogous to eqn (7.57),

$$\mathbf{Q} = \mathbf{p} + \frac{\partial W^{(s)}}{\partial \mathbf{P}}, \qquad \mathbf{p} = \mathbf{P} + \frac{\partial W^{(s)}}{\partial \mathbf{q}}, \tag{7.70}$$

in which the generator $W^{(s)}(\mathbf{P},\mathbf{q})$ is a homogeneous polynomial of degree s, starting with $s = 3$, while the Hamiltonian, $H(\mathbf{p},\mathbf{q})$ is also assumed to be expanded in a sequence of homogeneous terms

$$H(\mathbf{p},\mathbf{q}) = H_0^{(2)}(\mathbf{p},\mathbf{q}) + H^{(3)}(\mathbf{p},\mathbf{q}) + \ldots. \tag{7.71}$$

By analogy with eqns (7.60)–(7.61) the identity

$$H\left(\mathbf{P} + \frac{\partial W^{(s)}}{\partial \mathbf{q}}, \mathbf{q}\right) = K\left(\mathbf{P}, \mathbf{q} + \frac{\partial W^{(s)}}{\partial \mathbf{P}}\right) \tag{7.72}$$

is expanded in a Taylor series about $\mathbf{P}$ and $\mathbf{q}$, and the following sequence of equations is obtained by collecting terms of equal degree i:

$$H^{(i)}(\mathbf{P}, \mathbf{q}) = K^{(i)}(\mathbf{P}, \mathbf{q}), \qquad i < s \tag{7.73}$$

$$\hat{D}W^{(s)}(\mathbf{P}, \mathbf{q}) = K^{(s)}(\mathbf{P}, \mathbf{q}) - H^{(s)}(\mathbf{P}, \mathbf{q}), \qquad i = 2, \tag{7.74}$$

where the Liouville operator $\hat{D}$ is given by the equivalent expressions

$$\hat{D} = \sum_{\nu=a,b}\left(P_\nu \frac{\partial}{\partial q_\nu} - q_\nu \frac{\partial}{\partial P_\nu}\right) = \mathrm{i}\sum_{\nu}\left(a_\nu^\dagger \frac{\partial}{\partial a_\nu^\dagger} - a_\nu \frac{\partial}{\partial a_\nu}\right), \tag{7.75}$$

in which a_ν and $a_\nu^\dagger$ are the classical equivalents of the creation and annihilation operators,

$$a_\nu(P_\nu, q_\nu) = 2^{-1/2}(q_\nu - \mathrm{i}P_\nu), \qquad a_\nu^\dagger(P_\nu, q_\nu) = 2^{-1/2}(q_\nu + \mathrm{i}P_\nu), \tag{7.76}$$

although the dependence on (q_ν, P_ν) will be suppressed to simplify the notation.

Equation (7.73) ensures that the transformation of degree s has no effect on the terms of degree $i < s$, while eqn (7.74) serves to separate $H^{(s)}$ into its *null space* and *range space* components,

$$H^{(s)} = N^{(s)} + R^{(s)}. \tag{7.77}$$

To appreciate the value of the second expression for $\hat{D}$ in eqn (7.75), suppose that the real perturbation $H^{(s)}(\mathbf{P}, \mathbf{q})$ contains a term

$$H_{\mathbf{lm}}^{(s)}(\mathbf{P}, \mathbf{q}) = (\Phi_{\mathbf{l},\mathbf{m}} + \Phi_{\mathbf{m},\mathbf{l}}) \tag{7.78}$$

where

$$\Phi_{\mathbf{lm}} = \prod_\nu (a_\nu)^{l_\nu} \left(a_\nu^\dagger\right)^{m_\nu}, \tag{7.79}$$

and $\sum_\nu (l_\nu + m_\nu) = s$. It is readily verified that $\Phi_{\mathbf{lm}}$ is an eigenfunction of $\hat{D}$, with eigenvalue $\mathrm{i}\delta_{\mathbf{lm}} = \mathrm{i}(m_a + m_b - l_a - l_b)$. Two possibilities therefore arise. Either

$\delta_{\mathbf{lm}} = -\delta_{\mathbf{ml}} = 0$, in which case, by comparison with eqn (7.69), $H^{(s)}_{\mathbf{lm}}(\mathbf{P}, \mathbf{q})$ belongs to the null space $N^{(s)}$. Equation (7.74) is therefore satisfied by setting $K^{(s)}_{\mathbf{lm}}(\mathbf{P}, \mathbf{q}) = H^{(s)}_{\mathbf{lm}}(\mathbf{P}, \mathbf{q})$, with $W^{(s)}_{\mathbf{lm}}(\mathbf{P}, \mathbf{q}) = 0$. In fact any choice which makes $W^{(s)}_{\mathbf{lm}}(\mathbf{P}, \mathbf{q})$ proportional to $H^{(s)}_{\mathbf{lm}}(\mathbf{P}, \mathbf{q})$ is equally satisfactory because $\hat{D}W^{(s)}_{\mathbf{lm}}(\mathbf{P}, \mathbf{q}) = 0$. The second possibility is that $\delta_{\mathbf{lm}} \neq 0$, in which case the choice

$$W^{(s)}_{\mathbf{lm}}(\mathbf{P}, \mathbf{q}) = \mathrm{i}\lambda\delta^{-1}_{\mathbf{lm}}(\Phi_{\mathbf{l},\mathbf{m}} - \Phi_{\mathbf{m},\mathbf{l}}) \tag{7.80}$$

eliminates $H^{(s)}_{\mathbf{lm}}(\mathbf{P}, \mathbf{q})$ from eqn (7.74) and leaves $K^{(s)}_{\mathbf{lm}}(\mathbf{P}, \mathbf{q}) = 0$.

The final step in the scheme presented by Swimm and Delos (1979) is to back substitute from $(a_\nu, a_\nu^\dagger)$ to (P_ν, q_ν) and to use $W^{(s)}$ to accommodate contributions from higher terms of the Taylor expansion of eqn (7.72), by means of the following algorithm for $i > s$:

$$\begin{aligned} H^{(i)} = K^{(i)} &+ \sum_{\mathbf{j}} \frac{1}{(j_a!)(j_b!)} \\ &\times \sum_{\nu} \left[\left(\frac{\partial W^{(s)}}{\partial q_\nu} \right)^{j_\nu} \left(\frac{\partial^{j_\nu} H^{(l)}}{\partial P_\nu^{j_\nu}} \right) - \left(\frac{\partial W^{(s)}}{\partial P_\nu} \right)^{j_\nu} \left(\frac{\partial^{j_\nu} K^{(l)}}{\partial P_\nu^{j_\nu}} \right) \right], \end{aligned} \tag{7.81}$$

where $\mathbf{j}$ is a vector with components (j_a, j_b), and terms in the sum are constrained by the identity

$$l - |j| + |j|(s-1) = i \tag{7.82}$$

subject to the inequalities $l \geqslant 2$, $s \geqslant 3$ and $1 < |j| < l < i$.

Practical details of the transformation steps, which inevitably become quite complicated, are best performed by an algebraic computational package (see, for example, Fried and Ezra 1988). Applications to the Henon and Heiles (1964) Hamiltonian,

$$H = \frac{1}{2}(p_a^2 + p_b^2 + q_a^2 + q_b^2) - \lambda q_b \left(q_a^2 - \frac{1}{3} q_b^2 \right), \tag{7.83}$$

are of particular interest. The main results are due to Swimm and Delos (1979) and Jaffé and Reinhardt (1982), both of whom used variants of Birkhoff–Gustavson perturbation theory. It is sufficient, for illustrative purposes, to quote the following fourth-order normal form, although sixth-order results are also available:

$$\begin{aligned} K = \pi_0 &+ \frac{\lambda^2}{12}(7\pi_2^2 - 5\pi_0^2) + \frac{\lambda^4}{432}(-67\pi_0^3 - 21\pi_0\pi_2^2 \\ &-504\pi_0^2\pi_1 + 504\pi_1\pi_2^2 + 672\pi_1^3). \end{aligned} \tag{7.84}$$

Jaffé and Reinhardt (1982) combine eqns (7.67) and (7.84) to obtain

$$K = I + \frac{\lambda^2}{12}(7J_2^2 - 5I^2) + \frac{\lambda^4}{432}[-67I^3 - 21IJ_2^2 + 168(I^2 - J_2^2)^{3/2}\cos 6\alpha_2], \tag{7.85}$$

whereas Swimm and Delos (1979) used alternative angle–action expressions for the π_γ polynomials which fail to bring out the full three-fold symmetry of the problem (see Fig. 7.5). The interesting features of eqn (7.85) are that the degeneracy between states with different magnitudes of the vibrational angular momentum, $|J_2|$, is lifted in second order, but that the bifurcation into librational and precessional trajectory types, which arises from the term in $\cos 6\alpha_2$, does not occur until fourth order in λ. Just as in the local mode case (see Section 7.3), one type of mode corresponds to motion above and the other to motion below the barrier in eqn (7.25), so now the three vibrational modes are trapped in the wells of the cosine potential while the precessional modes ride above the barriers.

Table 7.4 lists the eigenvalues obtained by quantizing the sixth-order equivalent of eqn (7.85) by extension of the uniform semiclassical procedure discussed in Section 7.3 (Jaffé and Reinhardt 1982). The first notable feature is that good results are obtained even for the A_1, A_2 splittings, not addressed by the primitive EBK quantization methods of Section 7.2. Secondly, and perhaps more remarkably, reliable results are also obtained for states in which the classical motion is chaotic. Table 7.4 also lists the yet more accurate results obtained by Uzer et al. (1983b) who replaced the variables a_v and $a_v^\dagger$ by their quantum mechanical equivalents, so that the π_γ polynomials become operators $\hat{\pi}_\gamma$, of which $\hat{\pi}_0$ and $\hat{\pi}_2$ are diagonal in the basis $|n, l\rangle$, where l denotes the vibrational angular momentum, while the off-diagonal coupling is concentrated in a term

Table 7.4 Energy levels for the Henon–Heiles system in the classically chaotic regime

n	$\pm l$	Quantum	Uniform†	EBK‡	Uniform§
7	±1	7.6595	7.6674	7.6550	7.6556
7	±3	7.6977 7.7369	7.6953 7.7329	7.7178	7.7179
7	±5	7.8327	7.8293	7.8289	7.8281
7	±7	8.0094	8.0059	8.0054	8.0038
8	0	8.5541	8.5638	8.4919	8.4919
8	±2	8.5764	8.5881		8.5965
8	±4	8.6779	8.6730		8.6707
8	±6	8.8113 8.8152	8.8080 8.8132	8.8084	8.8085
8	±8	9.0217	9.0193	9.0151	9.0510
9	±1	9.444	9.4575		9.4276
9	±3	9.467 9.552	9.4850 9.5486		9.5279
9	±5	9.629	9.6254		9.623
9	±7	9.794	9.7943		9.760
9	±9	10.035 10.036	10.0350 10.0351		10.028
10	0	10.305	10.3238		10.194
10	±2	10.318	10.3393		10.358
10	±4	10.463	10.4591		10.453
10	±6	10.573 10.590	10.5703 10.5893		10.577
10	±8	10.774	10.7785		10.762
10	±10	11.050	11.0536		11.040

† Jaffé and Reinhardt (1982) ‡ Noid et al. (1977) § Uzer et al. (1983b)

$$\hat{\pi}_5 = 4\hat{\pi}_1^3 + 3(\hat{\pi}_2^2 - \hat{\pi}_0^2)\hat{\pi}_1 \tag{7.86}$$

with the property that

$$\hat{\pi}_5 |n, l\rangle = \frac{A}{2}(|n, l+6\rangle - |n, l-6\rangle), \tag{7.87}$$

where

$$A = \{(n-l)(n+l+6)[n^2 - (l+4)^2][n^2 - (l+2)^2]\}^{1/2}. \tag{7.88}$$

Not surprisingly one can show (Uzer et al. 1983b) that the classical function corresponding to the operator $\hat{\pi}_5$ is proportional to the term $(I^2 - J_2^2)^{3/2} \cos 6\alpha_2$ in eqn (7.85), but the quantum mechanical matrix element A gives a more accurate representation of the coupling than does the uniform semiclassical quantization procedure imposed by the six-fold barrier.

7.6 Adiabatic switching

A number of authors (Solov'ev 1978; Grozdanov and Solov'ev 1982; Johnson 1985; Skodje et al. 1985) have used the conceptually attractive Ehrenfest (1916) adiabatic hypothesis as a basis for so-called adiabatic quantization; the underlying assumptions have been explored at length by Skodje and Cary (1988). The idea is simply to follow the energy of an initially separable quantized system as the non-separable part of the Hamiltonian is slowly switched on. The meaning of the word 'slow' is of some importance but, when carefully handled, the method can be applied even to apparently chaotic systems in a way that illuminates the transition from regularity to chaos (Skodje et al. 1985).

To see the meaning of the word slow, it is convenient first to consider the case of a one-dimensional oscillator governed by the parametrically dependent Hamiltonian

$$H = \frac{1}{2m}p^2 + V(q; \xi), \tag{7.89}$$

so that, if the parameter ξ were held constant, the transformation to angle–action variables (I, α), such that

$$H = E(I; \xi), \tag{7.90}$$

would be induced by the generating function

$$W(q, I; \xi) = \int p(q', E; \xi)\, \mathrm{d}q'. \tag{7.91}$$

Suppose now that ξ is allowed to vary with time; the corresponding time-dependent Hamiltonian may be expressed as (Percival and Richards 1982)

$$\tilde{H} = E(I; \xi) + (\mathrm{d}W/\mathrm{d}t) = E(I; \xi) + G(I, \alpha; \xi)(\mathrm{d}\xi/\mathrm{d}t), \tag{7.92}$$

where

$$G(I, \alpha; \xi) = (\partial W / \partial \xi)_{I,q}, \tag{7.93}$$

in which the subscript q is taken to imply $q(I, \alpha; \xi)$. The time dependence of the action is therefore given by

$$\mathrm{d}I/\mathrm{d}t = -(\partial \tilde{H}/\partial \alpha) = -(\partial G/\partial \alpha)(\mathrm{d}\xi/\mathrm{d}t). \tag{7.94}$$

This means that the net change in I over the time T required to switch ξ from say zero to unity is given by

$$\Delta I = -\int_0^T \left(\frac{\mathrm{d}\lambda}{\mathrm{d}t}\right)\left(\frac{\partial G}{\partial \alpha}\right) \mathrm{d}t. \tag{7.95}$$

Now the function $\partial G/\partial \alpha$ may always be expanded as a Fourier series in the local angle $\alpha(\xi)$,

$$\partial G/\partial \alpha = \sum_k a_k \mathrm{e}^{\mathrm{i}k\alpha(\xi)}, \tag{7.96}$$

and the basis of the adiabatic theorem (see Landau and Lifshitz 1976) is that the oscillations due to the time dependence of $\alpha(\xi)$ at frequency $\omega(\xi)$ will cause almost complete cancellation in the integrand of eqn (7.95), provided first that $\mathrm{d}\xi/\mathrm{d}t \simeq 1/T \ll \omega(\xi)$ and secondly that ξ is switched on and off sufficiently smoothly to avoid end effects; Skodje et al. 1985 recommend the following switching function:

$$\xi(t) = \frac{t}{T} - \frac{1}{2\pi} \sin\left(\frac{2\pi t}{T}\right). \tag{7.97}$$

The quantization procedure is therefore to set the initial conditions such that $I = (n + \delta)\hbar$ and to take the final energy as the value of $E(I; \xi)$ when ξ reaches unity.

The extension of this procedure to systems with several degrees of freedom is complicated by the possible occurrence of resonances. The generalizations of eqns (7.95) and (7.96) are

$$\Delta I_v = -\int_0^T \frac{\mathrm{d}\xi}{\mathrm{d}t} \frac{\partial G}{\partial \alpha_v} \mathrm{d}t \tag{7.98}$$

and

$$(\partial G/\partial \alpha_v) = \sum_{\mathbf{k}} a_{\mathbf{k}}^v \exp[\mathrm{i}(k_a \alpha_a + k_b \alpha_b + \ldots)], \tag{7.99}$$

which means that the relevant frequencies for comparison with $\mathrm{d}\xi/\mathrm{d}t$ are not simply the frequencies ω_v but all combinations

$$\frac{\mathrm{d}}{\mathrm{d}t}(k_a \alpha_a + k_b \alpha_b + \ldots) = k_a \omega_a + k_b \omega_b + \ldots. \tag{7.100}$$

Thus any persistent resonance condition

$$k_a\omega_a + k_b\omega_b + \ldots = 0 \tag{7.101}$$

will spoil the adiabatic evolution—persistent over a timescale comparable with the switching time T (Berry 1983). For example, the variation of λ may cause the system to cross a variety of weak high-order resonances in a time short compared with T without perceptibly changing the actions I_v. On the other hand the motion would be very far from adiabatic if the trajectory spent half its time on one side of the separatrix between different types of motion and half on the other. This means in practice that the zeroth-order separable Hamiltonian must be taken in different forms in order to quantize different types of motion.

The study by Skodje et al. (1985) of the Henon–Heiles system offers an illustrative example. The Hamiltonian was expressed as

$$H = H_0 + \xi(t)H_1, \tag{7.102}$$

with the switching function $\xi(t)$ given by (7.97). Two divisions of H were then adopted according to the type of motion to be quantized. The polar forms

$$\begin{aligned} H_0 &= \tfrac{1}{2}(p_r^2 + r^2 + \tfrac{1}{r^2}p_\varphi^2) + f(r) \\ H_1 &= -\tfrac{1}{3}\lambda r^3 \cos 3\varphi - f(r) \end{aligned} \tag{7.103}$$

were chosen for the processing trajectories, with initial quantization conditions $p_\varphi = l$ and

$$\oint p_r \mathrm{d}r = 2\pi\left(n_r + \frac{1}{2}\right). \tag{7.104}$$

The alternative choice for the librating states was

$$\begin{aligned} H_0 &= \tfrac{1}{2}(p_x^2 + p_y^2 + x^2 + y^2) + g(x) \\ H_1 &= \lambda x\left(y^2 - \tfrac{1}{3}x^2\right) - g(x), \end{aligned} \tag{7.105}$$

with the zero-order trajectories quantized by the Noid et al. (1977) rules (see Section 7.2)

$$\begin{aligned} &\oint p_x\,\mathrm{d}x + \oint p_y\,\mathrm{d}y = 2\pi(n+1) \\ &\oint p_y\,\mathrm{d}y = \frac{2\pi l}{3}, \end{aligned} \tag{7.106}$$

designed to take partial account of the quantum mechanical splitting of the three-fold classical degeneracy. The choices $f(r) = 0$ and $g(x) = -0.19x^2$ for the resonance breaking are seen in Table 7.5 to give excellent results, even for the classically chaotic states. Note that the adiabatic energy levels are reported as a mean of the spread

Table 7.5 Adiabatically switched eigenvalues for the Henon–Heiles system (Skodje et al., 1985).

n	l	Type†	$E^{\ddagger}_{\text{exact}}$	$E^{\S}_{\text{adia}}$	$E^{\parallel}_{\text{EBK}}$	$\Delta E^{\P}_{\text{rms}}$
0	0	S	0.9986	0.9947	0.9947	
1	1	P	1.9901	1.9862	1.9863	6(−5)
2	0	S	2.9562	2.9506	2.9506	
2	2	P	2.9853	2.9814	2.9818	9(−5)
3	1	P	3.9260	3.9225	3.9233	7(−4)
3	3	P	3.9841*	3.9801	3.9803	1(−4)
4	0	S	4.8702	4.8573	4.8573	
4	2	P	4.8997	4.8954	4.8954	7(−4)
4	4	P	4.9863	4.9819	4.9821	1(−4)
5	1	L	5.8170	5.8167	5.8160	5(−6)
5	3	P	5.8742*	5.8709	5.8713	7(−4)
5	5	P	5.9913	5.9867	5.9869	1(−4)
6	0	S	6.7379	6.7077	6.7078	
6	2	P	6.7649	6.7667	6.7709	2(−3)
6	4	P	6.8354	6.8487	6.8500	8(−4)
6	6	P	6.9991*	6.9941	6.9958	2(−4)
7	1	L	7.6595	7.6556	7.6550	1(−5)
7	3	P	7.7173*	7.7179	7.7178	2(−3)
7	5	P	7.8327	7.8281	7.8289	9(−4)
7	7	P	8.0094	8.0038	8.0054	3(−4)
8	0	S	8.5541	8.4919	8.4919	
8	2	L	8.5764	8.5965	—	2(−4)
8	4	P	8.6779	8.6707	—	2(−3)
8	6	P	8.8132*	8.8085	8.8084	9(−4)
8	8	P	9.0217	9.0510	9.051	6(−4)
9	1	L	9.444	9.4276	—	3(−4)
9	3	P	9.509*	9.5279	—	5(−3)
9	5	P	9.629	9.623	—	3(−3)
9	7	P	9.794	9.790	—	3(−3)
9	9	P	10.035*	10.028	10.028	7(−4)

†Trajectory type: straight (S), librational (L), precessional (P).
‡ Noid as reported by Swimm and Delos (1979).
* Mean value of close doublet. § Skodje et al. (1985) ‖ Noid et al. (1977)
¶ From 9 precessing or 25 librating trajectories.

obtained from a random set of starting points on the zeroth-order trajectory (9 for precessing and 25 for librating modes) and that the root-mean-square deviation, ΔE_{rms}, tabulated in the final column, gives an estimate of the accuracy.

As an added bonus, the adiabatic method gives insight into how the classical trajectory distorts as chaos develops. Figure 7.11 shows the evolution of the (y, p_y) Poincaré section for the (11, 7) Henon–Heiles state starting from a regular distribution of initial trajectories. The shape of the invariant torus has changed quite markedly at the 29th crossing of the sectioning plane, but it appears to remain quite regular until the 33rd crossing when tendrils begin to appear, before growing rapidly in length and complexity. Nevertheless the overall shape can still be recognized, and it may well be that the success of the algebraic method of Section 7.5 in the chaotic regime is due to averaging out the tendril structure (Skodje et al. 1985).

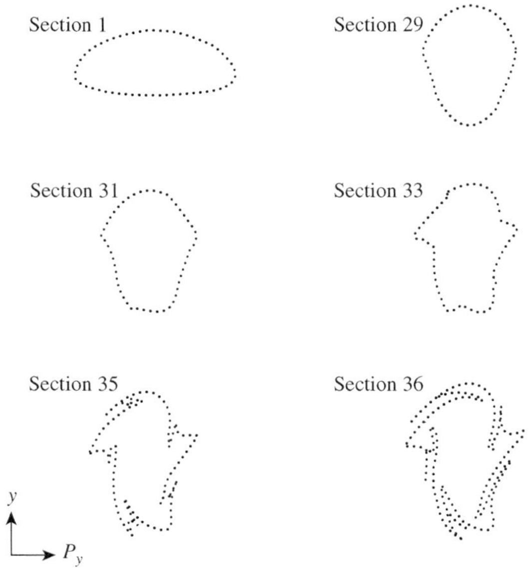

Fig. 7.11 Stages in evolution of the Poincaré section of an ultimately chaotic trajectory, as the coupling is adiabatically switched on. (Reprinted with permission from Skodje et al. (1985). Copyright 1985, AIP Publishing LLC.)

7.7 Periodic orbit quantization

Quantization procedures based on the properties of purely periodic orbits were suggested by Gutzwiller (1967, 1969, 1970, 1971, 1990) and later addressed by Balian and Bloch (1972), Berry and Mount (1972), Miller (1975) and Berry and Tabor (1976, 1977). The exposition given below follows Berry and Tabor (1976) for classically regular systems and Gutzwiller (1971) for the quantization of irregular ones. Littlejohn (1990) gives an illuminating discussion of the geometrical significance of trace formulae of this type.

Regular systems

The state density of a classically regular system is assumed to be consistent with the EBK quantization rule. Thus

$$\rho(E) = \sum_{\mathbf{n}} \delta\{E - H[(\mathbf{n} + \boldsymbol{\delta})\hbar]\}, \tag{7.107}$$

where $\mathbf{n}$ denotes the set of quantum numbers and $\boldsymbol{\delta}$ the associated Maslov parameters. Consequently, by use of the Poisson sum rule,

$$\rho(E) = \sum_{\mathbf{M}} \rho_{\mathbf{M}}(E) \exp(-2\pi i \boldsymbol{\delta} \,.\, \mathbf{M}), \tag{7.108}$$

where

$$\rho_{\mathbf{M}}(E) = \hbar^{-f} \int_{I_i \geq 0} \delta\{E - H(\mathbf{I})\} \exp(2\pi \mathrm{i} \mathbf{M} \,.\, \mathbf{I}/\hbar) \mathrm{d}^f I. \tag{7.109}$$

The components of $\mathbf{I}$ are the true but unknown action variables and the sum is taken over all integer combinations $(M_1, M_2, \ldots, M_n)$, which will in due course correspond to the numbers of cycles of the angle variables $(\alpha_1, \alpha_2, \ldots, \alpha_f)$, except that the special term with $\mathbf{M} = 0$ provides a Thomas–Fermi density equivalent to allocating a phase volume $\hbar^f$ to each quantum state (Berry and Mount 1972).

To obtain the periodic orbit sum, the integrals in (7.109) are approximated by invoking a transformation to the variables $(\xi_0, \xi_1, \ldots, \xi_{f-1})$ of which ξ_0 is normal to the energy surface, $H(\mathbf{I}) = E$, and the ξ_i with $i = 1, 2, \ldots f - 1$ lie in it. Hence, on integrating over ξ_0,

$$\rho_{\mathbf{M}}(E) = \hbar^{-f} \int \frac{1}{|\boldsymbol{\omega}[\mathbf{I}(\xi)]|} \exp[2\pi \mathrm{i} \mathbf{M} \,.\, \mathbf{I}(\xi)/\hbar] d^{f-1}\xi, \tag{7.110}$$

where the components of

$$\boldsymbol{\omega}(\mathbf{I}) = \nabla_{\mathbf{I}} H \tag{7.111}$$

are the frequencies. The term in $|\boldsymbol{\omega}|^{-1}$ may be seen to come from the Jacobian because, if ξ_0 is identified with H and the ξ_i are taken to be mutually perpendicular, the first row of $\partial(\xi_0, \xi_1, \ldots, \xi_{f-1})/\partial(I_1, I_2, \ldots, I_{f-1})$ contains the components of $\boldsymbol{\omega}$ while the remaining rows are orthogonal to $\boldsymbol{\omega}$ and to each other. The inverse Jacobian is therefore equal to $|\boldsymbol{\omega}|$ times the determinant of an orthogonal matrix, which is equal to unity.

The next step is to evaluate the integral in (7.110) by stationary phase, noting that the stationarity conditions

$$\mathbf{M} \,.\, (\partial \mathbf{I}/\partial \xi_i) = 0, \qquad i = 1, 2, \ldots, f - 1 \tag{7.112}$$

pick out points $\xi^{\mathbf{M}}$ corresponding to particular vectors $\mathbf{I}^{\mathbf{M}}$ in action space, whose significance becomes apparent in the light of the identities

$$\frac{\partial H}{\partial \xi_i} = \nabla_{\mathbf{I}} H. \left(\frac{\partial \mathbf{I}}{\partial \xi_i}\right) = \omega. \left(\frac{\partial \mathbf{I}}{\partial \xi_i}\right) = 0, \quad i = 1, 2, \ldots, f - 1, \tag{7.113}$$

in which the zero arises because the ξ_i $(i = 1, 2, \ldots, f - 1)$ lie on the energy surface. Comparison between (7.112) and (7.113) shows that the vector $\mathbf{I}^{\mathbf{M}}$ (which specifies a particular torus) must be chosen in such a way that the frequency ratios ω_1: ω_2: ω_3: …: ω_f are rationally related in the form M_1:M_2:M_3:…: M_f. In other words, the orbits on the torus $\mathbf{I}^{\mathbf{M}}$ must be periodic, with closure after M_1 cycles of α_1, M_2 cycles of α_2, etc. Moreover, the stationary value of the exponent in (7.110) is the classical action

$$W(\mathbf{M}) = 2\pi(M_1 I_1^M + M_2 I_2^M + \ldots + M_f I_f^M) = 2\pi \mathbf{M} \,.\, \mathbf{I}^{\mathbf{M}}, \tag{7.114}$$

which may be evaluated as the closed contour integral $\oint \mathbf{p}\,.\,d\mathbf{q}$ in any convenient system of variables. Repetitions of any given orbit are conveniently taken into account by writing

$$\mathbf{M} = q\boldsymbol{\mu} \tag{7.115}$$

and

$$W(\mathbf{M}) = qW(\boldsymbol{\mu}). \tag{7.116}$$

The reduced vectors $\boldsymbol{\mu}$ are termed by Berry and Tabor (1976) the contributing rays to points in the topological lattice that specify the relative winding numbers for the different component degrees of freedom.

The stationary phase approximation to the integral in eqn (7.110) is expressed in terms of the scalar curvature, $K(\mathbf{I})$, of the energy shell (Berry and Tabor 1976), as defined by the equation

$$|\boldsymbol{\mu}|^{f-1}K(\mathbf{I}) = \det\left|\boldsymbol{\mu}\cdot\frac{\partial^2\mathbf{I}}{\partial\xi_i\partial\xi_j}\right|. \tag{7.117}$$

The final expression for the full state density takes the form (Berry and Tabor 1976)

$$\rho(E) = \rho_0(E) + \frac{2}{\hbar^{(f+1)/2}}\sum_{\mu_i\geq 0}\frac{1}{|\mu|^{(f-1)/2}|\boldsymbol{\omega}(\mathbf{I}^{\boldsymbol{\mu}})||K(\mathbf{I}^{\boldsymbol{\mu}})|^{1/2}}$$
$$\times\sum_{q=1}^{\infty}\frac{\cos\{q[W(\boldsymbol{\mu})/\hbar - \boldsymbol{\delta}.\boldsymbol{\mu}] + \pi\beta(\boldsymbol{\mu})/4\}}{q^{(f-1)/2}}, \tag{7.118}$$

where $|\mu|$ is the magnitude of the ray vector $\boldsymbol{\mu}$ and $\beta(\boldsymbol{\mu})$ is the excess of positive over negative eigenvalues of the matrix with elements $\boldsymbol{\mu}\cdot\partial^2\mathbf{I}/\partial\xi_i\partial\xi_j$.

Equation (7.118) is the central result, but it should be noted that the derivation requires non-zero curvature $K(\mathbf{I}^{\mu})$ for all contributing rays $\boldsymbol{\mu}$. This means in particular that (7.118) is invalid for harmonic systems, because $K(\mathbf{I}) = 0$ for all $\mathbf{I}$. Breakdown also occurs for rays close to the caustics at which $K(\mathbf{I}) = 0$ if ever the curvature of the energy shell changes sign. The quantization procedure will therefore fail for particular states whose dominant contributing rays $\boldsymbol{\mu}$ lie close to the caustic, but it is unfortunately impossible to tell in advance where these lie because $H(\mathbf{I})$ (whose curvature is required) is the very function to be determined. Nevertheless there may be situations where some zeroth-order approximation gives adequate information on $K(\mathbf{I})$, and Berry and Tabor (1976) have suggested a uniform approximation to overcome the weaknesses of the stationary phase result.

It is also useful to remember that a typical quantizable EBK torus is incommensurate. The necessary constructive interference in eqn (7.118) therefore requires that many tori $\boldsymbol{\mu}$ must satisfy equations of the form

$$W(\boldsymbol{\mu})/\hbar - 2\pi\boldsymbol{\delta}\cdot\boldsymbol{\mu} \simeq 2\pi k_{\boldsymbol{\mu}}, \tag{7.119}$$

where $k_{\boldsymbol{\mu}}$ is an integer, a condition that can be understood on the basis that the contributing rays $\boldsymbol{\mu}$ correspond to periodic tori that cluster ever more closely around the true EBK torus. To see this, assume for simplicity two degrees of freedom, and suppose that the EBK torus has a frequency ratio 3.14159 ... and quantum numbers n_1 and n_2. Provided the curvature of $H(I_1, I_2)$ is sufficiently strong there will be a neighbouring 3:1 periodic orbit with $k = 3n_1 + n_2$, another with ratio 22:7 and $k = 22n_1 + 7n_2$, a third with ratio 31:10 and $k = 31n_1 + 10n_2$, and so on. The closer the ratio approaches the true value, the smaller will be the inequality between the two sides of (7.119) and the larger the contribution to the constructive sum. Hence it may well be that a single periodic orbit satisfying (7.119) has an energy close to that of the true torus, but the converse is not the case. The existence of a single periodic orbit satisfying or almost satisfying an equality in (7.119) is necessary for quantization, but not sufficient; there must be a whole family with related frequency ratios (i.e. related rays μ) that also satisfy (7.119) with appropriately related k values. (Note that the Sorbie and Handy (1976) procedure, based on a single almost-periodic orbit, overcomes this difficulty by also requiring carefully defined partial action integrals, as specified by (7.8).) As a further point of interest this picture of ever more tightly nested orbits also explains the failure of (7.118) to handle harmonic systems; either the frequency ratios are non-commensurate, in which case there are no periodic orbits, or they are commensurate, in which case all are periodic so that (7.119) will determine the energy, but not the individual quantum numbers.

A final comment concerns the rate of convergence of the sum in (7.118), which can be very slow, in view of the large number of ever longer and longer periodic orbits. Following Balian and Bloch (1972), Berry and Tabor (1976) recommend a smoothing procedure, consistent with the assumption of complex energy values, $E + \mathrm{i}\gamma$. A new state density is defined as

$$\rho_\gamma(E) = \frac{1}{\pi}\mathrm{Im}\sum_m \frac{1}{E_m - E - \mathrm{i}\gamma} = \sum_m \frac{\gamma/\pi}{(E_m - E)^2 + \gamma^2}, \tag{7.120}$$

and each action term in (7.118) is shown to acquire an additional imaginary term

$$2\pi\mathrm{i}\gamma\mathbf{M}\,.\left(\frac{\partial\mathbf{I}}{\partial E}\right) = \mathrm{i}\gamma\sum_i \frac{2\pi M_i}{\omega_i} = \mathrm{i}\gamma T(\mathbf{M}), \tag{7.121}$$

where $T(\mathbf{M})$ is the time to traverse the orbit $\mathbf{M}$. The net effect on (7.118) is to insert a factor $\exp[-\gamma T(\mathbf{M})/\hbar]$ in front of each cosine term, which can greatly increase the rate of convergence by eliminating the longest running orbits. Of course resolution is lost in the process, but the resulting peaks can be assigned to particular clustering frequency ratios (or rays), by reference to (7.119); refinement can then proceed with a smaller γ value by excluding rays far from those already identified.

Irregular systems

Following Gutzwiller (1990) the semiclassical quantization of irregular systems is also based on the properties of periodic orbits, which are the only recognizable classical entities in strongly chaotic situations. The density of states is now derived from the trace of the Green's function

$$\rho(E) = \sum_j \delta(E - E_j) = -\frac{1}{\pi}\operatorname{Im} g(E) = -\frac{1}{\pi}\operatorname{Im}\int \mathrm{d}\mathbf{q}G(\mathbf{q},\mathbf{q};E), \tag{7.122}$$

where the response function, $g(E)$, is given in the light of eqn (C.121) by

$$g(E) \simeq 2\pi\left(\frac{1}{2\pi\mathrm{i}\hbar}\right)^{(f+1)/2}\sum_{\mathrm{traj}}\int \mathrm{d}\mathbf{q}\,[D]^{1/2}\,\mathrm{e}^{\mathrm{i}W(\mathbf{q}''\mathbf{q}';E)/\hbar-\mathrm{i}\mu\pi/2}, \tag{7.123}$$

in which

$$W(\mathbf{q}''\mathbf{q}';E) = \int_{\mathbf{q}'}^{\mathbf{q}''}\mathbf{p}(\mathbf{E})\cdot\mathrm{d}\mathbf{q}$$

$$D = \left(\dot{q}_1'\dot{q}_1''\right)^{-1}\det\left(-\frac{\partial^2 W}{\partial\tilde{\mathbf{q}}''\partial\tilde{\mathbf{q}}'}\right)_E, \tag{7.124}$$

and the coordinates are chosen such that q_1 lies along the trajectory $\mathbf{q}(t)$, while the components $\tilde{\mathbf{q}}= (q_2, q_3, \cdots q_f)$ are perpendicular to it. Note that the trajectory end-points $\mathbf{q}'$ and $\mathbf{q}''$ are treated as formally distinct although the form of eqn (7.122) requires that a trajectory from $\mathbf{q}' = \mathbf{q}$ must return to $\mathbf{q}'' = \mathbf{q}$. Moreover, the stationary phase conditions

$$\frac{\partial W}{\partial\mathbf{q}} = \left[\frac{\partial W}{\partial\mathbf{q}''} + \frac{\partial W}{\partial\mathbf{q}'}\right]_{\mathbf{q}'=\mathbf{q}=\mathbf{q}''} = \mathbf{p}'' - \mathbf{p}' = 0 \tag{7.125}$$

ensure that the dominant contributions to the sum come from periodic orbits. The index μ in eqn (7.123) normally counts the number of sign changes of D along a multidimensional trajectory, although additional contributions may come from points at which $\dot{\mathbf{q}} = 0$ (Gutzwiller 1990).

Detailed attention is limited for simplicity to two degrees of freedom. The integral along the orbit is trivial because the stationary value

$$\Phi(E) = \frac{1}{\hbar}W(\mathbf{q}\mathbf{q};E) = \frac{1}{\hbar}\oint\mathbf{p}(\mathbf{E})\cdot\mathrm{d}\mathbf{q} \tag{7.126}$$

is independent of q_1 and $\mathrm{d}q_1/\dot{q}_1 = \mathrm{d}T$. There is however a minor complication in the stationary phase evaluation with respect to $\tilde{q}$, because second-order variations with respect to both $\tilde{q}'$ and $\tilde{q}''$ must be included. Hence

$$g(E) \simeq 2\pi\left(\frac{1}{2\pi\mathrm{i}\hbar}\right)^{3/2}\sum_{\mathrm{po}} T_0\left[-\frac{2\pi\mathrm{i}\hbar W_{q''q'}}{\left(W_{q''q''} + 2W_{q''q'} + W_{q'q'}\right)}\right]^{1/2}$$
$$\times\exp\left[\mathrm{i}\left(\Phi(E) - \mu\pi/2\right)\right], \tag{7.127}$$

in which $W_{q''q'}$ is a shorthand for $\left(\partial^2 W/\partial q''\partial q'\right)$ and the ratio in the pre-exponent is related by eqn (F.17) and table F.1 to the eigenvalues of the monodromy matrix (see

Appendix F). The sum over periodic orbits is taken to include all repetitions of the orbit with simple period T_0.

Two cases will be considered, according to whether the orbit is stable (elliptic) or unstable (hyperbolic). The contribution from the repetitions of a single stable orbit is found to be (Gutzwiller 1990; Miller 1975)

$$\rho(E) = \frac{1}{\pi}\,\mathrm{Im}\left[\frac{T_0(E)}{\hbar}\sum_{n=1}^{\infty}\frac{\exp\left[\mathrm{i}n\left(\Phi(E)-\mu\pi/2\right)\right]}{2\sin\left[nv(E)/2\right]}\right], \tag{7.128}$$

where $\mathrm{i}v(E)$ is the Lyapunov exponent of the orbit, which is related to the transverse frequency $\omega_\perp(E)$ in the form $v(E) = \omega_\perp(E)T_0 = \hbar\omega_\perp(E)\Phi'(E)$, and μ is the number of classical turning points for one cycle of the orbit. Miller (1975) uses the Poisson sum formula to reduce eqn (7.128) to a sum of delta functions at energies given by the quantization condition

$$\begin{aligned}\Phi(E_{n,m}) &- (m+1/2)\hbar\omega_\perp(E_{n,m})\Phi'(E)\\ &\simeq \Phi\left[E_{n,m}-(m+1/2)\hbar\omega_\perp(E_{n,m})\right] = 2\pi(n+\mu/4),\end{aligned} \tag{7.129}$$

where n and m are the quantum numbers for longitudinal and transverse motion respectively. The transverse term has the expected harmonic character because displacements from the orbit have been limited to quadratic terms. Note that ambiguities of 2π in the value of v must be resolved by accumulating the number of transverse oscillations in time T_0, in order to obtain a proper value for $\omega_\perp(E)$.

In the case of an unstable orbit, eqn (7.128) is replaced by (Gutzwiller 1990)

$$\begin{aligned}\rho(E) &= \frac{1}{\pi}\,\mathrm{Im}\left\{\frac{T_0(E)}{\hbar}\sum_{n=1}^{\infty}\frac{\exp\left[\mathrm{i}n\left(\Phi(E)-\mu\pi/2\right)\right]}{2\mathrm{i}\sinh\left[nu(E)/2\right]}\right\}\\ &\simeq \frac{1}{\pi}\,\mathrm{Im}\left\{-\mathrm{i}\frac{T_0(E)}{\hbar}\sum_{n=1}^{\infty}\exp\left[\mathrm{i}n\left(\Phi(E)-\mu\pi/2+\mathrm{i}u(E)/2\right)\right]\right\}.\end{aligned} \tag{7.130}$$

The geometric series in the second line can be summed to a complex cotangent function, which is then expanded in partial fractions to yield the following sequence of Lorentzian peaks:

$$\begin{aligned}\rho(E) &= \frac{T_0(E)}{\hbar}\sum_{n'=-\infty}^{\infty}\frac{u(E)/2}{\left[\Phi(E)-2\pi(n'+\mu/4)\right]^2+u^2(E)/4}\\ &= \frac{1}{\pi}\sum_{n'}\frac{\Gamma_{n'}/2}{(E-E_{n'})^2+\Gamma_{n'}^2/4}.\end{aligned} \tag{7.131}$$

The peak positions, $E_{n'}$, are quantized according to the formula

$$\Phi(E_{n'}) = \frac{1}{\hbar}\oint \mathbf{p}(E'_{\mathbf{n}})\cdot\mathbf{dq} = 2\pi(n'+\mu/4) \tag{7.132}$$

and the widths are given by $\Gamma_{n'} = u(E_{n'})/\Phi'(E_{n'})$.

A comparison between eqn (7.129) and (7.132) shows a common JWKB-like longitudinal contribution to the quantization condition, which ensures constructive interference between the numerators in (7.128) and (7.130). The difference is that the denominator in (7.128) gives rise to side bands arising from bound states of transverse vibrational motion, whereas the exponential spreading around the unstable orbit causes a sequence of broadened Lorentzian peaks. Of course this is an over-simplified picture, because attention has been restricted to a single transverse degree of freedom, and no account has been taken of bifurcations in the number and nature of the periodic orbits, as the energy increases. The density of states of a strongly chaotic system will depend on interference between contributions from a large number of individual orbits. The reader is referred to more specialized texts (Gutzwiller 1990; Ozorio de Almeida 1988) for further details.

Periodic orbit scars

To conclude this topic, we note that Heller (1984) has observed a remarkable 'scarring' phenomenon in the highly excited states of the ergodic stadium potential. Fig. 7.12(a)

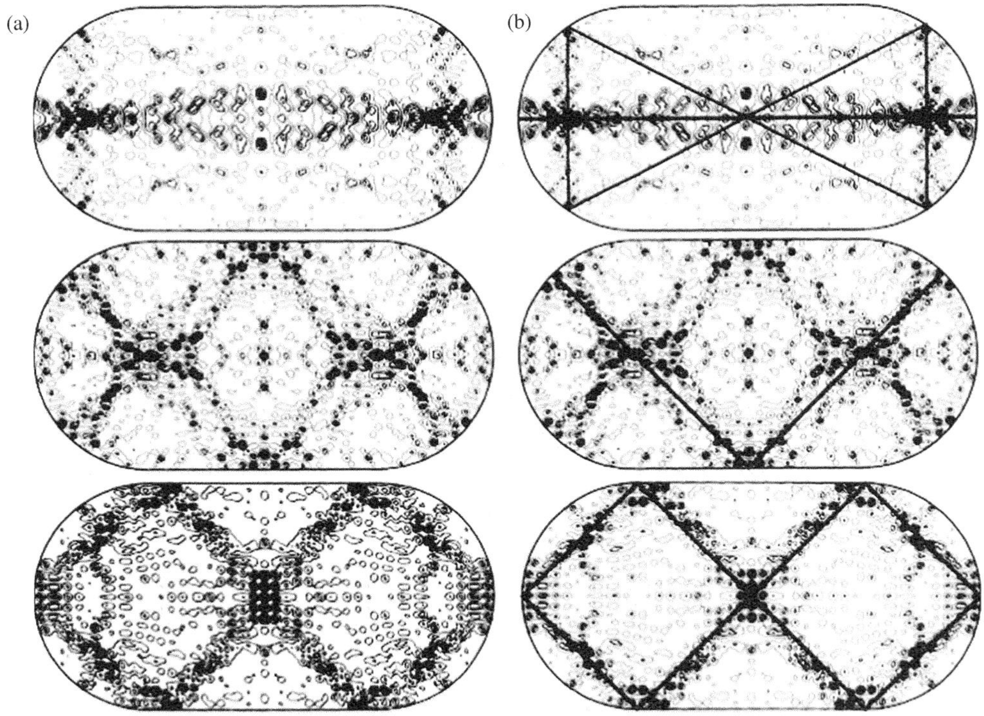

Fig. 7.12 (a) Three scarred wavefunctions of the stadium potential. (b) Associated isolated unstable periodic orbits. (Reprinted with permission from Heller (1984). Copyright 1984, AIP Publishing LLC.)

shows three examples of such wavefunctions, on which the tracks of three isolated unstable periodic orbits have been superimposed in 7.12(b). To understand this behaviour, imagine that the width, $\Gamma_{n'}$, of the Lorentzian state density in eqn (7.131) is small compared with the peak separation; $\Gamma_{n'} = \lambda\hbar \ll \hbar\omega_{\|}$, where $\omega_{\|} = 2\pi/T_0$ is the longitudinal frequency of the orbit and λ is the corresponding scaled level width. Suppose also that the spectral quantization procedure of Section 8.3 has been used to generate a localized wavefunction by propagation over a single cycle of the orbit. As discussed by Heller (1984) and Gutzwiller (1990), propagation of the wavefunction for long periods will lead to peaks in the auto-correlation function, with separation T_0, whose amplitudes will fall off according to the ratio λ/ω, while the power spectrum in eqn (8.68) will recover the Lorentzian form in eqn (7.131). The exponential decay responsible for the broadening is, however, a local phenomenon, based on a quadratic expansion around the orbit, and the amplitude that spreads away eventually populates other states of the stadium potential, all of which are strictly bound. The background density arising from the recipient states will typically be distributed over the entire stadium, which allows the concentrated density around the parent orbit to stand out in Fig. 7.12. Further discussion of this interesting scarring phenomenon is given by Heller (1987) and Gutzwiller (1990).

7.8 Problems

7.1. (i) Compute surfaces of section at $y = 0$ and $p_y > 0$ at energies $E = 1$ and $E = 5$ for the Hamiltonian

$$H = \frac{1}{2}(p_x^2 + p_y^2) + D(1 - \mathrm{e}^{-ax})^2 + \frac{1}{2}\omega_y^2 y^2 + \lambda x y^2,$$

with parameter values $\omega_y = 0.45, D = 12.5, a = (2D)^{-1/2}$ and $\lambda = 0.01$, choosing initial variables $x < 0$, $y = 0$, $p_x = 0$ and $p_y = \{2[E - V(x, y)]\}^{1/2}$.

(ii) Compute Lissajous figures corresponding to the sections plotted in part (i), and note that surfaces of section for trajectories 'within the resonance' at $E = 5$ would be invalid at constant x, but can be taken at constant η of the parabolic (ξ, η) system introduced in problem 4.2.

(iii) Locate three fixed points in the section at $E = 5$ and investigate their stabilities by means of the Newton–Raphson algorithm (see Section D.2)

$$\mathbf{r}^{\mathbf{k}+1} = \mathbf{r}^{\mathbf{k}} - (\mathbf{A} - \mathbf{I})^{-1}(\mathbf{R}^{\mathbf{k}} - \mathbf{r}^{\mathbf{k}}),$$

where $\mathbf{r}$ is an initial point (x, p_x) in the $y = 0, p_y > 0$ section, $\mathbf{R}$ is the position on returning to the section, and $\mathbf{A}$ is the matrix with elements $A_{X_p} = (\partial X/\partial p_x) \simeq (\Delta X/\Delta p_x)$, etc. The point is stable if the eigenvalues of $\mathbf{A}$ are complex and unstable if they are real (Percival and Richards 1982).

7.2. The degenerate harmonic oscillator described by eqns (4.29)–(4.37) is subject to a perturbation

$$H' = \frac{1}{2}\mu\delta^2(x^2 - y^2),$$

which imparts a frequency difference between the x and y modes. Show, by substituting from (4.35) and averaging over a cycle of α_I, that the mean effect of the perturbation is

$$\langle H' \rangle = J(\delta^2/2\omega),$$

and hence that the induced time dependence of α_J causes x and y, given by (4.35), to oscillate at frequencies $\omega \pm (\delta^2/2\omega)$. Use eqns (4.49) and (4.50) to analyse the effect of the same perturbation on the variables I_ϕ and α_ϕ which would otherwise be constants of the motion.

7.3. Employ the following procedure to obtain EBK eigenvalues of the coupled Fermi resonance Hamiltonian (Noid et al. 1979):

$$H = \frac{1}{2}(p_x^2 + p_y^2 + 4\omega^2 x^2 + \omega^2 y^2) + \lambda x(y^2 - \alpha x^2),$$

with parameter values $\omega = 0.7$, $\lambda = -0.04$ and $\alpha = 0.04$. Obtain appropriate Poincaré sections by noting that the relevant trajectories fluctuate about parabolic periodic orbits, which may be identified as lines of constant ξ or η in the parabolic system defined by $x = (\xi^2 - \eta^2)/2$ and $y = \xi\eta$.

(i) Plot Lissajous figures in the (x, y) and (ξ, η) planes for trajectories initiated at $(x, y, p_x, p_y) = (\mp 0.93, 0.00, \pm 1.18, \pm 1.79)$, roughly appropriate to the $(v_\xi, v_\eta) = (3, 0)$ and $(0, 3)$ states.

(ii) By obtaining fixed points of the separation constant, K, in problem 4.2 with respect to either (ξ, p_ξ) or (η, p_η), show that the periodic parabolic orbits of the uncoupled system correspond to $\xi_0 = \left[2E/3\omega^2\right]^{1/4}$ for $K < 0$, while $\eta_0 = \left[2E/3\omega^2\right]^{1/4}$ for $K > 0$.

(iii) Use the results in problem 4.2 to verify the reverse transformations

$$\xi = \sqrt{2r}\cos(\phi/2), \qquad \eta = \sqrt{2r}\sin\phi/2,$$
$$p_\xi = \xi p_x - \eta p_y, \qquad p_\xi = \eta p_x + \xi p_y,$$

where (r, ϕ) are polar coordinates in the (x, y) plane.

(iv) Use Poincaré sections at either $[\xi = \xi_0,\ p_\xi > 0]$ and $[\eta = 0,\ p_\eta > 0]$, or $[\xi = 0,\ p_\xi > 0]$ and $[\eta = \eta_0,\ p_\eta > 0]$ to obtain EBK eigenvalues for the states listed in Table 7.6.

Table 7.6 Quantum mechanical, E_{q}, and semiclassical, E_{sc}, eigenvalues taken from Noid et al. (1979).

v_ξ	v_η	E_{q}	E_{sc}	v_ξ	v_η	E_{q}	E_{sc}
0	0	1.050	1.050	0	1	1.747	1.750
0	2	2.442	2.421	2	0	2.471	2.467
0	3	3.100	3.099	3	0	3.184	3.185
0	4	3.789	3.769	4	0	3.906	3.908
2	2	3.844	3.850				

Table 7.7 Quantum mechanical eigenvalues of the Hamiltonian in problem 7.1.

v_x	v_y	E	v_x	v_y	E
0	0	0.722	1	3	3.096
0	1	1.174	2	1	3.054
1	0	1.687	0	6	3.417
0	3	2.074	1	4	3.514
1	2	2.149	2	2	3.570
0	5	2.970	3	0	3.487

7.4. Use the adiabatic switching method to determine semiclassical eigenvalues of the Hamiltonian in problem 7.1.

(i) Show that motions for $\lambda = 0$ satisfy the resonance condition, $\omega_x = 2\omega_y$ at $v_x = 2$. Thus states inside and outside the resonance correspond roughly to $v_x < 2$ and $v_x > 2$ respectively, although the precise boundary depends on the coupling strength.

(ii) Generate initial trajectories for the 'outside states' from the uncoupled Hamiltonian, with $\lambda = 0$, with coordinates and momenta given in Appendices E.1 and E.2.

(iii) For those 'inside the resonance', the initial trajectory should be chosen appropriate to the separable Fermi resonance Hamiltonian,

$$H_{\text{Ferm}} = \frac{1}{2}[p_x^2 + p_y^2 + \omega_y^2(4x^2 + y^2)].$$

Selected quantum mechanical eigenvalues are given in Table 7.7.

8
Wavepackets

The discussion in previous chapters has centred on the use of trajectory-based information to add quantum corrections to the classical picture, in a mainly time-independent formulation. This chapter takes an explicitly time-dependent standpoint by examining the relevance of various types of wavepacket to classical mechanics and vice versa; Littlejohn (1975) gives a much fuller mathematical account, while Tannor (2007) sets the subject in the context of time-dependent quantum mechanics. Intuition suggests that a unit-normalized wavepacket corresponds most nearly to a single quantum mechanical particle, but the associated uncertainty relations actually imply a closer connection with a classical ensemble. Analysis of this connection is one of the main concerns of the first half of the chapter.

The construction of such packets from time-independent functions is first illustrated for the case of free motion in Section 8.1, while Section 8.2 deals with the Gaussian wavepackets in general, with coherent harmonic oscillator states and with implications for the treatment of anharmonic systems. One central aspect concerns the spreading of free-motion and harmonic oscillator wavepackets, which is shown to have an exact classical analogue in the spreading of a classical ensemble taken from the Wigner (1932a) transform of the wavefunction. Closer analysis shows, however, that this harmonic classical analogue disguises special phase coherences, and that the spreading under anharmonic potentials is accompanied by interferences which disturb the shape of the packet. This leads to the introduction of a frozen Gaussian swarm technique, pioneered by Heller (1981), which incorporates classical spreading without sacrificing the quantum mechanical interference. A complex time path treatment of interferences (Huber and Heller 1987) is also referenced.

Subsequent sections of the chapter deal with applications of frozen Gaussian wavepacket techniques to bound state (Section 8.3) and photon-initiated processes (Section 8.4). The bound state discussion centres on construction of 'seeded Gaussian' wavefunctions for non-separable systems (Davis and Heller 1981) and 'spectral quantization' by Fourier analysis of the wavepacket autocorrelation function (DeLeon and Heller 1983, 1984). The photon-initiated section extends the discussion to include Franck–Condon transition intensities. Finally, the chapter concludes with an introduction to the unitary Herman–Kluk propagation procedure (Herman and Kluk 1984), as modified to handle weakly chaotic systems (Walton and Manolopoulos 1996).

8.1 The free-motion Gaussian wavepacket

A typical wavepacket is a square-integrable solution of the time-dependent Schrödinger equation

$$H\Psi = \mathrm{i}\hbar\frac{\partial \Psi}{\partial t}, \tag{8.1}$$

which may, in principle, be expressed as a linear combination of time-independent solutions,

$$\Psi(\mathbf{x}, t) = \sum_n \psi_n(\mathbf{x}) \exp(-\mathrm{i}E_n t/\hbar), \tag{8.2}$$

where $\psi_n(\mathbf{x})$ and E_n satisfy

$$H\psi_n = E_n\psi_n. \tag{8.3}$$

Other representations are, however, more revealing.

To construct a localized free-motion wavepacket at time $t = 0$, it is instructive to start with a superposition of the plane-wave states, $(2\pi)^{-1/2}\mathrm{e}^{\mathrm{i}kx}$, with energies $E = k^2\hbar^2/2m$. Since the index k is continuous, the superposition appears as an integral,

$$\Psi(x, 0) = (2\pi)^{-1/2} \int_{-\infty}^{\infty} a(k) \exp(\mathrm{i}kx)\mathrm{d}k, \tag{8.4}$$

and different choices for $a(k)$ define different types of wavepacket; the commonest form,

$$a(k) = \left(\frac{2\beta}{\pi}\right)^{1/4} \exp[-\beta(k - k_0)^2], \tag{8.5}$$

defines a Gaussian packet with mean momentum $p_0 = k_0\hbar$ and uncertainty $\Delta p = \beta^{-1/2}(\hbar/2)$. The corresponding coordinate representation at time $t = 0$ is given by

$$\begin{aligned}\Psi(x, 0) &= (\beta/2\pi^3)^{1/4} \int_{-\infty}^{\infty} \exp[-\beta(k - k_0)^2 + \mathrm{i}kx]\mathrm{d}k, \\ &= (\beta/2\pi^3)^{1/4} \int_{-\infty}^{\infty} \exp\left[-\beta\left(k - k_0 + \frac{\mathrm{i}x}{2\beta}\right)^2 + \mathrm{i}k_0x - \frac{x^2}{4\beta}\right] \mathrm{d}k, \\ &= (2\pi\beta)^{-1/4} \exp\left(-\frac{x^2}{4\beta} + \mathrm{i}k_0x\right);\end{aligned} \tag{8.6}$$

it is centred at $x = 0$ with uncertainty $\Delta x = \beta^{1/2}$.

The time dependence is obtained, according to (8.2), by augmenting $a(k)$ with a term

$$\exp(-\mathrm{i}E_k t/\hbar) = \exp(-\mathrm{i}k^2\hbar t/2m), \tag{8.7}$$

so that (8.6) goes over to

$$\Psi(x,t) = (\beta/2\pi^3)^{1/4} \int_{-\infty}^{\infty} \exp\left(-\beta(k-k_0)^2 + \mathrm{i}kx - \mathrm{i}\frac{k^2\hbar t}{2m} \right) \mathrm{d}k, \tag{8.8}$$

which reduces after completing the square to

$$\Psi(x,t) = \left(\frac{\beta}{2\pi\gamma^2(t)} \right)^{1/4} \exp\left(-\frac{(x-v_0t)^2}{4\gamma(t)} + \mathrm{i}k_0x - \mathrm{i}\frac{k_0^2\hbar t}{2m} \right), \tag{8.9}$$

where $v_0 = k_0\hbar/m$ and

$$\gamma(t) = \left(\beta + \mathrm{i}\frac{t\hbar}{2m} \right). \tag{8.10}$$

Equation (8.9) describes the desired free-motion Gaussian wavepacket, whose square modulus evolves as

$$|\Psi(x,t)|^2 = \left(\frac{\beta}{2\pi|\gamma(t)|^2} \right)^{1/2} \exp\left(-\frac{\beta(x-v_0t)^2}{2|\gamma(t)^2|} \right). \tag{8.11}$$

As expected, the centre moves with the classical velocity equivalent to k_0, but the width, in coordinate space, spreads with time according to the equation

$$\Delta x(t) = \beta^{-1/2}|\gamma(t)| = \beta^{-1/2}(\beta^2 + t^2\hbar^2/2m)^{1/2}, \tag{8.12}$$

while the momentum distribution, $a(k)$, remains unchanged. Thus the initial wavepacket, with minimum uncertainty $\Delta x(0)\Delta p = h/4\pi$, degrades with time.

Paradoxically this spreading, which presents at first sight a complete contrast to classical mechanics, has a revealing classical interpretation in relation to the evolution of a classical ensemble with the same initial uncertainties, Δp and Δx. To see this, assume for consistency with $\Psi(x,0)$, an initial classical Wigner (1932a) distribution

$$\begin{aligned} P(x,k;0) &= \frac{1}{2\pi} \int_{-\infty}^{\infty} \exp(\mathrm{i}k\xi)\Psi^*\left(x + \frac{1}{2}\xi, 0 \right) \Psi\left(x - \frac{1}{2}\xi, 0 \right) \mathrm{d}\xi \\ &= (2\pi)^{-3/2}\beta^{-1/2} \int_{-\infty}^{\infty} \exp\left(-\frac{x^2}{2\beta} - 2\beta(k-k_0)^2 - \frac{1}{8\beta}[\xi - 4\mathrm{i}\beta(k-k_0)]^2 \right) \mathrm{d}\xi \\ &= \frac{1}{\pi} \exp\left(-\frac{x^2}{2\beta} - 2\beta(k-k_0)^2 \right), \end{aligned} \tag{8.13}$$

which is designed to have the same coordinate and momentum probability distributions as the quantum mechanical state. Now a particle starting from (x,k) will reach

$$(x', k') = (x + vt, k) \tag{8.14}$$

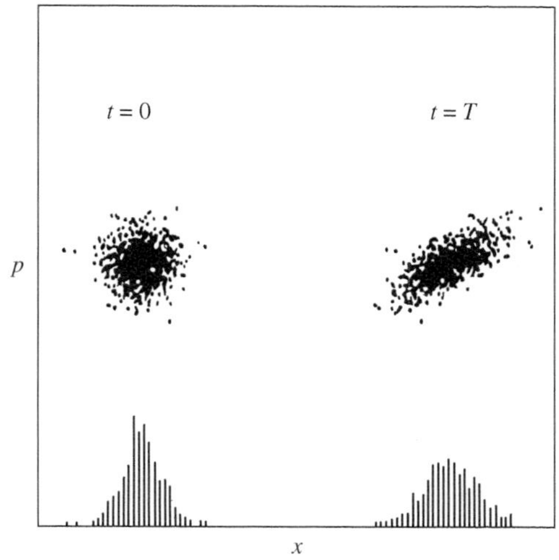

Fig. 8.1 Classical spreading of a free-motion Gaussian classical ensemble in phase space. The histograms show projections on to the coordinate axis.

at time t, where $v = k\hbar/m$. Hence the new distribution evolves to

$$P(x', k'; t) = \frac{1}{\pi} \exp\left(-\frac{(x' - vt)^2}{2\beta} - 2\beta(k' - k_0)^2\right), \tag{8.15}$$

which may be verified to have mean position and momentum

$$\langle x' \rangle = k_0 t\hbar/m = v_0 t, \quad \langle p' \rangle = k_0 \hbar, \tag{8.16}$$

and uncertainties

$$\begin{aligned} \Delta x' &= \left\langle (x' - v_0 t)^2 \right\rangle^{1/2} = (\beta + t^2\hbar^2/2m\beta)^{1/2} = \beta^{-1/2}|\gamma(t)|, \\ \Delta p' &= \left\langle (k' - k_0)^2 \right\rangle \hbar = \beta^{-1/2}(\hbar/2). \end{aligned} \tag{8.17}$$

One contribution to $\Delta x'$ (which is seen to be identical with the quantum result (8.12)) comes from the initial uncertainty $\beta^{1/2}$ and the other from the fact that v in (8.15) depends on k. In other words the ensemble spreads with time simply because some of its members are moving faster than others, as illustrated pictorially in Fig. 8.1. In other words the spreading comes solely from the initial quantum mechanical uncertainties and not from anything new in the quantum mechanical equations of motion.

8.2 Gaussian wavepackets and coherent harmonic oscillator states

The Gaussian form in eqn (8.9) suggests an analysis of the general form

$$\Psi(x, t) = \left(\frac{2\alpha(0)}{\pi\hbar}\right)^{1/4} \exp\left(\frac{1}{\hbar}\{-\alpha(t)[x - q(t)]^2 + ip(t)[x - q(t)] + c(t)\}\right), \tag{8.18}$$

where the pre-exponential factor ensures normalization to unity at time $t = 0$. Following Heller (1975), it may be shown that $\Psi(x,\ t)$ satisfies the time-dependent Schrödinger equation, provided that the potential takes the quadratic form:

$$V(x) \simeq V(q) + V'(q)(x-q) + \frac{1}{2}V''(x-q)^2. \tag{8.19}$$

The required conditions on $\alpha(t)$, $q(t)$, $p(t)$ and $c(t)$ are readily established by comparing the equations

$$H\Psi = \left(\frac{p^2}{2\mu} + V(x) + \frac{\alpha\hbar}{\mu} + \frac{2\mathrm{i}\alpha p}{\mu}(x-q) - \frac{2\alpha^2}{\mu}(x-q)^2\right)\Psi \tag{8.20}$$

and

$$\mathrm{i}\hbar\frac{\partial\Psi}{\partial t} = (p\dot{q} + \mathrm{i}\dot{c} + (2\mathrm{i}\alpha\dot{q} - \dot{p})(x-q) - \mathrm{i}\dot{\alpha}(x-q)^2)\Psi, \tag{8.21}$$

where a dot implies differentiation with respect to t. Comparison between the coefficients of various powers of $(x-q)$ leads to the following conditions on the guiding trajectory $[q(t),\ p(t)]$:

$$\frac{p^2}{2\mu} + V(q) = E, \quad \mu\dot{q} = p, \quad \dot{p} = -V'(q), \tag{8.22}$$

together with the following subsidiary conditions:

$$\dot{c} = \mathrm{i}\left(L - \frac{\alpha\hbar}{\mu}\right), \tag{8.23}$$

$$\dot{\alpha} = \frac{\mathrm{i}}{2}\left(V'' - \frac{4\alpha^2}{\mu}\right), \tag{8.24}$$

where L is the Lagrangian

$$L = p\dot{q} - E = \frac{1}{2}\mu\dot{q}^2 - V(q). \tag{8.25}$$

Equations (8.22) are of course the classical equations of motion under the potential $V(q)$, while the contribution $\mathrm{i}\int_0^t L dt$ to $c(t)$ is the classical action term, responsible for semiclassical interferences. The nature and validity of the full solution depend, however, on also satisfying (8.23) and (8.24), which allow a number of possibilities.

Suppose, for example, that $V'' = 0$, so that (8.24) yields

$$\frac{1}{\alpha(t)} = \frac{1}{\alpha(0)} + \frac{2\mathrm{i}t}{\mu}. \tag{8.26}$$

As expected, eqn (8.18) then goes over to (8.9) under the substitutions $\gamma(t) = \hbar/4\alpha(t)$ and $\beta = \hbar/4\alpha(0)$.

Cases when $V'' \neq 0$ allow more variety, and it is illuminating to follow Lee and Heller (1982) and Huber and Heller (1987) by relating α to auxiliary variables Z and P_z according to the equations

$$\alpha = -\frac{\mathrm{i}}{2} P_z / Z, \tag{8.27}$$

$$\begin{pmatrix} \dot{Z} \\ \dot{P}_z \end{pmatrix} = \begin{pmatrix} 0 & \mu^{-1} \\ -V'' & 0 \end{pmatrix} \begin{pmatrix} Z \\ P_z \end{pmatrix}, \tag{8.28}$$

where the so-called monodromy matrix in (8.28) determines the stability of the motion (Percival and Richards 1982); it is stable or unstable according to the positive or negative sign of the determinant, or equivalently of V'' (see Appendix F). The harmonic oscillator solution for $V'' > 0$ yields

$$\begin{aligned} Z &= X_+ \exp(\mathrm{i}\omega t) + X_- \exp(-\mathrm{i}\omega t), \\ P_z &= \mathrm{i}\mu\omega[X_+ \exp(\mathrm{i}\omega t) - X_- \exp(-\mathrm{i}\omega t)], \end{aligned} \tag{8.29}$$

where

$$\omega = (V''/\mu)^{1/2}. \tag{8.30}$$

Equation (8.29) may be rearranged in terms of the value α_0 at $t = 0$ to read

$$\alpha(t) = \alpha_* \left(\frac{\alpha_0 \cos \omega t + \mathrm{i}\alpha_* \sin \omega t}{\alpha_* \cos \omega t + \mathrm{i}\alpha_0 \sin \omega t} \right), \tag{8.31}$$

where

$$\alpha_* = \mu\omega/2. \tag{8.32}$$

Finally, if $V'' < 0$, eqns (8.27) and (8.28) allow only imaginary solutions for $\alpha(t)$ and the Gaussian form loses its localized significance.

The consequences of eqn (8.31) are of interest both in themselves and in relation to the classical picture outlined below. A special case arises if $\alpha_0 = \alpha_*$ because α remains constant (as may be verified from (8.24) or (8.31)). Consequently

$$c(t) = \mathrm{i} \int_0^t L \, \mathrm{d}t - \mathrm{i}\delta\hbar\omega \tag{8.33}$$

in eqn (8.18), with the parameter value $\delta = \frac{1}{2}$ which may be recognized as the Maslov index. Moreover, on the assumption that $V'(0) = 0$, so that $V'(q) = V''q$ in (8.22),

$$q(t) = q_0 \cos \omega t, \quad p(t) = -\mu\omega q_0 \sin \omega t, \tag{8.34}$$

which means, as seen in Fig. 8.2(a), that $|\Psi(x,t)|^2$ oscillates without change of shape at the natural harmonic frequency. More remarkably, the same is true of a packet with

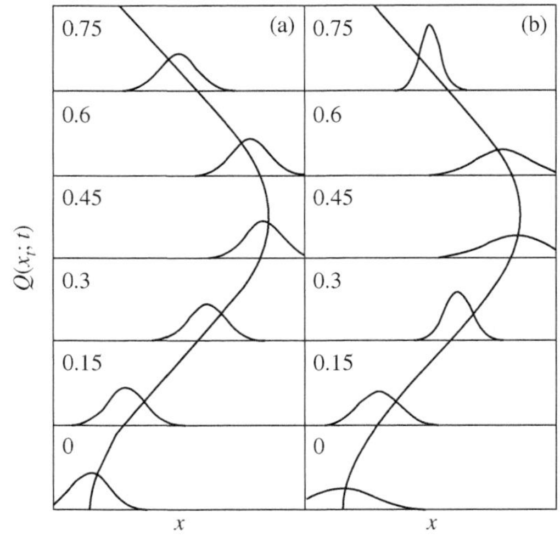

Fig. 8.2 Oscillations of a harmonically driven Gaussian wavepacket (a) with the initial co-ordinate width matched to the oscillator, and (b) with a larger initial width. Note the changes in profile in case (b).

the shape of any harmonic oscillator wavefunction, provided again that the overall Gaussian width is matched to the characteristic parameter α_* (see problem 8.1). Such special wavepackets may be termed strictly coherent states.

For a more general choice of α_0, $q(t)$ still follows (8.34) but (8.31) may be rearranged to read

$$\alpha(t) = \alpha_* \left(\frac{\alpha_0 \alpha_* + \mathrm{i}(\alpha_*^2 - \alpha_0^2) \sin \omega t \cos \omega t}{\alpha_*^2 \cos^2 \omega t + \alpha_0^2 \sin^2 \omega t} \right), \tag{8.35}$$

which implies that

$$|\Psi(x,t)|^2 = [\pi \rho(t)]^{-1/2} \exp\left(-\frac{1}{\rho(t)} (x - q_0 \cos \omega t)^2 \right), \tag{8.36}$$

where the notation

$$\rho(t) = \frac{\hbar(\alpha_*^2 \cos^2 \omega t + \alpha_0^2 \sin^2 \omega t)}{2\alpha_0 \alpha_*^2} \tag{8.37}$$

has been chosen for convenient comparison with the classical discussion given below. The difference in normalization between (8.18) and (8.36) may be verified to come from the imaginary part of $\alpha(t)$, via eqn (8.23). Equation (8.36) shows that the packet again follows the harmonic guiding path, but with a width that increases and decreases in the manner depicted in Fig. 8.2(b).

Turning to the classical picture, it is tempting, by analogy with the discussion in Section 8.1, to visualize the packet $\Psi(x,t)$ in relation to the evolution of the equivalent classical ensemble, given by the Wigner (1932a) distribution

$$\begin{aligned} P(x,p;0) &= \frac{1}{2\pi\hbar}\int_{-\infty}^{\infty} \exp(ip\xi/\hbar)\Psi^*\left(x+\frac{1}{2}\xi,0\right)\Psi\left(x-\frac{1}{2}\xi,0\right)\mathrm{d}\xi \\ &= \frac{1}{\pi\hbar}\exp\left[-\frac{1}{\hbar}\left(2\alpha_0(x-q_0)^2+\frac{p^2}{2\alpha_0}\right)\right]. \end{aligned} \tag{8.38}$$

A given phase point will evolve under a harmonic Hamiltonian according to the equations

$$\begin{aligned} x(t) &= \left(\frac{2I}{\mu\omega}\right)^{1/2}\cos(\omega t+\varphi), \\ p(t) &= -(2I\mu\omega)^{1/2}\sin(\omega t+\varphi), \end{aligned} \tag{8.39}$$

where (I,φ) are the initial angle–action variables (see Section 4.1, except that φ is now preferred to α as the angle variable, to avoid notational confusion). Equation (8.39) means that the point (x,p) at time $t=0$ will evolve to $(x_t,\ p_t)$ at time t, related to it by the equation

$$\begin{pmatrix} x \\ p \end{pmatrix} = \begin{pmatrix} \cos\omega t & -(2\alpha_*)^{-1}\sin\omega t \\ (2\alpha_*)\sin\omega t, & \cos\omega t \end{pmatrix}\begin{pmatrix} x_t \\ p_t \end{pmatrix}, \tag{8.40}$$

where α_* is given by (8.32). It follows (with some effort) after substitution for $(x,\ p)$ in (8.38) that

$$P(x_t,p_t;t) = \frac{1}{\pi\hbar}\exp\left(-\frac{1}{\hbar}\Omega(x_t,p_t;t)\right), \tag{8.41}$$

where

$$\begin{aligned} \Omega(x_t,p_t;t) &= \sigma(t)(x_t-q_0\cos\omega t)^2+\rho(t)(p_t+2\alpha_*q_0\sin\omega t)^2 \\ &\quad + 2\tau(t)(x_t-q_0\cos\omega t)(p_t+2\alpha_*q_0\sin\omega t), \\ \sigma(t) &= (2\hbar/\alpha_0)(\alpha_0^2\cos^2\omega t+\alpha_*^2\sin^2\omega t), \\ \rho(t) &= (\hbar/2\alpha_0\alpha_*^2)(\alpha_*^2\cos^2\omega t+\alpha_0^2\sin^2\omega t), \\ \tau(t) &= \hbar[(\alpha_*^2-\alpha_0^2)/\alpha_*\alpha_0]\sin\omega t\cos\omega t. \end{aligned} \tag{8.42}$$

Consequently, on completing the square in (8.41) and integrating over ρ_t, the coordinate space projection of $P(x_t,\rho_t;t)$ is found to be given by

$$Q(x_t;t) = [\pi\rho(t)]^{-1/2}\exp\left(-\frac{1}{\rho(t)}(x-q_0\cos\omega t)^2\right), \tag{8.43}$$

which agrees exactly with the quantum mechanical result (8.36).

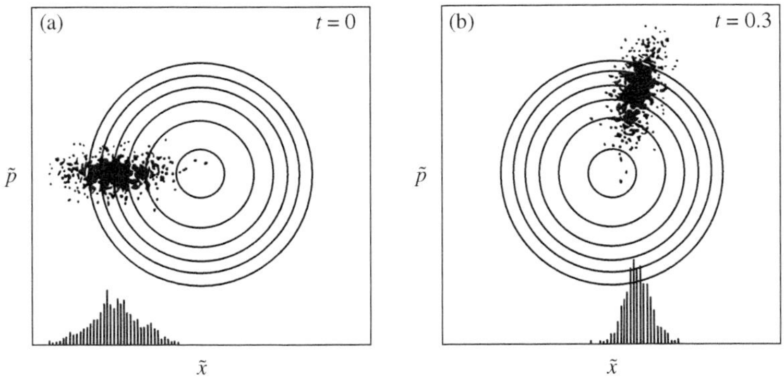

Fig. 8.3 Evolution of an initially elliptical classical ensemble from (a) $t = 0$ to (b) a later time. Note the sharpening of the coordinate projection, shown as a histogram.

The classical description also provides the striking pictorial explanation for the fluctuating width of the distributions shown in Fig. 8.3. It is simply that each initial phase point follows a circular orbit in the scaled phase space

$$(\tilde{x}, \tilde{p}) = [(\mu\omega)^{1/2}x, (\mu\omega)^{-1/2}p], \tag{8.44}$$

with a common angular frequency ω. Hence a general minimum uncertainty Gaussian packet starts with an elliptical distribution in the $(\tilde{x}, \tilde{y})$ plane, which rotates with one of its principal axes directed towards the origin, thereby offering a projection on to the coordinate axis whose width fluctuates with time. Only in the special case $\alpha_0 = \alpha_*$ is the initial distribution circular, so that the coordinate projection has a constant width (Fig. 8.2(a)). The same picture applies to all initially displaced harmonic oscillator states, even with $v \neq 0$, because the appropriate Wigner (1932a) distribution always has circular symmetry in an $[x(\mu\omega_0)^{-1/2}, p(\mu\omega_0)^{1/2}]$ projection (see problem 8.2).

This is not the whole story, however, because it applies only to pure displaced harmonic oscillator states. For mixed harmonic oscillator states there is a semiclassical correspondence in the sense that every classical phase point must return to its initial position at intervals of the period $T = 2\pi/\omega$, while in quantum mechanics any function $\Psi(x, t)$ may be expanded as

$$\Psi(x, t) = \Sigma c_n \psi_n(x) \exp(-\mathrm{i}E_n t/\hbar), \tag{8.45}$$

so that the equal energy spacing, $E_n = (n + \frac{1}{2})\hbar\omega$, ensures that $|\Psi(x, t)|$ returns to its original form at times $t = 2\pi k/\omega$. Thus an arbitrary probability distribution $|\Psi(x, t)|^2$ and its classical ensemble counterpart will both refocus at integral multiples of T. There is, however, no reason to expect any correspondence between the shapes of the classical and quantum mechanical distributions at intermediate times, even for a harmonic oscillator.

The behaviour of an initial Gaussian packet

$$\Phi(x, 0) = \pi^{-1/4} \exp\left[-\frac{1}{2}(x - x_0)^2\right] \tag{8.46}$$

under the scaled Morse oscillator Hamiltonian

$$H = \frac{1}{2}\left\{p^2 + a^2[1 - \exp(-x/a)]\right\} \tag{8.47}$$

is quite revealing from this point of view, because the quantum mechanical energy levels (Landau and Lifshitz 1965)

$$E_v = \left(v + \frac{1}{2}\right) - \frac{1}{2a^2}\left(v + \frac{1}{2}\right)^2 \tag{8.48}$$

imply a non-uniform level spacing or classical vibrational frequency

$$\omega_v = \mathrm{d}E/\mathrm{d}v \;=\; 1 - \left(v + \frac{1}{2}\right)\Big/\left(v_{\max} + \frac{1}{2}\right). \tag{8.49}$$

Consequently the anharmonicity in E_v and ω_v prevent any possibility of refocussing. Thus Fig. 8.4, which is drawn for $x_0 = 3$ and $a^2 = 15.5$, shows that both the wavepacket $\Phi(x,t)$ and the equivalent classical ensemble begin to spread immediately after they are launched, because in classical terms the lower v components of the ensemble have a higher frequency than those with larger v. Another significant feature is that $|\Phi(x,t)|^2$ develops interference structure as soon as the front of the classical ensemble passes the first turning point, but there are no equivalent oscillations in the classical distribution. The spreading is classical in origin but the interference is a quantum mechanical effect.

Insight into the origin of this interference may be obtained by a device due to Heller (1975) which also plays an important part in Section 8.4. It is designed to handle the spreading, despite the failure of any simple Gaussian representation wherever $V'' < 0$ (see above), in a way that also retains the phase information. The trick is to decompose the initial Gaussian term into a sum or 'swarm' of initial wavepackets, distributed around the appropriate phase space orbit, and to allow their centres to spread under classical mechanics, while holding their widths fixed (frozen) in order to prevent a breakdown in the representation. Interferences are then taken into account by summing the amplitudes at time t.

The justification for such a representation may be appreciated by reference to the identities

$$\begin{aligned}\psi_n(x) &= \pi^{-1/4}(2^n n!)^{-1/2} H_n(x)\exp(-x^2/2),\\ \exp(-z^2 + 2xz) &= \sum_n \frac{1}{n!} H_n(x) z^n,\end{aligned} \tag{8.50}$$

which together imply that

$$\exp\left(-z^2 + 2xz - \frac{1}{2}x^2\right) = \pi^{1/4}\sum_n (2^n/n!)^{1/2}\psi_n(x) z^n. \tag{8.51}$$

The substitution

$$z = \frac{1}{2}x_n \mathrm{e}^{-\mathrm{i}t}, \tag{8.52}$$

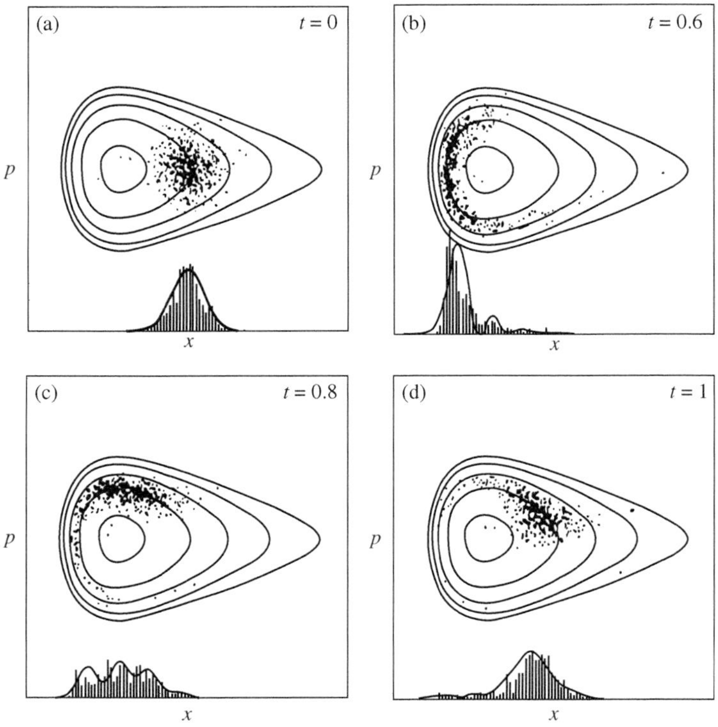

Fig. 8.4 Time evolution of an initially Gaussian classical ensemble moving under a Morse potential. Times are given as fractions of the period for small-amplitude vibrations of the Morse oscillator. The dashed envelopes show the evolution of the quantum mechanical wavepacket.

with $x_n = (2n+1)^{1/2}$, therefore casts (8.51) into a Fourier series, from which

$$\psi_n(\mathrm{x}) = \pi^{-1/4}\frac{a}{2\pi}\int_0^{2\pi} \exp\left\{ -\frac{1}{2}[x-q(t)]^2 + \mathrm{i}p(t)[x-q(t)] + \mathrm{i}\gamma(t)\right\} \mathrm{d}t, \tag{8.53}$$

where

$$q(t) = x_n \cos t, \quad p(t) = -x_n \sin t, \quad \gamma(t) = -\frac{1}{2}x_n^2 \sin t \cos t + (n+1/2)t, \tag{8.54}$$

and

$$a = \pi^{-1/4}\left(\frac{n!}{\left(n+\frac{1}{2}\right)^n}\right)^{1/2} \exp(x_n^2/4). \tag{8.55}$$

Equation (8.53) is exact, and the seeded approximation simply consists in approximating the integral by the N-point trapezoidal formula

$$\psi_n(x) \simeq \left(\frac{a}{N}\right)\sum_{k=0}^{N-1} \pi^{-1/4} \exp\left\{ -\frac{1}{2}[x-q(t_k)]^2 + \mathrm{i}p(t_k)[x-q(t_k)] + \mathrm{i}\gamma(t_k)\right\}, \tag{8.56}$$

where $t_k = 2\pi k/N$.

The generalization to the non-scaled case takes the form

$$\psi_n(r) \simeq \sum_{k=0}^{N-1} \left(\frac{a}{N}\right) g_k(r),$$

where

$$\begin{aligned}
g_k(r) &= \left(\frac{2\alpha}{\pi\hbar}\right)^{1/4} \exp\left(\frac{1}{\hbar}[-\alpha(x-q_k)^2 + \mathrm{i}p_k(x-q_k) + \mathrm{i}\gamma_k]\right) \\
\alpha &= (\mu\omega/2) \\
q_k &= [(2n+1)\hbar/\mu\omega]^{1/2} \cos\omega t_k \\
p_k &= -[(2n+1)\hbar\mu\omega]^{1/2} \sin\omega t_k \\
\gamma_k &= \int_0^{t_k} p_r \dot{r} \mathrm{d}t - \frac{1}{2}\hbar\omega t_k \\
t_k &= 2\pi k/N\omega.
\end{aligned} \tag{8.57}$$

Equations (8.56) or (8.57) describe the initial state in Fig. 8.4, which has a 'frozen Gaussian' time evolution described by

$$\Psi_n(r,t) \simeq \frac{a}{N} \sum_{k=0}^{N-1} g_k(r,\tau), \tag{8.58}$$

where $g_k(r,\tau)$ differs from (8.57) in allowing $q_k(\tau)$ and $p_k(\tau)$ to follow classical paths under the Morse Hamiltonian, while

$$\gamma_k(\tau) = \gamma_k(0) + \int_0^{\tau} (p_k\dot{q}_k - E_k)\mathrm{d}\tau. \tag{8.59}$$

This procedure is exact (as $N \to \infty$) for a displaced harmonic oscillator, because the phases of different members of the swarm may be verified to evolve in such a way as to preserve the Gaussian shape in any projection (see problem 8.3). Results for the Morse oscillator, shown for $a^2 = 15.5$ and $x_0 = 3$ in Fig. 8.5, indicate that the interference is well (but not perfectly) taken into account even for this quite strongly anharmonic system.

Huber and Heller (1987) and the Huber et al. (1988) have suggested another interesting semiclassical interpretation of this interference by noting that any initial Gaussian form

$$\Phi(x,0) = \exp[-\alpha(x-q)^2 + \mathrm{i}p(x-q) + c] \tag{8.60}$$

is invariant under the substitutions

$$(q,p,c) \to (q',p',c') \tag{8.61}$$

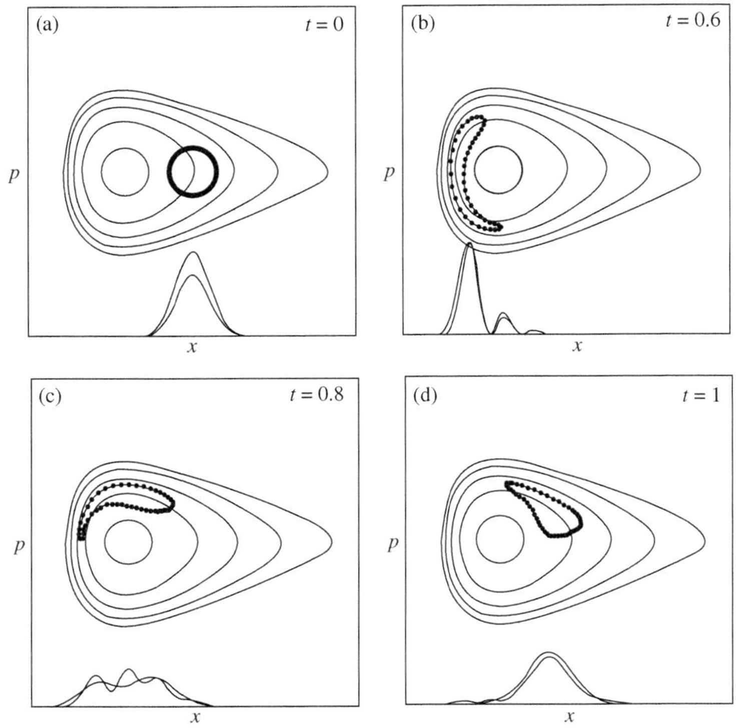

Fig. 8.5 Time evolution of a seeded Gaussian swarm moving under a Morse potential. Times are given as fractions of the period for small-amplitude vibrations of the Morse oscillator.

provided that

$$\begin{aligned} 2\alpha q' + \mathrm{i}p' &= 2\alpha q + \mathrm{i}p, \\ \alpha q'^2 + \mathrm{i}p'q' - c' &= \alpha q^2 + \mathrm{i}pq - c. \end{aligned} \tag{8.62}$$

Hence it is unduly restrictive to suppose that the centre of the packet (8.60) follows a classical orbit from the real phase point $(x,\ p) = (x_0,\ 0)$; it could equally well have been initiated from any equivalent complex point (x', p') consistent with (8.62). The generalized Gaussian wavepacket study by Huber and Heller (1987), based on this idea, shows that the structure seen in Figs. 8.3 and 8.4 can be quantitatively described in terms of interference between the main real branch of the packet and one dominant complex (classically forbidden) branch, which reaches the same coordinate point q at the same real time $t(q)$.

8.3 Seeded Gaussian wavefunctions and spectral quantization

The term seeded Gaussian wavefunction is used here to denote an approximation to a time-independent function $\psi_n(\mathbf{x})$ in terms of frozen Gaussian terms distributed along

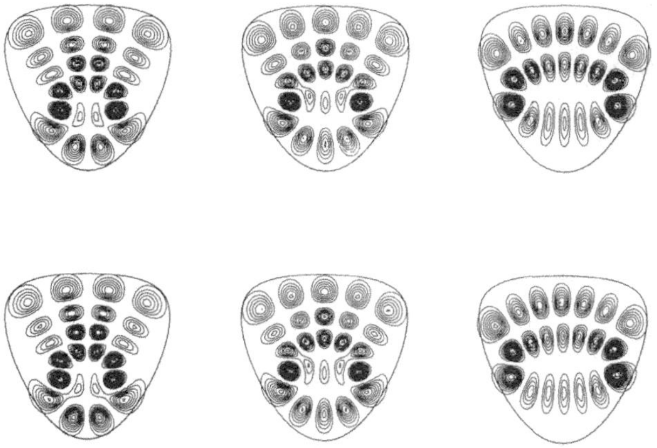

(a)

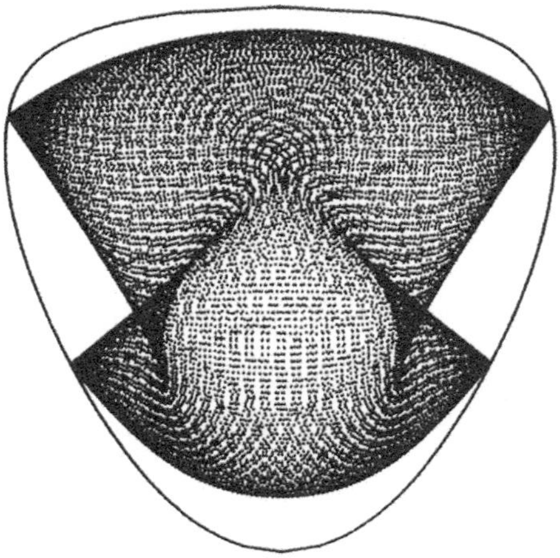

(b)

Fig. 8.6 (a) Seeded Gaussian wavefunctions (lower row) compared with their quantum mechanical counterparts (upper row). (b) The trajectory that quantizes the centre state in (a). (Reprinted with permission from Davis and Heller (1981). Copyright 1981, AIP Publishing LLC.)

an appropriate classical trajectory, $[\mathbf{q}(t),\ \mathbf{p}(t)]$ (Davis and Heller 1981). The method is a generalization of eqns (8.56) and (8.57), in which α is arbitrary within reasonable limits; the $(q_k,\ p_k)$ generalize to $[\mathbf{q}(t_k),\ \mathbf{p}(t_k)]$, with the points t_k equally spaced in time along a suitable quantizing trajectory, obtained by one of the methods of Section 7.2. Finally, the term $\hbar\omega/2$ in γ_k goes over to $\boldsymbol{\delta} \cdot \boldsymbol{\omega}\hbar$, where $\boldsymbol{\delta}$ denotes the proper set of Maslov indices. A comparison between plots of the resulting semiclassical forms and the exact wavefunctions for a coupled Morse oscillator system is shown in Fig. 8.6.

Subsequently DeLeon and Heller (1983, 1984) found a different procedure that eliminates the need to search for a proper quantizing trajectory, because several eigenvalues and the corresponding wavefunctions can, in fact, be extracted by propagation of a single d-dimensional frozen Gaussian function of the form

$$\Phi(x,t)=\left(\frac{2\alpha}{\pi\hbar}\right)^{d/4}\exp\left(\frac{1}{\hbar}\left\{-\alpha[\mathbf{x}-\mathbf{q}(t)]^2+\mathrm{i}\mathbf{p}(t)\cdot[\mathbf{x}-\mathbf{q}(t)]+\mathrm{i}\gamma(t)\right\}\right), \quad (8.63)$$

where

$$\gamma(t)=\int_0^t \mathbf{p}\cdot\mathbf{q}\,\mathrm{d}t-Et-\boldsymbol{\delta}\cdot\boldsymbol{\omega}t, \quad (8.64)$$

with the components of $\boldsymbol{\delta}$ being the Maslov indices.

To understand the reasoning behind this 'spectral quantization' procedure (DeLeon and Heller 1983, 1984; DeLeon and Mehta 1988) it is convenient to regard $\Phi(x,t)$ as an approximation to some true wavepacket, decomposable in the form

$$\Psi(\mathbf{x},t)=\sum_n c_n\psi_n(\mathbf{x})\exp(-\mathrm{i}E_nt/\hbar). \quad (8.65)$$

The form of the corresponding autocorrelation function,

$$\langle\Psi(0)|\Psi(t)\rangle=\sum_n |c_n|^2\exp(-\mathrm{i}E_nt/\hbar), \quad (8.66)$$

shows that the energy levels E_n could be extracted from peaks at

$$\omega_n=E_n/\hbar \quad (8.67)$$

in the power spectrum

$$P(\omega)=\lim_{T\to\infty}\frac{1}{T}\int_0^T\langle\Psi(0)|\Psi(t)\rangle\exp(\mathrm{i}\omega t)\mathrm{d}t. \quad (8.68)$$

Equally, once ω_n is known the wavefunction $\psi_n(\mathbf{x})$ may be determined as the Fourier transform

$$\lim_{T\to\infty}\frac{1}{4\pi T}\int_{-T}^{T}\Psi(\mathbf{x},t)\exp(\mathrm{i}\omega_nt)\mathrm{d}t=c_n\psi_n(\mathbf{x}). \quad (8.69)$$

Equivalent equations are shown below to apply to $\Phi(x,t)$ given by (8.63), to within the accuracy of the linear approximation

$$H(\mathbf{I})=H(\mathbf{I}_0)+\boldsymbol{\omega}\cdot(\mathbf{I}-\mathbf{I}_0), \quad (8.70)$$

where the components of $\mathbf{I}_0$ are the actions defined by the torus followed by the chosen arbitrary trajectory. To see this correspondence, notice that the autocorrelation function $\langle\phi(0)|\phi(t)\rangle$ constructed from (8.63) is a product of the phase factor $\exp[\mathrm{i}\gamma(t)/\hbar]$ and a quasi-periodic function, $B(t)$ say, which may be expanded in the Fourier series

$$B(t) = \sum_{\mathbf{n}} B_{\mathbf{n}} \exp(-\mathrm{i}\mathbf{n} \cdot \boldsymbol{\omega} t), \tag{8.71}$$

where $\mathbf{n}$ denotes a set of integers. Furthermore, in terms of the true but unknown angle–action variables, eqn (8.64) would read

$$\gamma(t) = [(\mathbf{I}_0 - \boldsymbol{\delta}\hbar) \cdot \boldsymbol{\omega} - H(I_0)]t/\hbar, \tag{8.72}$$

because the actions, which play the role of momenta, are constant and the angles, which act as coordinates, vary linearly with time at frequencies $\boldsymbol{\omega}$. Consequently the power spectrum of $\langle\phi(0)|\phi(t)\rangle$ has Fourier components at frequencies

$$\omega_{\mathbf{n}} = \mathbf{n} \cdot \boldsymbol{\omega} - [(\mathbf{I}_0 - \boldsymbol{\delta}\hbar) \cdot \boldsymbol{\omega} - H(I_0)]/\hbar, \tag{8.73}$$

which may also be expressed, with the help of (8.70), as

$$\omega_{\mathbf{n}} = H(\mathbf{I}_{\mathbf{n}})/\hbar, \tag{8.74}$$

where $\mathbf{I}_{\mathbf{n}}$ are the quantized actions

$$\mathbf{I}_{\mathbf{n}} = (\mathbf{n} + \boldsymbol{\delta})\hbar. \tag{8.75}$$

Equations (8.74) and (8.75) express the desired correspondence with (8.67), which may be attributed to the fact that the term $\gamma(t)$, responsible for the phase coherence of the packet, also provides a modulation on the classical frequencies, which brings the peaks in the power spectrum into coincidence with the semiclassical eigenvalues. Naturally the number of detectable eigenvalues depends on the choice of quantizing trajectory; for example, there would be only one if the actions $\mathbf{I}_0$ happened to satisfy (8.75). More commonly DeLeon and Heller (1984) report three or four eigenvalues per trajectory, as seen from the results in Table 8.1 for a model problem in two degrees of freedom. Their accuracy depends of course on a correct choice of Maslov indices in the definition of $\gamma(t)$ by (8.64), which may differ from trajectory to trajectory, but the necessary corrections for a change of indices is readily introduced in the form

$$\omega'_{\mathbf{n}} = \omega_{\mathbf{n}} + (\boldsymbol{\delta}' - \boldsymbol{\delta}) \cdot \boldsymbol{\omega}, \tag{8.76}$$

with $\omega_{\mathbf{n}}$ being given by (8.73) and $\boldsymbol{\omega}$ being calculable from the winding numbers in different degrees of freedom until the trajectory almost closes (DeLeon and Heller 1983, 1984). Quantization is achieved, however, only at the EBK quantization level, because no attempt is made to include the uniform corrections discussed in Section 7.4.

Table 8.1 Spectral quantization eigenvalues for the model of DeLeon and Heller (1984).

Traj no.	n_x	n_y	$E^{\dagger}_{\text{qm}}$	$E^{\ddagger}_{\text{sc}}$	$E^{\S}_{\text{av}}$
1	0	4	5.026	5.046	5.115
1	2	2	5.505	5.506	5.513
2	3	6	10.622	10.637	10.649
2	4	5	10.866	10.867	10.864
2	5	4	11.091	11.097	11.083
3	7	2	11.474	11.484	11.474
2	8	1	11.634	11.634	11.608
3	9	0	11.770	11.779	11.750
4	7	5	14.468	14.487	14.469
4	8	4	14.662	14.667	14.634
4	9	3	14.812	14.842	14.801
4	10	2	14.998	15.019	14.968
5‖	—	—	14.947	—	15.056
5‖	—	—	15.430	—	15.427
5‖	—	—	15.876	—	15.822
6	6	7	15.224	15.215	15.236
6	7	6	15.458	15.440	15.442
6	8	5	15.655	15.665	15.642
7	5	9	15.975	15.905	15.962
7	6	8	16.172	16.210	16.213
7	4	11	16.535	16.530	16.507
7	5	10	16.839	16.830	16.778

† Diagonalized quantum mechanical value
‡ Spectral quantization, eqn (8.74)
§ Semiclassical mean value, eqn (8.78)
‖ Trajectory within 3:1 resonance; no attempt at spectral quantization

To complete the analysis, the wavefunction corresponding to $\omega_{\mathbf{n}}$ may be generated by the Fourier transform (DeLeon and Heller 1984)

$$\phi_{\text{n}}(\mathbf{x}) \simeq \lim_{T\to\infty} \frac{1}{2T} \int_{-T}^{T} \phi(\mathbf{x}, t) \exp(\mathrm{i}\omega_{\mathbf{n}} t) \mathrm{d}t. \tag{8.77}$$

A check on the eigenvalues derived from the power spectrum is then offered by comparison with the mean values

$$\bar{E}_{\mathbf{n}} = \langle \phi_{\mathbf{n}} | H | \phi_{\mathbf{n}} \rangle \, / \, \langle \phi_{\mathbf{n}} / \phi_{\mathbf{n}} \rangle \, . \tag{8.78}$$

The results given in Table 8.1 show excellent general agreement between the two types of estimate, and with the exact eigenvalues. An exception occurs, however, in the case of trajectory 5, which lies inside a 3:1 resonance, for which no attempt was made to extract eigenvalues from the power spectrum.

8.4 Franck–Condon transitions

A time-dependent theory of Franck–Condon processes is readily formulated by imagining that a sudden transition from a state $\psi_0(\mathbf{x})$ on one potential surface generates a wavepacket

$$\Psi'(\mathbf{x},t) = \sum_n \langle \psi_n' |\mu| \psi_0 \rangle \, \psi_n'(\mathbf{x}) \exp(-\mathrm{i}E_n t/\hbar) \tag{8.79}$$

on another, where μ denotes the transition operator. Hence, if $\Psi'(\mathbf{x},t)$ were known, the state-to-state transition probability, $|\langle \psi_0 |\mu| \psi_n' \rangle|^2$, could be obtained as the Fourier transform

$$\varepsilon(\omega) = \lim_{T\to\infty} \frac{1}{2T} \int_{-T}^{T} \langle \Psi_0(\mathbf{x},0) |\mu| \Psi'(\mathbf{x},t) \rangle \exp[\mathrm{i}(\hbar\omega + E_0)/\hbar] \mathrm{d}t, \tag{8.80}$$

where the angle brackets imply integration over $\mathbf{x}$. With this definition $\varepsilon(\omega)$ necessarily peaks at $\omega = (E_n' - E_0)/\hbar$ and $|\langle \psi_0 |\mu| \psi_n' \rangle|$ corresponds to the value of $\varepsilon(\omega)$ at the peak. The following argument, due to Heller (1981), shows how this integral may be approximated by the use of frozen Gaussian wavepackets.

Assume for simplicity two degrees of freedom (x, y) and an initial separable harmonic oscillator ground state

$$\psi_0(x,y) = g_x(0,0;x) g_y(0,0;y), \tag{8.81}$$

where

$$g_x(q_x, p_x; x) = (m_x\omega_x/\pi\hbar)^{1/4} \exp\left(\frac{1}{\hbar}\left[\frac{1}{2} m_x\omega_x (x-q_x)^2 + \mathrm{i}p_x(x-q_x)\right]\right), \tag{8.82}$$

and similarly for $g_y(q_y, p_y; y)$. The computational technique is to project this Gaussian product on to the upper state surface and to follow its subsequent motion. To do this Heller (1981) creates a Gaussian swarm by decomposing $g_x(0,0;x)$ and $g_y(0,0;y)$ into frozen seeded Gaussians whose centres spread out under the classical equations of motion. Hence, on generalizing eqns (8.57), the upper state wavepacket is decomposed as

$$\begin{aligned} \Psi(x,y;0) &= \mu(x,y)\psi_0(x,y) \\ &\simeq \sum_{j=1}^{N} \sum_{k=1}^{N} \mu_{jk} c_j d_k g_x(q_{xj}, p_{xj}; x) g_y(q_{yk}, p_{yk}; y) \exp[\mathrm{i}\gamma_{jk}(0)], \end{aligned} \tag{8.83}$$

where

$$\begin{aligned} c_j &= N^{-1} \exp\left(\frac{1}{4} + \frac{\mathrm{i}}{2\hbar} q_{xj} p_{xj}\right), \\ q_{xj} &= (\hbar/m_x\omega_x)^{1/2} \cos(2\pi j/N), \\ p_{xj} &= -(\hbar m_x \omega_x)^{1/2} \sin(2\pi j/N), \end{aligned} \tag{8.84}$$

with similar equations for d_k, q_{yk} and p_{yk}. Also

$$\mu_{jk} = \mu(q_{xj}, q_{yk})$$

and

$$\gamma_{jk}(t) = \frac{1}{\hbar} \int_0^t [p_{xj}^2(t) + p_{yk}^2(t) - \bar{E}_{jk}(t)] \mathrm{d}t - (\delta_a \omega_a' + \delta_b \omega_b')t, \tag{8.85}$$

where ω_a' and ω_b' are the frequencies of the assumedly quasi-periodic motion on the upper state potential, δ_a and δ_b are the Maslov terms, and $\bar{E}_{jk}(t)$ is the (weakly time-dependent) mean value of the upper state Hamiltonian in the jkth Gaussian term.

The procedure now is to allow the individual packet centres, given by $[q_{xj}(t), q_{yk}(t), p_{xj}(t), p_{yk}(t)]$ to follow classical trajectories on the upper state surface, thereby allowing each subpacket to perform its own individual spectral quantization (see Section 8.3). Such trajectory runs may be quite time consuming but the remaining computations are trivial because the integrations over (x, y) in (8.80) can be evaluated in closed form, with the result

$$\langle \Psi(x, y; 0)|\mu|\Psi'(x, y; t)\rangle = N^{-2} \sum_{j,k} \mu_{jk} c_j d_k \alpha_{jk}(t) \exp[\mathrm{i}\gamma_{jk}(t)], \tag{8.86}$$

where

$$\alpha_{jk}(t) = \exp\left\{ -\frac{m_x\omega_x}{4\hbar}\left[q_{xj}^2(t) + \left(\frac{p_{xj}(t)}{m_x\omega_x}\right)^2\right] - \frac{m_y\omega_y}{4\hbar}\left[q_{yk}^2(t) + \left(\frac{p_{yk}(t)}{m_y\omega_y}\right)^2\right]\right\}. \tag{8.87}$$

All that is left is to perform the Fourier transform of (8.86) and pick out the peak positions and intensities.

Results obtained for a model problem by Gray et al. (1985) are given in Table 8.2. Since spectral quantization was seen in Section 8.3 to yield three or four eigenvalues per trajectory, it is not surprising that the procedure converges quite rapidly with respect to the number of subpackets employed. The results shown were derived from a 4×4 grid, and increasing the grid size to 10×10 caused changes of 0.4 per cent in the peak frequencies and 10 per cent in their heights, rising to 15 per cent for the smallest peaks.

Similar ideas have been extended to resonance Raman spectroscopy (Lee and Heller 1979) and photodissociation (Lee and Heller 1982).

Table 8.2 Frozen Gaussian transition frequencies and Franck–Condon factors for the model of Gray et al. (1985).

n_1	n_2	$E_{\mathrm{q}}^{\dagger}$	$E_{\mathrm{fg}}^{\ddagger}$	$\lvert\langle\psi_0\lvert\psi_n'\rangle\rvert_{\mathrm{q}}^2$	$\lvert\langle\psi_0\lvert\psi_n'\rangle\rvert_{\mathrm{fg}}^2$
0	1	2.278	2.31	0.081	0.05
1	1	2.958	2.98	0.084	0.08
0	2	3.545	3.56	0.127	0.12
2	1	3.635	3.63	0.045	0.04
1	2	4.216	4.23	0.105	0.10
0	3	4.804	4.81	0.123	0.14
1	3	5.460	5.49	0.073	0.08
0	4	6.046	6.08	0.079	0.08

† Quantum mechanical values
‡ Frozen Gaussian approximation. The eigenvalues correspond to peaks in $\varepsilon(\omega)$.

8.5 The Herman–Kluk propagator

The sections above focus on the evolution and manipulation of time-dependent quantum mechanical wavefunctions, with emphasis for the later applications on working in a frozen Gaussian basis. However no account is taken of changes to the normalization integral arising from the spreading of the underlying classical ensemble, particularly for anharmonic systems. The quantum mechanical propagator approach, as outlined in Appendix C.3, offers an alternative formulation, which takes explicit account of these normalization difficulties. Thus a wave function $\Psi(\mathbf{x}_0, 0)$ at an initial position $\mathbf{x}_0$ and time zero, evolves under the propagator $K(\mathbf{x}_t, \mathbf{x}_0; t)$ to $\Psi(\mathbf{x}_t, t)$ at position $\mathbf{x}_t$ and time t according to either of the equivalent equations

$$\begin{aligned}\Psi(\mathbf{x}_t, t) &= \int_{-\infty}^{\infty} K(\mathbf{x}_t, t; \mathbf{x}_0, t_0)\Psi(\mathbf{x}_0, 0)\mathrm{d}\mathbf{x}_0 \\ \langle \mathbf{x}_t, t \,|\Psi\rangle &= \int_{-\infty}^{\infty} \mathrm{d}\mathbf{x}_0 \, \langle \mathbf{x}_t, t|\, \mathrm{e}^{-\mathrm{i}Ht/\hbar} \,|\mathbf{x}_0, 0\rangle \, \langle \mathbf{x}_0, 0 \,|\Psi\rangle \,. \end{aligned} \tag{8.88}$$

Moreover, the appropriate semiclassical Van Vleck (1928) form, which is derived in Appendix C.3, is given by

$$\langle \mathbf{x}_t, t|\, \mathrm{e}^{-\mathrm{i}H_{\mathrm{sc}}t/\hbar} \,|\mathbf{x}_0, 0\rangle = \left[\left(\frac{1}{2\pi\mathrm{i}\hbar}\right)^f \det\left(-\frac{\partial^2 S_{\mathrm{cl}}}{\partial \mathbf{x}_t \partial \mathbf{x}_0}\right)\right]^{1/2} \exp\left[\mathrm{i}S_{\mathrm{cl}}(\mathbf{x}_t, \mathbf{x}_0)/\hbar\right], \tag{8.89}$$

where

$$S_{\mathrm{cl}}(\mathbf{x}_t, \mathbf{x}_0) = \int_0^t \left[\mathbf{p}\cdot\dot{\mathbf{x}} - H(p, x)\right] \mathrm{d}t = \int_{\mathbf{x}_0}^{\mathbf{x}_t} \mathbf{p}\cdot\mathbf{dq} - Et, \tag{8.90}$$

in which the prefactor ensures unitarity by taking proper account of the local trajectory density (Gutzwiller 1990). The explicit dependence on t has been dropped on the left-hand side to simplify the notation. It follows, for future reference, that

$$\left(\frac{\partial S_{\mathrm{cl}}}{\partial \mathbf{x}_t}\right)_{\mathbf{x}_0} = \mathbf{p}_t\left(\mathbf{x}_t, \mathbf{x}_0\right), \qquad \left(\frac{\partial S_{\mathrm{cl}}}{\partial \mathbf{x}_0}\right)_{\mathbf{x}_t} = -\mathbf{p}_0\left(\mathbf{x}_t, \mathbf{x}_0\right), \tag{8.91}$$

where $\mathbf{p}_0(\mathbf{x}_t, \mathbf{x}_0)$ and $\mathbf{p}_t(\mathbf{x}_t, \mathbf{x}_0)$ are the initial and final momenta required to ensure that the trajectory from the point $\mathbf{x}_0$ reaches $\mathbf{x}_t$ in time t. The difficulty, as discussed in Chapter 10, is that eqn (8.91) involves a trajectory root search for the appropriate initial momentum. Moreover, the determinant in the prefactor is known to diverge at the caustics of the classical motion, which raises difficulties in following the proper branch of the square root in eqn (8.89).

The aim of this section is to introduce an alternative propagator, due to Herman and Kluk (1984), which is designed to overcome these difficulties. It is set up, like the Heller frozen Gaussian wavepacket, in a coherent state basis, with components

$$\langle \mathbf{x}|\mathbf{z}\rangle = \left(\frac{\gamma}{\pi}\right)^{f/4} \exp\left[-\frac{\gamma}{2}(\mathbf{x}-\mathbf{q})^2 + \frac{\mathrm{i}}{\hbar}\mathbf{p}\cdot(\mathbf{x}-\mathbf{q})\right]. \tag{8.92}$$

Thus

$$\langle \mathbf{x}_t | \, \mathrm{e}^{-\mathrm{i}\hat{H}_{\mathrm{HK}}t/\hbar} \, |\mathbf{x}_0\rangle = \left(\frac{1}{2\pi\hbar}\right)^f \iint \mathrm{d}\mathbf{p}_0 \mathrm{d}\mathbf{q}_0 \langle \mathbf{x}_t | \mathbf{z}_t \rangle R_t \mathrm{e}^{\mathrm{i}S_t(\mathbf{q}_t,\mathbf{q}_0)/\hbar} \langle \mathbf{z}_0 | \mathbf{x}_0 \rangle$$
$$= \frac{1}{2\pi\hbar}\left(\frac{\gamma}{\pi}\right)^{f/2} \iint \mathrm{d}\mathbf{p}_0 \mathrm{d}\mathbf{q}_0 R_t \mathrm{e}^{\mathrm{i}\phi_t(\mathbf{q}_0,\mathbf{p}_0)/\hbar}, \tag{8.93}$$

where

$$\phi_t(\mathbf{q}_0,\mathbf{p}_{0.}) = \mathbf{p}_t\cdot(\mathbf{x}_t-\mathbf{q}_t) - \mathbf{p}_0\cdot(\mathbf{x}_0-\mathbf{q}_0) + S_t(\mathbf{q}_t,\mathbf{q}_0) + \frac{\mathrm{i}\hbar\gamma}{2}[(\mathbf{x}_t-\mathbf{q}_t)^2 + (\mathbf{x}_0-\mathbf{q}_0)^2], \tag{8.94}$$

and the aim is to choose a prefactor R_t such that stationary phase evaluation of the integral recovers the Van Vleck form.

The relevant stationary phase conditions

$$\begin{aligned}
\left(\frac{\partial S_t}{\partial \mathbf{q}_0}\right)_{\mathbf{q}_t} &= \left(\frac{\partial \mathbf{p}_t}{\partial \mathbf{q}_0}\right)_{\mathbf{q}_t}(\mathbf{x}_t - \mathbf{q}_t) - \left[\left(\frac{\partial \mathbf{p}_0}{\partial \mathbf{q}_0}\right)_{\mathbf{q}_t} + \mathrm{i}\hbar\gamma\right](\mathbf{x}_0 - \mathbf{q}_0) = 0 \\
\left(\frac{\partial S_t}{\partial \mathbf{q}_t}\right)_{\mathbf{q}_0} &= \left[\left(\frac{\partial \mathbf{p}_t}{\partial \mathbf{q}_t}\right)_{\mathbf{q}_0} - \mathrm{i}\hbar\gamma\right](\mathbf{x}_t - \mathbf{q}_t) - \left(\frac{\partial \mathbf{p}_0}{\partial \mathbf{q}_t}\right)(\mathbf{x}_0 - \mathbf{q}_0) = 0
\end{aligned} \tag{8.95}$$

require that $(\mathbf{q}_0, \mathbf{q}_t) = (\mathbf{x}_0, \mathbf{x}_t)$ and $\phi_t(\mathbf{q}_0, \mathbf{p}_{0.}) = S_t(\mathbf{x}_t, \mathbf{x}_0)$, while the second derivative matrix at this point may be expressed in terms of blocks of the second partial derivative matrix, such as $S_{0t} = \left(\partial^2 S/\partial \mathbf{q}_0 \partial \mathbf{q}_t\right)$, in the form

$$\Phi = \begin{pmatrix} \phi_{00} & \phi_{0t} \\ \phi_{t0} & \phi_{tt} \end{pmatrix} = \begin{pmatrix} -S_{00} + \mathrm{i}\hbar\gamma & -S_{0t} \\ -S_{t0} & -S_{tt} + \mathrm{i}\hbar\gamma \end{pmatrix}. \tag{8.96}$$

The remaining step is to use eqn (F.18) to relate the blocks of the partial derivative matrix, S_{0t} etc., to those of the monodromy matrix

$$\begin{pmatrix} M_{qq} & M_{qp} \\ M_{pq} & M_{pp} \end{pmatrix} = \begin{pmatrix} (\partial \mathbf{q}_t/\partial \mathbf{q}_0)_{\mathbf{p}_0} & (\partial \mathbf{q}_t/\partial \mathbf{p}_0)_{\mathbf{q}_0} \\ (\partial \mathbf{p}_t/\partial \mathbf{q}_0)_{\mathbf{p}_0} & (\partial \mathbf{p}_t/\partial \mathbf{p}_0)_{\mathbf{q}_0} \end{pmatrix}, \tag{8.97}$$

whose properties are discussed in Appendix F. The symmetries, such as $S_{00} = S_{00}^T$, $S_{0t} = S_{t0}^T$, etc., allow several alternative forms, the most convenient of which is

$$\Phi = \begin{pmatrix} \mathrm{i}\hbar\gamma - M_{qq}^T \left(M_{qp}^T\right)^{-1} & -M_{qp}^{-1} \\ -\left(M_{qp}^T\right)^{-1} & \mathrm{i}\hbar\gamma - M_{pp}M_{qp}^{-1} \end{pmatrix}. \tag{8.98}$$

Further simplification is obtained by multiplying on the right by $A^{-1}A$, where A is the block diagonal matrix

$$A = \begin{pmatrix} \left(M_{qp}^T\right)^{-1} & 0 \\ 0 & M_{qp}^{-1} \end{pmatrix}. \tag{8.99}$$

Thus

$$\Phi = \begin{pmatrix} i\hbar\gamma M_{qp}^T - M_{qq}^T & -I \\ -I & i\hbar\gamma M_{qp} - M_{pp} \end{pmatrix} A, \tag{8.100}$$

from which the symplectic properties in eqns (F.12)–(F.14) allow the following identities:

$$\begin{aligned} &\det\Phi \\ &= \det\left[-\hbar^2\gamma^2 M_{qp}M_{qp}^T - i\gamma\hbar M_{qp}M_{qq}^T - i\gamma\hbar M_{pp}M_{qp}^T + M_{pp}M_{qq}^T - I\right]\det A \\ &= \det\left[-\hbar^2\gamma^2 M_{qp}M_{qp}^T - i\gamma\hbar M_{qq}M_{qp}^T - i\gamma\hbar M_{pp}M_{qp}^T + M_{pq}M_{qp}^T\right]\det A \\ &= (-i\hbar\gamma)^f \det\left[M_{qq} + M_{pp} - i\hbar\gamma M_{qp} - (i\hbar\gamma)^{-1}M_{pq}\right](\det M_{qp})^{-1}. \end{aligned} \tag{8.101}$$

Now the stationary phase approximation to eqn (8.93) takes the form

$$\begin{aligned} &\langle \mathbf{x}_t|\, e^{-i\hat{H}_{\mathrm{HK}}t/\hbar}\,|\mathbf{x}_0\rangle \\ &\simeq \frac{1}{2\pi\hbar}\left(\frac{\gamma}{\pi}\right)^{1/2}\left[-\left(\frac{1}{2i\pi\hbar}\right)^{2f}\det\Phi\right]^{-1/2} R_t e^{-iS(\mathbf{x}_0,\mathbf{x}_t)/\hbar}(\det M_{qp})^{-1}, \end{aligned} \tag{8.102}$$

where the factor $(\det M_{qp})^{-1}$ is the Jacobian of the integral variable transformation $(d\mathbf{p}_0, d\mathbf{q}_0) \to (d\mathbf{q}_t, d\mathbf{q}_0)$. Hence on substituting for $\det\Phi$ and recognizing that $M_{qp}^{-1} = -\left(\partial^2 S/\partial\mathbf{q}_0\partial q_t\right)$, it may be verified that eqn (8.102) coincides with the Van Vleck form in eqn (8.89), provided that

$$R_t = \det\left[\frac{1}{2}\left(\left[M_{qq} + M_{pp} - i\gamma\hbar M_{qp} - (i\gamma\hbar)^{-1}M_{pq}\right]\right)\right]^{1/2}. \tag{8.103}$$

Kay (1994a) was the first to derive the result in this way, and other derivations from the Van Vleck expression have also been given. For example Miller (2002) obtains the same result by Filinov (1976) smoothing of the Van Vleck form, which raises questions about the theoretical status of the Herman–Kluk propagator. However, Kay (2006) set the question to rest by deriving eqn (8.93), and (8.103) directly from the Schrödinger equation. The difference between the two forms is therefore a question of representation. The Van Vleck and Herman–Kluk expressions are equivalent semiclassical approximations to the true propagator, presented in a plane wave and a coherent state basis respectively.

As mentioned at the outset the propagator is designed to follow the evolution of an initial wavepacket $|\psi(0)\rangle$ in the form

$$\begin{aligned} |\psi(t)\rangle &= e^{-i\hat{H}_{\mathrm{HK}}t/\hbar}|\psi(0)\rangle \\ &= \frac{1}{(2\pi\hbar)^f}\iint d\mathbf{p}_0 d\mathbf{q}_0 |\mathbf{z}_t\rangle R_t e^{iS_t(q_t\cdot\mathbf{q}_0)/\hbar}\langle\mathbf{z}_0|\psi(0)\rangle \end{aligned} \tag{8.104}$$

where, as in the earlier frozen Gaussian methods, the integral is approximated by a sum over an initial swarm of coherent states $|\mathbf{z}_0\rangle$, chosen to mimic $|\psi(0)\rangle$. The family of trajectories $[\mathbf{q}_t, \mathbf{p}_t]$ is conveniently propagated along with the monodromy matrix elements, $M_{\mu\nu}$, by use of a symplectic integrator given by Brewer et al. (1997). As distinct from Van Vleck propagation, by use of eqn (8.89), there is no question of a root search for relevant initial phase points. Moreover the pre-exponential factor R_t can never vanish identically, due to the constraint $M_{pp}M_{qq}^T - M_{pq}M_{qp}^T = I$ on the monodromy matrix elements. For example the vanishing of R_t in the one-dimensional case would require, according to eqn (8.103), that $M_{qq} = -M_{pp}$ and $M_{pq} = (\hbar\gamma)^2 M_{qp}$, leading to $\det M < 0$. Extension of this proof to the f-dimensional case is given in Appendix C of Kay (1994a). There is, however, a minor difficulty, arising from choice of the branch of the square root in eqn (8.103) when the sign of the real part of R_t changes; the proper choice is the branch on which the imaginary part is continuous. The propagation is completed in a scattering application by projecting $|\psi(t)\rangle$ onto the possible product states in an appropriate asymptotic region. Alternatively in a spectroscopic context a Fourier transform of the autocorrelation function yields the eigenvalues and eigenfunctions along the lines of eqns (8.68) and (8.69).

Applications of this well-defined Herman–Kluk method have been reviewed, for example, by Tannor and Garaschuk (2000) and Kay (2005). The technique works well for scattering applications and for the simulation of regular or weakly chaotic systems. Serious convergence difficulties are, however, encountered in strongly chaotic situations, due to exponential growth in some of the monodromy matrix elements (matched by exponential decay in others, so that $\det(M) = 1$). The difficulty is that the modulus of the prefactor, R_t, becomes very large, and that the action differences ΔS_t between the neighbouring trajectories of any finite swarm become large enough to cause wild fluctuations in the real and imaginary parts of R_t. Chance contributions from a few particular trajectories that happen to lie at a maximum of these fluctuations may therefore dominate the outcome, so that the sum in eqn (8.104) fails to converge as the number of trajectories is increased. Fig. 8.7 shows an example taken from Brewer et al. (1997). The autocorrelation function is well behaved for $t < 60$ time units, but becomes increasingly unreliable at longer times.

Kay (1994b) suggested overcoming this difficulty by simply rejecting trajectories with very large pre-exponents $|R_t|$, although these may amount to as many as 90% of the total in strongly chaotic situations. A more elegant procedure is to apply the following Filinov (1976) smoothing technique, or cellular dynamics algorithm (Heller 1991), which is designed to diminish the contributions to the integral from rapidly oscillatory regions, and to enhance those from regions of stationary phase.

The Filinov method recognizes that the dominant contribution to integrals such as

$$I = \int_{-\infty}^{\infty} \mathrm{d}q \exp[\mathrm{i}\phi(q)] \tag{8.105}$$

come from regions of stationary phase; and that knowledge of the trajectory $[q(t), p(t)]$ and its associated monodromy matrix provides information on the value and first and second derivatives of $\phi(q)$ at every point on the trajectory. To avoid errors in I

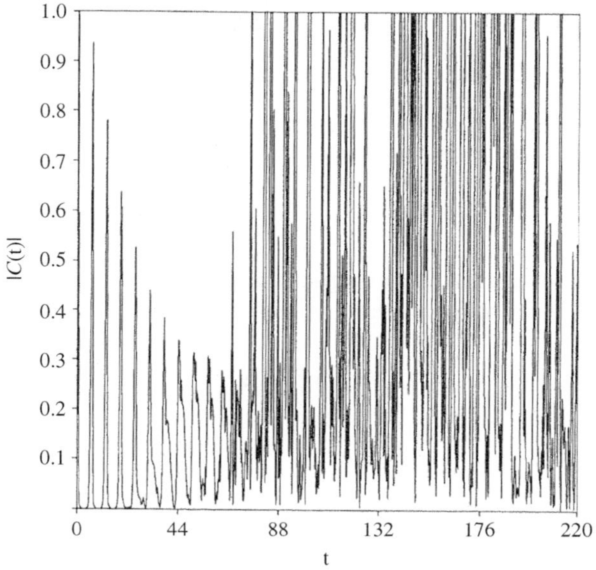

Fig. 8.7 Modulus of the autocorrelation function $|C(t)|$ for a chaotic two-dimensional Henon–Heiles system, calculated from 2047 trajectories. (Reprinted from Walton and Manolopoulos (1996), by permission of Taylor and Francis.)

arising from a relatively sparse Monte Carlo sum over regions of rapidly varying phase, Filinov (1976) proposes inserting unity into the integrand (in the form of a complete Gaussian integral in x), expanding $\phi(q)$ to quadratic terms in $(q-x)$ and performing the integral over q to obtain

$$\begin{aligned} I &= \sqrt{\frac{\alpha}{\pi}} \int_{-\infty}^{\infty} \int_{-\infty}^{\infty} \mathrm{d}q\mathrm{d}x \exp[\mathrm{i}\phi(q)/\hbar - \alpha(q-x)^2] \\ &\simeq \sqrt{\frac{\alpha}{\pi}} \int_{-\infty}^{\infty} \int_{-\infty}^{\infty} \mathrm{d}q\mathrm{d}x \exp[\mathrm{i}\phi(x)/\hbar + \mathrm{i}\phi'(x)(q-x)/\hbar + [\frac{\mathrm{i}}{2\hbar}\phi''(x) - \alpha](q-x)^2] \\ &\simeq \int_{-\infty}^{\infty} \mathrm{d}x \sqrt{\frac{\alpha}{\alpha - \mathrm{i}\phi''(x)/2}} \exp\left(\mathrm{i}\phi(x)/\hbar - \frac{[\phi'(x)^2]}{4\hbar[\alpha\hbar - \mathrm{i}\phi''(x)/2]}\right). \end{aligned} \tag{8.106}$$

It is evident that the approximation reverts to the original form in (8.105) in the limit $\alpha \to \infty$, and that the second term in the exponent damps out rapid oscillations in regions where $\phi'(x)$ is large, for suitably chosen finite values of α. The aim is to ensure a rapid onset of damping, without seriously curtailing the width of the stationary phase region.

Walton and Manolopoulos (1996) have extended this technique to the multidimensional Herman–Kluk autocorrelation integral

$$C(t) = \left\langle \psi_0 | e^{-i\hat{H}_{HK}t/\hbar} | \psi_0 \right\rangle$$
$$\simeq \frac{1}{(2\pi\hbar)^f} \iint d\mathbf{q}_i d\mathbf{p}_i \langle \psi_0 | \mathbf{z}_t \rangle R_{\mathbf{p}_i \mathbf{q}_i t} e^{iS_{\mathbf{p}_i \mathbf{q}_i t}/\hbar} \langle \mathbf{z}_i | \psi_0 \rangle, \tag{8.107}$$

using Gaussian filters with exponents α and β centred on $\mathbf{q}_i$ and $\mathbf{p}_i$ respectively. Note that $\mathbf{q}_i$ and $\mathbf{p}_i$ are chosen as integration variables, to conform with the paper, in which $\mathbf{q}_0$ and $\mathbf{p}_0$ relate to the properties of $|\psi_0\rangle$. The additional integrals over the Gaussian variables, $\mathbf{q}$ and $\mathbf{p}$ say, are approximated by expanding $S_{\mathbf{p}_i \mathbf{q}_i t}$ to quadratic terms, and $\mathbf{q}_t$ and $\mathbf{p}_t$ to linear terms, in $\mathbf{q}$-$\mathbf{q}_i$ and $\mathbf{p}$-$\mathbf{p}_i$. In view of the length and complexity of the derivation, it suffices to quote the final result:

$$C(t) = \left(\frac{1}{2\pi\hbar}\right)^n \iint d\mathbf{q}_i d\mathbf{p}_i \langle \psi_0 | \mathbf{z}_t \rangle R_{\mathbf{p}_i \mathbf{q}_i t} e^{iS_{\mathbf{pq}t}/\hbar} \langle \mathbf{z}_i | \psi_0 \rangle$$
$$\times \left(\frac{[\alpha\beta]^f}{\det[\mathbf{A}_{pqt}]}\right) \exp\left(\frac{1}{4}\mathbf{b}_{pqt}^T \mathbf{A}_{pqt}^{-1} \mathbf{b}_{pqt}\right). \tag{8.108}$$

The components of eqn (8.109) include a $(2n \times 2n)$ positive definite symmetric matrix $\mathbf{A}$ of the form

$$\mathbf{A} = \begin{pmatrix} [\gamma/4 + \alpha]\mathbf{I} & 0 \\ 0 & [(1/4\gamma\hbar^2) + \beta]\mathbf{I} \end{pmatrix} + \mathbf{\Delta A}, \tag{8.109}$$

where $\mathbf{\Delta A}$ depends on products of the blocks of the monodromy matrix $\mathbf{M}$ – see Walton and Manolopoulos (1996). The $(2n \times 1)$ vector $\mathbf{b}_{pqt}$ involves products of the monodromy matrix elements and displacements between the trajectory start point ($\mathbf{q}$,$\mathbf{p}$), the end point ($\mathbf{Q}$,$\mathbf{P}$) and the origin ($\mathbf{q}_0$,$\mathbf{p}_0$) of the initial wavepacket. It was also shown that the pre-exponent in the second line remains continuous even at the classical caustics.

Equation (8.109) shows that the elements of $\mathbf{A}$ are of order α or β, which means the leading term in the second exponent vanishes in the limit $\alpha \to \infty$, $\beta \to \infty$, while $\det[\mathbf{A}_{pqt}] \to [\alpha\beta]^f$. Thus eqn (8.108) reverts to the Herman–Kluk form as expected, because the Filinov Gaussian factors then reduce to delta functions. Walton and Manolopoulos (1996) also show that the Filinov form correctly reduces to $C(0) = 1$ in the limit $t \to 0$, for finite α and β, and that eqn (8.109) is exact for quadratic potentials.

As test of the reliability of eqn (8.108), the authors performed numerical convergence tests for a variety of two-dimensional and three-dimensional Henon–Heiles-based problems previously investigated by Heller (1981). The integral over initial phase space, which was evaluated by a Sobol sampling technique (Press et al., 1992), was found to converge for a sample of typically 2000 or 4000 trajectories in the two-dimensional or three-dimensional case respectively. Since the quadratic parts of the test Hamiltonians were taken as scaled harmonic oscillators, it was natural to take equal values of the two Filinov parameters α and β. The most interesting conclusion was that the relatively small values $\alpha = \beta = 100$ were found to be adequate for the regular or weakly chaotic test cases, while values of order $\alpha = \beta = 10^5$ were required in more strongly chaotic situations, which was rationalized on the basis that the higher parameter values allowed higher resolution in probing the complex dynamics of the chaotic phase space.

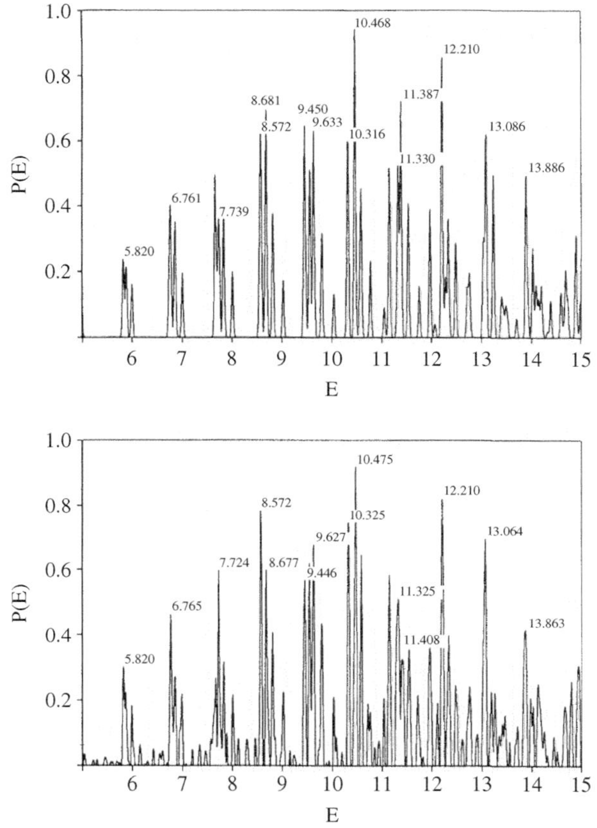

Fig. 8.8 Comparison between the quantum mechanical (top) and modified Herman–Kluk (bottom) power spectra for trajectories initiated in a strongly chaotic region of a two-dimensional Henon–Heiles system. (Reprinted from Walton and Manolopoulos (1996) by permission of Taylor and Francis.)

Fig. 8.8 shows close agreement with the accurate quantum mechanical power spectrum even in the strongly chaotic case – a clear demonstration that the Filinov correction has removed the wild fluctuations in the auto-correlation function seen in Fig. 8.7.

The method was extended in a later paper to the simulation of Ar_nI^- ($n = 2$–6) clusters with up to 15 vibrational degrees of freedom. The computational effort was found to increase approximately linearly with the number of degrees of freedom. Satisfactory agreement was obtained with experimental photo-detachment spectra for $n = 2$–4 (Yourshaw et al. 1996). Ideally one would also compare with accurate quantum mechanical calculations, but there is at present no computational method for obtaining accurately converged excited energy levels for these larger clusters.

8.6 Problems

8.1. Verify by means of the substitutions (Kerner 1958)

$$\xi(x,t) = x - u(t)$$
$$\Psi(x,t) = \phi(\xi,t)\exp[xv(t) + w(t)],$$

that the time-dependent forced harmonic oscillator equation

$$\left(-\frac{\hbar^2}{2\mu}\frac{\partial^2}{\partial x^2} + \frac{1}{2}kx^2 - xF(t)\right)\Psi = \mathrm{i}\hbar\left(\frac{\partial \Psi}{\partial t}\right)_x$$

goes over to

$$\left(-\frac{\hbar^2}{2\mu}\frac{\partial^2}{\partial \xi^2} + \frac{1}{2}k\xi^2\right)\phi = \mathrm{i}\hbar\left(\frac{\partial \phi}{\partial t}\right)_\xi,$$

provided that

$$\mu\frac{\mathrm{d}^2u}{\mathrm{d}t^2} + ku(t) = F(t)$$
$$v(t) = \mathrm{i}\frac{\mu}{\hbar}\frac{\mathrm{d}u}{\mathrm{d}t}$$
$$w(t) = -\frac{\mathrm{i}}{\hbar}\int_0^t \left[\frac{\mu}{2}\left(\frac{\mathrm{d}u}{\mathrm{d}t}\right)^2 - \frac{1}{2}ku^2\right]\mathrm{d}t.$$

8.2. Prove that the Wigner (1932a) transform

$$P_n(x,k) = \frac{1}{2\pi}\int_{-\infty}^{\infty} \exp(\mathrm{i}k\xi)\psi_n(x+\xi/2)\psi_n(x-\xi/2)\mathrm{d}\xi,$$

where $\psi_n(x)$ is any scaled harmonic oscillator function, depends only on $(x^2 + k^2)$. [Hint: use the generating function in eqn (8.51).]

8.3. Verify, by means of the substitutions

$$q = x_n\cos(t+\tau)$$
$$p = -x_n\sin(t+\tau)$$
$$\gamma = -\frac{1}{2}x_n^2\sin(t+\tau)\cos(t+\tau) - \frac{1}{2}(t+\tau)$$

in (8.53), that $|\psi_n(x)|^2$ is independent of τ.

8.4. Show that Herman–Kluk propagator in eqn (8.93) yields the exact form for a harmonic oscillator (Feynman and Hibbs 1965)

$$\left\langle x_t|\mathrm{e}^{-\mathrm{i}Ht/\hbar}|x_0\right\rangle$$
$$= \left(\frac{m\omega}{2\pi\mathrm{i}\hbar\sin\omega t}\right)^{1/2} \exp\left\{\frac{im\omega}{2\hbar\sin\omega t}\left[(x_0^2+x_t^2)\cos\omega t - 2x_0x_t\right]\right\}.$$

[Hints: Note, by using Hamilton's equations, that $p\dot{q} - H = (p\dot{q} + \dot{p}q)/2$. Moreover, the prefactor R_t is designed to reduce to the Van Vleck form in (8.89) for a strictly quadratic phase function.]

8.5. Multidimensional applications of Herman–Kluk theory require knowledge of the following connections between the elements of the monodromy matrix in Appendix F and partial derivatives of the action integral $S\left[\mathbf{q}_1, \mathbf{q}_2(\mathbf{p}_1, \mathbf{q}_1)\right]$ with respect to the initial variables $(\mathbf{p}_1, \mathbf{q}_1)$.

(i) Show that

$$\left(\frac{\partial S}{\partial q_{1i}}\right)_{p_1} = -p_{1i} + \sum_j p_{2j} M_{q_j q_i}, \qquad \left(\frac{\partial S}{\partial p_{1i}}\right)_{q_1} = \sum_j p_{2j} M_{q_j p_i}$$

and, to the extent that derivatives of the monodromy matrix are ignored,

$$\left(\frac{\partial^2 S}{\partial p_{1k}\partial q_{1i}}\right) = -\delta_{ki} + \sum_j M_{p_j p_k} M_{q_j q_i}, \qquad \left(\frac{\partial^2 S}{\partial q_{1k}\partial q_{1i}}\right) = \sum_j M_{p_j q_k} M_{q_j q_i}.$$

(ii) Express the full results in the matrix forms

$$\left(\frac{\partial S}{\partial q_1}\right)^T = \left(-p_1^T + p_2^T M_{qq}\right), \qquad \left(\frac{\partial S}{\partial p_1}\right)^T = p_2^T M_{qp}$$

and

$$\begin{pmatrix} S_{qq} & S_{qp} \\ S_{pq} & S_{pp} \end{pmatrix} = \begin{pmatrix} M_{pq}^T M_{qq} & M_{pq}^T M_{qp} \\ -I + M_{pp}^T M_{qq} & M_{pp}^T M_{qp} \end{pmatrix},$$

where the elements of S_{qp}, for example, are given by $\left(\partial^2 S/\partial q_{1i}\partial p_{1j}\right)$. In addition, M_{pq}^T is the transpose of the block M_{pq}, with elements $\left(M_{pq}^T\right)_{kj} = M_{p_j q_k}$.

9
Atom–atom scattering

Atomic collisions are normally termed elastic or inelastic. The former lead simply to momentum transfer between the collision partners, or equivalently to a change in the direction of their motions; the latter involve a simultaneous transfer of energy between internal motion and translation. This chapter lays the foundations of the theory in the elastic context and indicates, in the final section, how it generalizes to the inelastic case. The theory stems from the seminal papers of Ford and Wheeler (1959) which precipitated the recent revival of interest in semiclassical mechanics. The review by Berry and Mount (1972) was also very influential.

The chapter begins with a brief introduction to the classical and quantum mechanical theories, which are constructed around assumed knowledge of the classical deflection function, $\chi(E, l)$, and quantum phase shift, $\eta(E, l)$, respectively. It is then shown how the central semiclassical identity

$$\chi(E, l) = 2(\partial\eta/\partial l) \tag{9.1}$$

links the two theories. The resulting semiclassical approach is applied to the interpretation of a variety of experimentally observable interference phenomena including rainbow and glory scattering, diffraction and exchange oscillations. Resonance effects are also discussed both in a direct manner and by means of a complex angular momentum approach which has been reviewed by Connor (1980). Finally, the semiclassical origins of the Stückelberg (1932) interferences associated with inelastic scattering are discussed in Section 9.4. Comprehensive reviews of the elastic scattering theory are given by Bernstein (1966) and Pauly (1979), while the two-state non-adiabatic transition problem is discussed in more detail by Child (1979).

9.1 The classical and quantum mechanical limits

The classical theory of elastic scattering is based on assumed knowledge of the classical deflections, χ, experienced by different trajectories, each specified by an initial velocity, v, and impact parameter, b, as shown in Fig. 9.1(a). This motion may be followed by noting that the local angular velocity, $\dot{\varphi}$, at a point along the trajectory may be expressed in terms of the angular momentum by the equation

$$L = mvb = mR^2\dot{\varphi}, \tag{9.2}$$

while the radial velocity, $\dot{R}$, is determined by the energy conservation equation

$$E = \frac{1}{2}mv^2 = \frac{1}{2}m\dot{R}^2 + V_l(R), \tag{9.3}$$

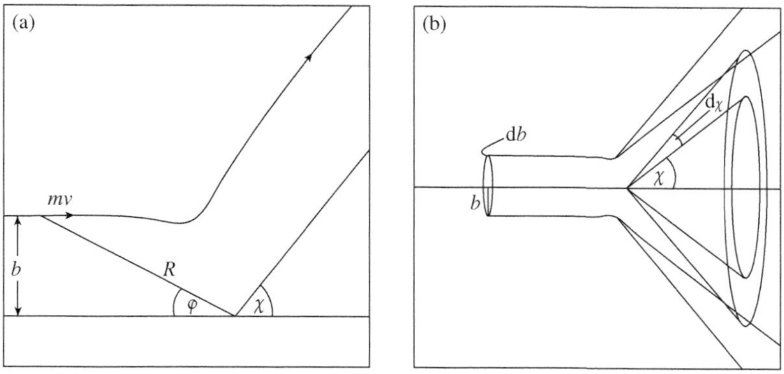

Fig. 9.1 (a) Notation employed, where b is the impact parameter; (b) surfaces and solid angles defined by scattering at impact parameters b and $b + \mathrm{d}b$.

where

$$V_l(R) = V(R) + (l + 1/2)^2\, \hbar^2/2mR^2, \tag{9.4}$$

with l related to the classical angular momentum by

$$L = (l + 1/2)\, \hbar. \tag{9.5}$$

Thus l is the semiclassical angular momentum quantum number, treated below as a continuous variable, and $V_l(R)$ is termed the centrifugally corrected potential function, whose typical form is shown for representative l values in Fig. 9.2.

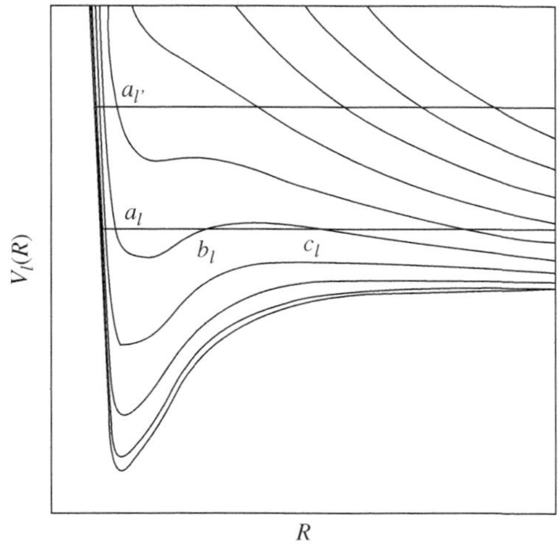

Fig. 9.2 Centrifugally corrected potential curves, $V_1(R)$.

It follows, on combining eqns (9.2)–(9.5) that

$$\frac{\mathrm{d}\varphi}{\mathrm{d}R} = \frac{\mathrm{d}\varphi}{\mathrm{d}t} \Big/ \frac{\mathrm{d}R}{\mathrm{d}t} = \pm \frac{1}{R^2} \frac{(l+1/2)\,\hbar}{\{2m[E - V_l(R)]\}^{1/2}}, \tag{9.6}$$

where the positive and negative signs apply to the outward and inward parts of the motion respectively. Integration of (9.6) leads to the following formula for the classical deflection, $\chi(E,l)$, at energy E and angular quantum number l:

$$\begin{aligned} \chi(E,l) &= \pi - \varphi(\infty) \\ &= \pi - \frac{(2l+1)\hbar}{(2m)^{1/2}} \int_{a_l}^{\infty} \frac{\mathrm{d}R}{R^2[E - V_l(R)]^{1/2}} \\ &= \pi - (2l+1) \int_{a_l}^{\infty} \frac{\mathrm{d}R}{R^2[k^2 - U(R) - (l+1/2)^2/R^2]^{1/2}}, \end{aligned} \tag{9.7}$$

where

$$k^2 = 2mE/\hbar^2 \tag{9.8}$$

and

$$U(R) = 2mV(R)/\hbar^2. \tag{9.9}$$

The notation in eqn (9.7) assumes that the energy E lies above the centrifugal barrier maximum in Fig. 9.2; otherwise the lower integration limit must be replaced by the outer turning point c_l. The energy has been replaced by k^2 in the final line in order to facilitate comparison with the quantum mechanical theory.

The classical theory itself is completed by noting that particles which enter an annulus of area $\mathrm{d}\sigma = 2\pi b\mathrm{d}b$, appropriate to an impact parameter range $b \rightarrow b + \mathrm{d}b$ in Fig. 9.1(b), will be scattered into an element of solid angle $\mathrm{d}\Omega = 2\pi|\sin\chi\,\mathrm{d}\chi|$. Hence the angular intensity per unit solid angle is given by

$$\frac{\mathrm{d}\sigma}{\mathrm{d}\Omega} = \sum \frac{b\mathrm{d}b}{|\sin\chi\,\mathrm{d}\chi|}, \tag{9.10}$$

where the sum is included to allow for the possibility that scattering at several different impact parameters leads to the same physical angle $\theta = |\chi|$, defined to lie in the range $0 < \theta < \pi$. Alternatively eqn (9.10) may be expressed as

$$\frac{\mathrm{d}\sigma}{\mathrm{d}\Omega} = \frac{1}{k^2} \sum \frac{(l+1/2)}{\sin\theta|\mathrm{d}\chi/\mathrm{d}l|}, \tag{9.11}$$

because, according to eqns (9.2), (9.3) and (9.8),

$$l + 1/2 = kb. \tag{9.12}$$

The final step in this brief classical exposition is to define the integral cross-section

$$\sigma(E) = 2\pi \int_0^{\pi} \frac{\mathrm{d}\sigma}{\mathrm{d}\Omega} \sin\theta\,\mathrm{d}\theta, \tag{9.13}$$

so that both $d\sigma/d\Omega$ and $\sigma(E)$ are completely determined by knowledge of the deflection function $\chi(E, l)$. Discussion of the implications of eqns (9.10) and (9.13) for the nature of the classical scattering is deferred until later in the section, when comparison can be made with the quantum mechanical theory. Readers seeking a more detailed exposition of the classical approach are referred to Child (1974b).

Turning to the quantum mechanical theory, the combined effect of the scattering at all angular momenta (or impact parameters) is taken into account by expressing the wavefunction as

$$\Psi(\mathbf{R}) = \frac{1}{R}\sum_{l=0}^{\infty} A_l \psi_l(R) P_l(\cos\theta). \tag{9.14}$$

Each partial wave, $\psi_l(R)$, then satisfies a radial equation

$$\left(\frac{d^2}{dR^2} + k^2 - U(R) - \frac{l(l+1)}{R^2}\right)\psi_l(R) = 0, \tag{9.15}$$

whose regular ($\psi_l(0) = 0$) solution is conventionally represented in the asymptotic region as (Child 1974b)

$$\psi_l(R) \overset{R\to\infty}{\approx} \sin(kR - l\pi/2 + \eta_l), \tag{9.16}$$

where the term $-l\pi/2$ is attributed to the centrifugal term in eqn (9.15) and the phase shift η_l to the effect of the reduced potential function $U(R)$, defined by eqn (9.9).

The second essential step is to choose the coefficients A_l in eqn (9.14) in such a way that

$$\Psi(\mathbf{R}) \overset{R\to\infty}{\approx} \exp(ikZ) + f(\theta)\exp(ikR)/R, \tag{9.17}$$

which corresponds to an incoming plane wave plus a purely outgoing scattered term. The appropriate choice is found to be (Taylor 1972; Child 1974b)

$$A_l = (2l+1)k^{-1}\exp[i(l\pi/2 - \eta_l)]. \tag{9.18}$$

Rearrangement of eqns (9.14), (9.16), (9.17) and (9.18) then leads to the following expression for the scattering amplitude:

$$f(\theta) = \frac{1}{2ik}\sum_{l=0}^{\infty}(2l+1)[\exp(2i\eta_l) - 1]P_l(\cos\theta), \tag{9.19}$$

from which the differential and total cross-sections may be derived in the forms

$$\frac{d\sigma}{d\Omega} = |f(\theta)|^2 \tag{9.20}$$

and

$$\sigma = 2\pi\int_0^{\pi}\frac{d\sigma}{d\Omega}\sin\theta d\theta = \frac{4\pi}{k^2}\sum_{l=0}^{\infty}(2l+1)\sin^2\eta_l. \tag{9.21}$$

Equations (9.19)–(9.21) assume that the collision partners are physically distinguishable. Mott and Massey (1965) show that the complications due to nuclear spin statistics, if the two particles are identical, are taken into account by generalizing (9.19) and (9.21) to read

$$f_{\pm}(\theta) = \frac{1}{2ik} \sum_{l=0}^{\infty} w_l^{\pm}(2l+1)[\exp(2i\eta_l) - 1]P_l(\cos\theta) \tag{9.22}$$

and

$$\sigma_{\pm} = \frac{4\pi}{k^2} \sum_{l=0}^{\infty} w_l^{\pm}(2l+1)\sin^2\eta_l. \tag{9.23}$$

The upper and lower signs in these two equations apply to Bose–Einstein and Fermi–Dirac statistics, respectively. The weighting factors are given by

$$w_l^{\pm} = 1 \pm (-1)^l, \tag{9.24}$$

which means that the sums in (9.22) and (9.23) are taken over either only even or only odd l values. This completes our résumé of the quantum mechanical theory.

The semiclassical connection between the above classical and quantal viewpoints will be shown below to rest on a close functional relationship between the deflection function, $\chi(E,\, l)$, and the JWKB approximation to the phase shift, the latter being given, according to (2.54) and (9.16), by

$$\eta(E, l) = \lim_{R\to\infty}\left[\int_{a_l}^{R} k_l(R)\mathrm{d}R - kR + \left(l+\frac{1}{2}\right)\frac{\pi}{2}\right], \tag{9.25}$$

where $k_l(\mathrm{R})$ is the Langer-corrected wavenumber function,

$$k_l(R) = \left[k^2 - U(R) - \frac{(l+1/2)^2}{R^2}\right]^{1/2}. \tag{9.26}$$

Hence, on comparing with (9.7),

$$\chi(E, l) = 2(\partial\eta/\partial l). \tag{9.27}$$

The implications of this result are explored at length in later sections of the chapter.

Before proceeding to details it is convenient to note some of the major differences between the classical and quantum mechanical pictures. Figure 9.3 shows a comparison between the classical and quantum mechanical differential cross-sections for scattering under a Lennard–Jones potential,

$$V(R) = 4\varepsilon[(R_0/R)^{12} - (R_0/R)^6], \tag{9.28}$$

in circumstances roughly appropriate to the thermal scattering of two Ar atoms. Here the behaviour of the classical cross-section may be understood in the light of eqn (9.11)

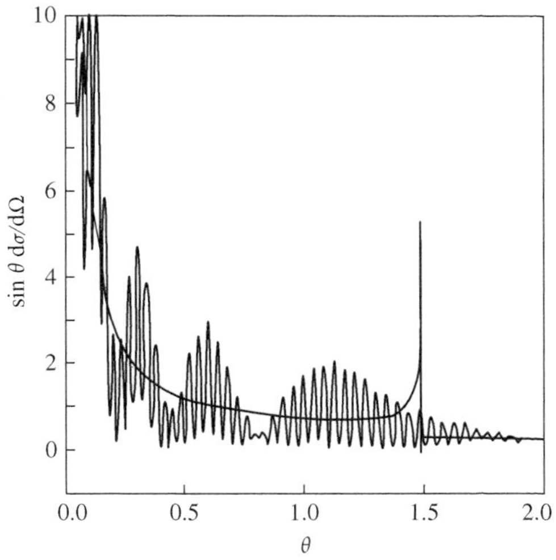

Fig. 9.3 Quantum mechanical and classical differential cross-sections for the Lennard–Jones potential (9.28) at a reduced energy $E/\varepsilon = 1.6$ and $\lambda_0 = (2mE)^{1/2}R_0/\hbar = 60$. The cross-section is given in units of R_0^2. Note the 'fast' diffraction oscillations imposed on slower 'rainbow' fluctuations, and the classical singularity at the rainbow angle $\theta_\mathrm{r} \simeq 1.5$.

by reference to the shape of the deflection function shown in Fig. 9.4; it shows a singularity at the 'rainbow' angle $\theta_\mathrm{r} = -\chi_\mathrm{r}$ where $\chi(E,l)$ passes through a minimum with respect to l and also diverges as $\theta \to 0$ because $\chi \to 0$ and $\partial\chi/\partial l \to 0$ as $l \to \infty$. On the other hand the quantum mechanical cross-section oscillates around the classical one (in a way that is analysed in detail in Section 9.2), but remains finite at all θ, even at $\theta = 0$.

The nature of the classical singularity at $\theta = 0$ also has important consequences for the total cross-section, $\sigma(E)$, which is infinite in classical theory for any potential with long-range form $V(R) \approx -C_n/R^n$, while the quantum mechanical cross-section remains finite for $n > 2$. To demonstrate this difference, Bernstein (1966) shows that

$$\eta(E,l) \overset{l\to\infty}{\approx} a_n l^{1-n}, \tag{9.29}$$

so that, in the light of (9.27),

$$\chi(E,l) \overset{l\to\infty}{\approx} -2(n-1)a_n l^{-n}, \tag{9.30}$$

where

$$a_n = \sqrt{\pi}\left[\Gamma\left(\tfrac{1}{2}n - \tfrac{1}{2}\right)/\Gamma\left(\tfrac{1}{2}n\right)\right](C_n/4E)k^n. \tag{9.31}$$

Hence, according to eqns (9.11) and (9.30),

$$\sin\theta(\mathrm{d}\sigma/\mathrm{d}\Omega)_\mathrm{cl} = \text{constant} \times \theta^{-1-1/n}, \tag{9.32}$$

Turning to the mathematical analysis of $f(\theta)$, it is convenient to note that eqn (9.19) reduces to

$$f(\theta) = \frac{1}{2ik} \sum_{l=0}^{\infty} (2l+1) \exp(2i\eta_l) P_l(\cos\theta), \tag{9.38}$$

for $\theta \neq 0$, because (Abramowitz and Stegun 1965)

$$\sum_{l=0}^{\infty} (2l+1) P_l(\cos\theta) = 0 \quad \text{for } \theta \neq 0. \tag{9.39}$$

Physically this expresses the fact that the term -1, which is lost in passing from (9.19) to (9.38), arises from the plane-wave contribution in (9.19), which has zero amplitude except in the forward direction, $\theta = 0$. Secondly it may be noted (see problem 2.2) that

$$P_l(\cos\theta) \simeq \left(\frac{2}{\pi\lambda\sin\theta}\right)^{1/2} \cos(\lambda\theta - \pi/4), \tag{9.40}$$

where

$$\lambda = l + 1/2, \tag{9.41}$$

provided that $|\lambda \sin\theta| \gg 1$. Finally it may be assumed that the number of significant terms in (9.38) is sufficiently large to justify replacement of the sum by an integral (Ford and Wheeler 1959; Landau and Lifshitz 1965); the range $0 < l < 100$ is quite typical under semiclassical conditions. The upshot, on combining these approximations, is that

$$\begin{aligned} f(\theta) &\simeq \frac{1}{2ik} \int_0^{\infty} 2\lambda \left(\frac{2}{\pi\lambda\sin\theta}\right)^{1/2} \exp[2i\eta(\lambda)] \cos(\lambda\theta - \pi/4) d\lambda \\ &\simeq \int_0^{\infty} g(\theta,\lambda)\{\exp[i\phi^+(\theta,\lambda)] + \exp[i\phi^-(\theta,\lambda)]\} d\lambda, \end{aligned} \tag{9.42}$$

where

$$g(\theta,\lambda) = -\frac{i}{k}\left(\frac{\lambda}{2\pi\sin\theta}\right)^{1/2} \tag{9.43}$$

and

$$\phi^{\pm}(\theta,\lambda) = 2\eta(\lambda) \pm (\lambda\theta - \pi/4). \tag{9.44}$$

Note that some authors (e.g. Pauly (1979) and Connor (1980)) first transform (9.38) to an exact (but infinite) sum of integrals with respect to λ, and then argue that a single term equivalent to (9.42) dominates in the semiclassical case. The present approach is adopted here to avoid undue notational complexity. The glory scattering case $\theta \to 0$, for which (9.40) is invalid, is covered by problems 9.2–9.4.

A first approximation to the integral in eqn (9.42) may be obtained by assuming that the phase terms $\phi_{\pm}(\theta,\lambda)$ are large and rapidly varying with λ. Hence the form of the integrand will lead to constructive interference, at any chosen scattering angle θ, only around particular λ values at which $\partial\phi_\pm/\partial\lambda = 0$. On combining this stationary phase condition with the semi-classical identity (9.27), between $2(\mathrm{d}\eta/\mathrm{d}\lambda)$ and the classical deflection $\chi(\lambda)$, one finds that

$$\chi(\lambda) = 2(\mathrm{d}\eta/\mathrm{d}\lambda) = \mp\theta. \tag{9.45}$$

In other words the dominant contribution to the scattering comes from λ values at which the classical trajectory would reach the chosen angle θ.

Figure 9.5 shows that there are typically three such λ values (labelled λ_a, λ_b and λ_c) if $\theta < \theta_\mathrm{r}$ and one value, λ_c, if $\theta > \theta_\mathrm{r}$; the a and b trajectories, for which $\chi = -\theta$, are shown in Fig. 9.6 as passing around one side of the scattering centre while the c branch trajectory passes around the other. See Section 11.2 for further discussion of this nearside–farside behaviour. There is also a special 'glory' scattering trajectory, labelled by λ_g in Fig. 9.5, which suffers no net deflection; the corresponding impact parameter

$$b_\mathrm{g} = k\lambda_\mathrm{g} \tag{9.46}$$

defines an effective hard-sphere radius for the scattering centre at any given energy, such that trajectories with $\lambda < \lambda_\mathrm{g}$ reach a given θ around one side and those with $\lambda > \lambda_\mathrm{g}$ around the other. This classical picture explains the existence of two distinct types of contribution to the interference pattern in Fig. 9.3. Since the a- and b-type trajectories follow quite similar paths, their associated phase difference will vary slowly with θ giving rise to the widely spaced 'rainbow' oscillatory structure. The phase

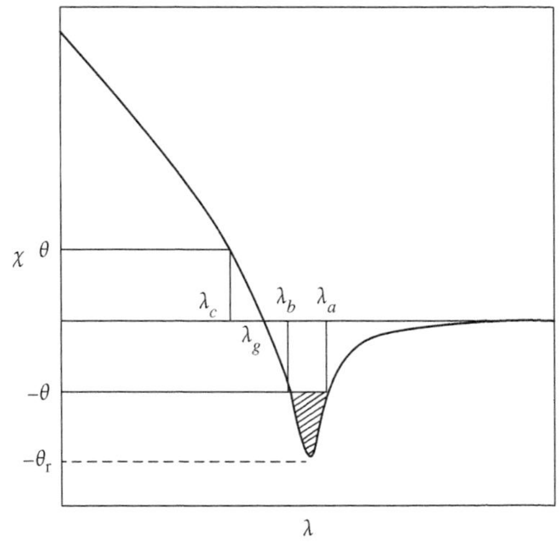

Fig. 9.5 The classical deflection function, drawn to illustrate the rainbow angle, θ_r, the glory angular momentum λ_g, and typically three angular momenta λ_a, λ_b and λ_c leading to scattering at $|\chi| = \theta$ for $\theta < \theta_\mathrm{r}$. The shaded area determines the phase of the rainbow oscillations in Fig. 9.3.

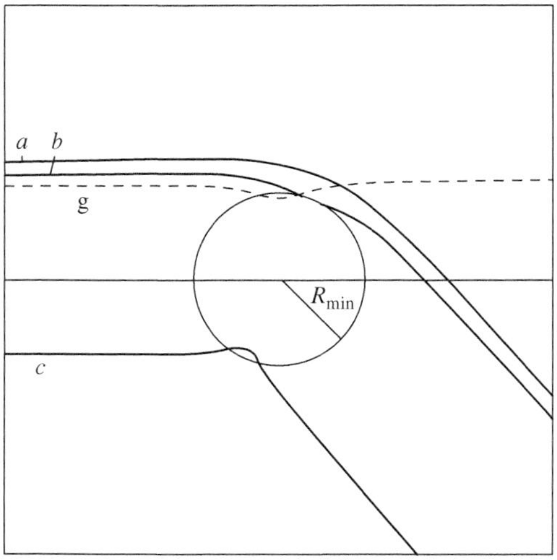

Fig. 9.6 Paths of the three trajectories leading to the same scattering angle θ. The dashed line labelled g is the glory trajectory, and the circle denotes the radius of the potential energy minimum.

difference between the c branch wave and the mean $(a,\ b)$ resultant varies much more rapidly with θ, however, giving rise to the much faster 'diffraction' oscillations, which are shown below to have a spacing determined by the glory impact parameter b_g.

Primitive semiclassical approximation

To put this discussion on a mathematical footing it is convenient first to evaluate the integral in (9.42) by stationary phase, thereby ignoring possible contributions due to coalescence of the a and b branch trajectories as $\theta \to \theta_r$. To do this the phase terms are expanded about the points $\lambda_\nu, \nu = a, b$ or c, in the obvious shorthand notation

$$\phi_\nu(\lambda) \simeq \phi_\nu + \frac{1}{2}\phi_\nu''(\lambda - \lambda_\nu)^2, \tag{9.47}$$

where, according to eqns (9.44) and (9.45),

$$\begin{aligned} \phi_\nu &= 2\eta_\nu(\lambda_\nu) + \lambda_\nu\theta - \pi/4, \quad \nu = a \text{ or } b \\ &= 2\eta_\nu(\lambda_\nu) - \lambda_c\theta + \pi/4, \quad \nu = c \end{aligned} \tag{9.48}$$

and

$$\phi_\nu'' = (\mathrm{d}\chi/\mathrm{d}\lambda)_{\lambda=\lambda_\nu}. \tag{9.49}$$

The identity (Dwight 1961)

$$\int_{-\infty}^{\infty} \exp(\pm\, \mathrm{i}\alpha^2 x^2)\mathrm{d}x = \frac{\sqrt{\pi}}{\alpha} \exp\left(\pm\frac{\mathrm{i}\pi}{4}\right) \tag{9.50}$$

then yields the primitive semiclassical (PSC) result

$$f_{\rm PSC}(\theta) \simeq \sum_{\nu=a,b,c} [I_\nu(\theta)]^{1/2} \exp[{\rm i}\phi_\nu(\theta) + {\rm i}\delta_\nu], \tag{9.51}$$

where

$$I_\nu(\theta) = \lambda_\nu / (k^2 \sin\theta |{\rm d}\chi/{\rm d}\lambda|) \tag{9.52}$$

and

$$\delta_a = \pi/4, \quad \delta_b = \delta_c = -\pi/4, \tag{9.53}$$

because ${\rm d}\chi/{\rm d}\lambda$ is positive on the a branch and negative on the b and c branches.

The interesting feature of eqn (9.51) is that the term $I_\nu(\theta)$ is identical with the classical contribution to the classical differential cross-section, as given by (9.11). Hence

$$\begin{aligned}\left(\frac{{\rm d}\sigma}{{\rm d}\Omega}\right)_{\rm PSC} &= |f_{\rm PSC}(\theta)|^2 \\ &= \sum_\nu I_\nu + 2\sum_{\nu' \neq \nu} (I_\nu I_{\nu'})^{1/2} \cos\,(\phi_\nu + \delta_\nu - \phi_{\nu'} - \delta_{\nu'}) \\ &= \left(\frac{{\rm d}\sigma}{{\rm d}\Omega}\right)_{\rm cl} + \text{interference terms}.\end{aligned} \tag{9.54}$$

It is seen from Fig. 9.7 that the interference pattern does indeed fluctuate around the classical form for $\theta < \theta_{\rm r}$, and that the primitive semiclassical form quantitatively reproduces the pattern of slow 'rainbow' undulations with superimposed faster 'diffraction' oscillations, provided that $\theta << \theta_{\rm r}$. There is, however, an inevitable spurious classical singularity at $\theta_{\rm r}$, plus a less drastic defect in the absence of diffraction oscillations for $\theta > \theta_{\rm r}$.

Uniform Airy approximation

Both these deficiencies are remedied by the following uniform Airy approximation, due to Berry (1966), which takes proper account of the coalescence between the stationary phase points λ_a and λ_b as $\theta \to \theta_{\rm r}$. The mathematical situation is similar to that encountered in Section 2.2, in relation to the form of the wavefunction at a classical turning point, and details of the derivation are given in a general context in Appendix B. The salient points for the present discussion are first that the Airy functional form is dictated by the occurrence of two stationary phase points of $\phi^+(\theta, \lambda)$ with respect to λ. Secondly, the resulting equation for $f_{\rm uni}(\theta)$ involves precisely the same stationary phase quantities I_ν and ϕ_ν as those required to calculate $f_{\rm PSC}(\theta)$, but in the following more sophisticated form:

$$f_{\rm uni}(\theta) = f_{\rm r}(\theta) + f_{\rm c}(\theta), \tag{9.55}$$

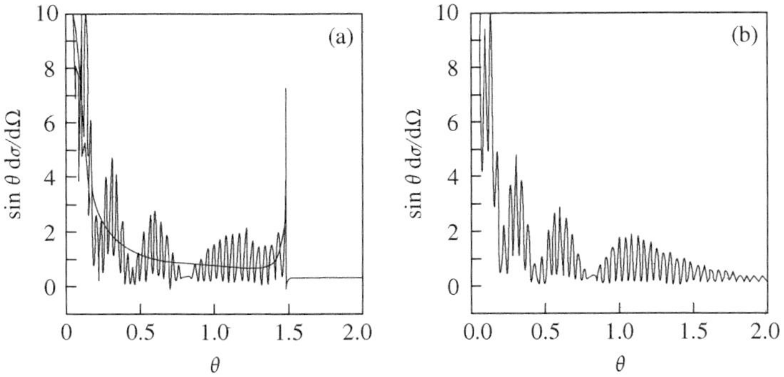

Fig. 9.7 (a) Primitive semiclassical and (b) uniform Airy approximations to the differential cross-section in Fig. 9.3. The Airy approximation is graphically indistinguishable from the partial-wave sum, except for $\theta \gg \theta_r$ where the analytical continuation becomes inaccurate.

where $f_c(\theta)$ is the c branch contribution to (9.51) and $f_r(\theta)$ is the rainbow term

$$\begin{aligned} f_r(\theta) = \quad & \sqrt{\pi} e^{i\bar{\phi}} [(I_a^{1/2} + I_b^{1/2})\zeta^{1/4} Ai(-\zeta) \\ & -i(I_a^{1/2} - I_b^{1/2})\zeta^{-1/4} Ai'(-\zeta)], \end{aligned} \tag{9.56}$$

with

$$\bar{\phi} = \frac{1}{2}(\phi_a + \phi_b) \tag{9.57}$$

and

$$\zeta = \left[\frac{3}{4}(\phi_b - \phi_a)\right]^{2/3}. \tag{9.58}$$

Note, by virtue of eqns (9.27) and (9.48), that the phase term $(\phi_b - \phi_a)$ in (9.58) has a pictorial interpretation as the shaded area in Fig. 9.5;

$$\begin{aligned} \phi_b - \phi_a \quad & = 2\eta_b - 2\eta_a + (\lambda_b - \lambda_a)\theta \\ & = \int_{\lambda_b}^{\lambda_a} [-\theta - \chi(\lambda)] d\lambda, \end{aligned} \tag{9.59}$$

which carries echoes of the general phase space interpretation of semiclassical interference seen elsewhere in Sections 2.2, 2.3, 3.5, 5.3, 9.4 and Appendix B.

One strength of the uniform approximation is that it is guaranteed by the asymptotic identities (Abramowitz and Stegun 1965)

$$\begin{aligned} Ai(-\zeta) & \overset{\zeta \gg 1}{\approx} \pi^{-1/2}\zeta^{-1/4} \sin\left(\tfrac{2}{3}\zeta^{3/2} + \pi/4\right) \\ Ai'(-\zeta) & \overset{\zeta \gg 1}{\approx} -\pi^{-1/2}\zeta^{1/4} \cos\left(\tfrac{2}{3}\zeta^{3/2} + \pi/4\right) \end{aligned} \tag{9.60}$$

to go over to the primitive semiclassical form for $\theta \ll \theta_r$. Secondly, the divergence of $(I_a^{1/2} + I_b^{1/2})$ at the rainbow angle is exactly compensated by the vanishing of $\zeta^{1/4}$ in (9.56) because $\phi_a - \phi_b \to 0$ as $\theta \to \theta_r$. Similarly the divergence of $\zeta^{-1/4}$ is exactly cancelled by the disappearance of $I_a^{1/2} - I_b^{1/2}$, and these cancellations are in fact so complete that $|f_{\rm uni}(\theta)|^2$ is graphically indistinguishable from the exact partial-wave sum shown in Fig. 9.3. Another virtue of eqn (9.56) is that $f_r(\theta)$ can be analytically continued to cover angles $\theta > \theta_r$, which are classically inaccessible to the a and b branches, in the sense that eqn (9.45) has no real solutions. To see this suppose that $\chi(\lambda)$ is expanded in the form

$$\chi(\lambda) \simeq -\theta_r + q(\lambda - \lambda_r)^2, \tag{9.61}$$

so that the a and b branch solutions of (9.45) are given by

$$\lambda_{a,b} = \lambda_r \pm [(\theta_r - \theta)/q]^{1/2}, \tag{9.62}$$

which are real for $\theta < \theta_r$ and complex for $\theta > \theta_r$. It follows, on inverting (9.27), that

$$\begin{aligned} \phi_\nu &= 2\eta_\nu + \lambda_\nu\theta - \pi/4 \\ &= 2\eta_r + \int_{\lambda_r}^{\lambda_\nu} \chi(\lambda)\mathrm{d}\lambda + \lambda_\nu\theta - \pi/4 \\ &= 2\eta_r + \lambda_r\theta \pm \tfrac{2}{3}q^{-1/2}(\theta_r - \theta)^{3/2} - \pi/4, \end{aligned} \tag{9.63}$$

where the upper and lower signs in the final line apply to the b and a branches respectively. Consequently

$$\begin{aligned} \bar{\phi} &= 2\eta_r + \lambda_r\theta - \pi/4 \\ \phi_b - \phi_a &= \frac{4}{3}q^{-1/2}(\theta_r - \theta)^{3/2}, \end{aligned} \tag{9.64}$$

while in light of (9.52), (9.58) and (9.64),

$$\begin{aligned} \left(I_a^{1/2} + I_b^{1/2}\right) &= (2\lambda_r/\sin\theta)^{1/2}q^{-1/3}k^{-1} \\ \left(I_a^{1/2} - I_b^{1/2}\right) &= 0 \\ \zeta &= q^{-1/3}(\theta_r - \theta). \end{aligned} \tag{9.65}$$

Equations (9.65) apply only in the immediate vicinity of θ_r, where the quadratic approximation (9.61) is valid, but they allow analytic continuation of (9.56) to the dark side of the rainbow, $\theta > \theta_r$, where the contributing angular momenta λ_a and λ_b are complex and the exponential decrease of the Airy functions for $\zeta < 0$ leads, by interference with the $f_c(\theta)$ branch, to the oscillatory tail of $|f(\theta)|^2$ in Fig. 9.7(b).

Glory scattering

Additional problems arise from divergence of the inverse sine functions in the forward direction, $\theta = 0$. The behaviour of the a branch contribution has been discussed in eqns (9.29)–(9.33). The behaviour of the 'glory' contribution

$$f_g(\theta) = f_b(\theta) + f_c(\theta) \tag{9.66}$$

arising from interference between the b and c branches was first discussed by Berry (1975) and later by Connor (2004). Details of the derivation are set as an exercise in problems 9.2–9.4. The result takes the form

$$\begin{aligned} f_g(\theta) = -\mathrm{e}^{\mathrm{i}A(\theta)+\mathrm{i}\pi/4}\left[\frac{\pi\zeta(\theta)}{2}\right]^{1/2} & \left\{\left[I_c^{1/2}(\theta) + I_b^{1/2}(\theta)\right] J_0[\zeta(\theta)]\right. \\ & \left. +\mathrm{i}\left[I_c^{1/2}(\theta) - I_b^{1/2}(\theta)\right] J_1[\zeta(\theta)]\right\}, \end{aligned} \tag{9.67}$$

where $J_0(\zeta)$ and $J_0(\zeta)$ are Bessel functions (Abramowitz and Stegun 1965) and

$$\begin{aligned} A(\theta) &= \frac{1}{2}[2\eta_b + 2\eta_c + \frac{1}{2}(\lambda_b - \lambda_c)\theta], \\ \zeta(\theta) &= \frac{1}{2}[2\eta_b - 2\eta_c + \frac{1}{2}(\lambda_b + \lambda_c)\theta]. \end{aligned} \tag{9.68}$$

This means that $\zeta(\theta) \to \lambda_g\theta$ as $\theta \to 0$, because $\lambda_b \to \lambda_c \to \lambda_g$ and $\eta_b - \eta_c \to 0$. Consequently the products $\zeta(\theta)I_{b,c}(\theta)$ remain finite at $\theta = 0$.

Classically forbidden contributions

So far it has been assumed that the number of partial waves that contribute to the scattering is large, in which case the picture may be loosely described as classical plus simple quantum mechanical interference. The next stage is to move to a more strongly quantal situation, by decreasing the assumed reduced mass of the collision partners. Suppose for illustrative purposes that the scattering potential is represented in the two-parameter form

$$V(R) = \varepsilon W(R/R_0), \tag{9.69}$$

with strength ε and range R_0. Equations (9.7) and (9.25) for the deflection function $\chi(\lambda)$ and phase shift $\eta(\lambda)$ respectively may then be expressed in the scaled forms

$$\chi(\lambda) = \tilde{\chi}(\tilde{\lambda}) = \pi - 2\tilde{\lambda}\int_{\tilde{\alpha}_\lambda}^{\infty} \frac{\mathrm{d}\rho}{\rho^2[1 - W(\rho)/K - \tilde{\lambda}^2/\rho^2]^{1/2}} \tag{9.70}$$

and

$$\eta(\lambda) = \lambda_0\tilde{\eta}(\tilde{\lambda}) = \lim_{\rho\to\infty} \lambda_0\left(\int_{\tilde{\alpha}_\lambda}^{\rho} [1 - W(\rho)/K - \tilde{\lambda}^2/\rho^2]^{1/2}\mathrm{d}\rho - \rho - \tilde{\lambda}\pi/2\right), \tag{9.71}$$

where

$$\rho = R/R_0, \quad K = E/\varepsilon, \quad \tilde{\lambda} = \lambda/\lambda_0, \tag{9.72}$$

and

$$\lambda_0 = kR_0 = K^{1/2}(2m\varepsilon)^{1/2}R_0/\hbar. \tag{9.73}$$

Equation (9.70) shows that the classical deflection function, $\tilde{\chi}(\tilde{\lambda})$, takes the same scaled form at given reduced energy K, regardless of the reduced mass. By contrast the phase shift and the phase terms ϕ_ν discussed above all scale (at given $\tilde{\lambda}$ and $\tilde{\chi}(\tilde{\lambda})$) with λ_0, and hence depend on $m^{1/2}$. Consequently the oscillation wavelengths in Figs. 9.3 and 9.7 increase in proportion to $m^{-1/2}$ until one ultimately reaches the situation shown in Fig. 9.8 ($\lambda_0 = 10$) in which the first rainbow peak covers the entire angular region, $0 < \theta < \pi$, and the only visible interference structure is attributable to diffraction, or alternatively to glory scattering because the period of the oscillations is well represented by eqn (9.67). At the same time the amplitudes of the oscillations in the region $\theta > \theta_r$ are much larger than of those in Fig. 9.7 ($\lambda_0 = 100$), indicating an increase in the non-classical (complex angular momentum) contribution to the scattering.

Zahr and Miller (1975) have given further emphasis to the connection between the number of significant terms in the partial-wave sum (9.19) and the strength of the non-classical contribution to the scattering, by considering the case of an exponentially decreasing phase shift

$$\eta(\lambda) = \eta_0 \exp(-\alpha\lambda). \tag{9.74}$$

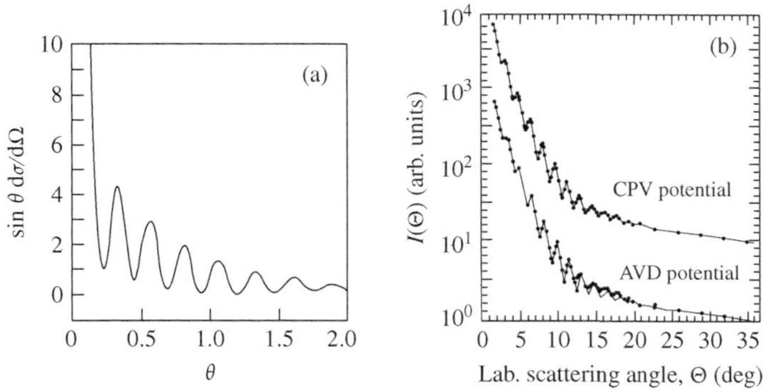

Fig. 9.8 (a) Pure diffraction oscillations for Lennard–Jones scattering at $E = 1.6$ and $\lambda_0 = 10$. Compare Fig. 9.3 at the same reduced energy but $\lambda_0 = 60$. (b) Experimentally resolved high-frequency oscillations in the elastic scattering of Ne and Ar at $E = 65.5\,\text{meV}$. Points indicate the experimental data; the two curves are theoretical reconstructions for different assumed potential functions. (Reprinted with permission from Beneventi et al. (1986). Copyright 1986, AIP Publishing LLC.)

The stationary phase condition (9.45) then yields

$$\chi(\lambda) = 2(\mathrm{d}\eta/\mathrm{d}\lambda) = -2\alpha\eta_0 \exp(-\alpha\lambda) = \mp\theta, \tag{9.75}$$

with a classically allowed solution for $\chi(\lambda) = -\theta$, namely

$$\lambda = \alpha^{-1} \ln(\theta/2\alpha\eta_0), \tag{9.76}$$

and a string of classically forbidden solutions for $\chi(\lambda) = \theta$, namely

$$\lambda = \alpha^{-1}[\ln(\theta/2\alpha\eta_0) \pm (2m+1)\pi\mathrm{i}]. \tag{9.77}$$

It is appropriate to take the lower sign in eqn (9.77), when performing the stationary phase approximation, on the grounds that the integration contour in (9.42) should strictly terminate not on the real axis but with a large imaginary component to λ in order that $\exp(\mathrm{i}\phi^-) \to 0$. Equations (9.74)–(9.77) imply in the notation of (9.44) that

$$\exp(\mathrm{i}\phi^+) = (\theta/2\alpha\eta_0)^{\mathrm{i}(\theta/\alpha)} \exp[\mathrm{i}(\theta/\alpha)], \tag{9.78}$$

while

$$\exp(\mathrm{i}\phi^-) = \exp(-\mathrm{i}\phi^+) \exp[-(2m+1)(\pi\theta/\alpha)]. \tag{9.79}$$

Consequently the classically forbidden ϕ^- branch contribution to the scattering varies as $\exp[-(\pi\theta/\alpha)]$, meaning that it increases with increasing values of α. However, large values of α imply, according to eqn (9.74), a decreasing number of terms in the partial-wave sum and hence larger quantum corrections to the normal classical picture. The point made by Zahr and Miller (1975) is that these quantum corrections are also contained within the classical equations of motion, but hidden away in the complex angular momentum plane.

Complex angular momentum (Regge pole) theory

Connor (1980) and his collaborators have followed a more systematic complex angular momentum approach, which is in principle applicable to both classically allowed and classically forbidden events, by working in terms of so-called Regge poles (Alfaro and Regge 1965). A similar approach has also been extended to inelastic processes (Thylwe and Connor 1988). The idea is that the phase shift function $\eta(E, l)$ has an underlying resonant structure which may be revealed either, as in eqns (9.35)–(9.37), by varying E at constant l or, as in the present context, by fixing the energy and varying the angular momentum, the Regge pole being the complex angular momentum $l = l_n + \mathrm{i}\gamma_n/2$ at the resonance. Consequently, close to a resonance,

$$\eta(l) \simeq \eta^0(l) + \arctan\left(\frac{\gamma_n}{2(l-l_n)}\right), \tag{9.80}$$

where the energy dependence has been suppressed to avoid notational congestion. The equivalent form

$$\exp[2\mathrm{i}\eta(l)] \simeq \exp[2\mathrm{i}\eta^0(l)] \left(\frac{l-l_n^*}{l-l_n}\right) \tag{9.81}$$

shows that $\exp[2i\eta(l)]$ diverges at the complex pole positions $l = l_n$. The Regge pole formulation is designed to exploit these divergences when evaluating the partial-wave sum for $f(\theta)$. To do this the sum in eqn (9.19) is converted into a contour integral by means of a Watson transformation (see Boas and Stutz 1971):

$$f(\theta) = -\frac{1}{2k}\int_c \frac{(1+\frac{1}{2})[\exp(2i\eta(l) - 1]P_l(\cos\theta)}{\cos[\pi\,(l+\frac{1}{2})]}\mathrm{d}l. \tag{9.82}$$

The contour is taken as shown in Fig. 9.9, from which it follows by Cauchy's theorem that

$$f(\theta) = f_{\mathrm{A}}(\theta) + f_{\mathrm{B}}(\theta) + f_{\mathrm{C}}(\theta) + f_{\mathrm{p}}(\theta), \tag{9.83}$$

where the labels A, B and C denote parts of the contour and p is the Regge pole contribution. Of these parts, $f_{\mathrm{p}}(\theta)$ is given by

$$f_{\mathrm{p}}(\theta) = -\frac{\mathrm{i}\pi}{k}\sum_n \frac{\left(l_n+\frac{1}{2}\right)r_n}{\cos\left[\pi\left(l_n+\frac{1}{2}\right)\right]}P_{l_n}(\cos\theta), \tag{9.84}$$

where according to (9.81)

$$r_n = (l_n - l_n^*)\exp[2\mathrm{i}\eta^0(l_n)], \tag{9.85}$$

while $f_{\mathrm{A}}(\theta)$ and $f_{\mathrm{C}}(\theta)$ cancel to zero, and Connor and Mackay (1978) argue that $f_{\mathrm{B}}(\theta)$ is well approximated by the short-range core contribution to the scattering, as given by $f_{\mathrm{c}}(\theta)$ in eqn (9.51).

The implementation of this approach requires an algorithm for the pole positions, $l_n + \mathrm{i}\gamma_n$, and residues, r_n. The comparison with numerically converged quantum mechanical values in Table 9.1 shows that extension of the connection formula approach of Section 3.5 by Connor et al. (1976) yields an excellent approximation. Secondly, it is important to include a sufficient number of terms in the sum in eqn (9.84) to

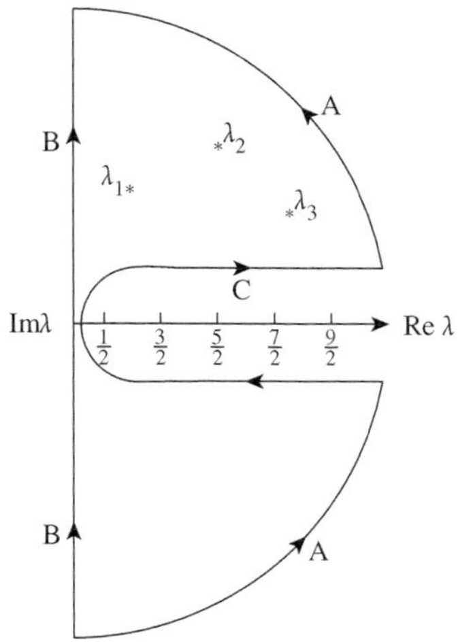

Fig. 9.9 Contours in the complex λ plane for the Watson transformation.

Table 9.1 Quantum mechanical and semiclassical Regge pole positions and residues for a Lennard–Jones (12,6) model (Connor et al. 1976).

	Position				Residue	
	Quantum		Semiclassical		Quantum	Semiclassical
n	$\mathrm{Re}\, l_n$	$\mathrm{Im}\, l_n$	$\mathrm{Re}\, l_n$	$\mathrm{Im}\, l_n$	$\|r_n\|^\dagger$	$\|r_n\|^\dagger$
0	180.012	21.219	180.015	21.218	0.734(6)	0.682(6)
1	179.239	24.035	179.242	24.034	0.634(7)	0.617(7)
2	178.522	26.890	178.526	26.889	0.252(8)	0.248(8)
3	177.866	29.780	177.869	29.779	0.617(8)	0.610(8)
4	177.272	32.700	177.275	32.699	0.106(9)	0.105(9)
5	176.742	35.645	176.745	35.644	0.138(9)	0.138(9)
10	175.074	50.561	175.176	50.560	0.341(8)	0.340(8)
20	176.187	79.644	176.189	79.643	0.151(5)	0.151(5)
40	187.625	128.784	187.627	128.783	0.910(0)	0.912(0)
60	201.095	168.240	201.097	168.239	0.119(0)	0.119(0)
75	210.626	194.336	210.628	194.335	0.998(−1)	0.999(−1)

† Values in parentheses indicate the power of 10 by which the entries should be multiplied

ensure convergence. Figure 9.10, taken from Connor and Jakubetz (1978), indicates that diffraction-dominated situations may be well described by a single pole term plus the background contribution. However, the number of significant pole terms proliferates rapidly as the reduced mass increases and the rainbow structure becomes well developed. The Regge pole description still appears to be tractable in the high angular range $\theta_\mathrm{r} < \theta < \pi$ (Connor 1980) but the uniform Airy approximation is strongly preferred in the rainbow region, $0 < \theta < \theta_\mathrm{r}$.

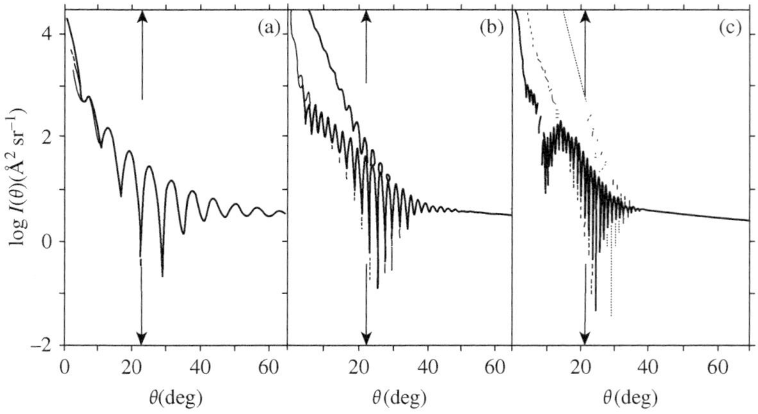

Fig. 9.10 Plots of log $I(\theta)$ against θ, where $I(\theta) = \mathrm{d}\sigma/\mathrm{d}\Omega$, for Lennard–Jones (18, 6) scattering at different values of $(2mE)^{1/2}R_0/\hbar$: (a) 29.13; (b) 82.383; and (c) 154.460. In each case the curves signify partial-wave sum (continuous), Regge theory with a single pole (dots) and Regge theory with several poles (dashed). (Reprinted from Connor and Mackay (1978) by permission of Taylor and Francis.)

9.3 The integral cross-section

The interesting features of the integral cross-section, given in terms of the phase shifts by eqns (9.21) and (9.23), include glory scattering, resonance structure and symmetry oscillations. The first is associated with the existence of a trajectory with the glory angular momentum λ_g in Fig. 9.5, which suffers zero net deflection in classical terms. Hence in an attenuation experiment (Levine and Bernstein 1985) the corresponding partial wave interferes at the detector with the 'unscattered' high angular momentum partial waves leading to undulations in the total cross-section governed by the energy dependence of the glory phase shift. Secondly, the resonant features are associated with temporary capture and hence to peaks in the total cross-section at particular resonant energies. Finally, the symmetry oscillations, which are particularly well developed for He–He scattering (Feltgen et al. 1978) arise from a combination of the nuclear spin statistics and the fact that

$$2(\partial\eta_l/\partial l) = \chi = \pi, \tag{9.86}$$

for scattering with angular momenta $l \simeq 0$. The nuclear spin statistics allow either only even or only odd l contributions to the partial-wave sums in (9.23) while (9.86) ensures that the low l contributions to these sums all arrive with constructive phases $\eta_0, \eta_0 + 2\pi, \eta_0 + 4\pi, \ldots,$ etc. or $\eta_0 + \pi, \eta_0 + 3\pi, \eta_0 + 5\pi, \ldots,$ etc.

The background cross-section

Before turning to these interference and resonance effects, it is convenient to note that the magnitude of the cross-section

$$\sigma = \frac{4\pi}{k^2}\sum_{l=0}^{\infty}(2l+1)\sin^2\eta_l \tag{9.87}$$

depends, as discussed in Section 9.1, on the l value at which η_l suddenly drops below unity because the factor $\sin^2\eta_l$ merely oscillates around 0.5 for smaller l values. Hence a background term, σ_a, consistent with this drop-off may be estimated by assuming that the a branch formula (9.29) extends to all l,

$$\eta_l \simeq a_n l^{1-n}, \tag{9.88}$$

where a_n is given by (9.31) and n denotes the inverse power of the long-range potential $V(R) \approx -C_n/R^n$. The following approximation (Massey and Mohr 1934) for σ_a then follows by replacing the sum in eqn (9.87) by an integral with respect to η and then using the identity (Dwight 1961)

$$\int_0^{\infty}\eta^{-1-p}\sin^2\eta\,\mathrm{d}\eta = \frac{\pi 2^{p-2}}{p\sin(p\pi/2)\Gamma(p)}. \tag{9.89}$$

The result may be expressed in the form

$$\sigma_a = p(n)(C_n/\hbar v)^{2/(n-1)}, \tag{9.90}$$

where

$$p(n) = \pi^x \left[\Gamma\left(\frac{n-1}{2}\right) \Big/ \Gamma\left(\frac{n}{2}\right) \right]^y \left[\sin\left(\frac{\pi}{n-1}\right) \Gamma\left(\frac{2}{n-1}\right) \right]^{-1}, \tag{9.91}$$

with $x = (2n-1)/(n-1)$ and $y = 2/(n-1)$.

Glory oscillations

The glory correction to this background term is readily estimated by noting that $\chi(l)$ in Fig. 9.5 passes through zero with a negative derivative at $l_{\rm g}$.

Hence, on integrating (9.27),

$$2\eta_l \simeq 2\eta_{\rm g} - \frac{1}{2}|{\rm d}\chi/{\rm d}l|(l - l_{\rm g})^2. \tag{9.92}$$

Consequently, on expressing $\sin^2 \eta_l$ in terms of $\cos 2\eta_l$ and approximating the sum in (9.87) by an integral, one obtains the following glory term

$$\sigma_{\rm g} \simeq \frac{2\pi}{k^2} \int_{-\infty}^{\infty} l \cos 2\eta_l {\rm d}(l - l_{\rm g}) \simeq \frac{4\pi}{k} I_{\rm g}^{1/2} \sin(2\eta_{\rm g} - 3\pi/4), \tag{9.93}$$

where $I_{\rm g}$ is the classical term

$$I_{\rm g} = 2\pi l_{\rm g}/(k^2|{\rm d}\chi/{\rm d}l|). \tag{9.94}$$

The predicted velocity dependence of

$$\sigma = \sigma_{\rm a} + \sigma_{\rm g} \tag{9.95}$$

is shown in Fig. 9.11 for a Lennard–Jones potential function characterized by the dimensionless parameter $B = 2m\varepsilon R_0^2/\hbar^2 = 500$, and similar experimental variations are well attested. Although the zero-velocity limit cannot be shown on this log–log diagram, it is interesting to note that the number of glory oscillations is related to the number of $l = 0$ bound vibrational states supported by the potential $V(R)$ (Bernstein 1962, 1963). This connection may be understood on the basis that the glory phase shift, $\eta_{\rm g}$, increases monotonically as $v \to 0$, while the l value at the glory point tends to zero in this limit. Consequently there is an upper limit on $\eta_{\rm g}$, set by the s-wave zero-energy phase shift

$$\eta(0,0) \approx \lim_{R\to\infty} \left(\int_{a_0}^{R} k_0(R){\rm d}R - \pi/4 \right), \tag{9.96}$$

because $k_0(\infty) = 0$ at zero energy. Moreover, the highest $l = 0$ bound state is quantized according to the equation

$$\left(n_{\max} + \frac{1}{2} \right) \pi = \int_{a_0}^{b_0} k_0(R){\rm d}R, \tag{9.97}$$

with $b_0 < \infty$. It follows in this semiclassical approximation that

$$\eta(0,0) > \left(n_{\max} + \frac{1}{4} \right) \pi; \tag{9.98}$$

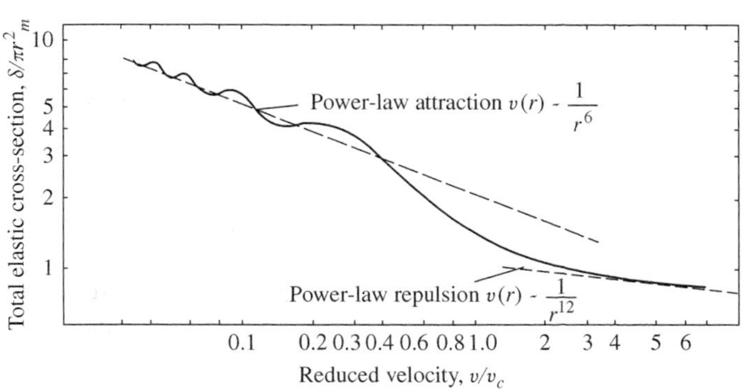

Fig. 9.11 Glory oscillations in the elastic scattering cross-section as a function of the reduced velocity $v/v_c = v\hbar/2\varepsilon R_0$. (Reprinted from Pauly (1979) with kind permission of Springer Science+Business Media B.V.)

more exact analysis based on Levinson's (1949) theorem (see Child 1974b) shows that $\eta(0,0)/\pi$ is precisely equal to the number of $l = 0$ bound states.

Orbiting resonances

The treatment of resonance scattering is quite straightforward because, in the light of eqns (9.21) and (9.35)–(9.37),

$$\sigma = \sigma_0 + \sigma_{\text{res}}, \tag{9.99}$$

where the background term is given in the notation of (9.90) and (9.93) by

$$\sigma_0 = \frac{4\pi}{k^2} \sum_{l=0}^{\infty} (2l+1) \sin^2 \eta_l \simeq \sigma_{\text{a}} + \sigma_{\text{g}} \tag{9.100}$$

and

$$\sigma_{\text{res}} = \frac{\pi}{k^2} \sum_{l=0}^{\infty} (2l+1) \frac{[\Gamma_{vl}^2 \cos 2\eta_l^0 + 2\Gamma_v (E_{vl} - E) \sin 2\eta_l^0]}{(E_{vl} - E)^2 + \Gamma_{vl}^2/4}. \tag{9.101}$$

E_{vl} and Γ_{vl} in these equations would be determined in the semiclassical approximation by eqns (3.108) and (3.111).

Figure 9.12 shows examples of the resonant structure predicted by eqn (9.101) superimposed on the glory oscillations described by eqn (9.93). There is of course an experimental lower limit, imposed by the experimental energy resolution, on the width of a detectable scattering resonance, but other narrower resonances in the same system may be observed as spectroscopic predissociation (Herzberg 1950).

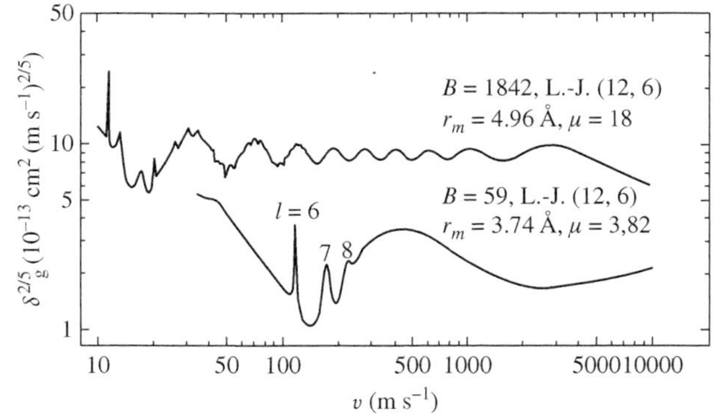

Fig. 9.12 Total elastic scattering cross-section showing glory oscillations and orbiting resonances. (Reprinted from Pauly (1979) with kind permission of Springer Science+Business Media B.V.)

Exchange oscillations

The final type of interference to be considered gives rise to the symmetry oscillations illustrated for He–He scattering in Fig. 9.13. The explanation for the monotonically decreasing ^3He–^4He cross-section is that the binding energy and reduced mass are too small to allow even vestigial glory oscillations. Hence the mutually out of phase undulations shown by ^4He–^4He and ^3He–^3He cannot be attributed to the glory effect;

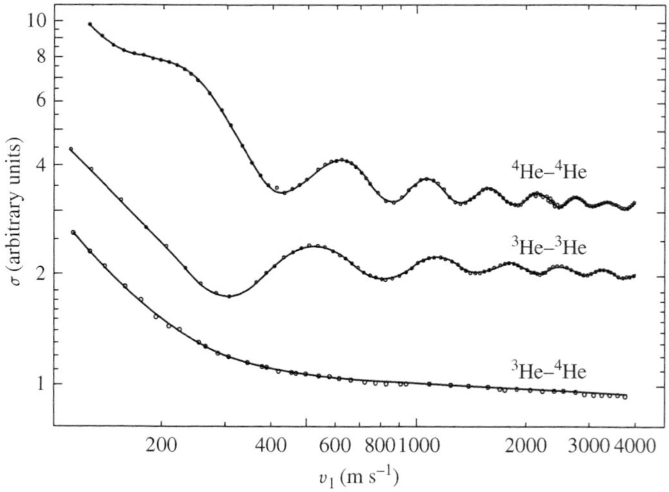

Fig. 9.13 Symmetry oscillations in He–He total scattering cross-sections. (Data of Feltgen et al. (1978). Reprinted from Pauly (1979) with kind permission of Springer Science+Business Media B.V.)

instead, they arise from the nuclear spin statistics which allow the occurrence of only even and only odd l values in the ^{4}He–^{4}He and ^{3}He–^{3}He cases respectively. To see this, note according to (9.86) that

$$\eta_l = \eta_0 + l\pi/2, \tag{9.102}$$

at least for small l values such that $\chi(l) = 2(\partial\eta/\partial l) \simeq \pi$. This means that

$$\begin{aligned}\sin^2\eta_l &= \sin^2\eta_0 \quad \text{for } l \text{ even} \quad \text{(Bose statistics)}\\ &= \cos^2\eta_0 \quad \text{for } l \text{ odd} \quad \text{(Fermi statistics)}.\end{aligned} \tag{9.103}$$

Hence the first few terms in the partial-wave sum (9.23) all contribute $\sin^2\eta_0$ or $\cos^2\eta_0$ in the symmetric cases as compared with $\sin^2\eta_0 + \cos^2\eta_0$ in the non-symmetric case. It follows that the undulations about the ^{3}He–^{4}He mean provide information about the energy dependence of $1 - \sin^2\eta_0$ in the ^{4}He–^{4}He case or of $1 - \cos^2\eta_0$ in the ^{3}He–^{3}He case. The resulting knowledge of $\eta_0(E)$ was shown in Section 6.4 to form a possible basis for direct determination of the scattering potential. Unfortunately, although the above discussion applies to the mutual scattering of any pair of identical particles, in heavy-atom cases eqn (9.103) applies to such a small fraction of the contributing partial waves that it is difficult to detect the symmetry fluctuations against the background of resonances and glory oscillations. The mutual scattering of helium atoms is therefore much the most favourable case.

9.4 Two-state non-adiabatic transitions

The analysis in the previous sections may also be extended to electronically non-adiabatic atom–atom scattering. To set the context it is convenient to give a brief outline of the quantum mechanical theory of inelastic atom–atom scattering. The Hamiltonian may be assumed to take the typical form

$$H = -\frac{\hbar^2}{2m}\nabla_R^2 + H_{\text{int}}(\xi) + V(\mathbf{R}, \xi), \tag{9.104}$$

where $H_{\text{int}}(\xi)$ determines the composite internal states, $\varphi_i(\xi)$, of the asymptotically separated partners, according to the equation

$$H_{\text{int}}(\xi)\varphi_i(\xi) = E_i\varphi_i(\xi). \tag{9.105}$$

$V(\mathbf{R}, \boldsymbol{\xi})$ is the interaction potential. The expansion

$$\phi(\mathbf{R}, \xi) = \sum_i \varphi_i(\xi)\Psi_i(\mathbf{R}) \tag{9.106}$$

therefore leads, by standard manipulations (Child 1974b), to the coupled equations

$$\left(-\frac{\hbar^2}{2m}\nabla_R^2 + E_i - E\right)\Psi_i(\mathbf{R}) = -\sum_j V_{ij}(\mathbf{R})\Psi_j, \tag{9.107}$$

where

$$V_{ij}(\mathbf{R}) = \int \varphi_i^*(\xi)V(\mathbf{R}, \xi)\varphi_j(\xi)\mathrm{d}\xi. \tag{9.108}$$

If the $V_{ij}(\mathbf{R})$ are assumed, for simplicity, to depend only on the magnitude of $\mathbf{R}$, the partial-wave expansions analogous to (9.14) lead to (Child 1974b)

$$\left(\frac{\mathrm{d}^2}{\mathrm{d}R^2} + k_i^2 - U_{ii}(R) + \frac{l(l+1)}{R^2}\right)\psi_{il}(R) = \sum_{j\neq i} U_{ij}(R)\psi_{jl}(R), \tag{9.109}$$

where

$$k_i^2 = 2m(E - E_i)/\hbar^2, \; U_{ij}(R) = 2mV_{ij}(r)/\hbar^2. \tag{9.110}$$

Of the $2N$ independent solutions of (9.6), the N that are regular (i.e. $\psi_{il}(0) = 0$) may be chosen in such a way that each has an incoming term in only one channel; thus

$$\begin{aligned}\psi_{il}^{(il)} &\overset{R\to\infty}{\approx} k_i^{-1/2}[\exp(-\mathrm{i}k_iR + \mathrm{i}l\pi/2) - S_{ii}(l)\exp(\mathrm{i}k_iR - \mathrm{i}l\pi/2)]\\ \psi_{jl}^{(il)} &\overset{R\to\infty}{\approx} k_j^{-1/2}[-S_{ij}(l)\exp(\mathrm{i}k_jR - \mathrm{i}l\pi/2)].\end{aligned} \tag{9.111}$$

Equation (9.111) defines the S matrix for the lth partial wave in a form which may be recognized as a direct generalization of the term exp $[2\mathrm{i}\eta(l)]$ in the elastic scattering theory. The upper index specifies the incoming channel and the lower the appropriate solution of (9.109). Hence by generalization of (9.19) the inelastic scattering amplitude $f_{ij}(\theta)$ takes the form (Child 1974b)

$$f_{ij}(\theta) = \frac{1}{2\mathrm{i}k_i}\sum_{l=0}^{\infty}(2l+1)[S_{ij}(l) - \delta_{ij}]P_l(\cos\theta). \tag{9.112}$$

The equations up to this point are exact; the semiclassical theory is used to obtain the S matrix elements and to approximate the sum in eqn (9.112). The example chosen for this section is the two-state non-adiabatic curve-crossing problem illustrated in Fig. 9.14(a), while Fig. 9.14(b) shows the associated phase space orbits. The functional form of the S matrix elements are readily obtained by use of the connection formulae given in Section 3.2. Suppose, as in Fig. 9.14(a), that the lower and upper adiabatic curves correlate with asymptotic states 1 and 2 respectively and that the general asymptotic behaviour in channel 1 is expressed as

$$\begin{aligned}\psi_{1l}(R) &\overset{R\to\infty}{\approx} k_1^{-1/2}\left[Q''_{-}\mathrm{e}^{-\mathrm{i}\int_{R_x}^{R} k_{-l}(R)\mathrm{d}R} + Q'\mathrm{e}^{\mathrm{i}\int_{R_x}^{R} k_{-l}(R)\mathrm{d}R}\right]\\ &\overset{R\to\infty}{\approx} k_1^{-1/2}[Q''_{-}\exp(-\mathrm{i}k_1R + \mathrm{i}l\pi/2 - \mathrm{i}\delta_{-l})\\ &\qquad + Q'\exp(\mathrm{i}k_1R - \mathrm{i}l\pi/2 + \mathrm{i}\delta_{-l})]\end{aligned} \tag{9.113}$$

with a similar expression for $\psi_{2l}(R)$ in terms of k_2, Q'_{+}, Q''_{+} and δ_{+l}, where

$$\delta_{\pm l} = \lim_{R\to\infty}\left(\int_{R_x}^{R} k_{\pm l}(R)\mathrm{d}R - k_1^2R + l\pi/2\right), \tag{9.114}$$

$$k_{\pm l}(R) = [2m(E - V_{\pm}(R) - l(l+1)\hbar^2/2mR^2)]^{1/2}/\hbar, \tag{9.115}$$

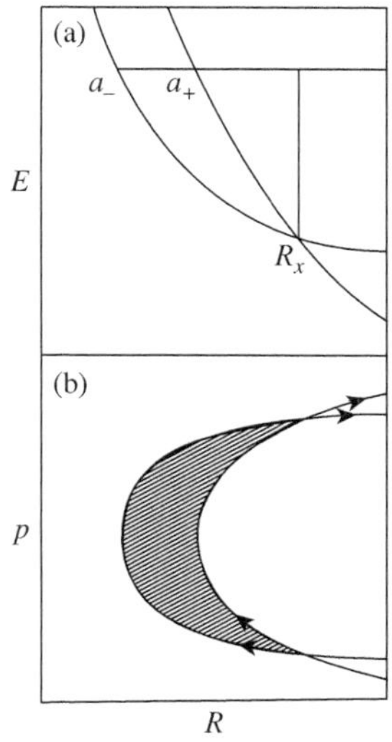

Fig. 9.14 Potential curves and phase space orbits for curve-crossing transitions. The shaded area determines the phase of the Stückelberg oscillations.

and

$$V_\pm(R) = \frac{1}{2}[V_1(R) + V_2(R) \pm \{[V_1(R) - V_2(R)]^2 + 4V_{12}^2(R)\}^{1/2}]. \tag{9.116}$$

The form of the standard channel 1 incoming boundary condition (9.111) then requires that

$$Q''_- = \exp(-\mathrm{i}\delta_{-l}) \quad \text{and} \quad Q''_+ = 0, \tag{9.117}$$

and that

$$S_{11} = -Q'_- \exp(\mathrm{i}\delta_{-l}) \quad \text{and} \quad S_{12} = -Q'_+ \exp(\mathrm{i}\delta_{+l}). \tag{9.118}$$

On the other hand eqns (3.59) imply the following connection between $Q'_\pm$ and $Q''_\pm$:

$$\begin{pmatrix} Q'_+ \\ Q'_- \end{pmatrix} = \begin{pmatrix} (1-x_l{}^2)^{1/2}\mathrm{e}^{\mathrm{i}\chi_l} & -x_l \\ x_l & (1-x_l{}^2)^{1/2}\mathrm{e}^{-\mathrm{i}\chi_l} \end{pmatrix}$$
$$\times \begin{pmatrix} \exp(2\mathrm{i}\alpha_+ - \mathrm{i}\pi/2) & 0 \\ 0 & \exp(2\mathrm{i}\alpha_- - \mathrm{i}\pi/2) \end{pmatrix} \tag{9.119}$$
$$\times \begin{pmatrix} (1-x_l{}^2)^{1/2}\mathrm{e}^{\mathrm{i}\chi_l} & x_l \\ -x_l & (1-x_l{}^2)^{1/2}\mathrm{e}^{-\mathrm{i}\chi_l} \end{pmatrix} \begin{pmatrix} Q''_+ \\ Q''_- \end{pmatrix}$$

where

$$\alpha_\pm = \int_{a_{\pm l}}^{R_x} k_{\pm l}(R)\mathrm{d}R, \tag{9.120}$$

and x and χ (whose l dependence has been suppressed in eqn (9.119)) are given in the Landau–Zener approximation by (see Section 3.2)

$$x_l = \exp(-\pi\nu_l) \simeq \exp(-\pi V_{12}^2/\hbar u_l \Delta F), \tag{9.121}$$

$$\chi_l = \arg\Gamma(\mathrm{i}\nu_l) - \nu_l \ln \nu_l + \nu_l + \pi/4, \tag{9.122}$$

ΔF being the difference in potential derivatives at the crossing point and u_l the radial classical velocity,

$$u_l = \left[\frac{2}{m}\left(E - E_x - l(l+1)\frac{\hbar^2}{2mR^2}\right)\right]^{1/2}. \tag{9.123}$$

It follows, on combining (9.116)–(9.118), that

$$\begin{aligned} S_{12}(l) &= -2\mathrm{i}x_l(1-x_l^2)^{1/2}\sin\tau_l \exp[\mathrm{i}(\eta_{1l}+\eta_{2l})], \\ S_{11}(l) &= [x_l^2 + (1-x_l^2)\exp(2\mathrm{i}\tau_l)]\exp(2\mathrm{i}\eta_l), \end{aligned} \tag{9.124}$$

where

$$\begin{aligned} \eta_{1l} &= \lim_{R\to\infty}\left(\int_{a_{+1}}^{R_x} k_{+l}(R)\mathrm{d}R + \int_{R_x}^{R} k_{-l}(R)\mathrm{d}R - k_1 R - \frac{l\pi}{2} + \frac{\pi}{4}\right), \\ \eta_{2l} &= \lim_{R\to\infty}\left(\int_{a_{-1}}^{R_x} k_{-l}(R)\mathrm{d}R + \int_{R_x}^{R} k_{+l}(R)\mathrm{d}R - k_2 R - \frac{l\pi}{2} + \frac{\pi}{4}\right), \end{aligned} \tag{9.125}$$

and

$$\tau_l = \int_{a_{-l}}^{R_x} k_{-l}(R)\mathrm{d}R - \int_{a_{+l}}^{R_x} k_{+l}(R)\mathrm{d}R - \chi_l. \tag{9.126}$$

Equations (9.124)–(9.126) have an appealing physical interpretation. As usual the probability $|S_{12}(l)|^2$ oscillates around a classical function $2x_l^2(1-x_l^2)$, which in this case may be attributed to the sum of two events—an initial diabatic transition, with probability x_l^2, followed by an adiabatic transition, with probability $(1-x_l^2)$, and vice versa. Moreover, the Stückelberg interference term depends on the phase difference τ_l loosely determined by the shaded area in Fig. 9.14(b), which corresponds to the classical action difference between the above two paths. It is also interesting to note that these Stückelberg oscillations have the same physical origin as those in the Franck–Condon matrix elements discussed in Section 5.3. Similar remarks apply to the diagonal probability $|S_{11}(l)|^2$ which is seen to involve a term x_l^4 attributable to two adiabatic transitions at the crossing point, a term $(1-x_l^2)^2$ attributable to two diabatic transitions, and an interference term again dependent on τ_l. The phases of $S_{ij}(l)$ also carry explicit information about the history of the motion. Thus the second term in $S_{11}(l)$ has a phase

$$\eta_{1l} + \tau_l = \lim_{R\to\infty}\left(\int_{a_{-l}}^{R} k_{-l}(R)\mathrm{d}R - k_1 R - \frac{l\pi}{2} + \frac{\pi}{4}\right) - \chi_l, \tag{9.127}$$

appropriate to motion under $V_-(R)$, plus a small phase correction. Similarly eqn (9.125) for the phase, $2\eta_{1l}$, of the first term in $S_{11}(l)$ suggests motion under V_-(R) for $R > R_x$ and under $V_+(R)$ for $R < R_x$, where the phase discontinuity at R_x may be attributed, according to the Stückelberg picture in Section A.5, to an excursion to the complex intersection point, R_c, where an energy-conserving transition from $V_-(R)$ to $V_+(R)$ can take place; Miller and George (1972) have even formulated a general theory of non-adiabatic transitions from this complex trajectory viewpoint (see also Lin et al. 1975; Lam and George 1980).

Turning to the scattering amplitude, the combination of (9.112) with (9.124) yields the partial-wave sum

$$f_{12}(\theta) = -2^{-1/2}k_1^{-1}\sum_{l=0}^{\infty}(2l+1)F_l^{1/2}\sin\tau_l \exp[\mathrm{i}(\eta_{1l}+\eta_{2l})]P_l(\cos\theta), \tag{9.128}$$

where F_l is the classical double passage probability

$$F_l = 2x_l^2(1-x_l^2). \tag{9.129}$$

Hence after using (9.40) to substitute for $P_l(\cos\theta)$ and replacing the sum by an integral with respect to $\lambda = l + \frac{1}{2}$,

$$f_{12}(\theta) \simeq \sum_{\nu}\int_0^{\infty} g(\theta,\lambda)\exp[\mathrm{i}\varphi_\nu(\theta,\lambda)]\mathrm{d}\lambda, \tag{9.130}$$

where the sum over v is taken over four branches with phases

$$\begin{aligned} \varphi_{a\pm}(\theta,\lambda) &= \eta_{1l}+\eta_{2l}+\tau_l \pm \lambda\theta \pm (-\pi/4) \\ &= 2\eta_a(\lambda) \pm \lambda\theta \pm (-\pi/4) \\ \varphi_{d\pm}(\theta,\lambda) &= \eta_{1l}+\eta_{2l}-\tau_l \pm \lambda\theta \pm (-\pi/4) \\ &= 2\eta_d(\lambda) \pm \lambda\theta \pm (-\pi/4), \end{aligned} \tag{9.131}$$

and

$$g(\theta,\lambda) = \frac{\mathrm{i}}{k_1}\left(\frac{\lambda F(\lambda)}{4\pi\sin\theta}\right)^{1/2}. \tag{9.132}$$

Here the notations α_a and η_a indicate an initial adiabatic transition at the crossing point followed by a diabatic one, while φ_d and η_d indicate transitions in the opposite order; in particular

$$\begin{aligned} \eta_a(\lambda) &= \tfrac{1}{2}(\eta_{1l}+\eta_{2l}) + \tau_l \\ \eta_b(\lambda) &= \tfrac{1}{2}(\eta_{1l}+\eta_{2l}) - \tau_l. \end{aligned} \tag{9.133}$$

Following the analysis in Section 9.2, it is now natural to use the equations

$$\chi_\nu = 2(\partial\eta_\nu/\partial\lambda), \quad \nu = a \quad \text{or} \quad d, \tag{9.134}$$

to define classical deflection functions for the above inelastic scattering paths. The previous stationary phase and uniform approximation arguments then follow through exactly, apart from the complication of four rather than two contributing branches to $f_{12}(\theta)$.

The simplest case occurs when $\chi_a(\lambda)$ and $\chi_d(\lambda)$ are both monotonically decreasing functions of λ, as illustrated in Fig. 9.15. The only stationary phase points then belong either to $\varphi_{a^-}(\theta,\lambda)$ or $\varphi_{d^-}(\theta,\lambda)$. Moreover, one sees that the diagram is closed because both scattering paths lead to π at $\lambda = 0$ and to a common deflection at the limiting angular momentum, λ_x, at which the crossing point becomes classically accessible. This closure has an interesting consequence, because it follows from eqns (9.133) and (9.134) that the term $2\tau(\lambda)$ that determines the oscillations of $|S_{12}(\lambda)|^2$ may be expressed as the integral

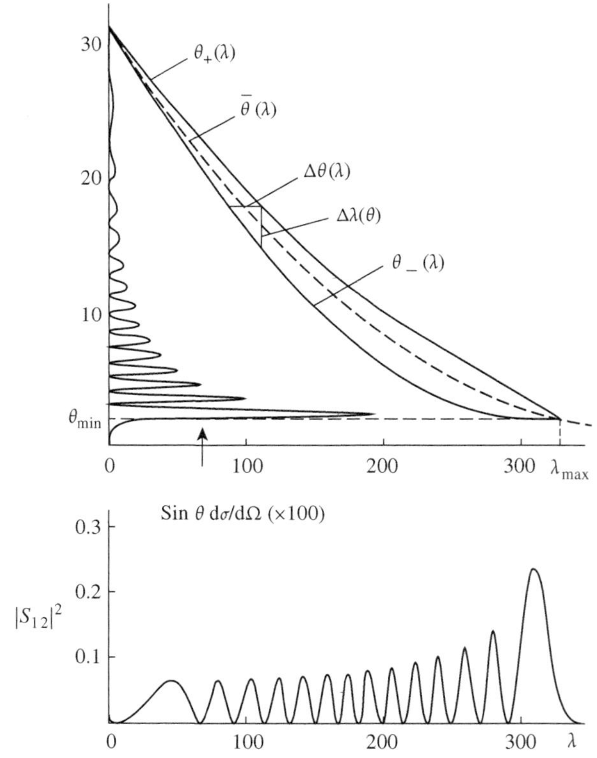

Fig. 9.15 Classical deflection functions and the 'reflection' relationship between the differential cross-section ($d\sigma/d\Omega$) and the oscillatory variation of $|S_{12}(\lambda)|^2$. (Reprinted from Child and Gerber (1979) by permission of Taylor and Francis.)

$$2\tau(\lambda) = 2\eta_a(\lambda) - 2\eta_d(\lambda) = \int_0^{\lambda} [\chi_a(\lambda) - \chi_d(\lambda)]\mathrm{d}\lambda, \tag{9.135}$$

which is taken between the two curves in Fig. 9.15. The same is true of the oscillations of $f_{12}(\theta)$ because the relevant phase term is given by

$$\begin{aligned}\varphi_a(\theta) - \varphi_d(\theta) &= 2\eta_a[\lambda_a(\theta)] - 2\eta_d[\lambda_d(\theta)] \\ &= \int_0^{\lambda_a(\theta)} \chi_a(\lambda)\mathrm{d}\lambda = \int_0^{\lambda_d(\theta)} \chi_d(\lambda)\mathrm{d}\lambda \\ &= \int_\theta^{\pi} [\lambda_a(\theta') - \lambda_d(\theta')]\mathrm{d}\theta',\end{aligned} \tag{9.136}$$

where the final line is obtained by integration by parts, with $\lambda_v(\theta)$ taken as the inverse function to $\chi_v(\lambda)$. Equations (9.135) and (9.136) mean that the number of oscillations in $|S_{12}(\lambda)|^2$ and in $|f_{12}(\theta)|^2$ must be equal because both are determined by the enclosed area in Fig. 9.15. Child and Gerber (1979) have based the first part of an inversion scheme of the form $|f_{12}(\theta)|^2 \to S_{12}(\lambda) \to V_{12}(R)$ on this phase conservation theorem. Readers interested in other applications of this two-state non-adiabatic theory are referred to Delvigne and Los (1973), Child (1979), Faist and Bernstein (1976) and Faist and Levine (1976). Olsen (1970) gives a complementary treatment of the integral atom–atom inelastic differential scattering cross-sections, while Tully and Preston (1971) and Tully (1976) extend the theory to surface-hopping atom–molecule collisions.

9.5 Problems

9.1. (i) Use eqn (9.7) and (9.25) to derive expressions for the classical deflection function, $\chi(\lambda)$, and the semiclassical phase shift, $\eta(\lambda)$, for scattering by a hard sphere of radius d at angular momentum $l + 1/2 = \lambda < kd$.

(ii) Show that the classical differential and total cross-sections are given by $(\mathrm{d}\sigma/\mathrm{d}\Omega)_{\mathrm{cl}} = d^2/4$ and $\sigma_{\mathrm{cl}} = \pi d^2$ respectively. Secondly, use the random phase approximation $\sin^2\eta(\lambda) \simeq 0.5$ to show that $\sigma_{\mathrm{semi}} = 2\pi d^2$ for $kd \gg 1$.

(iii) Compute the partial-wave sum for $\mathrm{d}\sigma/\mathrm{d}\Omega$ over the range $0 < \theta < \pi$ at $kd = 10.0$, using the above form for $\eta(\lambda)$. Truncate the sum at $\lambda = kd$ and estimate the ratio $\sigma/\pi d^2$.

9.2. Combine the random phase approximation to $f(\theta)$ with the approximation $P_l(\cos\theta) \simeq \sqrt{\theta/\sin\theta}J_0(\lambda\theta)$ from problem 2.7 to show that the small-angle hard sphere scattering amplitude may be estimated as

$$\begin{aligned}f(\theta) &\simeq \frac{1}{\mathrm{i}k}\sqrt{\frac{\theta}{\sin\theta}}\int_0^{kd} \lambda J_0(\lambda\theta)\mathrm{d}\lambda = \frac{d}{\mathrm{i}\sqrt{\theta\sin\theta}}J_1(kd\theta) \\ &\simeq \frac{kd^2}{2\mathrm{i}}\sqrt{\frac{\theta}{\sin\theta}}\exp\left[-\frac{(kd\theta)^2}{8}\right]\end{aligned}$$

over the angular region $0 < \theta < \pi/kd$, over which $P_l(\cos\theta) > 0$ for $\lambda \leqslant kd$. Estimate the small angle 'shadow' contribution to the integrated cross-section, and compare it with the discrepancy between the random phase value $2\pi d^2$ and the value πd^2 implied by equivalence between the classical and semiclassical differential cross-sections over the full angular region.
[Hints: $\int_0^a xJ_0(x)\mathrm{d}x = aJ_1(a) = (a^2/2)\left(1 - (a^2/8)\ldots\right) \simeq (a^2/2)\mathrm{e}^{-a^2/8}$ for $a < 3$.]

9.3. Show by means of the substitution in eqn (9.19) from problem 2.7, that

$$P_l(\cos\theta) \simeq (\theta/\sin\theta)^{1/2} J_0(\lambda\theta) = \left(\frac{\theta}{\sin\theta}\right)^{1/2} \frac{1}{\pi}\int_0^\pi \exp(-\mathrm{i}\lambda\theta\cos\varphi)\mathrm{d}\varphi.$$

Deduce, by stationary phase integration with respect to λ, that $f(\theta) \simeq f_a(\theta) + f_g(\theta)$, where $f_a(\theta)$ is the a branch contribution to (9.51) and $f_g(\theta)$ is the glory scattering amplitude

$$f_g(\theta) = \frac{\exp(-3\mathrm{i}\pi/4)}{k}\left(\frac{2\pi\theta}{\sin\theta}\right)^{1/2}\frac{1}{\pi} \times \int_0^\pi \frac{\lambda(\theta,\ \varphi)}{\sqrt{|\chi'[\lambda(\theta,\ \varphi)]|}} \exp\{2\mathrm{i}\eta[\lambda(\theta,\varphi)] - \mathrm{i}\lambda(\theta,\varphi)\theta\cos\varphi\}\mathrm{d}\varphi,$$

where $\chi'(\lambda) = \mathrm{d}\chi/\mathrm{d}\lambda$ and $\lambda(\theta,\ \varphi)$ is defined such that

$$\chi[\lambda(\theta,\varphi)] = 2(\partial\eta/\partial\lambda) = \theta\cos\varphi.$$

9.4. (i) The following mapping equation in the above integral for $f_g(\theta)$,

$$F(\theta,\psi) = 2\eta[\lambda(\theta,\ \varphi)] - \lambda(\theta,\ \varphi)\theta\cos\varphi = A(\theta) - \zeta(\theta)\cos\psi,$$

defines a function $\varphi(\psi)$. Use the approximation

$$G(\theta,\psi) = \frac{\lambda(\theta,\ \varphi)}{\sqrt{|\chi'[\lambda(\theta,\ \varphi)]|}}\left(\frac{\mathrm{d}\varphi}{\mathrm{d}\psi}\right) \simeq p(\theta) + q(\theta)\cos\psi$$

to obtain the uniform Bessel approximation (Berry 1975; Connor 2004)

$$f_g(\theta) \simeq \frac{\exp[\mathrm{i}A(\theta) - 3\mathrm{i}\pi/4]}{k}\left(\frac{2\pi\theta}{\sin\theta}\right)^{1/2}\{p(\theta)J_0[\zeta(\theta)] + \mathrm{i}q(\theta)J_0'[\zeta(\theta)]\},$$

where $J_0'(z) = \mathrm{d}J_0/\mathrm{d}z$.

(ii) Show that correspondence between the stationary phase points on the two sides of the mapping equation yields $\lambda_b(\theta) = \lambda(\theta,\ \pi)$ and $\lambda_c(\theta) = \lambda(\theta,\ 0)$, such the $\chi_b = -\theta$ and $\chi_c = \theta$ respectively, in terms of which

$$A(\theta) = \frac{1}{2}[2\eta_b + 2\eta_c + (\lambda_b - \lambda_c)\theta],$$

$$\zeta(\theta) = \frac{1}{2}[2\eta_b - 2\eta_c + (\lambda_b + \lambda_c)\theta],$$

$$p(\theta) = \frac{1}{2}\left(\frac{\zeta(\theta)}{\theta}\right)^{1/2}[(\lambda_c/|\chi_c'|)^{1/2} + (\lambda_b/|\chi_b'|)^{1/2}],$$

$$q(\theta) = \frac{1}{2}\left(\frac{\zeta(\theta)}{\theta}\right)^{1/2}[(\lambda_c/|\chi_c'|)^{1/2} - (\lambda_b/|\chi_b'|)^{1/2}].$$

[Hint: $\mathrm{d}\varphi/\mathrm{d}\psi$ may be deduced from the second derivative of the transformation identity (compare eqn (B.12)).]

9.5. Deduce that the results of problem 9.4 reduce under the approximation

$$\eta(\lambda) \simeq \eta_{\mathrm{g}} - \frac{1}{2}\eta_{\mathrm{g}}''(\lambda - \lambda_{\mathrm{g}})^2$$

to the limiting form

$$f_g(\theta) \simeq \frac{\exp(2\mathrm{i}\eta_{\mathrm{g}} - 3\mathrm{i}\pi/4)}{k}\left(\frac{2\pi\lambda_g^2\theta}{\sin\theta|\chi_{\mathrm{g}}'|}\right)^{1/2} J_0(\lambda_{\mathrm{g}}\theta) \quad \text{as} \quad \theta \to 0.$$

10 The classical S matrix

The classical S matrix is a natural extension of the semiclassical phase shift employed in Sections 9.1–9.3. A simple example appeared earlier in eqn (9.124), which assumed a tacit distinction between the treatment of translational and internal degrees of freedom. The more general approach, described below, treats all degrees of freedom equivalently, in a manner applicable to arbitrary non-separable processes, with specific reference to inelastic and reactive scattering and photodissociation.

Two different formulations have been adopted. One, due to Pechukas (1969) and Miller (1970, 1974, 1976), starts from the Feynman path integral approach to quantum mechanics (see Feynman and Hibbs 1965), as discussed in Appendix C. The other, due to Marcus (1970) and coworkers, which is based on a direct JWKB type of approximation for the scattering wavefunction, is expounded in Section 10.1 in a way that leads naturally to the uniform approximations that make the difference between an intuitively attractive and a quantitatively reliable theory. Reviews of the two approaches have been given by Miller (1974, 1976) and Child (1974b, 1976b).

The underlying physical idea is that any scattering wavefunction has a classical analogue in the family of trajectories emanating from the initial state in question, provided that each trajectory carries an associated phase determined by the classical action accumulated along its path. As derived in Section 10.1 the interference between members of the family, that determines any particular transition amplitude, is naturally expressed as an integral, which constitutes the fundamental representation of the S matrix. The classical connection is that the dominant contribution to this integral comes from the subset of trajectories that pass from the specified initial to the specified final state. As in earlier simpler contexts, the probabilities and classical actions associated with these root trajectories provide the basis both for primitive approximations and for various uniform improvements designed to remove spurious singularities. Details are expounded in Section 10.2, while Section 10.3 describes the use of complex trajectory methods to handle classically forbidden events. Extension of the theory to multidimensional inelastic situations in Section 10.5 shows that the well-attested 'rotational rainbow' pattern (Schinke and McGuire 1979; Schinke et al. 1982), seen in the rotationally inelastic differential cross-section, is in fact the simplest member of a family of generic diffraction patterns, which are discussed under the title 'waves and catastrophes' in Appendix B; relevant reviews have been given by Berry (1976) and Connor et al. (1976, 1990b). The 'Condon reflection' connection between the final state distribution and the form of the initial quantum mechanical wavefunction is also discussed.

Semiclassical Mechanics with Molecular Applications. Second Edition. M. S. Child

10.1 The integral representation

It is assumed for the purposes of exposition that the motion in a single translational variable is coupled to a single internal degree of freedom, as described by the Hamiltonian

$$H = \frac{1}{2m}P^2 + H_0(I) + V(R,\, I,\, \alpha), \tag{10.1}$$

where $(I,\, \alpha)$ are the angle–action variables for internal motion and $(P,\, R)$ are Cartesian translational variables. The proper motion, in the absence of any interaction $V(R,\, I,\, \alpha)$, is illustrated in Fig. 10.1(a) in terms of a spiral around an invariant tube (the analogue of bound state invariant torus); the corresponding wavefunction may be expressed, according to eqns (2.3) and (4.25), as

$$\Psi^0(R,\, \alpha) = (2\pi v_1)^{-1/2} \exp\{[(I_1 - \delta\hbar)\alpha + P_1 R]/\hbar\}, \tag{10.2}$$

where v_1 is the initial velocity, $P_1 = mv_1$, and the initial action I_1 is related to an integer initial quantum number n_1 in the form

$$I_1 = (n_1 + \delta)\hbar, \tag{10.3}$$

where δ is the Maslov index.

The effect of switching on the interaction is to distort the invariant tube, as shown in Fig. 10.1(b), so that different initial angles α_1 lead to different final actions I_2. The object of the theory is to follow the associated changes in the wavefunction in order

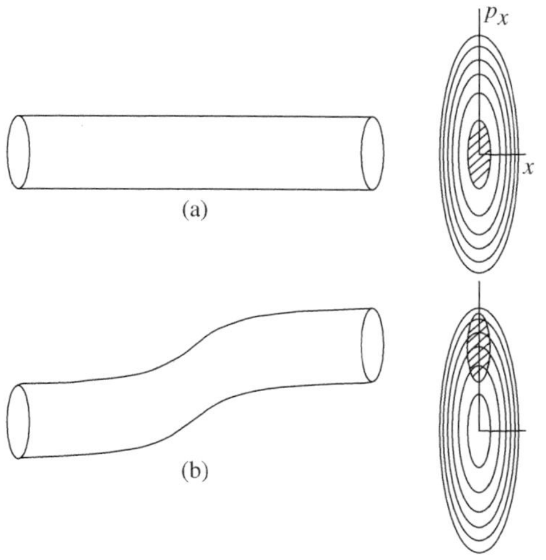

Fig. 10.1 A schematic view of (a) elastic and (b) inelastic scattering and related phase space transformations. The closed circle to the right of (b) shows the locus of final trajectory end points, initiated from the common initial state $v = 0$.

ultimately to extract the S matrix amplitude for scattering to an integer final quantum number n_2. It is essential for what follows to recognize that the final variables (I, α) are dependent, via an appropriate classical trajectory, on their initial values (I_1, α_1) or vice versa.

Following Marcus (1970), the wavefunction is expressed in the JWKB form

$$\Psi(R, \alpha) = \exp[\mathrm{i}\, W(\alpha, R)/\hbar], \tag{10.4}$$

and it is assumed that the Hamiltonian has a valid Taylor expansion over the range of actions of interest; for example

$$H_0(I) + V(R, I, \alpha) = \sum_k h_k(R, \alpha) I^k. \tag{10.5}$$

We also know, from eqn (4.23), that I has the associated semiclassical operator

$$\widehat{I} = -\mathrm{i}\hbar\frac{\partial}{\partial\alpha} + \delta\hbar, \tag{10.6}$$

from which it may be verified that

$$\begin{aligned}\widehat{I}^k \exp(\mathrm{i}W/\hbar) = &\left[\left(\frac{\partial W}{\partial\alpha} + \delta\hbar\right)^k\right.\\ &\left. -\mathrm{i}\hbar\frac{k(k-1)}{2}\left(\frac{\partial W}{\partial\alpha} + \delta\hbar\right)^{k-2}\frac{\partial^2 W}{\partial\alpha^2}\cdots\right]\exp(\mathrm{i}W/\hbar).\end{aligned} \tag{10.7}$$

A corresponding result applies, with $\delta = 0$, for the momentum operator, $\widehat{P} = -\mathrm{i}\hbar\partial/\partial R$, which of course appears only quadratically in the Schrödinger equation

$$\widehat{H}\Psi = \left(\frac{1}{2m}\widehat{P}^2 + H_0(\widehat{I}) + V(R, \widehat{I}, \alpha)\right)\Psi = E\Psi. \tag{10.8}$$

It follows from (10.4)–(10.8), on expanding W in powers of $\hbar$,

$$W = W_0 + \hbar W_1 + \hbar^2 W_2 + \ldots, \tag{10.9}$$

that the terms in W_0 lead, as in Section 2.1, to an equation of Hamilton–Jacobi form

$$\frac{1}{2m}\left(\frac{\partial W_0}{\partial R}\right)^2 + H_0\left(\frac{\partial W_0}{\partial\alpha} + \delta\hbar\right) + V\left(R, \frac{\partial W_0}{\partial\alpha} + \delta\hbar, \alpha\right) = 0 \tag{10.10}$$

but with the difference that eqn (10.10) cannot be solved in closed form. However, comparison with (10.1) suggests the identities

$$\frac{\partial W_0}{\partial\alpha} = I - \delta\hbar, \qquad \frac{\partial W_0}{\partial R} = P, \tag{10.11}$$

from which $W_0(R, \alpha)$ may be evaluated as the line integral

$$W_0(R, \alpha) = \int_{R_1}^{R} P\mathrm{d}R + \int_{\alpha_1}^{\alpha} (I - \delta\hbar)\mathrm{d}\alpha + \text{constant}, \tag{10.12}$$

taken along a classical trajectory appropriate to the Hamiltonian (10.1). The generalization to cover situations in which the Hamiltonian carries an explicit time dependence is readily verified, by solution of the time-dependent counterpart of (10.8), to be

$$S_0(R, \alpha, t) = \int^{t} [P\dot{R} + I\dot{\alpha} - H(t)]\mathrm{d}t. \tag{10.13}$$

In other words the phase of the semiclassical wavefunction is determined, as usual, by the classical action integral. Note, as discussed at greater length in Appendix C, that $S_0(R, \alpha, t) = W_0(R, \alpha) - Et$ for time-independent Hamiltonians.

Turning to terms in $\hbar$ in eqn (10.8), it may be verified by use of (10.7) that any term in $\widehat{I}^k$ in the Hamiltonian yields a term of order $\hbar$ involving the derivative $\mathrm{d}I^k/\mathrm{d}k = KI^{k-1}$; thus

$$\widehat{I}^k \exp(\mathrm{i}\, W/\hbar) = \left(I^k - \frac{\mathrm{i}\hbar}{2} \exp(-2\mathrm{i}\, W_1) \frac{\partial}{\partial \alpha} [\exp(2\mathrm{i}\, W_1) k I^{k-1}] \right) \exp(\mathrm{i}W/\hbar), \tag{10.14}$$

with a similar equation from the term $\hat{P}^2$. It follows from (10.5) that the condition for cancellation of terms in $\hbar$ may be expressed as

$$\frac{\partial}{\partial R}\left(\exp(2\mathrm{i}\, W_1) \frac{\partial H}{\partial P} \right) + \frac{\partial}{\partial \alpha}\left(\exp(2\mathrm{i}\, W_1) \frac{\partial H}{\partial I} \right) = 0, \tag{10.15}$$

or, on using Hamilton's equations to substitute for the partial derivatives,

$$\frac{\partial}{\partial R}(A^2 \dot{R}) + \frac{\partial}{\partial \alpha}(A^2 \dot{\alpha}) = 0, \tag{10.16}$$

where

$$A = \exp(\mathrm{i}\, W_1). \tag{10.17}$$

Equation (10.16) may be expressed in the more compact and more revealing form

$$\nabla \cdot \mathbf{i} = 0, \tag{10.18}$$

where $\mathbf{i}$ is the current vector $A^2\dot{\mathbf{q}}$. Hence, by Gauss' theorem, the net outflow from any region of space must be zero. Specializing to the area between neighbouring trajectories in (R, α) space, as illustrated in Fig. 10.2, this requires that

$$A_1^2 \dot{R}_1 \mathrm{d}\alpha_1 = A^2 \dot{R} \mathrm{d}\alpha, \tag{10.19}$$

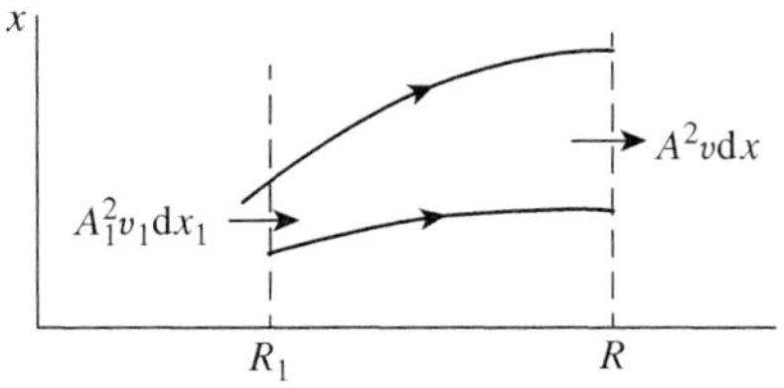

Fig. 10.2 Evolution of adjacent trajectories in (R, α) space.

because there is no outflow across the trajectories. Written in another way eqn (10.19) may be expressed as

$$A(R, \alpha) = A_1 \left[\frac{v_1}{v} \left(\frac{\partial \alpha_1}{\partial \alpha} \right) \right]^{1/2}, \tag{10.20}$$

where the precise partial derivative depends on the way that the family of trajectories is chosen—normally, for consistency with eqn (10.2), at a common value of n_1, or I_1.

In summary, eqns (10.13) and (10.20) give the terms in $W(\alpha, R)$ of order $\hbar^0$ and $\hbar^1$ respectively, while eqn (10.2) shows that $A_1 = (2\pi v_1)^{-1/2}$. Hence, to order $\hbar$ in the exponent of (10.4),

$$\Psi(R, \alpha) = (2\pi v)^{-1/2} [(\partial \alpha_1/\partial \alpha)_{n_1}]^{1/2} \exp[\mathrm{i} W_0(\alpha, R)/\hbar], \tag{10.21}$$

where

$$\begin{aligned} W_0(\alpha, R)/\hbar &= \int_{\alpha_1}^{\alpha} n \mathrm{d}\alpha + \int_{R_1}^{R} k \mathrm{d}R + n_1 \alpha_1 + k_1 R_1, \\ n &= I/\hbar - \delta, \quad k = P/\hbar. \end{aligned} \tag{10.22}$$

The constant in (10.12) has been chosen in such a way that (10.21) goes over to (10.2) at the start of the trajectory when $\alpha = \alpha_1$ and $R = R_1$. It is also important to remember that any trajectory at a given energy may be specified (at large enough R) by the initial variables (I_1, α_1) or (n_1, α_1). Hence the final variables (I, α) or (n, α) are dependent on this initial choice.

Before proceeding to the S matrix, it is convenient to introduce a significant modification to (10.21) by transforming in the asymptotic regions from the (α, R) to a new $(\bar{\alpha}, t)$ picture where

$$\bar{\alpha} = \alpha - \omega t = \alpha - \omega m R/P, \quad \omega = \partial H_0/\partial I. \tag{10.23}$$

This carries the practical advantage that $\bar{\alpha}$ is a constant of the asymptotic motion; it also removes the velocity dependence from the pre-exponent in (10.21). Details of the transformation are discussed in Appendix C, with the upshot, according to eqn (C.48), that $\Psi^0(R, \alpha)$, given by (10.2), goes over to

$$\Psi^0_{n_1 E}(\bar{\alpha}, t) = (2\pi)^{-1/2} \exp(\mathrm{i} n_1 \bar{\alpha} - \mathrm{i} Et/\hbar), \tag{10.24}$$

while (10.21) transforms to

$$\Psi_{n_1E}(\bar{\alpha}, t) \overset{t\to\infty}{\approx} (2\pi)^{-1/2}(\partial\bar{\alpha}_1/\partial\bar{\alpha})_{n_1}^{1/2}\exp\{\mathrm{i}[\chi(n_1, \bar{\alpha}) - Et/\hbar]\}, \tag{10.25}$$

where

$$\chi(n_1, \bar{\alpha}) = n\bar{\alpha} - \int_{n_1}^{n} \alpha\,\mathrm{d}n - \int_{k_1}^{k} R\,\mathrm{d}k. \tag{10.26}$$

Both the modified angle variable $\bar{\alpha}$ and the phase term $\chi(n_1, \bar{\alpha})$ are independent of the initial and final R values, provided these lie in the asymptotic region, because $\bar{\alpha}$, n and k are then constants of the motion. This representation is the classical analogue of the quantum mechanical interaction picture (Child 1974b).

The derivation of the S matrix is completed by comparing eqn (10.25) with the standard asymptotic form

$$\Psi_{n_1E}(\bar{\alpha}, t) \overset{t\to\infty}{\approx} (2\pi)^{-1/2}\sum_{n_2} S_{n_1n_2}\exp(\mathrm{i}n_2\bar{\alpha} - \mathrm{i}Et/\hbar); \tag{10.27}$$

thus, on projecting out the term in $\exp(\mathrm{i}n_2\bar{\alpha})$ and transforming from $\bar{\alpha}$ to $\bar{\alpha}_1$ as the integration variable,

$$S_{n_1n_2} = \frac{1}{2\pi}\int_0^{2\pi}\left(\frac{\partial\bar{\alpha}}{\partial\bar{\alpha}_1}\right)_{n_1}^{1/2}\exp[\mathrm{i}\Delta_{n_1n_2}(\bar{\alpha}_1)]\mathrm{d}\bar{\alpha}_1, \tag{10.28}$$

where

$$\begin{aligned}\Delta_{n_1n_2}(\bar{\alpha}_1) &= \chi[n_1, \bar{\alpha}(n_1, \bar{\alpha}_1)] - n_2\bar{\alpha}(n_1, \bar{\alpha}_1)\\ &= [n(n_1, \bar{\alpha}_1) - n_2]\bar{\alpha}(n_1, \bar{\alpha}_1)\\ &\quad - \int_{n_1}^{n(n_1, \bar{\alpha}_1)}\alpha\mathrm{d}n - \int_{k_1}^{k(n_1, \bar{\alpha}_1)} R\mathrm{d}k\Bigg].\end{aligned} \tag{10.29}$$

An alternative expression for the phase term, applicable to situations in which the trajectories are propagated in Cartesian, rather than angle–action, variables may be obtained by means of the substitution (Rankin and Miller 1971; Marcus 1973; see also eqn (C.110))

$$\int_{n_1}^{n(n_1, \bar{\alpha}_1)}\alpha\mathrm{d}n = \frac{1}{\hbar}\left(-\int_{x_1}^{x} p\mathrm{d}x + F_2(x, n) - F_2(x_1, n_1)\right), \tag{10.30}$$

where $F_2(x_i, n_i)$ are the classical generators for initial and final canonical transformations between angle–action and Cartesian variables (see Appendix C). An alternative path integral derivation of the general multidimensional Cartesian result is also given in this Appendix.

Equation (10.28), from which all other approximations will be derived, shows that the S matrix elements may be expressed as an integral over all trajectories specified by energy E and the appropriate initial internal action $(n_1+\delta_1)\hbar$, with each trajectory carrying a phase term determined by the classical action. A less obvious point is that this integral representation also enshrines certain symmetry selection rules. Consider, for example, the excitation of a centrosymmetric plane rotor, whose physical orientation is specified by the angle α. Clearly, the final variables $n(n_1, \bar{\alpha}_1)$ and $k(n_1, \bar{\alpha}_1)$ must be invariant to the substitution $\bar{\alpha}_1 \to \bar{\alpha}_1 + \pi$ while $\bar{\alpha}(n_1, \bar{\alpha}_1) \to \bar{\alpha}(n_1, \bar{\alpha}_1) + \pi$; hence the α-dependent terms in eqn (10.29) introduce a phase difference of $(n_1 - n_2)\pi$ between every such pair of symmetry-related trajectories. The integral from 0 to π in eqn (10.28) therefore exactly cancels that from π to 2π if $(n_1 - n_2)$ is odd. In other words eqn (10.28) automatically embodies the symmetry selection rule $\Delta n = 0, 2, 4, \ldots$. Extensions of this argument to more general rotationally inelastic situations have been given by McCurdy and Miller (1977) and Kreek et al. (1974, 1975)—see also problem 10.2.

Early numerical applications of the above 'initial value' integral representation may be noted (Miller 1970; Wong and Marcus 1971), but they suffer from two disadvantages. In the first place $S_{n_1 n_2}$ is not symmetric in n_1 and n_2, because all the trajectories that contribute to the integral are specified by the initial n value but relatively few may reach the neighbourhood of the final one. Secondly, rapid oscillations in the integrand may necessitate integration over a very fine $\bar{\alpha}_1$ grid, in order to handle the cancellations correctly, which could involve a prohibitive number of long trajectory runs. It is therefore preferable to employ the following stationary phase and uniform approximations which are specifically designed to exploit the existence of such rapidly oscillatory integrands.

10.2 Stationary phase and uniform approximations

The stationary phase reduction of the integral in eqn (10.28) proceeds in the usual way. The stationary phase points are given according to eqn (10.29) by

$$\partial \Delta_{n_1 n_2}/\partial \bar{\alpha}_1 = [n(n_1, \bar{\alpha}_1) - n_2](\partial \bar{\alpha}/\partial \bar{\alpha}_1) + (\partial n/\partial \bar{\alpha}_1)(\bar{\alpha} - \alpha) - (\partial k/\partial \bar{\alpha}_1)R, \quad (10.31)$$

a result that may be simplified by use of the identity

$$k^2(n, \bar{\alpha}_1) = 2m\{E - H_0[(n(\bar{\alpha}_1) + \delta)\hbar]\}/\hbar^2, \quad (10.32)$$

together with $I = (n + \delta)\hbar$ and $\omega = \partial H_0/\partial I$; hence

$$\partial k/\partial \bar{\alpha}_1 = -m(\partial H/\partial I)(\partial n/\partial \bar{\alpha}_1)/k\hbar = -m\omega(\partial n/\partial \bar{\alpha}_1)/P. \quad (10.33)$$

This means, in the light of eqn (10.23), that the two final terms in (10.31) cancel, so that the stationary phase condition reduces to the classical identity

$$n(n_1, \bar{\alpha}_1) = n_2. \quad (10.34)$$

In other words, as shown in Fig. 10.3, the stationary phase angles $\bar{\alpha}_1$ are those for which the final quantum number function $n(n_1, \bar{\alpha}_1)$ coincides with the target value n_2. Since $n(n_1, \bar{\alpha}_1)$ is periodic in $\bar{\alpha}_1$ the number of roots of (10.34) is even—typically two in simple applications.

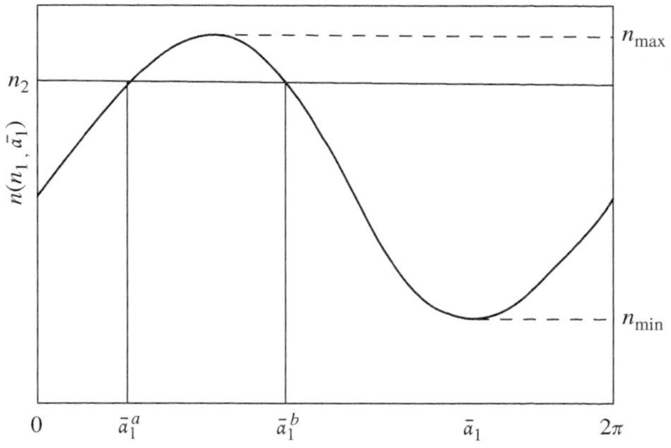

Fig. 10.3 Final quantum numbers $n(n_1, \bar{\alpha}_1)$ as a function of the modified initial angle $\bar{\alpha}_1$, for trajectories with a common initial quantum number n_1.

The approximation is completed by finding the second derivative,

$$\partial^2 \Delta_{n_1 n_2}/\partial \bar{\alpha}_1^2 = (\partial n/\partial \bar{\alpha}_1)_{n_1} (\partial \bar{\alpha}/\partial \bar{\alpha}_1)_{n_1}, \tag{10.35}$$

and using the standard integrals

$$\int_{-\infty}^{\infty} \exp(\pm \mathrm{i} a^2 x^2) \mathrm{d}x = (\pi)^{1/2} a^{-1} \exp(\pm \mathrm{i}\pi/4). \tag{10.36}$$

Hence, in the case of two roots labelled a and b for positive and negative values of $\partial^2 \Delta_{n_1 n_2}/\partial \bar{\alpha}_1^2$ respectively,

$$S_{n_1 n_2}^{\mathrm{PSC}} \simeq P_a^{1/2} \exp(\mathrm{i}\Delta_a + \mathrm{i}\pi/4) + P_b^{1/2} \exp(\mathrm{i}\Delta_b - \mathrm{i}\pi/4), \tag{10.37}$$

where

$$P_a = \left[\frac{1}{2\pi} \left(\frac{\partial \bar{\alpha}}{\partial \bar{\alpha}_1} \right)_{n_1} \Big/ \left(\frac{\partial^2 \Delta}{\partial \bar{\alpha}_1^2} \right) \right]_a = \left[\frac{1}{2\pi} \left(\frac{\partial \bar{\alpha}_1}{\partial n} \right)_{n_1} \right]_a \tag{10.38a}$$

and similarly

$$P_b = - \left[\frac{1}{2\pi} \left(\frac{\partial \bar{\alpha}_1}{\partial n} \right)_{n_1} \right]_b . \tag{10.38b}$$

Equations (10.38a) and (10.38b) assume that $(\partial \bar{\alpha}/\partial \bar{\alpha}_1)_{n_1}$ is positive and that $(\partial n/\partial \bar{\alpha}_1)_{n_1}$ is positive and negative at roots a and b respectively; if, on the other hand, $(\partial \bar{\alpha}/\partial \bar{\alpha}_1)_{n_1} < 0$ the a and b roots would have negative and positive values of $(\partial n/\partial \bar{\alpha}_1)_{n_1}$ respectively and the signs of P_a and P_b in the above expressions would be reversed. In either case P_a and P_b have classical interpretations as the limits of the fractional number of trajectories, $\delta\bar{\alpha}_1/2\pi$, falling into the quantum number range δn. It is readily verified that eqn (10.37) yields a primitive semiclassical probability

$$P^{\mathrm{PSC}}_{n_1 n_2} = |S_{n_1 n_2}|^2 = P_a + P_b + 2(P_a P_b)^{1/2} \sin(\Delta_b - \Delta_a) \tag{10.39}$$

which oscillates around the sum of these classical probabilities over the range of classically accessible final quantum numbers n_2. However, as usual, $P^{\mathrm{PSC}}_{n_1 n_2}$ diverges as n_2 approaches the classical thresholds $n_{\max}$ and $n_{\min}$, where $\partial n/\partial\bar{\alpha}_1 = 0$.

Connor and Marcus (1971) were the first to tackle these spurious singularities by use of a uniform Airy approximation similar to that used to describe rainbow scattering (see Section 9.2). The mathematical similarity, which is expounded at length in Appendix B, lies in the fact that any integral of the form

$$I(\gamma) = \int_{-\infty}^{\infty} g(x) \exp[\mathrm{i}f(x, \gamma)]\mathrm{d}x, \tag{10.40}$$

such that $f(x, \gamma)$ has two possibly coalescent stationary phase points, a minimum at x_a and a maximum at x_b, may be approximated as (see eqn (B.14))

$$\begin{aligned} I(\gamma) \simeq \sqrt{\pi} \exp[\mathrm{i}\bar{f}(\gamma)] \Big\{ &(p_a^{1/2} + p_b^{1/2})\zeta^{1/4}(\gamma) Ai[-\zeta(\gamma)] \\ &- \mathrm{i}(p_a^{1/2} - p_b^{1/2})\zeta^{-1/4}(\gamma) Ai'[-\zeta(\gamma)] \Big\}, \end{aligned} \tag{10.41}$$

where

$$\begin{aligned} p_a &= 2\pi g_a^2/f_a'', \qquad p_b = -2\pi g_b^2/f_b'', \\ \zeta(\gamma) &= \{3[f_b(\gamma) - f_a(\gamma)]/4\}^{2/3}, \quad \bar{f}(\gamma) = [f_a(\gamma) + f_b(\gamma)]/2. \end{aligned} \tag{10.42}$$

This approximation is therefore applicable to the S matrix integral in eqn (10.28), provided that the widths of the stationary phase regions are sufficiently small compared with 2π to justify formal extension of the integration range to $-\infty < x < \infty$. $g(x)$ is identified with $(\partial\bar{\alpha}/\partial\bar{\alpha}_1)^{1/2}_{n_1}$, $f(x, \gamma)$ with $\Delta_{n_1 n_2}(\bar{\alpha}_1)$, and n_2 with γ. Hence

$$S^{\mathrm{Airy}}_{n_1 n_2} \simeq \sqrt{\pi}\mathrm{e}^{\mathrm{i}\overline{\Delta}} \left\{ \left(P_a^{1/2} + P_b^{1/2}\right) \zeta^{1/4} Ai(-\zeta) - \mathrm{i}\left(P_a^{1/2} - P_b^{1/2}\right) \zeta^{-1/4} Ai'(-\zeta) \right\} \tag{10.43}$$

where P_a and P_b are given by (10.38a) and (10.38b) and

$$\overline{\Delta} = (\Delta_a + \Delta_b)/2, \quad \zeta = [3(\Delta_b - \Delta_a)4]^{2/3}. \tag{10.44}$$

As usual the uniform approximation depends on the same quantities, P_ν and Δ_ν, as the primitive semiclassical form but with the advantage that the singularities in $P_a^{1/2}$ and $P_b^{1/2}$ as $n_2 \rightarrow n_{\max}$ or $n_2 \rightarrow n_{\min}$ are exactly cancelled by the disappearance of $\zeta^{1/4}$; similarly the divergence of $\zeta^{-1/4}$ is compensated by the vanishing of $P_a^{1/2} - P_b^{1/2}$. Equation (10.44) may also be continued into the classically forbidden regions, with $\Delta_b - \Delta_a$ assigned the phase $\exp(3\mathrm{i}\pi/2)$ so that $\zeta = -|\zeta|$, but details are deferred until Section 10.5.

Two types of difficulty can, however, arise in applying the uniform Airy approximation. One is fundamental, in that it is not always valid to assume an effectively infinite integration range in eqn (10.28); the second, which is merely computational, concerns elimination of phase uncertainties of $(2n_i + 1)\pi$, in order to ensure that the phase difference $\Delta_b - \Delta_a$ vanishes at the trajectory coalescence points where the P_v have their singularities.

To tackle the fundamental difficulty first, Stine and Marcus (1973) build in the periodicity of the final quantum number function $n(n_1, \bar{\alpha}_1)$ in Fig. 10.3 by mapping the phase variation on to a canonical function of the form

$$f(x, \gamma) = \varphi(\gamma) - \xi(\gamma)\cos x - mx, \tag{10.45}$$

in place of the cubic form used to derive the Airy approximations (10.41) and (10.43). Details of the derivation are given in Section B.4. The result is termed the uniform Bessel approximation

$$S^{\text{Bessel}}_{n_1 n_2} \simeq (\pi/2)^{1/2} \exp(\mathrm{i}\bar{\Delta} - \mathrm{i}m\pi/2) \Big[(P_a^{1/2} + P_b^{1/2})(\xi^2 - m^2)^{1/4} J_m(\xi) + \mathrm{i}(P_a^{1/2} - P_b^{1/2})\xi(\xi^2 - m^2)^{-1/4} J'_m(\xi)\Big], \tag{10.46}$$

where $J_m(\xi)$ is the Bessel function (Abramowitz and Stegun 1965), with index $m = |n_2 - n_1|$ and argument ξ determined by the transcendental equation

$$(\xi^2 - m^2)^{1/2} - m\arccos(m/\xi) = (\Delta_b - \Delta_a)/2, \tag{10.47a}$$

if the transition is classically accessible, or

$$(m^2 - \xi^2)^{1/2} - m\,\text{arccosh}(m/\xi) = |\Delta_b - \Delta_a|/2 \tag{10.47b}$$

in the classically inaccessible case. The superiority of eqn (10.46) over (10.44) in 'near elastic' situations (where the overall phase difference between any pair of trajectories is small) is illustrated below in the context of the analytically soluble forced harmonic problem. Various other two-root uniform approximations have also been suggested: Laguerre (Child and Hunt 1977), harmonic (Hunt and Child 1978) and non-integer Bessel (Connor and Mayne 1979).

We turn now to the problem of phase ambiguities discussed, for example, by Fraser et al. (1975), which is particularly serious if the trajectories are propagated in Cartesian coordinates, because the classical generators $F_2(x_i, n_i)$, which appear in eqn (10.30) for the overall phase function, are ambiguous by multiples of $(2n_i + 1)\pi$. The difficulty may be removed by reference to Fig. 10.4, which shows the final trajectory end points, plotted as a contour C'_1, against the target orbit C_2 in the final phase space. Now Child (1978) argues (i) that the area enclosed by C'_1 is the same as that enclosed by the initial phase orbit C_1, namely $(n_1 + \frac{1}{2})\hbar$ for a quantized oscillator, and (ii) that the action difference $(\Delta_b - \Delta_a)$ that determines ξ may be interpreted as the smaller of the two areas into which C'_1 is divided by C_2 if $n_1 < n_2$, or C_2 is divided by C'_1 if $n_1 > n_2$. This choice of the smaller area may also be understood in terms of the sign of

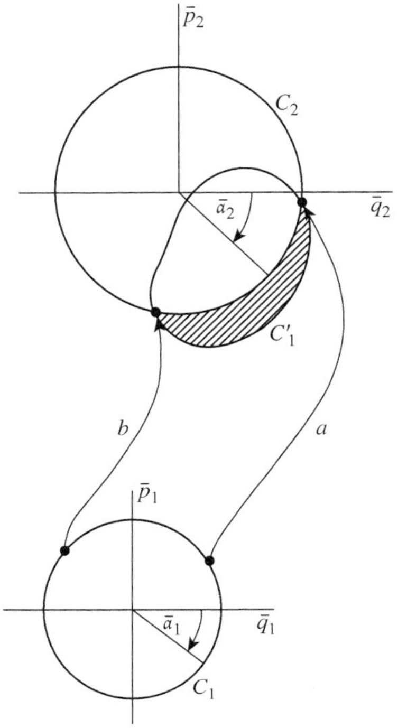

Fig. 10.4 A pictorial view of the dynamical phase space transformation. C_2 is the target quantized orbit and C_1' is the transform of the initial orbit C_1.

$(\partial\bar{\alpha}/\partial\bar{\alpha}_1)_{n_1}$. Thus the proper area is the shaded one in Fig. 10.4, but it is unnecessary to construct the diagram to compute its value; all that is necessary is to add or subtract multiples of $(2n_i + 1)\pi$ to the computed but ambiguous phase difference until

$$0 < \Delta_b - \Delta_a < \left(n_{\min} + \frac{1}{2}\right)\pi. \tag{10.48}$$

This correction is unnecessary at the primitive semiclassical level because (10.39) is invariant to substitution of $\Delta_b - \Delta_a$ by $(2n + 1)\pi - (\Delta_b - \Delta_a)$, but the uniform approximations require that the phase difference should vanish as the final quantum number approaches $n_{\min}$ or $n_{\max}$.

As an illustration of the strengths and weaknesses of the various approximations this section concludes with a discussion of the analytically soluble forced harmonic oscillator problem (Pechukas and Child 1976). The Hamiltonian takes the scaled form

$$H = \frac{1}{2}(p^2 + q^2) - f(t)q, \tag{10.49}$$

and the effect of an even forcing function, such that $f(t) = f(-t) \to 0$ as $t \to \pm\infty$, may be shown to cause a simple upward translation of the initial phase orbit in the modified $(n, \bar{\alpha})$ or $(\bar{p}, \bar{q})$ phase space, as defined by the equation

$$\bar{p} + i\bar{q} = (p + iq)e^{-it}. \tag{10.50}$$

In other words, as depicted in Fig. 10.5,

$$\bar{p} \to \bar{p} + \gamma, \quad \bar{q} \to \bar{q}, \tag{10.51}$$

where

$$\gamma = \int_{-\infty}^{\infty} f(t)e^{-it}dt. \tag{10.52}$$

It may be assumed without loss of generality that $n_1 < n_2$, in which case the geometry of Fig. 10.5 shows that the final variables $(n, \bar{\alpha})$ and the phase overlap area A are given by

$$\begin{aligned} n &= n_1 + \alpha^2/2 + \gamma(2n_1+1)^{1/2}\cos\bar{\alpha}_1 \\ (2n+1)^{1/2}\sin\bar{\alpha} &= (2n_1+1)^{1/2}\sin\bar{\alpha}_1 \\ A &= (2n_1+1)\bar{\alpha}_1 - (2n+1)\bar{\alpha} + \gamma(2n_1+1)^{1/2}\sin\bar{\alpha}_1. \end{aligned} \tag{10.53}$$

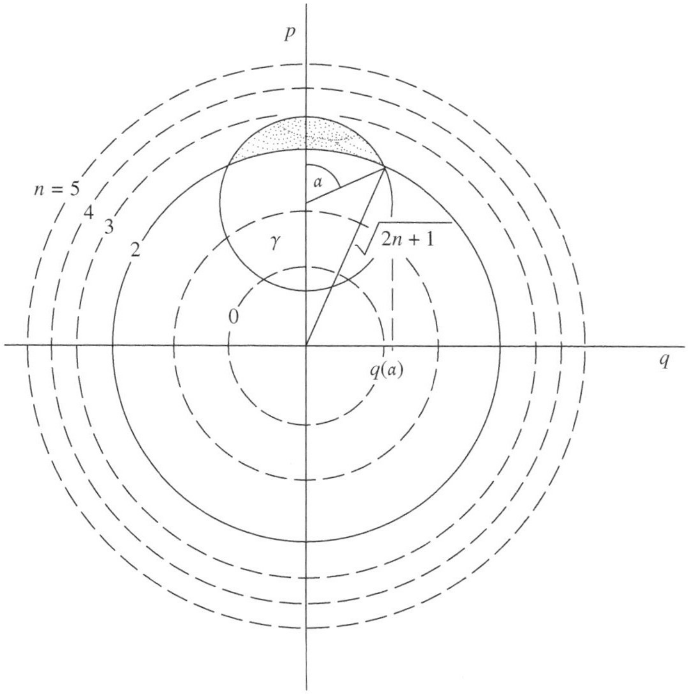

Fig. 10.5 The phase space transformation experienced by a symmetrically forced harmonic oscillator. The shaded area $A(\gamma)$ determines the phase difference between the root trajectories that contribute to the $0 \to 2$ transition. (Reprinted from Pechukas (1976), by permission of Taylor and Francis.)

Note that $\bar{\alpha}$ and $\bar{\alpha}_1$ have been measured, for geometrical convenience, from the $\bar{p}$ rather than the $\bar{q}$ axis and that the phase difference $\Delta_b - \Delta_a$ is equal either to the area A or, if A exceeds $(n_1 + 1/2)\pi$, to $(2n_1 + 1)\pi - A$. The physically relevant situations arise when $n = n_2$ in which case the classically allowed transition probabilities are given, according to eqns (10.38a) and (10.38b), by

$$P_a = P_b = |2\pi(\partial n/\partial\bar{\alpha}_1)_{n_1}|^{-1} = [2\pi\gamma(2n_1+1)^{1/2}\sin\bar{\alpha}_1]^{-1}. \tag{10.54}$$

Figure 10.6 shows a comparison between the primitive semiclassical, uniform Airy and uniform Bessel $0 \to 0$ and $0 \to 1$ transition probabilities and the exact forms given according to Pechukas and Light (1966) by

$$P_{00}^{\text{exact}} = \exp(-\gamma^2/2), \quad P_{01}^{\text{exact}} = (\gamma^2/2)\exp(-\gamma^2/2). \tag{10.55}$$

Notice that the analysis in Section 10.3 has been anticipated by including semiclassical probabilities for classically forbidden as well as classically allowed events. Two types of spurious singularity may be noted. The first occurs when $\gamma = (2n_2 + 1)^{1/2} \pm (2n_1 + 1)^{1/2}$ which implies, on setting $\bar{\alpha}_1$ equal to 0 or π in (10.53), that $n_2 = n_{\max}$ or $n_2 = n_{\min}$. These singularities arise from the divergence of $(\sin\bar{\alpha}_1)^{-1}$ in the probability terms given by (10.54) and they are removed by both forms of uniform correction, although there is evidence of overcompensation for the $0 \to 0$ transition in the Bessel case. The second type of singularity, which occurs in P_{00}^{Airy} as $\gamma \to 0$, arises from the term in γ^{-1} in eqn (10.55) which is correctly handled by the Bessel but not by the Airy uniform correction.

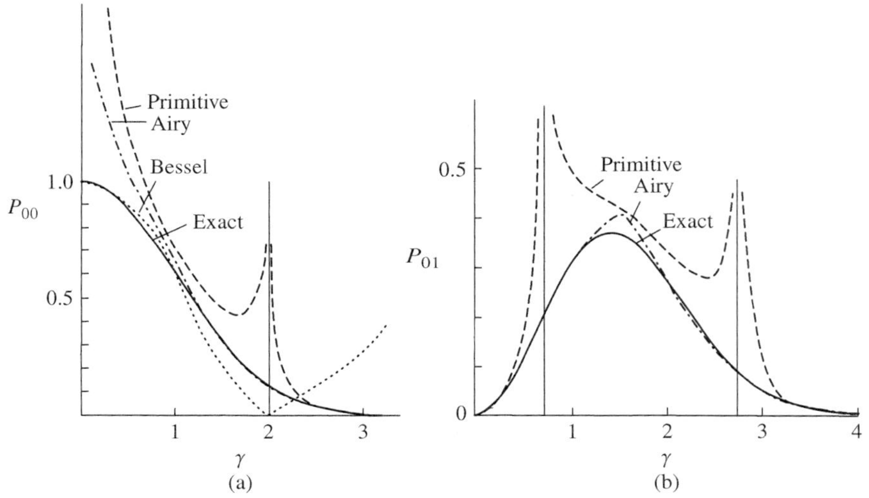

Fig. 10.6 (a) $0 \to 0$ and (b) $0 \to 1$ exact transition probabilities, as functions of the forcing parameter γ, compared with primitive, uniform Airy and uniform Bessel approximations. (Reprinted from Pechukas and Child (1976), by permission of Taylor and Francis.)

Further analysis of the $0 \to 0$ case is quite illuminating. Consider first the behaviour as $\gamma \to 0$. Equations (10.53) may be verified to have solutions $\bar{\alpha}_1 \simeq \pi/2 + \varepsilon$ and $\bar{\alpha} \simeq \pi/2 - \varepsilon$, where $\varepsilon \simeq \gamma/(2\sqrt{2})$, so that $A \simeq 2\gamma$ and $P_a \simeq \gamma^{-1}$. It follows from (10.44) that ζ varies as $\gamma^{2/3}$, with the result that $P_a^{1/2}\zeta^{1/4}$ diverges as $\gamma^{-1/3}$; consequently the Airy uniform approximation is invalid. On the other hand eqns (10.47a) and (10.47b), with $m = 0$, show that the Bessel parameter ξ is directly proportional to γ; hence $P_a^{1/2}(\xi^2 - m^2)^{1/4}$ remains finite as $\gamma \to 0$ and the Bessel approximation remains valid.

The second singularity in P_{00} may be handled by setting $\gamma = 2 - \delta$, in which case eqns (10.53) yield $\bar{\alpha}_1 \simeq \pi - \nu$ and $\bar{\alpha} \simeq \pi + \nu$ where $\nu^2 = \delta$. It also turns out that the complementary area $A' = \pi - A$, which determines the phase difference $\Delta_b - \Delta_a$ in this case, varies as $\delta\nu = \nu^3$. Hence $P_a^{1/2}$ varies as $\nu^{-1/2}$ while the Airy parameter ζ varies as $(A')^{2/3}$ or ν^2; consequently $P_a^{1/2}\zeta^{1/4}$ remains finite as $\nu \to 0$. By contrast the Bessel parameter ξ is directly proportional to A' and hence varies as ν^3; hence $(\xi^2 - m^2)^{1/4}$, with $m = 0$, varies as $\nu^{2/3}$, so that the term $P_a^{1/2}(\xi^2 - m^2)^{1/4}$ in eqn (10.46) vanishes as $\nu \to 0$. This failure of the Bessel approximation looks less spectacular than the divergences in Fig. 10.6 but it is equally serious from a fundamental point of view.

To summarize, this analysis shows that the uniform Airy approximation gives excellent results except for a catastrophic failure in the near elastic case, with similar failures expected for any $n \to n$ transition in weak interaction situations. Fortunately the Bessel approximation handles this near elastic situation with spectacular success, but it is probably best reserved for these special cases, both in view of its own weakness at the second singularity in Fig. 10.6 and because the transcendental nature of eqns (10.47a) and (10.47b) gives an awkward prescription for the parameter ξ. The point at which one decides to transfer from one approximation to the other is relatively unimportant because both give excellent results in regions away from the singularities. The reader is referred to Hunt and Child (1978) and Connor and Mayne (1979) in order to assess the merits of other 'uniform Laguerre' and 'non-integer uniform Bessel' approximations, respectively. Clarke and Dickinson (1973) and Duff and Truhlar (1975) report extensive tests for a number of standard vibrationally inelastic scattering models.

10.3 Classically forbidden events

The idea that classically forbidden probabilities can be handled by semiclassical methods first occurred in Section 2.1, where the probability $|\psi(x)|^2$ of finding a particle outside the classical region required the use of a 'non-classical' imaginary momentum. Similarly non-classical elastic scattering beyond the rainbow angle was handled in Section 9.2 by introduction of complex angular momenta. Here it will be shown how complex time solutions of the classical equations of motion may be used to treat quite general classically forbidden processes. The ideas were first formulated in this context by Stine and Marcus (1972) and Miller and George (1972).

Consider first the example of penetration through a quadratic potential barrier (Miller 1972):

$$V(x) = -\frac{1}{2}\lambda x^2. \tag{10.56}$$

The classical motion, at negative energy $-\Delta E$ and initial conditions $x < 0$ and $p > 0$ at $t \to -\infty$, follows the equations

$$x = -(2\Delta E/\lambda)^{1/2} \cosh \omega^* t, \qquad p = -(2m\Delta E)^{1/2} \sinh \omega^* t, \tag{10.57}$$

where

$$\omega^* = (\lambda/m)^{1/2}. \tag{10.58}$$

In other words, in accordance with normal classical experience, the coordinate x is restricted to negative values and the momentum changes sign on reflection from the barrier at time $t = 0$. Suppose, however, that the time is given an imaginary increment, $\mathrm{i}\pi/\omega^*$, during the motion. Substitution for the final time

$$t_2 = t_2' + \mathrm{i}\pi/\omega^* \tag{10.59}$$

in eqn (10.57) yields

$$\begin{aligned} x_2 &= -(2\Delta E/\lambda)^{1/2} \cosh(\omega^* t_2' + \mathrm{i}\pi) = (2\Delta E/\lambda)^{1/2} \cosh \omega^* t_2', \\ p_2 &= -(2m\Delta E)^{1/2} \sinh(\omega^* t_2' + \mathrm{i}\pi) = (2m\Delta E)^{1/2} \sinh \omega^* t_2'. \end{aligned} \tag{10.60}$$

The choice of this special time increment has caused the particle to pass through the barrier and to appear at positive x with positive momentum. From a mathematical viewpoint it is actually immaterial when the imaginary time increment is acquired, but it is physically most illuminating to suppose that it occurs between $t = 0$ and $t = \mathrm{i}\pi/\omega^*$, because on setting $t = \mathrm{i}t''$ in (10.57), with $0 < t'' < \pi$,

$$\begin{aligned} x &= -(2\Delta E/\lambda)^{1/2} \cosh(\mathrm{i}\omega^* t'') = -(2\Delta E/\lambda)^{1/2} \cos(\omega^* t''), \\ p &= -(2m\Delta E)^{1/2} \sinh(\mathrm{i}\omega^* t'') = -\mathrm{i}(2m\Delta E)^{1/2} \sin(\omega^* t''), \end{aligned} \tag{10.61}$$

showing that the imaginary time motion has been restricted to the barrier region $-(\Delta E/\lambda)^{1/2} < x < (\Delta E/\lambda)^{1/2}$ where the momentum is purely imaginary. This idea will be explored at greater length under the heading of 'instanton theory' in Section 11.4.

Finally, it is easy to verify from eqns (10.57) that the tunnelling path has an associated imaginary classical action

$$\begin{aligned} S &= \int_0^{\mathrm{i}\pi/\omega^*} p\dot{x}\mathrm{d}t = (2\Delta E/\omega^*) \int_0^{\mathrm{i}\pi/\omega^*} \sinh^2 \omega^* t \, \mathrm{d}t \\ &= \mathrm{i}\pi\Delta E/\omega^*. \end{aligned} \tag{10.62}$$

The corresponding semiclassical amplitude, $\exp(\mathrm{i}S/\hbar)$, therefore coincides exactly with the primitive semiclassical form $\exp(-\pi\Delta E/\hbar\omega^*)$. It is, however, difficult to adapt this complex time approach to obtain the correct uniform amplitude, $(1 + \kappa^2)^{-1/2}\mathrm{e}^{-\mathrm{i}\phi}$ implied by (3.48), although Miller (1974) has obtained the corresponding tunnelling probability by summing over complex time segments appropriate to repeated internal reflections inside the barrier.

For a second type of insight into the complex time solutions of the classical equations of motion, it is instructive to refer back to the exactly soluble forced harmonic oscillator model discussed in the previous section. One sees for example, from the first of eqns (10.53), that the final quantum number n is restricted to the range

$$n_1 + \gamma^2/2 - \gamma(2n_1+1)^{1/2} < n < n_1 + \gamma^2/2 + \gamma(2n_1+1)^{1/2} \qquad (10.63)$$

if the initial angle $\bar{\alpha}_1$ is chosen to be real. Classically forbidden real n values outside the above range may, however, be reached by starting from carefully chosen complex angles $\bar{\alpha}_1$; for example, the choice $\bar{\alpha}_1 = \mathrm{i}\bar{\alpha}_1'$ leads according to eqn (10.53) to

$$n = n_1 + \gamma^2/2 + \gamma(2n_1+1)^{1/2}\cosh\bar{\alpha}_1' > n_{\mathrm{max}}, \qquad (10.64a)$$

while the choice $\bar{\alpha}_1 = \pi + \mathrm{i}\bar{\alpha}_1'$ yields

$$n = n_1 + \gamma^2/2 - \gamma(2n_1+1)^{1/2}\cosh\bar{\alpha}_1' < n_{\mathrm{min}}. \qquad (10.64b)$$

Pechukas and Child (1976) used eqns (10.64a) and (10.64b), together with the corresponding variants of eqns (10.53) for the final angle $\bar{\alpha}$ and complex action A, to cover the non-classical regions in Fig. 10.6.

Of course few situations are analytically soluble, but eqns (10.64a) and (10.64b) suggest the general existence of lines in the complex initial angle plane, along which the final quantum number takes real values outside the classically accessible range. The aim of the complex time approach is to locate these lines and to compute the necessary classical probabilities P_ν and actions Δ_ν for use in the uniform Airy or uniform Bessel equations, (10.43) or (10.46) respectively. Equations (10.29), (10.30), (10.38a) and (10.38b) still apply but the probabilities P_ν are imaginary and the Airy and Bessel function arguments lie outside their classical ranges, $\zeta > 0$ and $|\xi| > m$ respectively. It is also important to choose the branch of $\Delta_b - \Delta_a$ in such a way that ζ or ξ is real; for example, $\arg(\Delta_b - \Delta_a) = 3\pi/2$ in the Airy case (see the comment after eqn (10.44)).

Computational problems raised by this complex trajectory approach have been discussed by Miller and George (1972) and Stine and Marcus (1972). The most stable algorithm is to initiate the trajectory with real n, P and R and a complex angle α_1, and then to follow a complex time path such that the translational coordinate remains real throughout. It is then a matter of interpolating for the α_1 values for which $n(n_1, \alpha_1)$ is real and finally for those for which n coincides with the desired value n_2. A less accurate, but fairly reliable, approach for transition probabilities close to the classical thresholds is to generate the classical probabilities and action differences by analytical continuation from the classical region (Miller 1970).

This complex trajectory approach was extended to reactive scattering by George and Miller (1972). The most interesting finding is that the real parts of tunnelling trajectories at energies below the saddle point tend to 'cut the corner' on the potential energy surface, in a manner which is the converse of the classical 'bobsled' effect; just as trajectories at energies above the barrier maximum are flung out by the curvature in the reaction path, those that have energies below it seem to be flung

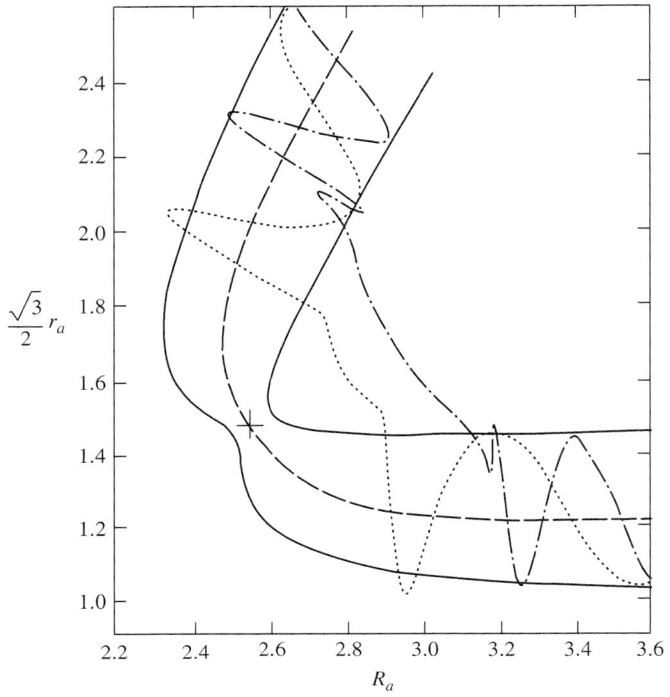

Fig. 10.7 Tunnelling trajectories for $H + H_2$, at $E = 0.2\,\text{eV}$ (dotted line) and $E = 0.02\,\text{eV}$ (chain line). (Reprinted with permission from George and Miller (1972). Copyright 1972, AIP Publishing LLC.)

inwards. Figure 10.7 shows an example for the collinear $H + H_2$ reaction (George and Miller 1972).

Such reactive complex trajectory calculations are extremely taxing, both with respect to time and to ingenuity. Marcus and Coltrin (1977) have therefore introduced an alternative vibrationally adiabatic approach, based on a natural collision coordinate system (u, v). The coordinate u is measured along the steepest descents path between reactants and products in the proper skewed axis system (Child 1974b), while v is taken perpendicular to this path. The procedure is then to define a set of adiabatic barrier functions, $W_n(u)$, by the semiclassical quantization condition

$$\frac{(2m)^{1/2}}{\hbar} \int_{a(u)}^{b(u)} [W_n(u) - V(u,v)]\,\mathrm{d}v = (n + 1/2)\hbar. \tag{10.65}$$

Finally, the Marcus and Coltrin (1977) tunnelling path is taken as the locus of inner turning points $a(u)$ in eqn (10.65), a prescription that is found to give excellent results for the $H + H_2$ reaction. Extensions of these ideas to higher-dimensioned systems have been given by Skodje et al. (1982), Garrett and Truhlar (1983) and Lynch et al. (1989).

10.4 Rotational rainbows and higher interference structures

The exposition so far has assumed the existence of a single physical variable—an angle in the case of elastic scattering, or a final quantum number if the scattering is inelastic. Formal extension of the theory to cover situations involving several final observables is quite straightforward in principle, and the nature of possible interference patterns raises some points of interest.

The simplest and best-studied example is the so-called rotational rainbow scattering of an atom from a rigid rotor in a prepared initial rotational state j_1. The final observables are the final rotational state j and the angle of deflection χ, each of which is regarded for the purpose of the theory as a function of two trajectory sampling variables—the initial orientation of the target, α, and the incident angular momentum l. One finds, by an obvious generalization of eqn (10.28), that the scattering amplitude for observation of final state j_2 at angle θ may be expressed in the canonical integral representation

$$f_{j_1 j_2}(\theta) = \frac{1}{2\pi}\int_0^\infty \int_0^{2\pi} g(\alpha,\, l)\exp[\mathrm{i}\Delta(j_1,\, j_2,\, \theta;\, \alpha,\, l)]\mathrm{d}\alpha\,\mathrm{d}l, \tag{10.66}$$

where $\Delta(j_1,\, j_2,\, \alpha;\, \theta,\, l)$ is defined in such a way that the stationary phase conditions $\partial\Delta/\partial\alpha = \partial\Delta/\partial l = 0$ yield the following pairs of root trajectory conditions:

$$j(\alpha,\, l) = j_2, \quad |\chi(\alpha,\, l)| = \theta, \tag{10.67}$$

$j(\alpha,\, l)$ and $\chi(\alpha,\, l)$ being the final observables arising from a trajectory specified by initial variables $(\alpha,\, l)$. For example, following Schinke and McGuire (1979) and Schinke et al. (1982), one finds within the infinite order sudden (IOS) fixed orientation approximation (Kouri 1979) that

$$f_{0\, j_2}(\theta) \simeq \frac{-(-1)^{j_2}}{8\pi k(\sin\theta)^{1/2}}\int_0^\infty \int_0^{2\pi} \sum_v \left[\left(l+\frac{1}{2}\right)\sin\alpha\right]^{1/2} \exp(\mathrm{i}\Delta_v)\mathrm{d}\alpha\,\mathrm{d}l, \tag{10.68}$$

where the sum is taken over all four sign choices of the function

$$\Delta_\nu(j_2,\, \theta;\, \alpha,\, l) = 2\eta(\alpha,\, l) \pm \left[\left(j_2+\frac{1}{2}\right)\alpha + \pi/4\right] \pm \left[\left(1+\frac{1}{2}\right)\theta - \pi/4\right]. \tag{10.69}$$

$\eta(\alpha,\, l)$ in this expression is the JWKB phase shift at the fixed orientation α;

$$\begin{aligned}\eta(\alpha,\, l) &= \lim_{R\to\infty}\left(\int_{a(\alpha,\, l)}^{R} k_l(\alpha,\, R)\mathrm{d}R + \left(l+\frac{1}{2}\right)\pi - kR\right),\\ k_l(\alpha,\, R) &= [2m(E - V(\alpha,\, R) - L^2/2mR^2)]^{1/2}/\hbar,\\ L^2 &= \left(l+\frac{1}{2}\right)^2 \hbar^2.\end{aligned} \tag{10.70}$$

Thus the stationary phase conditions yield

$$\theta = \pm\chi(\alpha,\, l), \quad j_2 = \pm j(\alpha,\, l) \tag{10.71}$$

where

$$\begin{aligned}
\chi(\alpha,\, l) &= 2(\partial\eta/\partial l) \\
&= \pi - 2L\int_a^\infty R^{-2}[2m(E - V(\alpha,\, R) - L^2/2mR^2)]^{-1/2}\mathrm{d}R, \\
j(\alpha,\, l) &= 2(\partial\eta/\partial\alpha) \\
&= -2m\int_a^\infty (\partial V/\partial\alpha)[2m(E - V(\alpha,\, R) - L^2/2mR^2)]^{-1/2}\mathrm{d}R.
\end{aligned} \tag{10.72}$$

The appearance of the final probability distribution $|f_{j_1 j_2}(\theta)|^2$ depends on the number and disposition of the roots of eqn (10.67). Some features may be predicted by symmetry. For example, as discussed above, the initial conditions $(\alpha,\, l)$ and $(\alpha + \pi,\, l)$ yield physically identical trajectories if the target has central symmetry but the associated root trajectory contributions to $f_{j_1 j_2}(\theta)$ differ in phase by $\exp[\mathrm{i}(j_1 - j_2)\pi]$. Hence the symmetry selection rule $\Delta j = 2k$ appears as a semiclassical interference effect. McCurdy and Miller (1977) and Kreek et al. (1974, 1975) discuss particular examples.

The nature and classification of other non-symmetry-determined interference features may be introduced by reference to Fig. 10.8, which is taken from Schinke et al. (1982). It shows contours of the functions $\chi(\alpha,\, l)$ and $j(\alpha,\, l)$ for two interaction

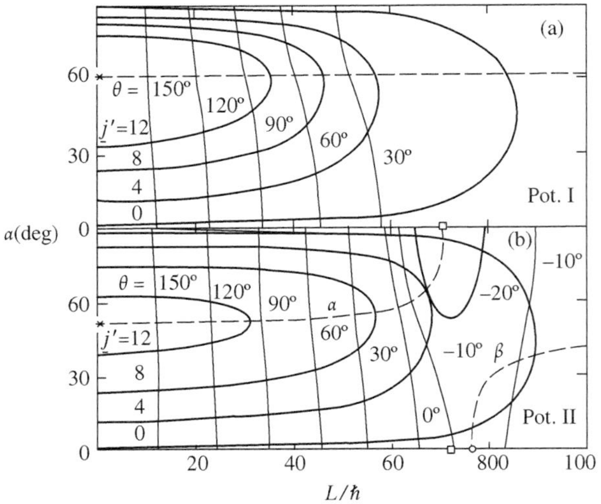

Fig. 10.8 Contours of the final angular momentum j' and scattering angle θ as functions of the initial angular momentum L and orientation γ for two potential models. (Reprinted with permission from Schinke et al. (1982). Copyright 1982, AIP Publishing LLC.)

potentials. Potential I, which is purely repulsive, allows at most two real solutions of (10.67), which appear as contour intersections in Fig. 10.8(a), whereas potential II has an isotropic attractive term which allows up to four real roots of (10.67) or contour intersections in Fig. 10.8(b). It is also evident that the existence or otherwise of real roots of (10.67), which correspond to classically allowed events, is limited by the choice of j and χ; for example, there is no intersection between the $\chi = 30°$ and $j = 8$ contours in Fig. 10.8(a). This means that the observation space (j, θ) may be divided into regions according to the number of real roots of the stationary phase conditions (10.67). Figure 10.9(a) shows that potential I gives rise to two such regions—a lower region that is accessible by two real trajectories and an upper one that is classically inaccessible except in a complex sense (see Section 10.3). Similarly, Fig. 10.9(b) is divided into three regions—a small region reached by four real trajectories, a larger one that is accessible by two and a third classically inaccessible region.

The dispositions of the dividing lines (or caustics) in Fig. 10.9 play a vital role in determining the qualitative form of the interference pattern. In the first place one expects oscillatory behaviour wherever two or more real trajectories interfere, and smooth exponential decay elsewhere. Secondly, the pattern will become more complicated, the larger the number of real trajectories. Berry (1976) and Connor et al. (1976) have demonstrated that the techniques of catastrophe theory (Thom 1975; Postern and Stewart 1978) are admirably suited for the classification of such interference patterns. Details of the mathematics are outlined in Appendix B, but it is sufficient for the general reader to note that the catastrophes are associated in a mathematical sense with coalescences between the stationary points of certain generic functions, and that the caustic boundaries in Fig. 10.9 correspond to coalescence of the roots of (10.67).

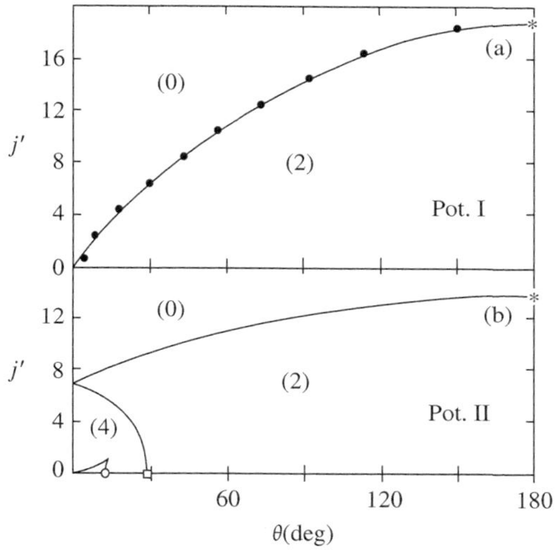

Fig. 10.9 Divisions of the observation space (θ, j') according to the number of real root trajectories, which appear as contour intersections in Fig. 10.8. (Reprinted with permission from Schinke et al. (1982). Copyright 1982, AIP Publishing LLC.)

Moreover, such coalescences aptly cause catastrophic singularities in the primitive semiclassical theory. The higher catastrophe types give rise to more and more complicated caustic boundary patterns and each such pattern has an associated generic diffraction integral, which at least in principle provides the basis for an appropriate uniform approximation.

The simplest situation is the one illustrated in Fig. 10.9(a)—a single caustic boundary with two (classically allowed) real stationary phase points on one side and two (forbidden) complex ones on the other. The catastrophe type is termed a 'fold' because the classically allowed region folds back at the caustic and the canonical diffraction integral is the Airy function, regardless of the dimensionality of the observation space (e.g. the two-dimensional (j, θ) space in Fig. 10.9(a)). Figures 10.10(a) and 10.10(b) confirm that the transition probabilities, taken either as functions of θ at constant j or vice versa, do indeed conform to this behaviour, with oscillations on the bright side of the caustic and exponential decay on the other. The term 'rotational rainbow' has been coined for this type of behaviour, although it arises in this simple form only if the interaction potential is predominantly repulsive whereas the elastic scattering

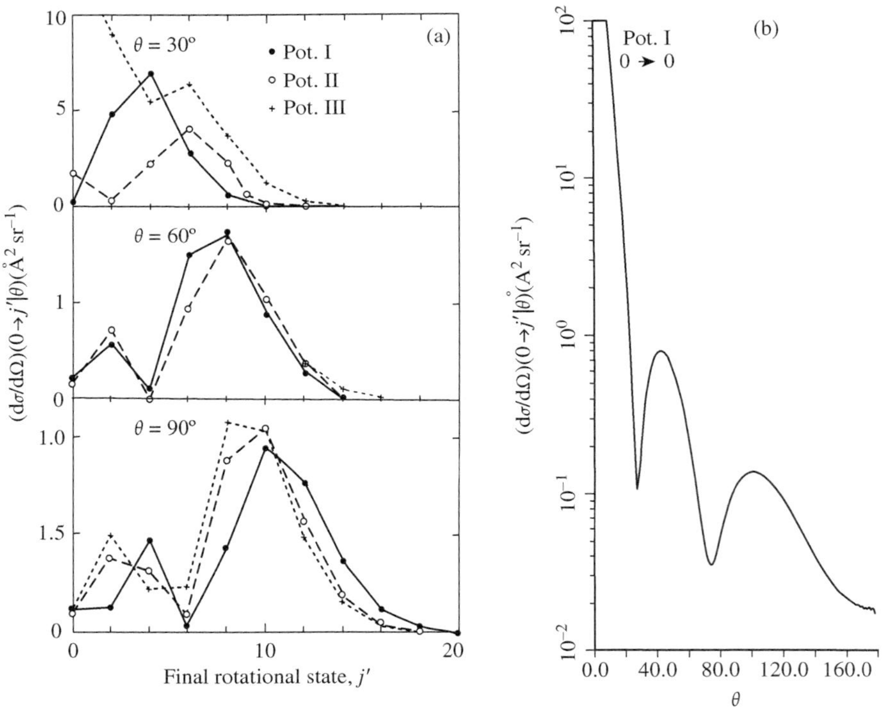

Fig. 10.10 (a) Rotational rainbow structure seen as a function of j' at fixed scattering angles θ. (b) Rotational rainbow structure as a function of θ at the fixed value $j' = 0$. (Taken from Schinke et al. (1982) with permission. Copyright 1982, AIP Publishing LLC.)

rainbow, which gave rise to the name, is attributed to attractive forces. The fundamental similarity is that both situations involve interference between just two stationary phase trajectories.

The situation in Fig. 10.9(b) is complicated from a mathematical point of view by the existence of four contributing trajectories, all of which are real inside the additional small cusp-shaped boundary. Although the region involved appears to be too small to have a striking effect, Wolf and Korsch (1984) have used a cusp-catastrophe-based theory to account for the interference pattern, which arises when more strongly attractive terms are added to the potential energy function. It should also be noted that Connor and Farrelly (1981) have applied a similar theory to the non-adiabatic scattering of Ne and He^+. A related uniform 'hyperbolic umbilic' theory has also been successfully applied to the photodissociation of CH_3I (Gray and Child 1984).

10.5 Condon reflection principles

The catastrophe-based diffraction patterns, discussed in the previous section, are designed to be locally accurate close to caustic boundaries of the appropriate complexity. However, they take no account of other possible boundaries between classically allowed and forbidden behaviour. The strength of this local validity is that two locally valid patterns are frequently well separated in phase, so that they can be joined by use of the stationary phase approximation; Fig. 10.6(b), which uses two different uniform Airy approximations, is a case in point. Such expedients can, however, break down, as shown for example by the superiority of the uniform Bessel approximation in Fig. 10.6(a).

A different type of more global approximation is also available in certain circumstances, which aims to relate the observed interference pattern to the nodal structure of the parent wavefunction. The term 'Condon reflection', suggested by analogy with the fluctuations in Franck–Condon factors (see Section 5.3), has been coined for approximations designed to bring out this connection. A simple example is the uniform harmonic approximation of Hunt and Child (1978), which exploits the limits imposed by flux conservation on the possible phase difference between a pair of interfering trajectories. It is given in the notation of eqns (10.29), (10.30), (10.38a) and (10.38b) by the equations (see eqn (B.105))

$$\begin{aligned} S^{\mathrm{Harm}}_{n_1 n_2} &= \left[\frac{\pi}{2}\right]^{1/2} \mathrm{e}^{\mathrm{i}\bar{\Delta}} \left[\left(P_a^{1/2} + P_b^{1/2}\right) \left(2n_1 + 1 - \eta^2\right)^{1/4} \psi_{n_1}(\eta) \right. \\ &\left. -\mathrm{i}\left(P_a^{1/2} - P_b^{1/2}\right) \left(2n_1 + 1 - \eta^2\right)^{-1/4} \left\{ \psi'_{n_1}(\eta) - \frac{1}{2\eta}[1 - (-1)^{n_1}]\psi_{n_1} \right\} \right], \end{aligned} \tag{10.73}$$

where it is assumed that $n_1 \leqslant n_2$. Here $\psi_n(\eta)$ is the harmonic oscillator wavefunction

$$\psi_n(\eta) = \pi^{-1/4}(2^n n!)^{-1/2} H_n(\eta) \exp(-\eta^2/2), \tag{10.74}$$

with an argument determined by the transcendental equation

$$\Delta_b - \Delta_a = \int_{-(2n_1+1)^{1/2}}^{\eta} (2n_1 + 1 - \eta^2)^{1/2} \mathrm{d}\eta, \tag{10.75}$$

where Δ_a and Δ_b are the phases of the contributing trajectories. The integral can of course be evaluated analytically according to the location of η, which lies in the classical range, $|\eta| < (2n_1 + 1)^{1/2}$, if $\Delta_b - \Delta_a$ is real, and outside it if $\Delta_b - \Delta_a$ is imaginary (see (B.99)–(B.101)).

The strength of this approximation is that it enshrines a direct relationship between the final state probability distribution and the form of the initial state wavefunction, subject to the existence of only two interfering trajectories—a collisional generalization of the spectroscopic Condon reflection approximation (Herzberg 1950). Figure 10.11 shows an example of this behaviour for inelastic scattering (Secrest and Johnson 1966). Notice that the reflection character applies either to the variation of $P_{n_1 n_2}(E)$ as a function n_2 at a fixed energy, or as a function of E at fixed n_2. To guide the eye n_2 has been taken as continuous for the purpose of drawing contours although the physically relevant values are of course integers. Child and Lefebvre (1978) obtain a similar diagram from reactive scattering probabilities obtained by Baer (1973).

The same idea may be extended to direct polyatomic photodissociation, in which case the parent wavefunction is spread over two or more coordinates. Child and Shapiro (1983) was able to demonstrate, by a numerically accurate quantum computation, that the nodal structure of an initial CH_3I vibrational wavefunction was reflected on the distribution of the photofragments over the translational and vibrational motion. Child et al. (1983) then demonstrated the semiclassical origin of this behaviour. The algorithm for the classical simulation initiated a family of trajectories from sets of coordinates and momenta $(\mathbf{Q}, \mathbf{P})$ appropriate to the initial vibrational state on the lower electronic energy surface, followed by a jump to the upper dissociative surface in such a way that position and momentum are conserved. The resulting trajectories were propagated out to dissociation and analysed the distribution of the energy between the translational and the various possible internal degrees of motion of the photodissociation fragments. Thus the classical dynamics provides a mapping between the initial coordinate and the final energy spaces, which was found to preserve the essential topology of the parent wavefunction. Subsequently Gray and Child (1984) demonstrated that a uniform classical S matrix approach is quantitatively as well as qualitatively accurate for this reaction. The rather different but computationally simpler frozen Gaussian semiclassical technique of Lee and Heller (1982), which is discussed in Section 8.3, is also appropriately noted at this point.

Finally, it is interesting to note that Schinke (1989) has devised yet another Condon reflection principle for direct photodissociation, which makes specific reference to the form of the reflecting mirror. The argument is that the cross-section for photodissociation into final state n at absorption frequency ν may be approximated as

$$\sigma_{\mathrm{cl}}(n|\nu) = \nu \int\int W_{\mathrm{gr}}(\mathbf{Q},\, \mathbf{P})\delta[H_{\mathrm{ex}}(\mathbf{Q},\, \mathbf{P}) - E]\delta[N(\mathbf{Q},\, \mathbf{P}) - (n+1/2)\hbar]\mathrm{d}\mathbf{Q}\,\mathrm{d}\mathbf{P}, \tag{10.76}$$

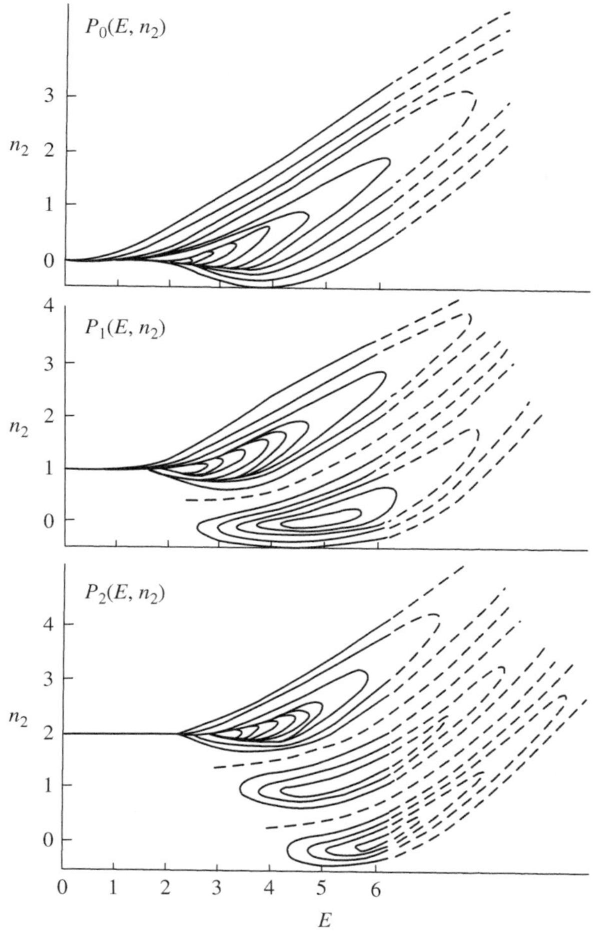

Fig. 10.11 Condon reflection structure associated with inelastic scattering. (Reprinted from Child (1978) by permission of Taylor and Francis.)

where $W_{gr}(\mathbf{Q}, \mathbf{P})$ is the quantum mechanical Wigner (1932a) distribution function for the ground state, which is expressible, in a harmonic oscillator model, in terms of the coordinate and momentum representations of the initial wavefunction, $\Psi_{gr}(Q_1, Q_2)$ and $\phi(P_1, P_2)$, respectively:

$$W_{gr}(Q_1, Q_2, P_1, P_2) = \Psi_{gr}^2(Q_1, Q_2)\phi_{gr}^2(P_1, P_2). \tag{10.77}$$

$H_{ex}(\mathbf{Q}, \mathbf{P})$ and $N(\mathbf{Q}, \mathbf{P})$ in eqn (10.76) respectively denote the upper state Hamiltonian and the fragment vibrational quantum number obtained by the classical photodissociation algorithm given above. The integral in eqn (10.76) may be evaluated by Monte Carlo sampling from an ensemble of trajectories, but convergence with a Wigner function weighting is notoriously slow; for example, Untch et al. (1989)

employed samples of 40 000 trajectories in order to obtain statistically reliable estimates for the classically forbidden events. On the other hand the calculation may be drastically simplified if the upper state potential that contributes to $H_{ex}(\mathbf{Q}, \mathbf{P})$ is extremely steep in the dissociative mode. The dominant contributions to the integral then come from the regions of the upper state classical turning points (compare eqns (5.61)–(5.73)), so that the initial momenta may be set equal to zero. This leaves only a double integral involving two delta functions, which are conveniently evaluated by adopting H_{ex} and N as the integration variables. The result reduces to

$$\begin{aligned}\sigma(n|\nu) &\simeq \mathrm{N}\Psi^2_{gr}[Q_1(n, \nu), Q_2(n, \nu)]|\partial(Q_1, Q_2)/\partial(H_{ex}, N)| \\ &\simeq \mathrm{N}\Psi^2_{gr}\{Q_1, [Q_2(Q_1)]\}(\partial V_{ex}/\partial Q_2)^{-1}_{Q}(\mathrm{d}N_v/\mathrm{d}Q_1)^{-1},\end{aligned} \tag{10.78}$$

where $Q_1(n, \nu)$ and $Q_2(n, \nu)$ are determined by the mapping equations

$$H_{ex}(Q_1, Q_2, 0, 0) = E, \quad N(Q_1, Q_2) = (n + 1/2)\hbar. \tag{10.79}$$

The energy E here obviously depends on the optical excitation frequency. The second line of eqn (10.78) is obtained by solving the first of eqns (10.79) for Q_2 as a function Q_1 and ν, allowing the second to be cast into the form $(n+1/2)\hbar = N_v(Q_1)$. Hence the Jacobian in the first of eqn (10.78) acts as the mirror that maps the nodal structure of the ground state wavefunction onto the product state distribution.

10.6 Problems

10.1. Consider a model in which $\bar{\alpha}_2 = \bar{\alpha}_1$ and

$$\Delta_{n_1^0 n_2^0}(\bar{\alpha}_1) = (n_1^0 - n_2^0)\bar{\alpha} - a\cos\bar{\alpha}_1.$$

(i) Confirm that eqn (10.28) reduces to

$$S_{n_1^0 n_2^0} = J_m(a),$$

where $m = |n_1^0 - n_2^0|$ and $J_m(a)$ denotes the Bessel function.

(ii) Compute the primitive semiclassical and uniform Airy transition probabilities for $a = 4.5$ and $-6 < n_1^0 - n_2^0 < 6$, and compare with the Bessel results.

10.2. The Hamiltonian for atom/rigid rotor scattering may be expressed in the reduced form (Kreek et al. 1974)

$$H' = \frac{1}{2}P^2 + \frac{l^2}{2R^2} + \frac{j^2}{2I} + V(R, \gamma),$$

where the geometrical angle γ may be expressed in terms of the angles α_l and α_j conjugate to l and j in the form (see problem 4.3)

$$\cos\gamma = \cos\alpha_1 \cos\alpha_j + [(j^2 - l^2 - j^2)/2lj]\sin\alpha_l \sin\alpha_j.$$

The phase function in the two-dimensional analogue of (10.28) takes the form

$$\Delta = (l^{\mathrm{f}} - l_2)\alpha_l^{-\mathrm{f}} + (j^{\mathrm{f}} - j_2)\alpha_j^{-\mathrm{f}} - \int_{l_1}^{l^{\mathrm{f}}} \alpha_l \mathrm{d}l - \int_{j_1}^{j^{\mathrm{f}}} \alpha_j \mathrm{d}j - \int_{k_1}^{k^{\mathrm{f}}} R\mathrm{d}k$$
$$+ \frac{\pi}{2}(l_1 + l_2 + 1),$$

where the superscript f implies a final variable dependent on the initial conditions and l_i and j_i $(i = 1, 2)$ are quantum numbers for the transition in question.

Deduce, by considering the dependence of the final variables on changes by π in the initial angle variables, that the semiclassical S matrix is subject to the parity selection rule that $l + j$ must remain either always even or always odd. Similarly, in the homonuclear case when $V(R, \gamma)$ is independent of the sign of cos γ, show that j is subject to the selection rule $|\Delta j| = 0, 2, 4, \ldots$.

10.3. The collisional excitation of a plane rotor is modelled in terms of impulsive collisions with a stationary hard ellipse, described by the equation

$$(x'/a)^2 + (y'/b)^2 = 1,$$

with an angle ϕ between $\mathbf{x}'$ and the space-fixed $\mathbf{x}$ direction. A collision partner with mass m and space-fixed velocity $(v_x, v_y) = (v, 0)$ strikes the target at the point $(x', y') = (r\cos\gamma, -r\sin\gamma)$ at which the tangent makes an angle ϵ with the x' axis.

(i) Derive expressions for the deflection angle, $\chi = 2\psi$, the initial and final orbital angular momenta, and the angular momentum transferred to the rotor.

(ii) Show that either the minimum deflection to obtain an angular momentum transfer Δj, or the maximum transfer at a given deflection, are given by $\sin\psi_{\min} = \Delta j/(2L\xi_{\max})$ or $(\Delta j)_{\max} = 2L\xi_{\max}\sin\psi$, where $\xi = \rho\cos(\gamma + \epsilon)$, $\rho = r/a$ and $L = mva$.

(iii) Show, by analogy with the hard sphere expressions in problem 9.1, that the classical action associated with scattering at mean angular momentum $\bar{l} = (l_i + l_f)/2$ from an ellipse with orientation $\alpha = -\phi$ may be expressed as

$$S\left(\bar{l}, \alpha\right) = 2\left[\bar{l}\arccos\left(\bar{l}/L\xi_\alpha\right) - \sqrt{\left(L\xi_\alpha\right)^2 - \bar{l}^2}\right],$$

where the suffix α indicates that ξ varies with the orientation angle α.

(iv) Verify, numerically or otherwise, that $(\partial S/\partial\alpha)_{\bar{l}}$ closely approximates your expression for Δj, assuming for example that $a = 3$, $b = 2$ and $L = 50$.

[Hints: The excluded free motion path comprises the combined projections of r onto the initial and final axes of motion.]

11
Reactive scattering

11.1 Definitions and working identities

Progress in the theory of chemically reactive scattering over the past twenty years has focussed mainly on the accurate quantum treatment of state-to-state dynamics. There are, however, some areas in which semiclassical ideas have proved fruitful, three of which are described below. The first is a so-called nearside–farside modification of the theory of elastic scattering in Section 9.2. Secondly, attention is given to the influence of geometric phase on chemical reactions, while the final section concerns the 'instanton' theory of chemical rate constants.

The systems of interest are restricted, for simplicity, to triatomic ones involving rearrangements of the constituent atoms from a reactant channel, say A + BC, to a product channel, AB + C or AC + B, along the lines laid out by Miller (1969), Zhang and Miller (1989) and Skouteris et al. (2000). The quantum mechanical labels include an arrangement index α, which might take values (a, b, c) according to the choice of free atom (A,B,C). Products are indicated below by a prime symbol, α' say. In addition, the reactant and product species have specified vibrational and rotational quantum numbers (v, j), of which the rotational ones combine with the orbital angular momentum quantum number ℓ to give a resultant J, subject to the normal triangular angular momentum constraints (Edmonds 1974). Expressions for the scattering amplitude and the cross-sections are simplified by transforming from this *orbital* or *space-fixed* representation to a *helicity* or *body-fixed* representation in which ℓ is replaced by the helicity quantum number K, equal to the projection of J onto the incoming or outgoing relative velocity vector, subject to the constraints $|K| \leqslant \min(j, J)$ and $|K'| \leqslant \min(j', J)$. The S matrix elements in the two representations are related in the notation of Edmonds (1974) by

$$\begin{aligned} &\langle \alpha' v' j' K' |\, S^J \,|\alpha v j K\rangle \\ &\quad = \sum_{\ell\ell'} \mathrm{i}^{\ell-\ell'} \langle j' K' J - K' |\, \ell' 0\rangle \langle \alpha' v' j' \ell' |\, S^J \,|\alpha v j \ell\rangle \langle \ell 0 \,| j K J - K\rangle , \qquad (11.1) \end{aligned}$$

where $\langle jJK - K|\, \ell 0\rangle$ is a Clebsch–Gordan coefficient, with $\langle \ell 0 \,| jKJ - K\rangle$ as its transpose. Notice that ℓ and ℓ', which are constrained by the triangular conditions, both have zero projection onto the body-fixed quantization axis, because there can be no orbital angular momentum component about the instantaneous direction of motion.

It is beyond the scope of this book to explore details of the numerical solution of the scattering equations, but readers are referred to the efficient code published by

Skouteris et al. (2000). The choice of chemical rearrangements is handled by working in a hyperspherical coordinate system (Pack and Parker 1987), which allows a parity factorization, $P = \pm 1$, according to the cyclic ordering of the three atoms in the chosen representation. The S matrix elements, $S^J_{n'n}$ where $|n\rangle = |\alpha v j K\rangle$, are then given by linear combinations of the elements of the parity-adapted scattering matrices $S^{J,P}$. The rotational states of homonuclear diatomic reactants and/or products also have a rotational parity, p, according to whether j is even or odd.

Following Zhang and Miller (1989), the scattering amplitudes are given by the partial-wave sum

$$f_{n'n}(\theta) = \frac{1}{2ik_{vj}} \sum_{J=J_{\min}}^{\infty} (2J+1) d^J_{K'K}(\theta) S^J_{n'n}, \tag{11.2}$$

in which $J_{\min} = \max(|K|, |K'|)$, θ is the angle between the relative velocity vectors of the reactant atom and the product molecule and $d^J_{K'K}(\theta)$ is a reduced Wigner rotation matrix element in the convention adopted by Edmonds (1974). Finally $k^2_{vj} = 2m(E - E_{vj})/\hbar^2$, where m is the reactant reduced mass, E is the total energy and E_{vj} is the energy of the reactant vibrational–rotational state. Consequently the differential cross-section is given by

$$I_{n'n}(\theta) = |f_{n'n}(\theta)|^2, \tag{11.3}$$

from which the fully state-selected integral cross-section becomes

$$\sigma_{n'n} = \sigma_{\alpha'v'j'K',\alpha vjK} = \frac{\pi}{k^2_{vj}} \sum_{J=J_{\min}}^{\infty} (2J+1) |S^J_{\alpha'v'j'K',\alpha vjK}|^2. \tag{11.4}$$

Situations often occur in which the helicity quantum numbers are unresolved, in which case one finds, with the help of eqn (11.1), that

$$\begin{aligned}\bar{\sigma}_{\alpha'v'j',\alpha vj} &= \frac{\pi}{(2j+1)k^2_{vj}} \sum_{K=-j}^{j} \sum_{K'=-j'}^{j'} \sum_{J=J_{\min}}^{\infty} (2J+1) |S^J_{\alpha'v'j'K',\alpha vjK}|^2 \\ &= \frac{\pi}{(2j+1)k^2_{vj}} \sum_{J=0}^{\infty} \sum_{\ell\ell'} (2J+1) |S^J_{\alpha'v'j'\ell',\alpha vj\ell}|^2,\end{aligned} \tag{11.5}$$

where the limits on ℓ are fixed by the triangular condition $|J - j| \leqslant \ell \leqslant J + j$, and similarly for ℓ'.

The final quantity of interest for subsequent sections is the thermally averaged chemical rate constant per unit volume, summed over all product states, as given by Miller (1975) and Seiderman and Miller (1992):

$$k_{\text{rate}}(T) = [2\pi\hbar Q_r(T)]^{-1} \int_0^{\infty} e^{-E/kT} N(E) dE, \tag{11.6}$$

where $Q_r(T)$ is the unit volume reactant partition function,

$$Q_r(T) = \left[\frac{2\pi mkT}{\hbar}\right]^{3/2} \sum_{vj} \mathrm{e}^{-E_{vj}/kT}, \tag{11.7}$$

and $N(E)$ is the cumulative reaction probability,

$$\begin{aligned} N(E) &= \sum_J \sum_{vjK} \sum_{v'j'K'} (2J+1)|S^J_{v'j'K',vjK}(E)|^2 \\ &= \sum_J \sum_{vj\ell} \sum_{v'j'\ell'} (2J+1)|S^J_{v'j'\ell',vj\ell}(E)|^2, \end{aligned} \tag{11.8}$$

or more generically

$$N(E) = \sum_{Jnn'} (2J+1)|S^J_{n'n}(E)|^2, \tag{11.9}$$

where n and n' include all reactant and product quantum numbers. While formally exact, equations (11.8) and (11.9) are seldom used in practice, except as points of reference in a proof, because $N(E)$ may be computed by using absorbing boundary conditions (Seiderman and Miller 1992), without reference to either initial or final states.

11.2 Nearside–farside interpretation of differential cross-sections

The nearside–farside theory of chemical reactions employs a technique derived from the nuclear physics literature, which has close connections with the Ford and Wheeler theory of elastic scattering in Chapter 9 (Dobbyn et al. 1999; McCabe et al. 2001; Connor 2004; Monks et al. 2006; Xiahou et al. 2011). Emphasis is, however, placed on the interpretation, rather than the semiclassical prediction, of the reactive cross-sections.

To emphasize the similarity between the elastic and reactive scattering formulations, we start by considering the special case in which the initial and final helicities are zero, in which case the reduced rotation matrix elements become Legendre polynomials. Thus equation (11.2) reduces to

$$f_{n'n}(\theta) = \frac{1}{2\mathrm{i}k_{vj}} \sum_{J=0}^{\infty} (2J+1) S^J_{n'n} P_J(\cos\theta) = \frac{1}{2\mathrm{i}k_{vj}} \sum_{J=0}^{\infty} a_J P_J(\cos\theta), \tag{11.10}$$

where a_J is an obvious contraction of the Jth term in the previous equality. Hence half the phase of $S^J_{n'n}$ plays the role of the elastic phase shift in equation (9.38), while $|S^J_{n'n}|$ modulates the scattering strength. Moreover there is now no unscattered term, independent of $S^J_{n'n}$, because the product states are independent of those of the reactants. Given the similarity between the elastic and reactive formulations, it is convenient to define a 'reactive deflection function'

$$\Theta(J) = 2\frac{\mathrm{d}\eta^{\mathrm{react}}_J}{\mathrm{d}J} = \frac{\mathrm{d}(\arg S^J_{n'n})}{\mathrm{d}J} = \mathrm{Im}\left(\frac{\mathrm{d}\ln S^J_{n'n}}{\mathrm{d}J}\right), \tag{11.11}$$

on the assumption that multiples of 2π are added to the principal values of $\arg S^J_{n'n}$ to ensure a smooth variation, and that the resulting points are interpolated to lie on a continuous curve. Note that it is sometimes convenient to 'resum' eqn (11.10) by factoring out polynomial expressions in $\cos\theta$, in order to eliminate the lowest-order coefficients in the sum itself (Totenhofer et al. 2010). The procedure is illustrated problem 11.1.

The essential step in nearside–farside theory is to express each Legendre polynomial as a sum of complex conjugate components (Dobbyn et al. 1999)

$$P_J(\cos\theta) = P_J^{(N)}\cos(\theta) + P_J^{(F)}(\cos\theta), \tag{11.12}$$

with the terms on the right-hand side usually given by

$$\begin{aligned} P_J^{(N,F)}(\cos\theta) &= \frac{1}{2}\left[P_J(\cos\theta) \pm \frac{2\mathrm{i}}{\pi}Q_J(\cos\theta)\right] \\ &\sim \left[2\pi(J+\frac{1}{2})\sin\theta\right]^{-1/2} \exp\left\{\mp\mathrm{i}\left[(J+\frac{1}{2})\theta - \frac{\pi}{4}\right]\right\}, \end{aligned} \tag{11.13}$$

where $Q_J(\cos\theta)$ is the standard irregular solution of the Legendre differential equation (Abramowitz and Stegun 1965), and the final line applies for $(J+1/2)\sin\theta \gg 1$. It follows that $f_{n'n}(\theta)$ divides into two parts

$$\begin{aligned} f_{n'n}(\theta) &= f^{(N)}_{n'n}(\theta) + f^{(F)}_{n'n}(\theta), \\ f^{(N,F)}_{n'n}(\theta) &= \frac{1}{2\mathrm{i}k_{vj}}\sum_{J=0}^{\infty} a_J P_J^{(N,F)}(\cos\theta), \end{aligned} \tag{11.14}$$

each of which may be evaluated exactly for any known set of S matrix elements. The significance of this division is readily understood by reference to the stationary phase argument in Section 9.2, according to which the dominant contributions to the components $f^{(N,F)}_{n'n}(\theta)$, at a given angle θ, come from angular momenta J such that

$$\Theta = \mp\theta - r\pi, \tag{11.15}$$

respectively, where r is a positive or negative integer chosen to ensure a smooth variation with J. Positive and negative deflections in this equation, when $r = 0$, are ones in which the scattered particle emerges on the 'nearside' or 'farside' of the scattering centre with respect to a detector in the upper half plane, as illustrated in Fig. 11.1. Since θ lies in the positive range $0 \leqslant \theta \leqslant \pi$, this means that $f^{(N)}_{n'n}(\theta)$ and $f^{(F)}_{n'n}(\theta)$ may be termed the 'nearside' and 'farside' components of the full scattering amplitude $f_{n'n}(\theta)$. The situation is, however, more complicated for large negative deflections, because, for example, a farside trajectory passes around the scattering centre to become a nearside one over the range $-2\pi < \Theta(J) < -\pi$.

One may also verify from eqns (11.13) and (11.14) that the logarithmic derivatives of $f^{(N,F)}_{n'n}(\theta)$ would correspond to 'stationary phase' angular momenta, $J^{(N,F)} + 1/2$, in cases involving only a single solution of equation (11.15). By extension it proves

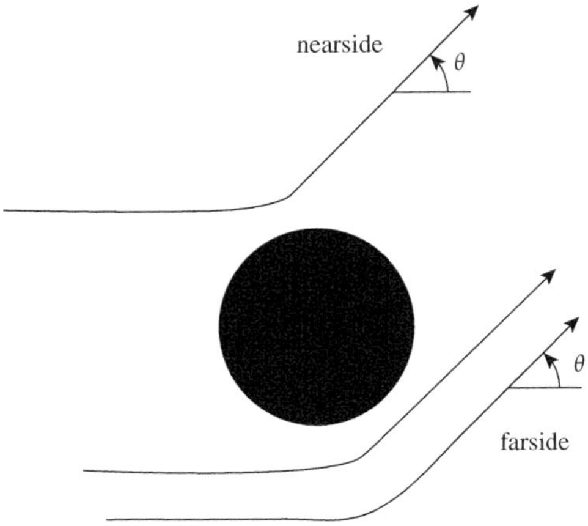

Fig. 11.1 Nearside and farside scattering trajectories. (Reproduced from Dobbyn et al. (1999) with permission The Royal Society of Chemistry.)

convenient to define 'nearside' and 'farside' *local angular momenta* by the equations (Monks et al. 2006)

$$LAM_{\mathrm{N,F}}(\theta) = \frac{\mathrm{d}\arg f^{(N,F)}}{\mathrm{d}\theta}, \tag{11.16}$$

where the arguments are again not necessarily principal values. Again $LAM_{\mathrm{N,F}}(\theta)$ may be evaluated exactly without reference to the stationary phase argument. A smooth variation of one or other of these $LAM(\theta)$ functions implies a monotonic variation with J in the appropriate part of the deflection function. In addition, Monks et al. (2006) have obtained the identity

$$\sigma(\theta)LAM(\theta) = \sigma_{\mathrm{N}}(\theta)LAM_{\mathrm{N}}(\theta) + \sigma_{\mathrm{F}}(\theta)LAM_{\mathrm{F}}(\theta) + C(\theta), \tag{11.17}$$

where, for consistency with the paper, we adopt the notation $\sigma(\theta)$ for the differential cross-section, $\sigma_{n'n}(\theta) = |f_{n'n}(\theta)|^2$ and similarly for $\sigma_{\mathrm{N,F}}(\theta)$ in terms of $|f_{n'n}^{(N,F)}(\theta)|^2$. Finally, $C(\theta)$ is an 'NF interference' term which may be shown to be given by (Monks et al. 2006)

$$C(\theta) = f^{(N)}(\theta) \otimes \mathrm{d}f^{(F)}/\mathrm{d}\theta + f^{(F)}(\theta) \otimes \mathrm{d}f^{(N)}/\mathrm{d}\theta, \tag{11.18}$$

where the product $z_1 \otimes z_2$ between two complex numbers, with arguments ϕ_1 and ϕ_2, is defined as

$$z_1 \otimes z_2 = \mathrm{Im}(z_1^* z_2) = |z_1||z_2|\sin(\phi_2 - \phi_1). \tag{11.19}$$

In addition it may be shown that

$$LAM_{\mathrm{N,F}}(\theta) = \frac{f^{(N,F)}(\theta) \otimes [\mathrm{d}f^{(N,F)}/\mathrm{d}\theta]}{|f^{(N,F)}(\theta)|^2}. \tag{11.20}$$

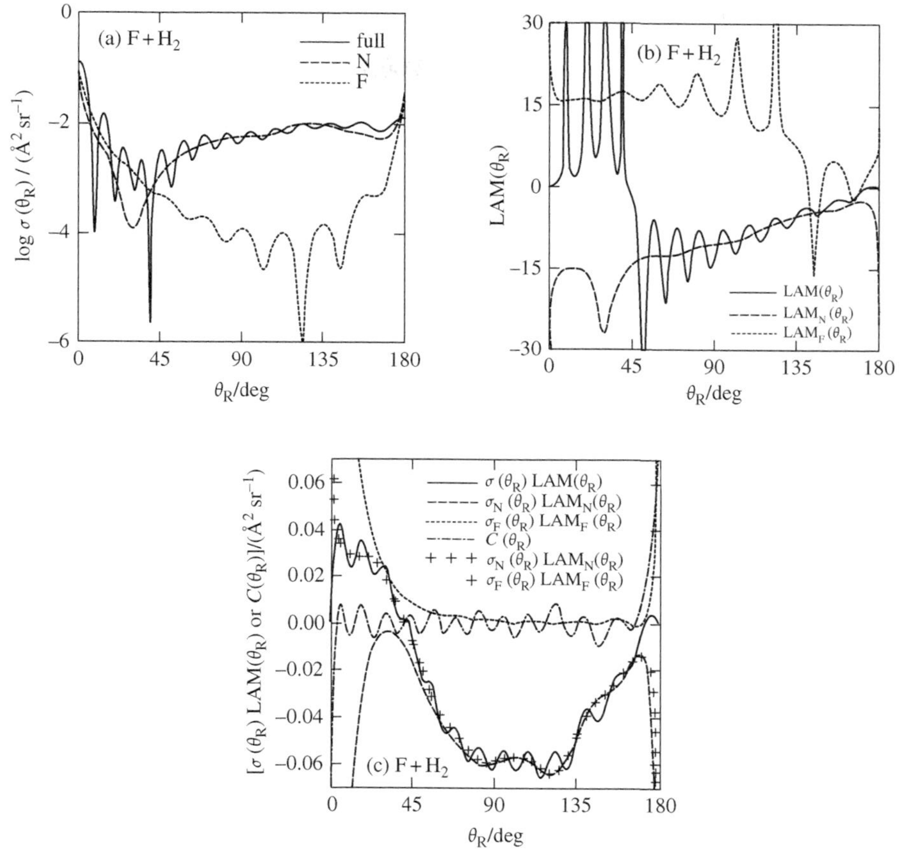

Fig. 11.2 (a) Plots of full (solid), nearside (dashed) and farside (dotted) differential cross-sections. (b) Local angular momenta derived from the full scattering amplitude (solid line) and from its nearside (dashed) and farside (dotted) components. (c) A plot of the left-hand side of eqn (11.17) (solid line) against the various terms on the right. (Reprinted with permission from Monks et al. (2006). Copyright 2006, AIP Publishing LLP.)

Figure 11.2 illustrates the significance of these results, by reference to the accurately computed differential cross-section for the quantum state-to-state reaction (Monks et al. 2006)

$$\mathrm{F} + \mathrm{H}_2(v = 0, j = 0, K = 0) \;\rightarrow\; \mathrm{FH}(v' = 3, j' = 3, K' = 0) + \mathrm{H},$$

with incident translational energy, $E_{\text{trans}} = 0.119$ eV. Figure 11.2(a) compares the full differential cross-section (solid line) with its nearside (dashed) and farside (dotted) analogues. The reaction is seen to be strongly N dominated (by up to two orders of magnitude) for $\theta_{\mathrm{R}} \gtrsim 43°$ and moderately F dominated at smaller scattering angles, although the two contributions have approximately equal magnitudes at small angles.

The main significance of this nearside–farside separation is that the two complements are much smoother than the full cross-section, which shows rapid oscillations over the full angular range. In particular the nearside contribution, $\sigma_{\rm N}(\theta)$, is smooth over all angles, while $\sigma_{\rm F}(\theta)$ is smooth in the forward direction, $0 < \theta_{\rm R} < 45^\circ$, and only begins to oscillate in regions where it is very small.

Figure 11.2(b) adds further information. Note first that the negative values of $LAM_{\rm N}(\theta)$ at small angles arise because the term $(J+1/2)\theta$ appears with a negative sign in the exponent of eqn (11.13). It is interesting to see that $LAM_{\rm F}(\theta) \simeq -LAM_{\rm N}(\theta) \simeq 15$ in the forward scattering direction $\theta_{\rm R} \simeq 0$, which implies that the deflection function $\Theta(J)$ passes through zero at $J \simeq 15$ to give a 'forward glory' (see below). Moreover the dominant contribution to the high-angle scattering is seen to arise from the low, $J \simeq -LAM_{\rm N}(\theta) < 15$, terms in the nearside partial wave sum. Given that the quantum calculation converges with $J_{\rm max} = 23$, this indicates that the weakly dominant farside scattering in the forward direction ($\theta_{\rm R} \leqslant 45^\circ$) comes from the highest angular momentum partial waves.

The final panel, Fig. 11.2(c), illustrates the relative magnitudes of the terms in eqn (11.17), bearing in mind that $\sigma_{\rm N}(\theta)LAM_{\rm N}(\theta)$ carries the same negative sign as $LAM_{\rm N}(\theta)$. Notice in particular that the magnitude of the interference term $C(\theta)$ is small compared to the remaining terms in this so-called CLAM plot (CLAM = Cross-section×LAM), except in the angular range where $\sigma(\theta)LAM(\theta)$ itself passes through zero. It is interesting to see that the products $\sigma_{\rm N}(\theta)LAM_{\rm N}(\theta)$ and $\sigma_{\rm F}(\theta)LAM_{\rm F}(\theta)$ are both smooth functions of θ, despite what appear as strong oscillations in separate functions $\sigma_{\rm F}(\theta)$ and $LAM_{\rm F}(\theta)$, of which the peaks in Fig. 11.2(a) correspond to the dips in Fig. 11.2(b). This farside behaviour may be understood, in the light of eqn (11.20), on the basis that the numerator on the right-hand side is much smoother than the denominator. Monks et al. (2006) illustrate similar behaviour for a variety of other reactions, but the origin of the relative smoothness of $f_{\rm N,F}(\theta) \otimes [f'_{\rm N,F}(\theta)]$ has not been analysed in detail.

It should also be mentioned that Connor (2004) has obtained a uniform Bessel approximation, which accounts well for the angular oscillations in Fig. 11.2(a) over the range $0 \leqslant \theta \lesssim 40^\circ$. The analogues of $I_\nu(\theta)$ and $\phi_\nu(\theta)$ in eqns (9.47) and (9.48) are written as

$$\sigma_\pm(\theta) = \frac{\lambda_\pm(\theta)|S_{\lambda_\pm}|^2}{k^2 \sin\theta |{\rm d}\Theta/{\rm d}\lambda_\pm|}, \qquad \beta_\pm(\theta) = \arg S_{\lambda_\pm} \pm \lambda_\pm(\theta)\theta, \tag{11.21}$$

where $\lambda_\pm(\theta)$ are the values of $J+1/2$ at which $\beta_\pm(\theta)$ are stationary with respect to λ at the chosen θ value. The resulting uniform Bessel approximation, which is designed to eliminate the $(\sin\theta)^{-1}$ divergence of $\sigma_\pm(\theta)$ as $\theta \to 0$ and $\lambda_\pm(\theta) \to \lambda_g$, takes the form

$$\begin{aligned}\sigma_{\rm UBA}(\theta) = \frac{\pi}{2}\zeta(\theta)\Big\{&\left[\sigma_+(\theta)^{1/2} + \sigma_-(\theta)^{1/2}\right]^2 J_0[\zeta(\theta)]^2 \\ &+ \left[\sigma_+(\theta)^{1/2} - \sigma_-(\theta)^{1/2}\right]^2 J_1[\zeta(\theta)]^2\Big\},\end{aligned} \tag{11.22}$$

where $J_n[\zeta(\theta)]$ are Bessel functions (Abramowitz and Stegun 1965) with argument

$$\zeta(\theta) = \frac{1}{2}[\beta_+(\theta) - \beta_-(\theta)] \simeq \lambda_g \theta \qquad \text{as } \theta \to 0. \tag{11.23}$$

Arbitrary helicity indices

This section is completed by outlining the theory for arbitrary helicity indices K and K' (Dobbyn et al. 1999; McCabe et al. 2001). The difference is that the Legendre polynomials in eqn (11.10) are replaced by the reduced rotation matrix elements in eqn (11.2). Thus

$$f_{K'K}(\theta) = \frac{1}{2ik_{vj}} \sum_{J=J_{\min}}^{\infty} a_J d^J_{K'K}(\theta). \tag{11.24}$$

The analogues of the Legendre functions $P_J(\cos\theta)$ and $Q_J(\cos\theta)$ are written in terms of Jacobi polynomials of the first and second kinds (Szegö 1975):

$$\begin{aligned} d^J_{K'K}(\theta) &= \qquad N^J_{K'K}[\sin(\theta/2)]^\alpha \cos[(\theta/2)]^\beta P_n^{(\alpha,\beta)}(\cos\theta) \\ e^J_{K'K}(\theta) &= (-1)^\alpha N^J_{K'K}[\sin(\theta/2)]^\alpha \cos[(\theta/2)]^\beta Q_n^{(\alpha,\beta)}(\cos\theta), \end{aligned} \tag{11.25}$$

where

$$n = J - K', \qquad \alpha = K' - K, \qquad \beta = K' + K \tag{11.26}$$

and

$$N^J_{K'K} = \left[\frac{(J-K')!(J+K')!}{(J-K)!(J+K)!}\right]^{1/2}. \tag{11.27}$$

In the equations that follow the symmetry relations obeyed by $d^J_{K'K}(\theta)$ (Edmonds 1974) have been used to ensure that $K' \pm K$ are positive integers or zero. The nearside and farside components of $f_{K'K}(\theta)$ in eqn (11.24) are then taken as

$$f^{(N,F)}_{K'K}(\theta) = \frac{1}{2ik_{vj}} \sum_{J=J_{\min}}^{\infty} a_J d^{(N,F)}_{JK'K}(\theta), \tag{11.28}$$

where it can be shown that

$$\begin{aligned} &d^{(N,F)}_{JK'K}(\theta) \\ &= \frac{1}{2}\left[d^J_{K'K}(\theta) \pm \frac{2i}{\pi} e^J_{K'K}(\theta)\right] \\ &\sim \left[2\pi(J+\frac{1}{2})\sin\theta\right]^{-1/2} \exp\left\{\mp i\left[(J+\frac{1}{2})\theta - (K'-K)\frac{\pi}{2} - \frac{\pi}{4}\right]\right\}, \end{aligned} \tag{11.29}$$

and an algorithm for evaluation of $d^J_{K'K}(\theta)$ and $e^J_{K'K}(\theta)$ has been given by McCabe et al. (2001).

The similarity between eqns (11.13) and (11.29) means that there is no fundamental difficulty in extending the theory to cases involving arbitrary changes in the helicity quantum number, except that the asymptotic forms in eqn (11.29), at a given angle θ, are valid only for integer values of J greater than

$$J_{\rm lim}^{(K',K)}(\theta) = {\rm int}\left[\frac{\sqrt{K^2 + K'^2 - 2KK'\cos\theta}}{\sin\theta}\right] + 1. \tag{11.30}$$

Consequently, Dobbyn et al. (1999) and McCabe et al. (2001) argue that restricted sums over the range $J \geqslant J_{\rm lim}^{(K',K)}(\theta)$ in eqn (11.28) will typically be valid except in the near forward and backward directions $\theta \simeq 0$ and $\theta \simeq \pi$ respectively.

11.3 The influence of geometric phase on reactive scattering

It is well known that divergence of the non-adiabatic coupling at a conical intersection between a pair of potential energy surfaces requires that the related adiabatic eigenfunctions, in a real representation, should change sign around a path sufficiently close to the said intersection (Longuet-Higgins 1975; Berry 1984; Domcke et al. 2003). The observable consequences of this electronic sign change arise from a compensating sign change in the nuclear wavefunction. Familiar examples occur in the bound state dynamics of Jahn–Teller systems (Englman 1972; Domcke et al. 2003; Bersuker 2006), but the treatment of scattering states is more complicated. Particular interest attaches to the influence of a well-known conical intersection on the chemical reaction $H + H_2$ reaction and its isotopic variants. The first theoretical work was reported by Kuppermann and Wu (2001), although their predictions of a large experimentally observable effect were later contradicted by Kendrick (2003). It is now generally accepted that the geometric phase boundary conditions on the nuclear wavefunction cause significant changes in the individual S matrix elements and in the differential cross-sections, but that these changes cancel out in the integral cross-section (Kendrick 2003; Juanes-Marcos et al. 2007; Bouakline et al. 2008). In addition, Althorpe (2006) has used an interesting symmetry argument to relate the geometric phase correction to particular types of classical trajectory.

The lowest adiabatic potential surface of the H_3 system (over which the reaction occurs) is connected in all equilateral geometries by a *hyperline* of conical intersections to what may be regarded at short range as the upper component of a Jahn–Teller coupled 2p(E) Rydberg state—a mechanism that actually provides the dominant contribution to the dissociative recombination reaction $e + H_3^+ \rightarrow H + H_2$ (Kokooline and Greene 2003). Before tackling the scattering problem, it may be helpful to recall some aspects of the related bound state system, with particular reference to the origin of the geometric phase and the nature of the adiabatic potential surface in the vicinity of the intersection.

To this end distortions from a particular equilateral configuration are modelled by the two-component electronic Hamiltonian matrix

$$\mathbf{W}_{\rm el} = \begin{pmatrix} W_{\rm diag}(r) + \lambda_1 x + \lambda_2(x^2 - y^2) & \lambda_1 y - 2\lambda_2 xy \\ \lambda_1 y - 2\lambda_2 xy & W_{diag}(r) - \lambda_1 x - \lambda_2(x^2 - y^2) \end{pmatrix}, \tag{11.31}$$

where $(x, y) = (r\cos\phi, r\sin\phi)$ are the coordinates of a degenerate nuclear vibration, while λ_1 and λ_2 are the first- and second-order coupling parameters that lift the degeneracy. The relative signs in the diagonal and off-diagonal elements are determined by symmetry (Englman 1972; Domcke et al. 2003; Bersuker 2006). The two resulting adiabatic potential surfaces take the forms

$$W_\pm(r, \phi) = W_{\text{diag}}(r) \pm \sqrt{\lambda_1^2 r^2 + 2\lambda_1\lambda_2 r^3 \cos 3\phi + \lambda_2^2 r^4}. \tag{11.32}$$

We focus initially on the first-order case, with $\lambda_2 = 0$, for which Fig. 11.3(a) illustrates the first-order splitting around a double cone. In addition, the form of the first-order eigenvector matrix,

$$U = \sqrt{\frac{1}{2}}\begin{pmatrix} \cos\phi/2 & -\sin\phi/2 \\ \sin\phi/2 & \cos\phi/2 \end{pmatrix}, \tag{11.33}$$

implies the expected sign change in the electronic eigenfunctions as $\phi \rightarrow \phi + 2\pi$. The single-valuedness of the total wavefunction is therefore most simply achieved by requiring a compensating sign change in the nuclear wavefunction: $\psi_{\text{nuc}}(r, \phi + 2\pi) = -\psi_{\text{nuc}}(r, \phi)$. It is, however, computationally more convenient to work in the phase-modified representation

$$\psi_{\text{nuc}}(r, \phi) = \mathrm{e}^{-\mathrm{i}\phi/2}\tilde{\psi}_{\text{nuc}}(r, \phi), \tag{11.34}$$

such that $\tilde{\psi}_{\text{nuc}}(r, \phi)$ is single-valued in ϕ, subject to a modification in the derivative operation

$$\frac{\partial\psi_{\text{nuc}}^{(-)}(r, \phi)}{\partial\phi} = \mathrm{e}^{-\mathrm{i}\phi/2}\left[\frac{\partial}{\partial\phi} - \frac{\mathrm{i}}{2}\right]\tilde{\psi}_{\text{nuc}}^{(-)}(r, \phi). \tag{11.35}$$

When written in general coordinates, normally in a higher-dimensional space, eqn (11.35) implies the vector potential substitution (Mead and Truhlar 1979)

$$\boldsymbol{\nabla}\psi_{\text{nuc}}^{(-)} \rightarrow \left[\boldsymbol{\nabla} - \frac{\mathrm{i}}{2}\mathbf{A}\right]\tilde{\psi}_{\text{nuc}}^{(-)} \tag{11.36}$$

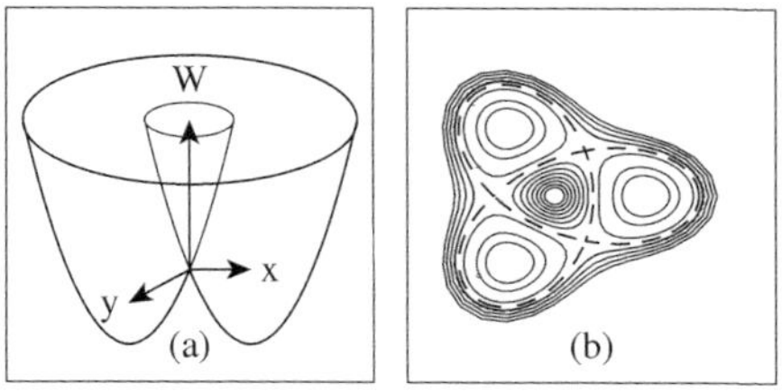

Fig. 11.3 (a) Adiabatic first-order Jahn–Teller potential surfaces, intersecting as a double cone. (b) Contour representation of the lower second-order Jahn–Teller potential surface with conical points at the centre, and at the three intersections of the dashed contour.

in the nuclear Schrödinger equation, where $\mathbf{A} = \boldsymbol{\nabla}\phi$. Note that $\mathbf{A}$ may in fact be taken as the gradient of any generalized angle that increases by 2π around the conical intersection (Berry 1984).

Moving on to the influence of the second-order terms in eqn (11.31), Fig. 11.3(b) shows a contour representation of the lower adiabatic potential surface for $\lambda_1\lambda_2 > 0$, in which the three intersections between the dashed contours, at $r_x = \lambda_1/\lambda_2$ and $\phi = \pm\pi/3$ and $\phi = \pi$, are additional conical intersections, at which $W_{\pm}(r, \phi)$ again coincide. Consequently any cycle of ϕ in the outer region $r > r_x$ passes round four conical intersections to undergo an overall sign change in the wavefunction of $(-1)^4 = 1$, which means that any geometric phase effects are restricted to inner parts of the wavefunction.

Figure 11.4 is the reactive analogue of Fig. 11.3(b) in a hyperspherical representation in which equilateral and linear geometries lie respectively at the centre and on the boundaries of the diagram. The saddle points (or transition states), marked by ‡, lie on paths around the outer conical points in Fig. 11.3(b) and the three entrance or exit valleys replace the outer minima in the earlier diagram. Geometric phase effects arise in classical terms from interference between trajectories that pass around opposite sides of the central conical intersection, as for example the solid clockwise trajectory that crosses one saddle point and the dashed anticlockwise one that crosses two saddle points to reach the same $H_B+H_AH_C$ product valley. The saddle points occur at 0.42 eV above the reactant asymptotes; the minimum energy on the conical intersection hyperline occurs at 2.4 eV. In the calculations reported below the state-to-state reaction probabilities and cross-sections were derived from nuclear wavefunctions Ψ_G and Ψ_N which were computed by respectively including and excluding the vector potential term in the Schrödinger equation (Juanes-Marcos et al. 2007; Bouakline et al. 2008).

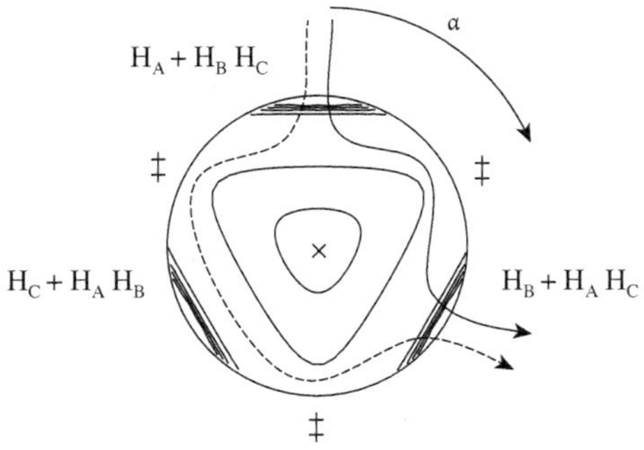

Fig. 11.4 Schematic representation of the reaction in a hyperspherical representation in which the central point X marks the conical intersection at an equilateral geometry, while the outer boundary corresponds to linear geometries, which are symmetrical at the three saddle points, ‡. The arrows indicate classical trajectories that cross one (solid line) or two (dashed line) transition states. (Reprinted with permission from Juanes-Marcos et al. (2007). Copyright 2007, AIP Publishing LLP.)

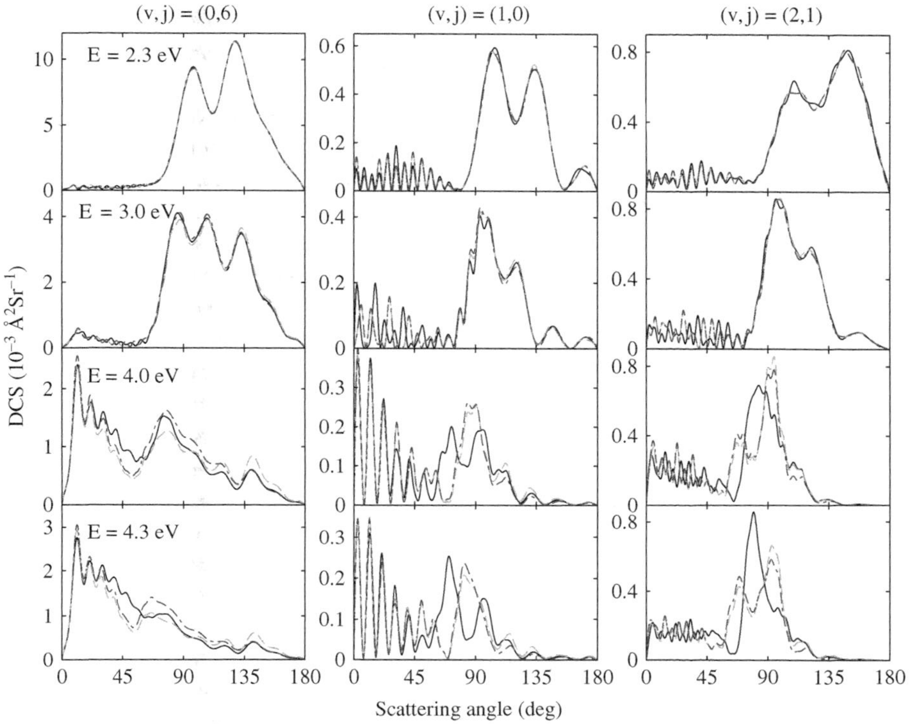

Fig. 11.5 Differential cross-sections for the $(v, j) = (0, 0) \to (0, 1)$ reactions at selected energies. Solid and dashed lines indicate normal and geometric phase boundary conditions. Dot-dash lines were obtained by including explicit coupling to the upper adiabatic surface. (Reprinted with permission from Bouakline et al. (2008). Copyright 2008, AIP Publishing LLP.)

To avoid any ambiguity, the relative signs of Ψ_G and Ψ_N were fixed by requiring them to have a common incoming component.

As an illustration of the expected increasing influence of geometric phase as the energy increases, Fig. 11.5 shows the differential cross-sections for different state-to-state reactions from $(v, j) = (0, 0)$ to various product combinations (Bouakline et al. 2008). The reported energies were chosen to span the range around the minimal value, $E_x = 2.74\,\text{eV}$, along the conical intersection hyperline. Thus the nuclear wavefunctions at energies for $E > E_x$ are in principle subject to the joint influence of both the upper and lower adiabatic potentials. However, little or no evidence of interaction with the upper surface could be detected. Thus the dashed and solid lines were derived from single-surface calculations with and without the vector potential term in eqn (11.36), while the dot-dash results were obtained by including explicit non-adiabatic coupling to the upper potential surface. The small differences from the dashed lines indicate minimal inter-surface coupling. The geometric phase effect is seen to be quite large, particularly at the two highest energies. It is also seen that geometric phase enhancement of the differential cross-section at certain angles is matched by reduction at neighbouring ones. It is therefore no surprise that the above differences cancel out in the integrated cross-section. There is, however, as yet no general physical explanation for the widespread observance of such cancellations.

On the other hand (Althorpe 2006) has employed an elegant symmetry argument to relate the changes in the differential cross-section to differences between the wavefunctions Ψ_N and Ψ_G that arise from the two types of trajectory in Fig. 11.4. For reasons of space we can only outline the main ideas, leaving the reader to refer to the paper for technical details. The first step is to express the difference in boundary conditions over the cycle $[0 : 2\pi]$ as a difference in symmetry over the extended double space $[0 : 4\pi]$. The resulting double group therefore has two operations, E and $R_{2\pi}$, such that Ψ_N and Ψ_G belong to the symmetric and antisymmetric irreducible representations, respectively. Secondly, new wavefunction components, Ψ_e and Ψ_o, are taken in such a way that $\Psi_o = R_{2\pi}\Psi_e$. Consequently, by use of the appropriate projection operator (Cotton 1971),

$$\Psi_N = \sqrt{\frac{1}{2}}\left(\Psi_e + \Psi_o\right), \qquad \Psi_G = \sqrt{\frac{1}{2}}\left(\Psi_e - \Psi_o\right) \tag{11.37}$$

and conversely

$$\Psi_e = \sqrt{\frac{1}{2}}\left(\Psi_N + \Psi_G\right), \qquad \Psi_o = \sqrt{\frac{1}{2}}\left(\Psi_N - \Psi_G\right). \tag{11.38}$$

As noted above, the imposition of common incoming boundary conditions removes any sign ambiguity between Ψ_N and Ψ_G, which suggests that the difference function Ψ_o carries the geometric phase effect. Althorpe (2006) goes on to relate Ψ_e and Ψ_o to the two types of path in Fig. 11.4 by reference to a Feynman path integral interpretation of the Aharanhoff–Bohm effect (Schulman 1981). The essential points are that paths with different winding numbers around the conical intersection belong to topologically distinct *homotopic* classes. Thus the Feynman propagator may be expressed as a sum over winding numbers, in a form that contains an arbitrary real parameter α,

$$K(\mathbf{x},\mathbf{x}_0|t) = \sum_{n=-\infty}^{\infty} \mathrm{e}^{in\alpha} K_n(\mathbf{x},\mathbf{x}_0|t). \tag{11.39}$$

A simple example is given in problem 11.2. To see the significance of the phase term α, note that increasing the angle ϕ around the conical intersection by 2π is equivalent to increasing the winding number. Thus $K_n \rightarrow K_{n+1}$, while overall $K \rightarrow \mathrm{e}^{\mathrm{i}\alpha}K$, which means that K has character $\mathrm{e}^{\mathrm{i}\alpha}$ under $R_{2\pi}$. The choices $\alpha = 0$ and $\alpha = \pi$ therefore correspond to normal and geometric phase boundary conditions. Moreover, in the latter case the odd and even n components of K have opposite signs. Thus the resulting propagators

$$K_N = K_e + K_o, \qquad K_G = K_e - K_o \tag{11.40}$$

generate the wavefunctions in equation (11.37), when applied to a common starting function χ say. To set these conclusions in the context of Fig. 11.4 the difference between the two types of trajectory is a topological $n = -1$ path that encircles the conical intersection in an anticlockwise sense. Wavefunction components associated with this composite path therefore contribute to Ψ_o, although for shorthand purposes Althorpe (2006) and Juanes-Marcos et al. (2007) refer to Ψ_e and Ψ_o as the single and double transition state ($1-TS$ and $2-TS$) components of Ψ_N and Ψ_G. The focus in understanding the geometric phase effect is now placed on Ψ_e and Ψ_o, each of which has its own S matrix, S_e or S_o (or S^{1-TS} or S^{2-TS}), and its own scattering amplitude, $f_e(\theta)$ or $f_o(\theta)$ [or $f^{1-TS}(\theta)$ or $f^{2-TS}(\theta)$]. Consequently

$$f_{N,G}(\theta) = \sqrt{\frac{1}{2}}\,[f_e(\theta) \pm f_o(\theta)]$$
$$\sigma_{N,G}(\theta) = \frac{1}{2}\left[|f_e(\theta)|^2 + |f_o(\theta)|^2 \pm 2\,\mathrm{Re}\,\{f_e(\theta)f_o(\theta)\}\right]. \tag{11.41}$$

Figure 11.6 illustrates the diagonal contributions, $|f_e(\theta)|^2$ and $|f_o(\theta)|^2$, to the differential cross-sections for the $(v, j) = (0, 0) \rightarrow (1, 0)$ reaction in the central column of

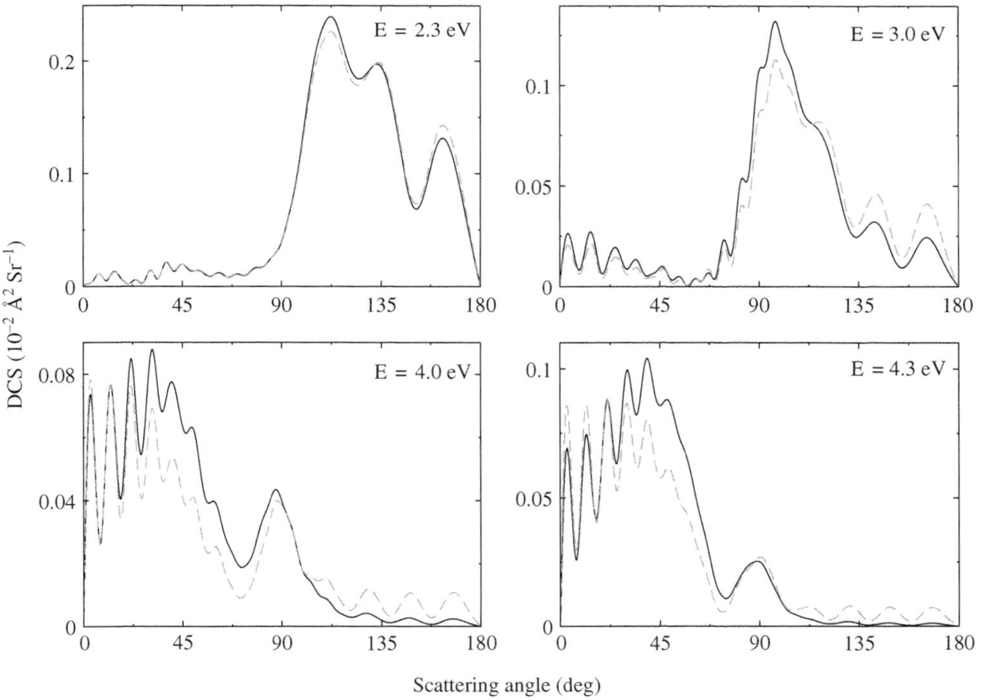

Fig. 11.6 Diagonal even (solid line) and odd (dashed line) contributions to the $(v, j) = 0, 0 \rightarrow 1, 0$ differential cross-section at $E = 2.3\,\mathrm{eV}$ and $4.3\,\mathrm{eV}$. (Reprinted with permission from Bouakline et al. (2008). Copyright 2008, AIP Publishing LLP.)

Fig. 11.5. The solid and dashed lines show that the even $(1 - TS)$ and odd $(2 - TS)$ parts of the wavefunction predominantly scatter into the forward $(\theta > 90^\circ)$ and backward $(\theta < 90^\circ)$ hemispheres respectively, and that $|f_o(\theta)|^2$ is negligibly small at the lower energy. It is therefore easily understood why the geometric phase effect is so small at $E = 2.3\,\text{eV}$ in Fig. 11.6 and why it is restricted to the region around $\theta = 90^\circ$ at $E = 4.3\,\text{eV}$, where the two contributions in Fig. 11.5 overlap.

Juanes-Marcos et al. (2007) have taken the semiclassical discussion further by performing quasiclassical trajectory calculations at $E = 2.3\,\text{eV}$. As a preliminary it was found that the phases of $f_e(\theta)$ and $f_o(\theta)$ respectively decrease and increase with increasing θ at $E = 2.3\,\text{eV}$, which means in the language of the previous section that the local angular momenta, $LAM_e(\theta)$ add $LAM_o(\theta)$, are respectively negative and positive. Now the trajectories responsible for the even or odd wavefunction components, Ψ_e or Ψ_o, correspond respectively to clockwise $(1 - TS)$ or anticlockwise $(2 - TS)$ ones in Fig. 11.4. Following the reasoning in the previous section, the signs of $LAM_e(\theta)$ and $LAM_o(\theta)$ imply that the corresponding trajectories should scatter in the nearside $(\theta > 0)$ and farside $(\theta < 0)$ directions respectively. These conclusions are fully confirmed by the results in Fig. 11.7. In addition, detailed examination of the different types of circuit indicate quite different reaction mechanisms for the two types of path. The three linear transition states may be labelled as ABC, ACB and BAC according to the choice of the central atom. The $1-TS$ path proceeds by the normal substitution mechanism whereby atom A approaches BC in the ABC geometry and causes C to depart from the opposite end. By contrast the $2 - TS$ trajectories necessarily proceed

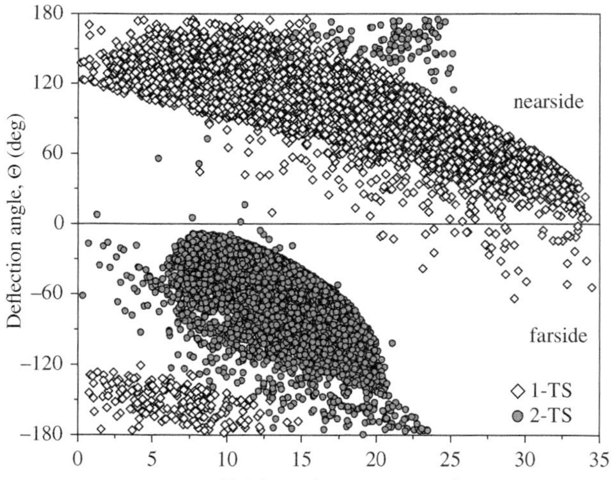

Fig. 11.7 Scatter plots showing the classical deflection angle as a function of total angular momentum arising from $1 - TS$ trajectories (open squares) and $2 - TS$ trajectories (closed circles). (Reprinted with permission from Juanes-Marcos et al. (2007). Copyright 2007, AIP Publishing LLP.)

by a substitution mechanism. Atom A passes around atom C, close to the ACB transition state, to attack atom B from the centre of the BC bond, near the BAC transition state. Consequently the product AB is formed with the A atom aligned towards the departing C atom.

In summary, it is firmly established that the geometric phase, arising from the presence of a conical intersection, affects the nature of the scattering wavefunction. In the test case of the H+H_2 reaction, increasingly significant effects on the state-to-state differential cross-sections were observed at higher energies, although they cancel out in the integrated cross-sections. An elegant symmetry argument was used to relate the geometric phase contributions to the forms of classical trajectory at the conical intersection.

11.4 Instanton theory of deep tunnelling

Instanton theory concerns the quantum mechanical tunnelling contributions to the chemical reaction rate. The topic was first addressed by Wigner (1932b), within a parabolic approximation to the reaction profile, which correctly predicts curvature in the Arrhenius plot but diverges catastrophically at very low temperatures (see below). The remainder of this section shows how anharmonic terms in the potential give rise to an 'instanton' correction to this divergence, first in the context of a one-dimensional Eckart barrier and then for a multidimensional reactive potential surface, along the lines pioneered by Miller (1975).

The parabolic model employs a potential of the form

$$V(q) = V_0 - \frac{1}{2} m\omega^{*2} q^2, \tag{11.42}$$

in which ω^* may be interpreted as the classical frequency in the upturned potential function, a situation that may be formalized by working with imaginary time, $t = -\mathrm{i}\tau$, and imaginary momentum, $p = \mathrm{i}\bar{p}$, giving rise to the Hamiltonian $H = -\bar{p}^2/2m - \bar{V}(q)$, where $\bar{V}(q) = -V(q)$. Hamilton's equations then revert to the canonical form

$$\frac{\mathrm{d}\bar{p}}{\mathrm{d}\tau} = \frac{\partial H}{\partial q} = -\frac{\mathrm{d}\bar{V}}{\mathrm{d}q}, \qquad \frac{\mathrm{d}q}{\mathrm{d}\tau} = -\frac{\partial H}{\partial \bar{p}} = \frac{\bar{p}}{m}, \tag{11.43}$$

with the frequency given by $m\ddot{q} = -(\mathrm{d}^2\bar{V}/\mathrm{d}q^2) = -m\omega^* q$.

The cumulative reaction probability, $N(E)$, for this parabolic model is given by the transmission probability $T(E)$ in eqn (3.54), for which the rate constant in eqn (11.6) may be evaluated in closed form (Gradsteyn and Ryyzhik 1980). Using the notation $\beta = 1/k_\mathrm{B}T$, where k_B denotes the Boltzmann constant,

$$\begin{aligned} k_\mathrm{rate}^\mathrm{parab} &= [2\pi\hbar Q_r]^{-1} \int_{-\infty}^{\infty} \left\{ 1 + \exp\left[\frac{-2\pi(E - V_0)}{\hbar\omega^*} \right] \right\}^{-1} \exp(-\beta E)\mathrm{d}E \\ &= \frac{\beta\hbar\omega^*}{2} \left[\sin\left(\frac{\beta\hbar\omega^*}{2} \right) \right]^{-1} \left[\frac{1}{2\pi\beta\hbar Q_r} \right] \mathrm{e}^{-\beta V_0}, \\ &\Rightarrow \frac{k_\mathrm{B}T}{h} Q_r^{-1} \mathrm{e}^{-\beta V_0} \qquad \text{for} \quad \beta^{-1} = k_\mathrm{B}T \gg \hbar\omega^*. \end{aligned} \tag{11.44}$$

The final line conforms to the familiar transition state theory formula in the high temperature limit (Atkins and de Paula 2010), while the product of the two leading factors in the second line accounts for a progressive increase in the rate constant at lower temperatures. However, the presence of the sine term in the denominator means that k_{rate} diverges when $\beta = 2\pi/\hbar\omega^* = \tau/\hbar$, where τ and ω^* are respectively the imaginary time period and the imaginary frequency of motion on the upturned parabolic barrier. The divergence itself is the outcome of an unphysical model, in which the potential function is unbounded below, but the temperature (or β value) at which it occurs will be seen to be important for the validity of the instanton correction. From henceforth temperatures will be measured in inverse β units, to avoid confusion between temperature and time.

Eckart barrier

The Eckart potential

$$V(q) = \frac{V_0}{\cosh^2 \alpha q} \tag{11.45}$$

is more realistic, in restricting energies to the physical range $E > 0$. The following equations are conveniently expressed in terms of the composite variables:

$$\tau_0 = \sqrt{\frac{2\pi^2 m}{\alpha^2 V_0}}, \qquad \gamma = \sqrt{\frac{2mV_0}{\alpha^2\hbar^2}}, \tag{11.46}$$

of which τ_0 is the period of small-amplitude motion in the upturned well, and $2\pi\gamma$ is the tunnelling integral at $E = 0$. The integral at intermediate energies, $0 < E < V_0$, may be verified to take the form

$$\begin{aligned}\frac{\bar{W}(E)}{\hbar} &= \frac{1}{\hbar}\oint \bar{p}(q,E)\mathrm{d}q = 2\int_{-a}^{a} \frac{\sqrt{2m\left[V_0 - E\cosh^2 \alpha q\right]}}{\hbar\cosh\alpha q}\mathrm{d}q \\ &= 2\pi\gamma\left[1 - \sqrt{\frac{E}{V_0}}\right],\end{aligned} \tag{11.47}$$

and its energy derivative is given in the light of eqn (11.43) by

$$\begin{aligned}\frac{1}{\hbar}\frac{\mathrm{d}\bar{W}}{\mathrm{d}E} &= \frac{1}{\hbar}\oint \frac{\partial \bar{p}}{\partial E}\mathrm{d}q = -\frac{1}{\hbar}\oint\left(\frac{\partial \bar{H}}{\partial \bar{p}}\right)^{-1}\mathrm{d}q = -\frac{1}{\hbar}\oint\left(\frac{\mathrm{d}\tau}{\mathrm{d}q}\right)\mathrm{d}q \\ &= -\frac{\tau(E)}{\hbar} = -\frac{\tau_0}{\hbar}\sqrt{\frac{V_0}{E}},\end{aligned} \tag{11.48}$$

in which $\tau(E)$ is time period of the motion in the upturned potential well, whose energy dependence will be shown below to eliminate the divergence shown by the parabolic model.

Taken together with eqns (3.55) and (11.6), eqn (11.47) implies the following expression for the rate constant:

$$k_{\text{rate}} = [2\pi\hbar Q_r]^{-1} \int_0^\infty \frac{\exp(-\beta E)}{1+\exp[\bar{W}(E)/\hbar]} \mathrm{d}E$$
$$\simeq [2\pi\hbar Q_r]^{-1} \int_0^\infty \exp[-\beta E - \bar{W}(E)/\hbar] \mathrm{d}E. \tag{11.49}$$

The validity of the approximation in the second line requires a negligible contribution to the integral from energies $E > V_0$, which leads in turn to the instanton approximation. Following a stationary phase argument, the dominant contribution to the integral comes from the range around E_β such that

$$\beta = \tau(E_\beta)/\hbar, \tag{11.50}$$

which may be compared with the above condition for divergence of the parabolic rate. However, the anharmonicity of the Eckart potential yields an energy dependence in eqn (11.50) such that the second derivative of the exponent is finite. The resulting instanton approximation to the rate constant is given by

$$k_{\text{rate}} Q_r = \sqrt{\frac{1}{2\pi\hbar^2}\left(-\frac{\mathrm{d}E_\beta}{\mathrm{d}\beta}\right)} \exp\left[-\beta E_\beta - \bar{W}(E_\beta)/\hbar\right],$$
$$= \sqrt{\frac{1}{2\pi\hbar^2}\left(-\frac{\mathrm{d}E_\beta}{\mathrm{d}\beta}\right)} \exp\left[-\bar{S}(\beta)/\hbar\right], \tag{11.51}$$

where $\bar{S}(\beta) = \bar{W}(E_\beta) + \beta\hbar E_\beta$ is the classical action for imaginary time propagation through the barrier. The rate given in this way remains finite, even at the critical 'cross-over temperature', such that $\beta_c = \tau_0/\hbar$ when the dominant contribution to the integral comes from the immediate vicinity of the barrier maximum.

The results in eqns (11.44) and (11.51) are compared in Fig. 11.8 to the β variation of the rate constant, computed by quadrature from the exact transmission probability (Landau and Lifshitz 1965), using parameter values $\beta_c V_0 = \pi\gamma = 12$, consistent with those employed by Voth et al. (1989). The parabolic approximation is seen to lie close to the exact rate at high temperatures ($\beta V_0 < 8$), but to diverge when $\beta V_0 = 12$. By contrast the instanton approximation remains finite at the cross-over point. Moreover, the deviations in the range $12 < \beta V_0 < 15$ may be removed by including a factor $\{1 + \exp[-\bar{W}(E_s)]\}^{-1}$ to compensate for the tacit assumption $\exp[-\bar{W}(E_s)] \gg 1$ implied by the above derivation. The resulting modest correction cannot however be extended to the range $\beta V_0 < 12$ because $\bar{W}(E_s)$ becomes negative, which invalidates the second line of eqn (11.49).

This completes the one-dimensional analysis. The essential elements, for the general theory, include the introduction of an energy-dependent imaginary time period for the tunnelling motion, coupled with a stationary phase argument to pick out the time period $\tau(E_s)$ appropriate to any required β value, above the critical cross-over value β_c.

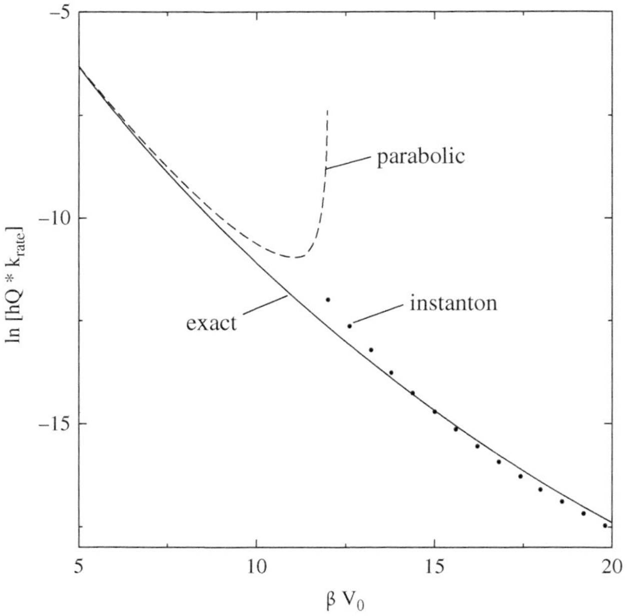

Fig. 11.8 Comparison between the exact Eckart barrier rate constant (solid line) and the parabolic (dashed line) and instanton (dots) approximation, for a model with $\pi\gamma = 12$.

Miller theory

We turn now to the more physically relevant case of a non-separable potential surface, along the lines pioneered by Miller (1975). The starting point is the following formally exact expression for the rate constant:

$$k_{\text{rate}} Q_r = tr\left[e^{-\beta\hat{H}}\hat{F}\hat{\mathcal{P}}\right] = \left\langle \hat{F}\hat{\mathcal{P}} \right\rangle, \tag{11.52}$$

in which $\hat{F} = \delta(q_1 - q_1^{\ddagger})\hat{p}_1/m$ is the flux operator at a point $q_1^{\ddagger}$ on the reaction coordinate, and $\hat{\mathcal{P}}$ is the projection operator that discriminates between reactive and non-reactive flow at $q_1 = q_1^{\ddagger}$. Within the spirit of transition state theory Miller expresses $\hat{\mathcal{P}}$ in terms of the Heaviside function such that

$$\hat{p}_1\hat{\mathcal{P}} \rightarrow p_1 h(p_1) = \frac{1}{2}(p_1 + |p_1|) = \begin{matrix} p_1 & \text{for} & p_1 > 0 \\ 0 & \text{for} & p_1 < 0 \end{matrix} . \tag{11.53}$$

The symbol $\left\langle \hat{A} \right\rangle$ denotes the Boltzmann weighted quantum mechanical mean value, and the trace of the operator product, which is independent of the representation, may be evaluated in a free motion basis, under semiclassical conditions, by the Wigner (1932a) prescription

$$\left\langle \hat{A} \right\rangle \simeq (2\pi\hbar)^{-f} \int \mathrm{d}\mathbf{p}_i \int \mathrm{d}\mathbf{q} \int \mathrm{d}q' A[\mathbf{p}, (\mathbf{q}_1 + \mathbf{q}')/2] \left\langle \mathbf{q} \right| \mathrm{e}^{-\beta H} \left| \mathbf{q}' \right\rangle$$
$$\times \exp\left[-\frac{\mathrm{i}}{\hbar} \mathbf{p}_i \cdot (\mathbf{q}' - \mathbf{q}) \right], \tag{11.54}$$

in which ($\mathbf{q}$,$\mathbf{p}$) and ($\mathbf{q}'$,$\mathbf{p}'$) denote initial and final phase variables, respectively, and $\mathbf{p}_i$ is an intermediate momentum. Moreover, the operator $\mathrm{e}^{-\beta H}$ may be regarded as a quantum propagator $\mathrm{e}^{-\mathrm{i}Ht/\hbar}$ in imaginary time, $t = -\mathrm{i}\tau = -\mathrm{i}\beta\hbar$. Consequently the matrix element may be approximated in the van Vleck form of Appendix C.3:

$$\left\langle \mathbf{q} \right| \mathrm{e}^{-\beta H} \left| \mathbf{q}' \right\rangle = [\det(-\partial^2 \bar{S}/\partial \mathbf{q}' \partial \mathbf{q})/(2\pi\hbar)^f]^{1/2} \exp[\text{-}\bar{S}(\mathbf{q}', \mathbf{q})]/\hbar \tag{11.55}$$

where

$$\begin{aligned} \bar{S}(\mathbf{q}', \mathbf{q}) &= \mathrm{i} \int_0^{-\mathrm{i}\beta\hbar} \mathrm{d}t \left\{ \frac{1}{2} \mathbf{p}(t) \cdot \dot{\mathbf{q}}(t) - V\left[\mathbf{q}(t)\right] \right\} \\ &= \int_0^{\beta\hbar} \mathrm{d}\tau \left\{ \frac{1}{2} \bar{\mathbf{p}}(\tau) \cdot \dot{\mathbf{q}}(\tau) + \bar{V}\left[\mathbf{q}(\tau)\right] \right\} \end{aligned} \tag{11.56}$$

is the classical action integral for imaginary time propagation, with the properties that

$$(\partial \bar{S}/\partial \mathbf{q}') = \bar{\mathbf{p}}'(\mathbf{q},\mathbf{q}') \quad \text{and} \quad (\partial \bar{S}/\partial \mathbf{q}) = -\bar{\mathbf{p}}(\mathbf{q},\mathbf{q}'), \tag{11.57}$$

in which the bold symbols imply a common result for all components. The notation $\bar{\mathbf{p}}'(\mathbf{q},\mathbf{q}')$ implies the final momentum such that the trajectory from $\mathbf{q}$ reaches $\mathbf{q}'$ in imaginary time $\tau = \beta\hbar$. We now see, by applying the stationary phase conditions to the integrand in eqn (11.54), that

$$\begin{aligned} \frac{\partial}{\partial \mathbf{q}} \left[\bar{S}(\mathbf{q}', \mathbf{q}) + \mathrm{i}\mathbf{p}_i \cdot (\mathbf{q}' - \mathbf{q}) \right] &= -\bar{\mathbf{p}} + \mathrm{i}\mathbf{p}_i = \mathbf{0} \\ \frac{\partial}{\partial \mathbf{q}'} \left[\bar{S}(\mathbf{q}', \mathbf{q}) + \mathrm{i}\mathbf{p}_i \cdot (\mathbf{q}' - \mathbf{q}) \right] &= \bar{\mathbf{p}}' - \mathrm{i}\mathbf{p}_i = \mathbf{0}, \\ \frac{\partial}{\partial \mathbf{p}_i} \left[\bar{S}(\mathbf{q}', \mathbf{q}) + \mathrm{i}\mathbf{p}_i \cdot (\mathbf{q}' - \mathbf{q}) \right] &= \mathbf{q} - \mathbf{q}' = \mathbf{0}, \end{aligned} \tag{11.58}$$

where the final line follows because $\bar{S}(\mathbf{q}', \mathbf{q}_1)$ is independent of $\mathbf{p}_i$. Notice that although the stationary points of $\mathbf{p}_i$ are imaginary, the integral in (11.54) is taken in the direction of the real $\mathbf{p}_i$ axis.

Equations (11.58) are central to the theory. Their solutions are restricted to energies at or below the saddle point on the potential energy surface. The saddle itself is a 'fixed point' (with $\bar{\mathbf{p}} = \mathrm{i}\mathbf{p}_i = \bar{\mathbf{p}}' = \mathbf{0}$) which is unstable with respect to displacements along the reaction coordinate, written as q_1, and stable with respect to displacements in the orthogonal vibrational coordinates, $\tilde{\mathbf{q}} = (q_2, q_3, \ldots, q_f)$, with frequencies ω_{0i}, $i = 2, 3, \ldots, f$. Alternatively the analogous point on the upturned potential is stable to displacements along q_1, with frequency ω_0^* and time period $\tau = 2\pi/\omega_0^*$. Moreover any displacement $\delta\tilde{\mathbf{q}}$ in the remaining modes is amplified or diminished over a cycle of q_1 by a factor $e^{\pm u_i}$, where the Lyapunov exponent u_i is given by the product of the time period of q_1 and the frequency of the ith transverse vibrational frequency, $u_i = \omega_{0i}\tau$ (Tabor 1989). At lower energies the saddle point solution of eqns (11.58) goes over to a 'periodic orbit' solution on the upturned potential, which will again be taken to lie along the (possibly curvilinear) coordinate q_1. The imaginary time period $\tau(E)$ typically increases as the energy decreases. Motions orthogonal to this tunnelling mode on the physical surface are stable with respect to small displacements, with energy-dependent frequencies $\omega_i(E)$, for $i \neq 1$, while those on the upturned surface are unstable, with Lyapunov exponents

$$u_i(E) = \omega_i(E)\tau(E). \tag{11.59}$$

These periodic orbit solutions provide the basis for a multidimensional generalization of the previous one-dimensional instanton approximation to the rate constant. The periodic orbit itself is readily determined, for collinear systems, by the surface of section technique of Section 7.1, while a ring polymer approach (see below) is appropriate in higher dimensions. The Lyapunov exponents are derived from the eigenvalues, $\{e^{\pm u_i}\}$, $i = 2 - f$, of the $2(f-1) \times 2(f-1)$-dimensional monodromy matrix M (see Appendix F), which is defined such that

$$\begin{pmatrix} \delta\tilde{\mathbf{q}}' \\ \delta\tilde{\mathbf{p}}' \end{pmatrix} = \begin{pmatrix} M_{\tilde{\mathbf{q}}'\tilde{\mathbf{q}}} & M_{\tilde{\mathbf{q}}'\tilde{\mathbf{p}}} \\ M_{\tilde{\mathbf{p}}'\tilde{\mathbf{q}}} & M_{\tilde{\mathbf{p}}'\tilde{\mathbf{p}}} \end{pmatrix} \begin{pmatrix} \delta\tilde{\mathbf{q}} \\ \delta\tilde{\mathbf{p}} \end{pmatrix}, \tag{11.60}$$

in which $(\delta\tilde{\mathbf{q}}', \delta\tilde{\mathbf{p}}')$ are a displacement in $(\tilde{\mathbf{q}}, \tilde{\mathbf{p}})$ at imaginary time $t = -i\tau$ arising from initial displacements $(\delta\tilde{\mathbf{q}}, \delta\tilde{\mathbf{p}})$ at $t = 0$. Symplectic algorithms for determining the elements of the monodromy matrix have been given by Brewer et al. (1997). The resulting numerical solutions take full account of the potential surface along the periodic orbit tunnelling mode, q_1, plus quadratic terms in the transverse modes $\tilde{\mathbf{q}}$. The motion is therefore treated as separable in the two types of motion at any particular energy. However, Fig. 11.9(a) shows that the location of the instanton orbit moves progressively away from the saddle point, as the energy decreases, and the resulting energy dependence of the transverse frequencies, $\omega_i(E)$, and Lyapunov exponents, $u_i(E)$, allows for changes in the intermode coupling at different energies.

Having established the conditions for the underlying periodic orbit at any given energy, it remains to evaluate the right-hand side of eqn (11.54), a procedure that is simplified by recognizing that the integral with respect to $\mathbf{p}$ is an f-dimensional delta function. Hence, after substituting from eqn (11.55) for $\langle \mathbf{q}| e^{-\beta H} |\mathbf{q}\rangle$, with $A(\mathbf{p}, \mathbf{q})$ given by eqns (11.52)–(11.53),

$$\begin{aligned} k_{\text{rate}} Q_r = \langle A \rangle &= \frac{1}{2m} \int d\mathbf{q} \delta(q_1 - q_1^{\ddagger})[p_1 + |p_1|] \langle \mathbf{q}| e^{-\beta H} |\mathbf{q}\rangle \\ &= \frac{|\dot{q}_1^{\ddagger}|}{(2\pi\hbar)^{f/2}} \int d\tilde{\mathbf{q}} \left[-\det \frac{\partial^2 \bar{S}}{\partial \mathbf{q}' \partial \mathbf{q}} \right]^{1/2}_{q_1 = q_1^{\ddagger}} e^{-\bar{S}(q', q; \beta)}, \end{aligned} \tag{11.61}$$

because contributions from p_1 cancel around the cycle, while those from $|p_1|$ are counted twice.

The remaining integrals are evaluated by an argument due to Gutzwiller (1971), as modified for imaginary time propagation by Althorpe (2011). Details are given in Appendix C.4. The Legendre transformation

$$\bar{S}(\mathbf{q},\mathbf{q},\tau) = \bar{W}(\mathbf{q},\mathbf{q},E) + E\tau = \oint \bar{\mathbf{p}}\cdot d\mathbf{q} + E\tau \tag{11.62}$$

is employed, where E is the energy of the orbit with period $\tau = \beta\hbar$, because the properties of the reduced classical action integral $\bar{W}(\mathbf{q},\mathbf{q},E)$ are more convenient than those of $\bar{S}(\mathbf{q},\mathbf{q},\tau)$. The following quadratic expansion for $\bar{S}(\mathbf{q},\mathbf{q},\tau)$ around the periodic orbit,

$$\bar{S}(\mathbf{q},\mathbf{q},\tau) \simeq \bar{W}(E) + E\tau + \frac{1}{2}\boldsymbol{\delta}\tilde{\mathbf{q}}^T\left(\frac{\partial^2\bar{W}}{\partial\tilde{\mathbf{q}}\partial\tilde{\mathbf{q}}} + 2\frac{\partial^2\bar{W}}{\partial\tilde{\mathbf{q}}'\partial\tilde{\mathbf{q}}} + \frac{\partial^2\bar{W}}{\partial\tilde{\mathbf{q}}'\partial\tilde{\mathbf{q}}'}\right)\boldsymbol{\delta}\tilde{\mathbf{q}}', \tag{11.63}$$

takes into account that the linear terms vanish, by virtue of eqn (11.57), and that the transverse displacement $\boldsymbol{\delta}\tilde{\mathbf{q}}$ at the beginning of an orbit cycle is physically indistinguishable from $\boldsymbol{\delta}\tilde{\mathbf{q}}'$at the end. The partial derivatives of $\bar{S}(\mathbf{q},\mathbf{q},\tau)$ and $\bar{W}(\mathbf{q},\mathbf{q},E)$ with respect to the transverse coordinates $\tilde{\mathbf{q}}$ are identical, while the full Van Vleck determinant of $\bar{S}(\mathbf{q},\mathbf{q},\tau)$ is given in eqn (C.118) as

$$\det\left(-\frac{\partial^2\bar{S}}{\partial\mathbf{q}'\partial\mathbf{q}}\right)_\tau = \left(-\dot{q}_1\dot{q}_1'\frac{\partial\tau}{\partial E}\right)^{-1/2}\det\left(-\frac{\partial^2\bar{W}}{\partial\tilde{\mathbf{q}}'\partial\tilde{\mathbf{q}}}\right)_E, \tag{11.64}$$

after taking into account that $\partial\bar{W}/\partial E = -\tau(E)$ on the inverted potential. It follows, by combining eqns (11.56), (11.63) and (11.64) with the identity $\tau = \beta\hbar$, that the stationary phase approximation to $\langle A\rangle$ in (11.61) is given by

$$\begin{aligned} k_{\text{rate}}Q_r = \langle A\rangle &\simeq \sqrt{\frac{1}{2\pi\hbar^2}\left(-\frac{\partial E_\beta}{\partial\beta}\right)}\mathrm{e}^{-\bar{W}(E_\beta)/\hbar - E_\beta\beta}\left[\det\left(-\frac{\partial^2\bar{W}}{\partial\tilde{\mathbf{q}}'\partial\tilde{\mathbf{q}}}\right)\right]^{1/2} \\ &\times\left[\det\left(\frac{\partial^2\bar{W}}{\partial\tilde{\mathbf{q}}\partial\tilde{\mathbf{q}}} + 2\frac{\partial^2\bar{W}}{\partial\tilde{\mathbf{q}}'\partial\tilde{\mathbf{q}}} + \frac{\partial^2\bar{W}}{\partial\tilde{\mathbf{q}}'\partial\tilde{\mathbf{q}}'}\right)\right]^{-1/2}, \end{aligned} \tag{11.65}$$

which differs from the one-dimensional result in eqn (11.51) by the ratio of the two $(f-1)\times(f-1)$-dimensional determinants.

Gutzwiller (1971) uses the argument in problem 11.3 to relate this ratio to the so-called residue of the monodromy matrix in eqn (11.60). In particular, if the periodic orbit is unstable to all $f-1$ transverse coordinate displacements,

$$\begin{aligned} &\det\left(\bar{W}_{\tilde{\mathbf{q}}\tilde{\mathbf{q}}} + \bar{W}_{\tilde{\mathbf{q}}\tilde{\mathbf{q}}'} + \bar{W}_{\tilde{\mathbf{q}}'\tilde{\mathbf{q}}} + \bar{W}_{\tilde{\mathbf{q}}'\tilde{\mathbf{q}}'}\right)/\det(\bar{W}_{\tilde{\mathbf{q}}\tilde{\mathbf{q}}'}) \\ &\quad = R = \det(M-I) = (-)^{f-1}\prod_i 4\sinh^2(u_i/2). \end{aligned} \tag{11.66}$$

Eqn (11.65) for the rate constant therefore reduces to

$$k_{\text{rate}}Q_r = \sqrt{\frac{1}{2\pi\hbar^2}\left(-\frac{\mathrm{d}E_\beta}{\mathrm{d}\beta}\right)}\mathrm{e}^{-\beta E_\beta - \bar{W}(E_\beta)/\hbar}Q^{\ddagger}_{\text{vib}}(E_\beta)$$
$$= \sqrt{\frac{1}{2\pi\hbar^2}\left(-\frac{\mathrm{d}E_\beta}{\mathrm{d}\beta}\right)}\mathrm{e}^{-\bar{S}(\beta)/\hbar}Q^{\ddagger}_{\text{vib}}(E_\beta), \tag{11.67}$$

where

$$Q^{\ddagger}_{\text{vib}}(E_\beta) = \prod_{i=2}^{f}\frac{1}{2\sinh\beta\hbar\omega_i(E_\beta)/2}, \tag{11.68}$$

which has the structure of a vibrational partition function, with energy-dependent frequencies, related to u_i by eqn (11.59).

Equations (11.67) and (11.68) give the (normally dominant) contribution to the rate constant arising from a single cycle of the instanton orbit. The extension to k cycles involves applying factors of k to both $\bar{W}(E_\beta)$ and $u_i = \beta\hbar\omega_i(E_\beta)$. To the extent that the longitudinal and transverse motions are treated as separable, Miller (1975) shows that the resulting sum over k may be expressed in the form

$$k_{\text{rate}}Q_r = \frac{1}{2\pi\hbar}\int \mathrm{d}E\mathrm{e}^{-\beta E}P(E)$$
$$P(E) = \sum_{\mathbf{n}}\frac{1}{1+\exp[\bar{W}(E_n)/\hbar]}, \tag{11.69}$$

where

$$E_n = E - \sum_{i=2}^{N}(n_i + 1/2)\hbar\omega_i \tag{11.70}$$

is the translational energy in the tunnelling mode, $\bar{W}(E_n)/\hbar$ is the tunnelling integral and the sum over $\mathbf{n}$ is taken over all partitions $(n_2, n_3, \ldots, n_N)$ between the vibrational modes for which $E_n < E$. In other words the cumulative reaction probability $P(E)$ is a sum of one-dimensional tunnelling probabilities, taken over all energetically accessible states of the activated complex. Serious doubts were cast, however, on the separability assumptions by poor agreement between quantum mechanically accurate cumulative probabilities for the collinear $H+H_2$ reaction and those given by eqn (11.69). However, substantial improvements were obtained by employing an iterative solution of eqn (11.70) to obtain an energetically dependent transverse frequency, $\omega_i(E)$, a procedure that led to agreement to within a factor of 2.0 for probabilities that varied over at least four orders of magnitude, depending on the chosen potential surface (Chapman et al. 1975).

Before moving on to extensions of this Miller–Gutzwiller formulation, the reader should note that the results following from eqn (11.67) are independent of any specific 'transition state', because reference to $q_1^{\ddagger}$ occurs only in the transition state velocity

$\dot{q}_1^\ddagger$ in eqn (11.61), which cancels with a factor from eqn (11.65) when the Van Vleck determinant is evaluated at $q_1^\ddagger$. The quantities $\bar{S}(\beta)$, $Q_{\text{vib}}^\ddagger(E_\beta)$ and $(\partial E/\partial\beta)$ all belong to the extended instanton orbit, which takes over the role of the classical transition state in the quantum mechanical tunnelling regime. The validity of eqn (11.67) cannot therefore be assessed by asking whether or not the relevant trajectories 'recross' a particular plane $q_1 = q_1^\ddagger$ (Pechukas 1976). Instead one must enquire into the nature of the dynamics on the upturned potential surface. The derivation above assumes a dispersive classical system governed by the properties of a single unstable periodic orbit, in which case the accuracy of eqn (11.61) is limited only by the quadratic expansion in eqn (11.63). There can, however, be interactions with other periodic orbit resonances, possibly leading to chaotic motion, which would clearly invalidate the present isolated instanton picture.

Kryvohuz modification

More recently, Kryvohuz (2011) has taken explicit account of the energy dependence of $Q_{\text{vib}}^\ddagger(E)$ in performing the integral in eqn (11.69). It is convenient for comparison with the paper to replace the Cartesian variables q_i by mass-reduced coordinates $X_i = \sqrt{m_i}q_i$. Consequently the second of eqn (11.56) is written as

$$\bar{S}_0(\beta) = E_0(\beta)\tau_\beta + \bar{W}_0(\beta) = \int_0^{\tau_\beta} \left\{\frac{1}{2}[\dot{\mathbf{X}}(\tau)]^2 + V([\mathbf{X}(\tau)])\right\} \mathrm{d}\tau. \tag{11.71}$$

The essential step in the subsequent argument is to incorporate the partition function into an effective phase term

$$\tilde{S}(\beta) = \bar{S}_0(\beta) + \sigma(\beta), \tag{11.72}$$

where

$$\sigma(\beta) = \hbar \sum_{i=2}^{N} \ln[2\sinh u_{i\beta}/2], \tag{11.73}$$

in which $u_{i\beta}$ are the Lyapunov exponents of the instanton orbit at the chosen β value. The resulting quasi one-dimensional theory employs the equations

$$\begin{aligned} \tilde{E}(\beta) &= \frac{\mathrm{d}\tilde{S}(\beta)}{\mathrm{d}\tau_\beta} = E_0(\beta) + \frac{\mathrm{d}\sigma}{\mathrm{d}\tau_\beta}, \\ \tilde{W}(\beta) &= \tilde{S}(\beta) - \tau_\beta\tilde{E}(\beta) = \bar{W}_0(\beta) - \tau_\beta^2\frac{\mathrm{d}}{\mathrm{d}\tau_\beta}\left(\frac{\sigma(\beta)}{\tau_\beta}\right) \end{aligned} \tag{11.74}$$

to define an effective energy, $\tilde{E}(\beta)$, and an effective tunnelling action, $\tilde{W}(\beta)$, for the orbit, which leads in turn to the critical value β_c at which $\tilde{W}(\beta_c) = 0$, and an effective saddle point energy $\tilde{V}_0 = \tilde{E}(\beta_c)$.

In the above so-called SI1 version of the theory, $\bar{S}_0(\beta)$ and $\sigma(\beta)$ are determined by the instanton orbit, whose path automatically minimizes the action integral $S_0(\beta)$

according to Hamilton's principle. There is also an iterative SI2 version, in which the shape of the path is optimized to minimize the modified action $\tilde{S}(\beta)$, taking into account that each new path generates a new set of transverse displacements and associated Lyapunov exponents. A Fourier representation for the path,

$$\mathbf{X}(\tau) = \sum_{j=1}^{J} \mathbf{C}_j \cos\left(\frac{2\pi j}{\hbar\beta}\tau\right), \tag{11.75}$$

is adopted for practical convenience, so that $\tilde{S}(\beta)$ can be minimized with respect to the elements of the $N \times J$ Fourier component matrix $\mathbf{C}$. The iterated functions $\tilde{S}(\beta)$ and $\sigma(\beta)$ are then used to define $\tilde{E}(\beta)$, $\tilde{W}(\beta)$ and new parameters $\tilde{V}_0$ and β_c as given by eqn (11.74).

The analogue of eqn (11.67) for $\beta > \beta_c$ was then taken, for either version of the theory, in the form

$$\begin{aligned} k_{\text{rate}} Q_r &= \sqrt{-\frac{\tilde{E}'(\beta)}{2\pi\hbar^2}} e^{-\tilde{S}(\beta)/\hbar} P\left\{[\tilde{V}_0 - \tilde{E}(\beta)]/\sqrt{-\tilde{E}'(\beta)}\right\} \\ &= \sqrt{-\frac{\tilde{E}'(\beta)}{2\pi\hbar^2}} e^{-S_0(\beta)/\hbar} \prod_{i=2}^{N} \frac{1}{2\sinh u_i/2} P\left\{[\tilde{V}_0 - \tilde{E}(\beta)]/\sqrt{-\tilde{E}'(\beta)}\right\}, \end{aligned} \tag{11.76}$$

where $P(z)$ is the probability integral from $-\infty$ to z. Equation (11.76) has the same essential structure as (11.67) except that $P(z)$ truncates the full stationary phase integral, for $\beta \simeq \beta_c$, in such a way that the value $P(0) = 1/2$ ensures 50% transmission and 50% reflection. At higher temperatures, with $\beta < \beta_c$, Kryvohuz (2011) employs an expression that was derived by Cao and Voth (1996) to interpolate between eqn (11.76) and the parabolic barrier formula (11.44), in a way that eliminates the discontinuity at $\beta = \beta_c$.

This theory was tested by Kryvohuz (2011) on seven different collinear symmetric and asymmetric atom transfer reactions involving a variety of relative mass combinations. Fig. 11.9 illustrates paths of the instanton orbits for the $H + H_2$ reaction. Panel (a) shows how the locations of the SI1 orbits move progressively away from saddle point on the surface as the temperature decreases from 400 K to 300 K and 200 K, while panel (b) illustrates a change in the shape of the 300 K orbit in going from the SI1 to the SI2 version of the theory. The rate constants given by eqn (11.76) together with the Cao and Voth (1996) extrapolation to higher temperatures were compared with numerically accurate quantum mechanical values over the temperature range 200 to 1500 K. Agreement was found within a factor of 2.0 for the SI1 method, improving to a factor of 1.5 for the SI2 approach.

Connection with ring polymer theory

It is interesting to complete this section by indicating a close connection between semiclassical instanton theory and the so-called Im F (imaginary free energy) version of ring polymer rate theory. The account below is taken from Althorpe (2011). The

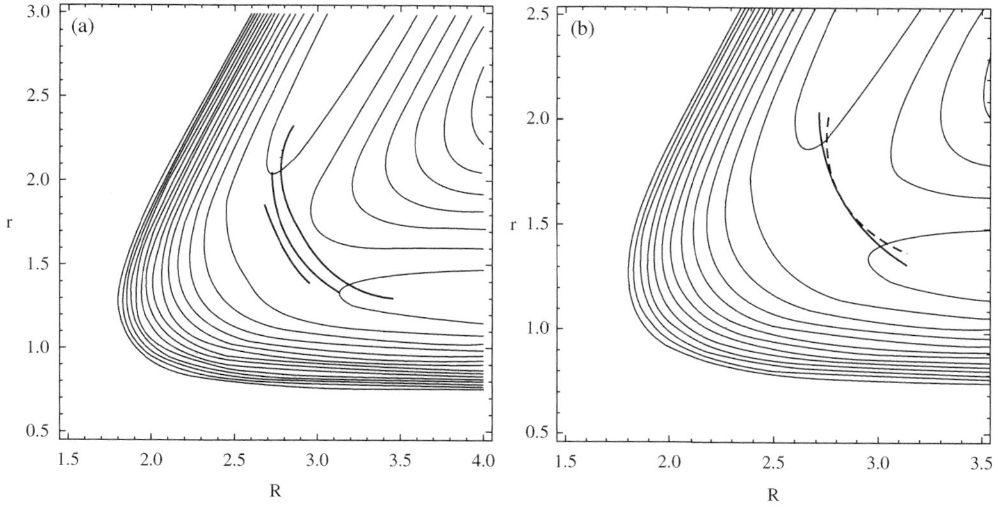

Fig. 11.9 Instanton orbits for the collinear H + H_2 reaction obtained by (a) the SI1 and (b) the SI2 versions of the effective action theory. The orbits in (a) are drawn for temperatures 400 K, 300 K and 200 K. Those in (b) are the SI1 (full) and SI2 (dashed) orbits at 300 K. (Reprinted with permission from Kryvohuz (2011). Copyright 2011, AIP Publishing LLP.)

ring polymer technique starts by evaluating the phase space average for the partition function $Q(\beta)$ along an imaginary time trajectory broken into N equal steps of duration $\epsilon_N = \beta\hbar/N$, which is taken around a loop from $\mathbf{x}_0 \to \mathbf{x}_1 \to \cdots \mathbf{x}_N$, with $\mathbf{x}_N = \mathbf{x}_0$; thus, in a mass-weighted Cartesian coordinate system,

$$Q_N(\beta) = \left(\frac{1}{2\pi_N\hbar}\right)^{fN/2} \int \mathrm{d}\mathbf{x}_0 \ldots \int \mathrm{d}\mathbf{x}_{N-1} \mathrm{e}^{-\bar{S}_N(\mathbf{x}_0,\mathbf{x}_1,\cdots,\mathbf{x}_{N-1},\mathbf{x}_N)/\hbar}, \tag{11.77}$$

where

$$\bar{S}_N(\mathbf{x}_0, \mathbf{x}_1, \cdots, \mathbf{x}_N; \beta) = \frac{\epsilon_N}{2}\left[V(\mathbf{x}_0) + V(\mathbf{x}_N)\right] + \epsilon_N \sum_{n=1}^{N-1} V(\mathbf{x}_n)$$
$$+ \frac{1}{2\epsilon_N} \sum_{i=1}^{f} \sum_{n=0}^{N} (x_{ni} - x_{(n+1)i})^2. \tag{11.78}$$

$Q_N(\beta)$ is then evaluated for bound systems by expanding $\bar{S}_N(\mathbf{x}_0, \mathbf{x}_1, \ldots, \mathbf{x}_{N-1}, \mathbf{x}_N; \beta)$ up to quadratic terms about its global minimum (written as $\{\mathbf{x}_0^{(e)}, \ldots, \mathbf{x}_{N-1}^{(e)}\}$) and performing the necessary Gaussian integrals—a procedure that becomes exact for finite β in the limit $N \to \infty$.

There are obvious analogies with the semiclassical instanton method, because $S(\mathbf{x}_0, \mathbf{x}_1, \ldots, \mathbf{x}_{N-1}, \mathbf{x}_0; \beta)$ may be recognized as a finite difference approximation to the classical action integral on the inverted potential. By the same token, the stationarity condition acts as a finite difference approximation to Newton's equation $\mathrm{d}^2\mathbf{x}/\mathrm{d}\tau^2 = \partial V/\partial \mathbf{x}$ on the upturned surface. There is also the complication, in a reactive situation, that the global stationary point is not in fact a minimum, which means that displacements along the reaction coordinates require separate treatment from those transverse to it. Since all the beads are equivalent, the coordinates analogous to $\mathbf{q}$ above will be written as $\mathbf{q}_0$, with displacements δq_{01} taken to lie along the chain, and the remaining displacements $\delta\tilde{\mathbf{q}}_0$ perpendicular to it.

The terminology 'imaginary free energy' or 'Im F' theory implies that the presence of a saddle point on the potential surface introduces an imaginary contribution to the partition function, in terms of which the rate constant is given by (Benderskii et al. 1994)

$$k_{\text{rate}}^{(imf)} = -\frac{2}{\hbar}\,\mathrm{Im}\,F = \left[-\frac{2}{\hbar\beta}\ln(\mathrm{Re}\,Q + \mathrm{i}\,\mathrm{Im}\,Q)\right] \simeq \frac{2}{\beta\hbar}\frac{\mathrm{Im}\,Q}{Q_r}, \tag{11.79}$$

for $\mathrm{Im}\,Q \ll \mathrm{Re}\,Q \simeq Q_r$. Details of the necessary analytic continuation of $Q(\beta)$ into the complex plane are beyond the scope of this section. We simply rearrange the result given by Althorpe (2011) to conform with the present notation. In the absence of a global minimum, the analogue of eqn (11.77) is evaluated by expanding $\bar{S}_N(\mathbf{x}_0, \mathbf{x}_1, \ldots, \mathbf{x}_{N-1}, \mathbf{x}_N)$ about a value $\bar{S}_N(\mathbf{x}_0, \mathbf{x}_N; \beta)$, consistent with prescribed values of $\mathbf{x}_0 = \mathbf{x}_N$. The remaining variables $(\mathbf{x}_1 \ldots \mathbf{x}_{N-1})$ are then taken to fluctuate quadratically about their equilibrium values $\left(\mathbf{x}_1^{(e)} \ldots \mathbf{x}_{N-1}^{(e)}\right)$ as constrained by the end point $\mathbf{x}_0 = \mathbf{x}_N$. In addition, the $\mathbf{x}_0$ variables are transformed to 'reaction coordinates' $\mathbf{q}_0$ of which δq_{01} is taken to lie along the chain, while the remaining displacements $\delta\tilde{\mathbf{q}}_0$ are perpendicular to it. The resulting Nf-dimensional integral is therefore evaluated as

$$\begin{aligned}
k_{\text{rate}}^{(imf)} Q_r &= 2\dot{q}_{01}\int \mathrm{d}\tilde{\mathbf{q}}_0 \det(\partial\mathbf{x}_0/d\mathbf{q}_0) F(\mathbf{x}_0, \mathbf{x}_N; \beta) \mathrm{e}^{-\bar{S}_N(\mathbf{x}_0,\mathbf{x}_N;\beta)/\hbar} \\
F(\mathbf{x}_0, \mathbf{x}_N; \beta) &= \lim_{N\to\infty}\left(\frac{1}{2\pi\epsilon_N\hbar}\right)^{fN/2}\int \mathrm{d}\mathbf{x}_1 \ldots \int \mathrm{d}\mathbf{x}_{N-1} \\
&\quad \times \exp\left[-\frac{\left(\mathbf{x}-\mathbf{x}^{(e)}\right)^T \mathbf{J}(\mathbf{x}_0,\mathbf{x}_N)\left(\mathbf{x}-\mathbf{x}^{(e)}\right)}{2\epsilon_N\hbar}\right] \\
&= \lim_{N\to\infty}\frac{1}{2\sqrt{(2\pi\epsilon_N\hbar)^f \det\mathbf{J}(\mathbf{x}_0,\mathbf{x}_N)}},
\end{aligned} \tag{11.80}$$

where $\det(\partial\mathbf{x}_0/d\mathbf{q}_0)$ is the Jacobian for the transformation between $\mathbf{q}_0$ and the Cartesian variables $\mathbf{x}_0$.

We also note that the constraints on $(\mathbf{x}_1 \ldots \mathbf{x}_{N-1})$ ensure that $\left(\partial\bar{S}_N/\partial x_{nj}\right) = 0$ at $\mathbf{x} = \mathbf{x}^{(e)}$ for $n \neq 0, N$, which means that

$$\frac{\partial\bar{S}_N}{\partial x_{0i}} = \frac{\epsilon_N}{2}\left(\frac{\partial V(\mathbf{x}_0)}{\partial x_{0i}}\right) + \frac{x_{0i} - x_{1i}^{(e)}}{\epsilon_N} \quad \text{and} \quad \frac{\partial^2\bar{S}_N}{\partial x_{0i}\partial x_{Nj}} = -\frac{1}{\epsilon_N}\frac{\partial x_{1i}^{(e)}}{\partial x_{Nj}}. \tag{11.81}$$

Similarly, the fluctuations of $\bar{S}_N(\mathbf{x}_0, \mathbf{x}_1, \ldots, \mathbf{x}_N; \beta)$ around the intermediate equilibrium points $\mathbf{x}_n^{(e)}(\mathbf{x}_0, \mathbf{x}_N; \beta)$ are controlled by the Hessian matrix with the banded block structure

$$\mathbf{J}(\mathbf{x}_0,\mathbf{x}_N)=\begin{pmatrix} \epsilon_N^2\mathbf{V}_{11}+2\mathbf{I} & -\mathbf{I} & 0 & 0 \\ -\mathbf{I} & \epsilon_N^2\mathbf{V}_{22}+2\mathbf{I} & -\mathbf{I} & 0 \\ 0 & -\mathbf{I} & \epsilon_N^2\mathbf{V}_{33}+2\mathbf{I} & -\mathbf{I} \\ 0 & 0 & -\mathbf{I} & \epsilon_N^2\mathbf{V}_{N-1,N-1}+2\mathbf{I} \end{pmatrix}, \tag{11.82}$$

in which each block has dimension $f \times f$, and $\mathbf{V}_{nn}$ is the matrix of second derivatives, $\partial^2 V/\partial x_{ni}\partial x_{ni'}$.

The key to the connection with the above semiclassical formulation lies in an algebraic identity between the $(N-1)f \times (N-1)f$ determinant $\det \mathbf{J}(\mathbf{x}_0,\mathbf{x}_N)$ in eqn (11.80) and the smaller $f \times f$ Van Vleck determinant $\det\left(\partial^2 \bar{S}_N/\partial\mathbf{x}_0\partial\mathbf{x}_N\right)$ (Gel'fand and Yaglom 1960; Althorpe 2011). One finds by taking partial derivatives of eqn (11.78) and substituting for $\mathbf{V}_{nn}$ in eqn (11.82) that

$$\mathbf{J}(\mathbf{x}_0,\mathbf{x}_N)\mathbf{A} = \mathbf{D}, \tag{11.83}$$

where $\mathbf{A}$ and $\mathbf{D}$ are stacked in $f \times f$-dimensional blocks $\mathbf{A}^{(n)}$ and $\mathbf{D}^{(n)}$, with elements $A_{ij}^{(n)} = \partial x_{ni}^{(e)}/\partial x_{Nj}$ and $D_{ij}^{(n)} = \delta_{nN-1}\delta_{ij}$. Thus, by comparison with eqn (11.81), $\mathbf{A}^{(1)} = -\partial^2 S/\partial\mathbf{x}_0\partial\mathbf{x}_N$ and $\mathbf{D}^{(n)}$ are null matrices apart from $\mathbf{D}^{(N-1)} = \mathbf{I}$. It follows by applying Cramer's rule in blocks that

$$(\epsilon_N)^f\,\mathbf{A}^{(1)} = -\partial^2 S/\partial\mathbf{x}_0\partial\mathbf{x}_N = (\epsilon_N)^f\,[\det(\mathbf{P})/\det(\mathbf{J})]\mathbf{I}, \tag{11.84}$$

where $\mathbf{P}$ is the matrix obtained by removing the first block row and the last block column of $\mathbf{J}$. Consequently

$$\det\left(-\frac{\partial^2\bar{S}}{\partial\mathbf{x}_0\partial\mathbf{x}_N}\right) = \lim_{N\to\infty}\det\left(-\frac{\partial^2\bar{S}_N}{\partial\mathbf{x}_0\partial\mathbf{x}_N}\right) = \lim_{N\to\infty}\frac{1}{(\epsilon_N)^f\det(\mathbf{J})}. \tag{11.85}$$

It remains to substitute for $\det(\mathbf{J})$ in eqn (11.80) and to use the identity

$$\det\left(-\frac{\partial^2\bar{S}}{\partial\mathbf{q}_0\partial\mathbf{q}_N}\right) = \det\left(\frac{\partial\mathbf{x}_0}{d\mathbf{q}_0}\right)\det\left(-\frac{\partial^2\bar{S}}{\partial\mathbf{x}_0\partial\mathbf{x}_N}\right)\det\left(\frac{\partial\mathbf{x}_0}{d\mathbf{q}_0}\right) \tag{11.86}$$

to see that in the limit $N \to \infty$

$$k_{\text{rate}}^{(imf)}Q_r = \frac{\dot{q}_{01}}{(2\pi\hbar)^{f/2}}\int d\tilde{\mathbf{q}}_0\left[\det\left(-\frac{\partial^2\bar{S}}{\partial\mathbf{x}_0\partial\mathbf{x}_N}\right)\right]^{1/2} e^{-\bar{S}(\mathbf{x}_0,\mathbf{x}_N;\beta)/\hbar}, \tag{11.87}$$

which is identical with the semiclassical result in eqn (11.61). It is also easy to see from this correspondence that equally close agreement would be obtained by combining the transition state conditions (11.52)–(11.53) with the ring polymer partition function in (11.77), instead of using the Im F formulation.

Moreover, the ring polymer method has significant computational advantages over the semiclassical technique, because the latter is severely hampered, even for three-dimensional systems, by the notorious difficulty of solving Hamilton's equations for unstable periodic orbits in multidimensional systems. Instead, the ring polymer approach iterates towards the path of the orbit by minimizing the discrete approximation to the action integral, $\bar{S}_N(\mathbf{x}_0, \mathbf{x}_1, \ldots, \mathbf{x}_{N-1}, \mathbf{x}_N)$, subject to constraining the transformation $\mathbf{x}_0 \to \mathbf{q}_0$ such that displacements δq_{01} lie in the direction of the reaction coordinate.

Before leaving this topic, the reader might take note of an alternative semi-classical approach, which builds on the classical perturbation theory in Section 7.5. Miller (1977) argues that the reaction dynamics close to the barrier may be set in an angle–action context, based on a zeroth-order model in which the tunnelling integral for a parabolic model acts as an imaginary time action variable, while the transverse vibrations are modelled as harmonic oscillators. Systematic corrections may then be added along the lines of Section 7.5. Applications of this idea have been given by Chapman et al. (1976).

11.5 Problems

11.1. (i) Assuming the Legendre polynomial recurrence relation (Abramowitz and Stegun 1965) in the form

$$xP_J(x) = [J|x|J+1]\, P_{J+1}(x) + [J|x|J-1]\, P_{J-1}(x),$$

where $[J|x|J+1] = (J+1)/(2J+1)$ and $[J|x|J-1] = J/(2J+1)$, show that

$$\sum_J a_J^{(0)} P_J(x) = (1+\beta x)^{-1} \sum_J a_J^{(1)} P_J(x),$$

where

$$a_J^{(1)} = a_J^{(0)} + \beta \left\{ a_{J-1}^{(0)} [J-1|x|J] + a_{J+1}^{(0)} [J+1|x|J] \right\}.$$

Conclude that the choice $\beta = -\left\{ a_0^{(0)}/a_1^{(0)} [1|x|0] \right\} = -3a_0^{(0)}/a_1^{(0)}$ eliminates the coefficient $a_0^{(1)}$.

(ii) Use the notation $[J|x^2|J'] = \sum_{J''} [J|x|J''][J''|x|J']$ to obtain equations for β_1 and β_2 such that $a_0^{(2)} = a_1^{(2)} = 0$ in the expansion

$$\sum_J a_J^{(0)} P_J(x) = [(1+\beta_1 x)(1+\beta_2 x)]^{-1} \sum_J a_J^{(2)} P_J(x).$$

11.2. (i) Show that the propagator for motion in a circle with moment of inertia I, $\hbar = 1$ and arbitrary boundary conditions $\psi(\phi + 2\pi, t) = \mathrm{e}^{\mathrm{i}\alpha}\psi(\phi, t)$ may be expressed as

$$K(\phi'', t''; \phi', t') = \sum_{m=-\infty}^{\infty} u_m(\phi'')u_m^*(\phi')\mathrm{e}^{-\mathrm{i}E_m(t''-t')}$$
$$= \frac{1}{2\pi}\sum_{m=-\infty}^{\infty} \exp\left[\mathrm{i}(m+\gamma)\phi - \mathrm{i}(m+\gamma)^2 T/2I\right],$$

where $\phi = \phi'' - \phi'$, $T = t'' - t'$ and $\gamma = (\alpha/2\pi)$.

(ii) Follow §23.1 of Schulman (1981) in recognizing the connection with the Jacobi theta function, $\theta_3(z, \tau)$, which satisfies the transformation identity

$$\theta_3(z,\tau) = \sum_{m=-\infty}^{\infty} e^{\mathrm{i}[m^2\pi\tau + 2nz]} = (-\mathrm{i}\tau)^{-1/2}\mathrm{e}^{(z^2/\mathrm{i}\pi\tau)}\theta_3\left(\frac{z}{\tau}, -\frac{1}{\tau}\right).$$

A proof is given in §21.51 of Whittaker and Watson (1962).

(iii) Hence derive the path integral form

$$K(\phi, T) = \sum_{n=-\infty}^{\infty} \mathrm{e}^{\mathrm{i}n\alpha}K_n(\phi, T)$$
$$= \left(\frac{I}{2\pi\mathrm{i}T}\right)^{1/2}\sum_{n=-\infty}^{\infty} \mathrm{e}^{\mathrm{i}n\alpha}\exp\left(\frac{\mathrm{i}I(\phi - 2n\pi)^2}{2T}\right),$$

in which $K_n(\phi, T)$ is the analogue of the free motion propagator in eqn (C.67) for n cycles of the angular motion.

11.3. (i) Use the identities $p'_\nu = \partial W/\partial q'$ and $p_\nu = -\partial W/\partial q$ to show that initial and final transverse displacements from a periodic orbit are related by

$$\begin{pmatrix} \delta\tilde{\mathbf{p}}' \\ \delta\tilde{\mathbf{p}} \end{pmatrix} = \begin{pmatrix} \mathbf{W}_{\tilde{\mathbf{q}}'\tilde{\mathbf{q}}'} & \mathbf{W}_{\tilde{\mathbf{q}}'\tilde{\mathbf{q}}} \\ -\mathbf{W}_{\tilde{\mathbf{q}}\tilde{\mathbf{q}}'} & -\mathbf{W}_{\tilde{\mathbf{q}}\tilde{\mathbf{q}}} \end{pmatrix}\begin{pmatrix} \delta\tilde{\mathbf{q}}' \\ \delta\tilde{\mathbf{q}} \end{pmatrix},$$

in which $(\mathbf{W}_{\tilde{\mathbf{q}}'\tilde{\mathbf{q}}})_{\nu\mu} = (\partial^2 W/\partial\tilde{q}'_\nu\partial\tilde{q}_\mu)$, etc.

(ii) Express the blocks of the monodromy matrix in eqn (11.60) in terms of those of the above Van Vleck matrix $\mathbf{W}$. Hence show that

$$R = \det(M - I) = \det\left[\mathbf{W}_{\tilde{\mathbf{q}}'\tilde{\mathbf{q}}} + \mathbf{W}_{\tilde{\mathbf{q}}'\tilde{\mathbf{q}}'} + \mathbf{W}_{\tilde{\mathbf{q}}\tilde{\mathbf{q}}} + \mathbf{W}_{\tilde{\mathbf{q}}\tilde{\mathbf{q}}'}\right]/\det(\mathbf{W}_{\tilde{\mathbf{q}}'\tilde{\mathbf{q}}}).$$

Appendix A
Phase integral techniques

A.1 The Stokes phenomenon

The behaviour of the JWKB wavefunction,

$$\psi_{\mathrm{JWKB}}(x) = k^{-1/2}(x)\left[A'\exp\left(\mathrm{i}\int_a^x k(x)\mathrm{d}x\right) + A''\exp\left(-\mathrm{i}\int_a^x k(x)\mathrm{d}x\right)\right], \quad \text{(A.1)}$$

in the complex plane raises some puzzling features, whose resolution relates to the choice of Maslov index for a given situation (Section 3.1), the structure of the connection formulae (Section 3.3) and the overall strengths and limitations of the theory.

The first puzzling point is that the classical turning points of the motion, normally taken as the integration limits a in eqn (A.1), are branch points of the function

$$k(x) = \{2m[E - V(x)]\}^{1/2}/\hbar, \quad \text{(A.2)}$$

although they are regular points with respect to the underlying Schrödinger equation

$$\frac{\mathrm{d}^2\psi}{\mathrm{d}x^2} + k^2(x)\psi = 0. \quad \text{(A.3)}$$

An arbitrary solution of eqn (A.3) must therefore be single valued with respect to encirclement of a turning point, whereas the sign change of $k(x)$ on crossing a branch cut (Phillips 1957) leads to an interchange of the two exponential factors in eqn (A.1). The encirclement must therefore be accompanied by a compensating interchange of A' and A''; the questions are where and how does this interchange take place?

Another feature is that solutions of eqn (A.3) may contain two exponential terms in one region of space but only one term in another. For example, the asymptotic part of the Airy function (see eqn (2.77)) consists of a real oscillatory combination in the classically allowed region where $k(x)$ is real, but only a single exponentially decreasing term in the forbidden region. Where and how in the complex plane does the second term disappear?

The answer to these questions lies in the Stokes phenomenon (Heading 1962; Jeffreys 1962; Froman and Froman 1967), which is associated with the fact that errors in an exponentially large asymptotic term, in a region where $k(x)$ is purely imaginary, are of the same order of magnitude as the whole exponentially small counterpart. Consequently the coefficient of the exponentially small, or subdominant, term is indeterminate in these regions and typically suffers changes proportional to that of the coefficient of the dominant term. The discussion of the behaviour of the Airy function in the complex plane by Jeffreys and Jeffreys (1956) is very illuminating on this point. This means that the puzzling coefficient changes are localized around the so-called Stokes lines, which emanate from the classical turning points in directions along which the exponents in eqn (A.1) are real, one being positive and the other negative. Conversely, directions in which the exponents are both imaginary are termed anti-Stokes directions. It follows that classically allowed and classically forbidden regions of the real axis correspond to anti-Stokes and Stokes segments respectively, but the Stokes phenomenon is by no means restricted to this real axis.

The consequences of the Stokes behaviour are best illustrated by the simple case in which

$$k(z) = (z - a)^{1/2}, \tag{A.4}$$

where z is taken to denote the general complex variable. The exponents in eqn (A.1) are then given by

$$\begin{aligned} \mathrm{i}\int_a^z k(z)\mathrm{d}z &= \frac{2}{3}\mathrm{i}(z-a)^{3/2} \\ &= \frac{2}{3}\rho^{3/2}\left[\cos\left(\frac{3\varphi}{2}+\frac{\pi}{2}\right) + \mathrm{i}\sin\left(\frac{3\varphi}{2}+\frac{\pi}{2}\right)\right], \end{aligned} \tag{A.5}$$

where ρ and φ are the modulus and argument of $(z - a)$ respectively. The Stokes lines are therefore given by $\varphi = (2n-1)\pi/3$, with the second term in eqn (A.1) dominant at $\varphi = \pi/3$ and $5\pi/3$, and the first term dominant at $\varphi = \pi$. Consequently the coefficient, A', of the first term may be expected to change in proportion to A'' over an interval $\pi/3 - \varepsilon < \varphi < \pi/3 + \varepsilon$; that is

$$A' \to A' + TA'', \qquad A'' \to A'', \tag{A.6}$$

where the Stokes proportionality constant, T, remains to be determined. The combination with these new constants then remains insensitive to φ until the line $\varphi = \pi$ is approached, where the first term in (A.1) is dominant. The crossing of this line leads to the further coefficient changes

$$A' + TA'' \to A' + TA'', \quad A'' \to A'' + U(A' + TA''), \tag{A.7}$$

with a new Stokes constant U.

The important point is that the Stokes constants T and U, etc., may depend on the positions of the Stokes lines and the sense in which they are crossed (Heading 1962) but not on the values of A' and A''. Hence knowledge of the locations of the Stokes lines and of their associated constants provides a complete picture of the asymptotic behaviour of the solutions of (A.3) without any necessity to examine the full general solution in each case.

The following sections illustrate the application of these ideas to isolated turning points (Section A.2), barrier penetration (Section A.3), the linear oscillator (Section A.4) and non-adiabatic transitions (Section A.5).

A.2 Isolated turning points

The general behaviour at, say, a left-hand turning point, $z = a$, is handled by setting

$$k(z) = (z-a)^{1/2} f(z), \tag{A.8}$$

where $f(z)$ is assumed to be slowly varying and single valued. The disposition of the corresponding (dashed) Stokes lines is shown in Fig. A.1. T, U and V are the so far unknown Stokes constants and the wavy line is a branch cut.

This branch cut has two important consequences. First it sets the range of $\arg(z-a)$ as lying between $-\pi/2$ and $3\pi/2$, which means that if $\psi(z)$ is written as

$$\psi(z) = A_+\psi_+(z) + A_-\psi_-(z), \tag{A.9}$$

where

$$\psi_\pm(z) = k^{-1/2}(z) \exp\left(\pm \mathrm{i} \int_a^z k(z)\mathrm{d}z\right), \tag{A.10}$$

then ψ_- is dominant at the first Stokes line because $\arg(z-a) \simeq \pi/3$ and

$$\arg\left(\mathrm{i}\int_a^z (z-a)^{1/2} f(z)\mathrm{d}z\right) = \pi. \tag{A.11}$$

Similarly ψ_+ is dominant on the other two lines because $\arg(z-a) \simeq \pi$ or $-\pi/3$ and

$$\arg\left(\mathrm{i}\int_a^z (z-a)^{1/2} f(z)\mathrm{d}z\right) = 2\pi \text{ or } 0. \tag{A.12}$$

Notice, however, that insertion of the branch cut in region 1 in Fig. A.1 would make $\arg(z-a) = 5\pi/3$ on the third Stokes line, so that the argument of the exponent would be 3π and ψ_- would be dominant. This shows that the Stokes constants are defined only with respect to a given disposition of the branch cut.

The second consequence of the branch cut is that $(z-a)$ acquires a factor of $\exp(-2\pi\mathrm{i})$ on crossing it in an anticlockwise sense; hence

$$\begin{aligned} k^{-1/2}(z) &= (z-a)^{-1/4}[f(z)]^{1/2} \\ &\to \exp(\mathrm{i}\pi/2)(z-a)^{-1/4}[f(z)]^{1/2} = \mathrm{i}k^{-1/2}(z), \end{aligned} \tag{A.13}$$

and

$$\begin{aligned} \int_a^z k(z)\mathrm{d}z &= \int_0^{z-a} (z-a)^{1/2} f(z)\mathrm{d}(z-a) \to \exp(-3\mathrm{i}\pi)\int_a^z k(z)\mathrm{d}z \\ &= -\int_a^z k(z)\mathrm{d}z. \end{aligned} \tag{A.14}$$

This means that

$$\psi_+(z) \to \mathrm{i}\psi_-(z), \qquad \psi_-(z) \to \mathrm{i}\psi_+(z), \tag{A.15}$$

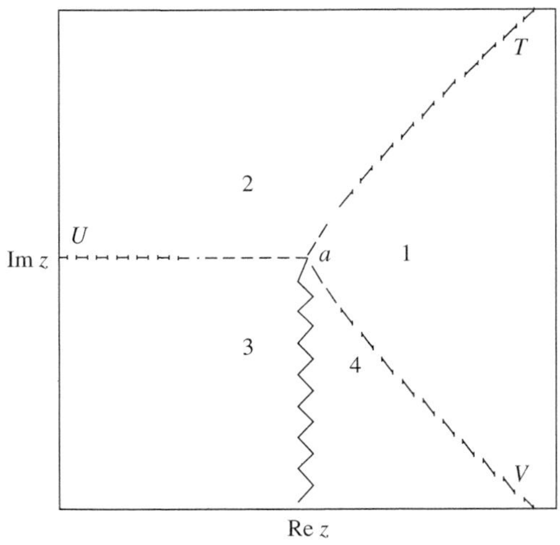

Fig. A.1 Stokes lines (dashed) and branch cut (zig-zag) at a classical turning point. T, U and V denote the Stokes constants.

so that the coefficients $A_\pm$ must suffer compensating changes to ensure the continuity of ψ:

$$A_+ \to -\mathrm{i}A_- \quad \text{and} \quad A_- \to -\mathrm{i}A_+. \tag{A.16}$$

The full sequence of coefficient changes starting from $(A_+, A_-) = (A', A'')$ in region 1 in Fig. A.1 is summarized in Table A.1.

The constants T, U and V are fixed by requiring that the function $\psi(z)$ must be single valued. In other words the two expressions in region 1 (given by the first and last lines of Table A.1) must be identical, regardless of the values of A' and A''. Hence, by equating coefficients of A' and A'',

$$\begin{aligned} 1 &= -\mathrm{i}U \\ 0 &= -\mathrm{i}(1 + TU) \\ 0 &= -\mathrm{i}(1 + UV) \\ 1 &= -\mathrm{i}(T + V + TUV). \end{aligned} \tag{A.17}$$

The solution of the first three equations, namely

$$T = U = V = \mathrm{i}, \tag{A.18}$$

Table A.1 Coefficient changes at a turning point (Fig. A.1).

Region	A_+	A_-
1	A'	A''
2	$A' + TA''$	A''
3	$A' + TA''$	$A'' + U(A' + TA'')$
1	$-\mathrm{i}[A'' + U(A' + TA'')]$	$-\mathrm{i}\{A' + TA'' + V[A'' + U(A' + TA'')]\}$

may also be verified to satisfy the fourth. Consequently, in this simple case, the Stokes constants are completely determined by the single-valuedness of $\psi(z)$.

One also sees that the requirement that $\psi(z)$ should be exponentially small in the classically forbidden region (i.e. on the real Stokes line between regions 2 and 3 in Fig. A.1) means that the coefficient of the dominant term ψ_+ must vanish:

$$A' + TA'' = A' + \mathrm{i}A'' = 0. \tag{A.19}$$

In other words

$$A' = \frac{A}{2}\exp(-\mathrm{i}\pi/4), \quad A'' = \frac{A}{2}\exp(+\mathrm{i}\pi/4), \tag{A.20}$$

so that

$$\begin{aligned}\psi(z) &= A\,k^{-1/2}(z)\cos\left(\int_a^z k(z)\mathrm{d}z - \frac{\pi}{4}\right)\\ &= A\,k^{-1/2}(z)\sin\left(\int_a^z k(z)\mathrm{d}k + \frac{\pi}{4}\right),\end{aligned} \tag{A.21}$$

showing that the Maslov term of $\pi/4$ may be attributed solely to the existence of the three Stokes lines and the branch cut in Fig. A.1.

A.3 Barrier penetration

The case of barrier penetration, with two turning points b and c as shown in Fig. A.2, implies that

$$k(z) = [(z-b)(z-c)]^{1/2}f(z), \tag{A.22}$$

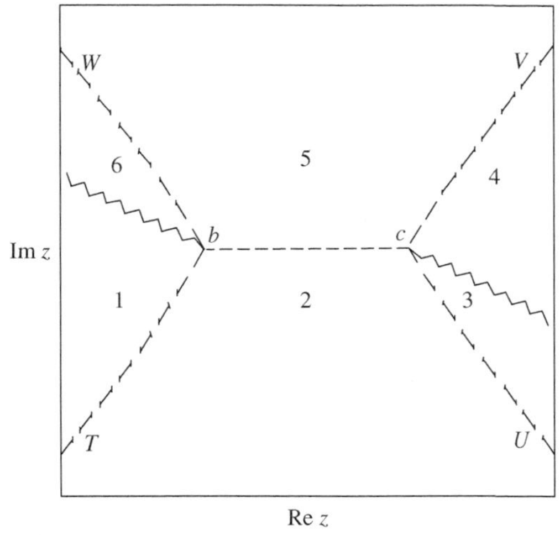

Fig. A.2 Stokes lines (dashed) and branch cuts (zig-zag) for energies below a potential maximum. T, U, V and W denote the Stokes constants.

and the wavefunction is conveniently considered in two forms; either

$$\psi(z) \approx k^{-1/2}(z)\left[B_+ \exp\left(\mathrm{i}\int_b^z k(z)\mathrm{d}z\right) + B_- \exp\left(-\mathrm{i}\int_b^z k(z)\mathrm{d}z\right)\right] \tag{A.23}$$

or

$$\psi(z) \approx k^{-1/2}(z)\left[C_+ \exp\left(\mathrm{i}\int_c^z k(z)\mathrm{d}z\right) + C_- \exp\left(-\mathrm{i}\int_c^z k(z)\mathrm{d}z\right)\right], \tag{A.24}$$

with the two sets of coefficients related by

$$C_\pm = B_\pm \exp\left(\pm\mathrm{i}\int_b^c k(z)\mathrm{d}z\right) = B_\pm \mathrm{e}^{\mp\delta}, \tag{A.25}$$

where δ is the positive barrier penetration integral

$$\delta = -\mathrm{i}\int_b^c k(z)\mathrm{d}z = \int_b^c [(c-z)(z-b)]^{1/2} f(z)\mathrm{d}z, \tag{A.26}$$

because, according to the disposition of the branch cuts in Fig. A.2, $(z-c) = \mathrm{e}^{\mathrm{i}\pi}(c-z)$ on the real axis between b and c. The way that the branch cuts are chosen also ensures that the B_+ term is dominant on the Stokes lines with constants T and W, while the C_- term is dominant on the lines with constants U and V. Also, by analogy with eqn (A.16),

$$B_+ \to -\mathrm{i}B_-, \qquad B_- \to -\mathrm{i}B_+ \tag{A.27}$$

on crossing the branch cut from region 6 to region 1, while

$$C_+ \to -\mathrm{i}C_-, \qquad C_- \to -\mathrm{i}C_+ \tag{A.28}$$

on crossing from region 3 to region 4.

The consequent coefficient changes on circling the diagram in an anticlockwise sense are summarized in Table A.2.

It follows, on substituting for (C', C'') in terms of (B', B'') in the final line of Table A.2 and comparing coefficients with the first line, that the single-valuedness of $\psi(z)$ requires the following identities between the Stokes constants:

$$\begin{aligned} 1 &= -[(1+TU\mathrm{e}^{2\delta})(1+VW\mathrm{e}^{2\delta})\mathrm{e}^{-2\delta} + TW\mathrm{e}^{2\delta}], \\ 0 &= -[U + (1+UV)W\mathrm{e}^{2\delta}] \\ 0 &= -[V + (1+UV)T\mathrm{e}^{2\delta}] \\ 1 &= -(\mathrm{e}^{2\delta} + UV\mathrm{e}^{2\delta}). \end{aligned} \tag{A.29}$$

Consequently

$$U = W, \qquad T = V, \qquad UV = TW = -(1+\mathrm{e}^{-2\delta}). \tag{A.30}$$

A second set of identities may be obtained by setting

$$B' = B'' = 1 \tag{A.31}$$

Table A.2 Coefficient changes at a potential barrier (Fig. A.2).

Region	B_+	B_-
1	B'	B''
2	B'	$B'' + TB'$
Region	C_+	C_-
2	$B'\mathrm{e}^{-\delta}$	$(B'' + TB')\mathrm{e}^{\delta}$
3	$B'\mathrm{e}^{-\delta} + U(B'' + TB')\mathrm{e}^{\delta}$	$(B'' + TB')\mathrm{e}^{\delta}$
4	$-\mathrm{i}(B'' + TB')\mathrm{e}^{\delta}$	$-\mathrm{i}[B'\mathrm{e}^{-\delta} + U(B'' + TB')\mathrm{e}^{\delta}]$
	$\equiv C'$	$\equiv C''$
5	$C' + VC''$	C''
Region	B_+	B_-
5	$(C' + VC'')\mathrm{e}^{\delta}$	$C''\mathrm{e}^{-\delta}$
6	$(C' + VC'')\mathrm{e}^{\delta}$	$C''\mathrm{e}^{-\delta} + W(C' + VC'')\mathrm{e}^{\delta}$

so that $\psi(z)$ given by eqn (A.23) is real on the real axis in region 1. Since the corresponding solution in region 4 must also be real on the real axis, it follows that $C'' = (C')^*$ or, on substituting for B' and B'',

$$\mathrm{i}(1 + T^*)\mathrm{e}^{\delta} = -\mathrm{i}[\mathrm{e}^{-\delta} + U(1 + T)\mathrm{e}^{\delta}]. \tag{A.32}$$

Taken together with eqns (A.30) this means that

$$T = -U^* = V = -W^* = (1 + \mathrm{e}^{-2\delta})^{1/2}\mathrm{e}^{\mathrm{i}\gamma}, \tag{A.33}$$

where only the phase term γ remains to be determined.

The resulting connection between the coefficients (B', B'') and (C', C'') bears a close resemblance to the standard barrier connection formula in Section 3.2,

$$\begin{pmatrix} C' \\ C'' \end{pmatrix} = \begin{pmatrix} (1 + \mathrm{e}^{2\delta})^{1/2}\exp(\mathrm{i}\gamma + \mathrm{i}\pi/2) & -\mathrm{i}\mathrm{e}^{\delta} \\ \mathrm{i}\mathrm{e}^{\delta} & (1 + \mathrm{e}^{2\delta})^{1/2}\exp(-\mathrm{i}\gamma - \mathrm{i}\pi/2) \end{pmatrix} \begin{pmatrix} B' \\ B'' \end{pmatrix}, \tag{A.34}$$

particularly when it is recognized that the tunnelling parameter δ is expressed in the convention of eqns (3.48)–(3.49) as

$$\delta = \int_b^c k(x)\mathrm{d}x = -\pi\varepsilon. \tag{A.35}$$

The final step is to determine the phase constant γ, which cannot be done by purely phase integral methods. Instead it is necessary to exploit the known asymptotic properties of the solutions of a suitable comparison equation, in this case the parabolic cylinder equation (Abramowitz and Stegun 1965)

$$\left(\frac{\mathrm{d}^2}{\mathrm{d}x^2} + \frac{1}{4}x^2 - a\right)y = 0. \tag{A.36}$$

The most convenient solution is $W(a,x)$ which is known to behave for $x^2 \gg 4a$ as

$$W(x,a) \overset{x>0}{\approx} \left(\frac{2\tau}{x}\right)^{1/2} \cos\left(\frac{1}{4}x^2 - a\ln x + \frac{\pi}{4} + \frac{1}{2}\arg\Gamma\left(\frac{1}{2} + \mathrm{i}a\right)\right),$$
$$W(-x,a) \overset{x<0}{\approx} \left(\frac{2}{\tau x}\right)^{1/2} \sin\left(\frac{1}{4}x^2 - a\ln x + \frac{\pi}{4} + \frac{1}{2}\arg\Gamma\left(\frac{1}{2} + \mathrm{i}a\right)\right), \tag{A.37}$$

where

$$\tau = [1 + \exp(2\pi a)^{1/2}] - \exp(\pi a). \tag{A.38}$$

The relationship to the phase integral form may be recognized by setting

$$k(x) = \frac{1}{2}(x^2 - 4a)^{1/2}, \tag{A.39}$$

and noting that

$$\int_{2\sqrt{a}}^{x} k(x)\mathrm{d}x = -\int_{-2\sqrt{a}}^{-x} k(x)\mathrm{d}x$$
$$\overset{x^2 \gg 4a}{\approx} \frac{1}{4}x^2 - a\ln x - \frac{1}{2}a + \frac{1}{2}a\ln|a|, \tag{A.40}$$

and

$$\delta = \int_{-2\sqrt{a}}^{2\sqrt{a}} k(x)\mathrm{d}x = \pi a. \tag{A.41}$$

The integration limits $\pm 2\sqrt{a}$ are of course the turning points b and c for this model. The coefficient choice

$$B_+ = B' = \frac{\mathrm{i}B}{2}\mathrm{e}^{\mathrm{i}\beta}, \quad B_- = B'' = \frac{-\mathrm{i}B}{2}\mathrm{e}^{-\mathrm{i}\beta}, \tag{A.42}$$

with

$$\beta = \frac{1}{2}a\ln|a| - \frac{1}{2}a - \frac{1}{2}\arg\Gamma\left(\frac{1}{2} + \mathrm{i}a\right) - \frac{\pi}{4}, \tag{A.43}$$

therefore casts eqn (A.23) into the form of (A.37). The final step is to find the phase term γ such that

$$C_+ = C' = \frac{1}{2}\tau B\mathrm{e}^{-\mathrm{i}\beta}, \quad C_- = C'' = \frac{1}{2}\tau B\mathrm{e}^{\mathrm{i}\beta}, \tag{A.44}$$

so that (A.24) is also consistent with the first of (A.37). To do this it is sufficient to note from Table A.2 that

$$C' = -\mathrm{i}(B'' + TB')\mathrm{e}^{\delta}, \tag{A.45}$$

so that on substitution from (A.33) and (A.44)

$$C' = \frac{1}{2}B[-(1 + \mathrm{e}^{2\delta})^{1/2}\mathrm{e}^{2\mathrm{i}\beta + \mathrm{i}\gamma} - \mathrm{e}^{\delta}]\mathrm{e}^{-\mathrm{i}\beta}, \tag{A.46}$$

which reduces to (A.38) and (A.44) if

$$\gamma = \pi - 2\beta = \arg\Gamma\left(\frac{1}{2} + \mathrm{i}a\right) - a\ln|a| + a - \frac{\pi}{2}. \tag{A.47}$$

The final connection formula, with if $a = -\varepsilon$ as required by (A.35) and (A.41), takes the form given in Section 3.2, namely

$$\begin{pmatrix} C' \\ C'' \end{pmatrix} = \begin{pmatrix} (1+\kappa^2)^{1/2}\mathrm{e}^{-\mathrm{i}\varphi} & -\mathrm{i}\kappa \\ \mathrm{i}\kappa & (1+\kappa^2)^{1/2}\mathrm{e}^{\mathrm{i}\varphi} \end{pmatrix} \begin{pmatrix} B' \\ B'' \end{pmatrix}, \tag{A.48}$$

where

$$\kappa = \mathrm{e}^{-\pi\varepsilon} \tag{A.49}$$

and

$$\varphi = \arg\Gamma\left(\frac{1}{2} + \mathrm{i}\varepsilon\right) - \varepsilon\ln|\varepsilon| + \varepsilon. \tag{A.50}$$

As presented, this derivation applies only for energies below the barrier (i.e. only for $\varepsilon < 0$), but similar arguments apply with precisely the same results for $\varepsilon > 0$. The disposition of the Stokes lines and the complex turning points, $x_\pm$, is as shown in Fig. A.3. The proper phase reference point, which appears in eqns (A.23) and (A.24) as a real integration limit, is the point x_0 at the intersection between the real axis and the Stokes line from x_+ to x_-.

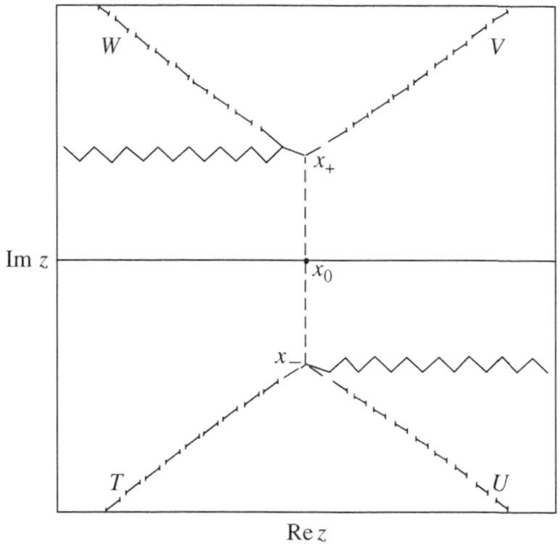

Fig. A.3 Stokes lines (dashed) and branch cuts (zig-zag) for energies above a potential maximum. T, U, V and W denote the Stokes constants. x_0 is the intersection between the Stokes line and the real axis.

A.4 The linear oscillator

The case of a linear oscillator with two real turning points a and b gives rise to the disposition of Stokes lines and branch cuts shown in Fig. A.4, and a similar argument to that used in Section A.3 yields the quantization condition.

It is assumed by analogy with (A.22)–(A.24) that

$$k(z) = [(z-a)(b-z)]^{1/2} f(z), \tag{A.51}$$

giving rise to

$$\psi(z) \approx k^{-1/2}(z)\left[A_+ \exp\left(\mathrm{i}\int_a^z k(z)\mathrm{d}z\right) + A_- \exp\left(-\mathrm{i}\int_a^z k(z)\mathrm{d}z\right)\right]$$

or

$$\psi(z) \approx k^{-1/2}(z)\left[B_+ \exp\left(\mathrm{i}\int_b^z k(z)\mathrm{d}z\right) + B_- \exp\left(-\mathrm{i}\int_b^z k(z)\mathrm{d}z\right)\right], \tag{A.52}$$

where

$$B_\pm = A_\pm \mathrm{e}^{\pm \mathrm{i}\alpha}, \qquad \alpha = \int_a^b k(z)\mathrm{d}z. \tag{A.53}$$

It may be verified that the term in B_+ is dominant on the lines with Stokes constants U and Z and subdominant on the line marked V. Similarly the A_+ term is dominant on Stokes line Y and subdominant on those marked W and X. Finally,

$$A_+ \to -\mathrm{i}A_-, \qquad A_- \to -\mathrm{i}A_+ \tag{A.54}$$

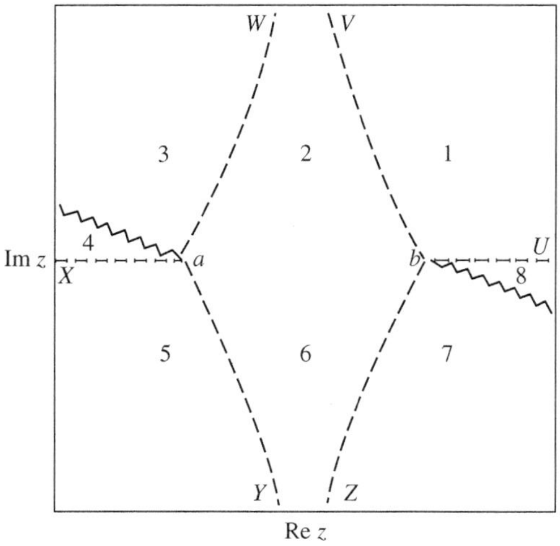

Fig. A.4 Stokes lines (dashed) and branch cuts (zig-zag) for a linear oscillator. U, V, W, X, Y and Z denote the Stokes constants.

on crossing from region 3 to 4, while

$$B_+ \to -\mathrm{i}B_-, \qquad B_- \to -\mathrm{i}B_+ \tag{A.55}$$

on crossing from region 7 to 8. The consequent coefficient changes during an anticlockwise encirclement from region 1 back to itself are listed in Table A.3.

Subsequent manipulations are assisted by defining

$$F = Y\mathrm{e}^{-\mathrm{i}\alpha} + Z\mathrm{e}^{\mathrm{i}\alpha}, \quad G = V\mathrm{e}^{-\mathrm{i}\alpha} + W\mathrm{e}^{\mathrm{i}\alpha}. \tag{A.56}$$

Thus, on comparing coefficients of B' and B'' in the two prescriptions for region 1, one finds that

$$\begin{aligned}
1 &= -(\mathrm{e}^{-2\mathrm{i}\alpha} + FX\mathrm{e}^{-\mathrm{i}\alpha}),\\
0 &= -(G\mathrm{e}^{-\mathrm{i}\alpha} + F\mathrm{e}^{\mathrm{i}\alpha} + FXG)\\
0 &= -(X + U\mathrm{e}^{-2\mathrm{i}\alpha} + FXU\mathrm{e}^{-\mathrm{i}\alpha})\\
1 &= -[\mathrm{e}^{2\mathrm{i}\alpha} + (FU + XG)\mathrm{e}^{\mathrm{i}\alpha} + FGUX + UG\mathrm{e}^{-\mathrm{i}\alpha}].
\end{aligned} \tag{A.57}$$

It follows after some manipulation that

$$F = G, \quad X = U, \quad FX = UG = -2\cos\alpha. \tag{A.58}$$

Now the boundary conditions for an acceptable bound state wavefunction are that the coefficients of the dominant (exponentially increasing) terms on the real axes between regions 8 and 1 and regions 4 and 5 should vanish. In other words, in the notation of Table A.3,

$$A'' = B' = 0, \tag{A.59}$$

Table A.3 Coefficient changes at a potential well (Fig. A.4).

Region	B_+	B_-
1	B'	B''
2	$B' + VB''$	B''
Region	A_+	A_-
2	$(B' + VB'')\mathrm{e}^{-\mathrm{i}\alpha}$	$B''\mathrm{e}^{\mathrm{i}\alpha}$
3	$(B' + VB'')\mathrm{e}^{-\mathrm{i}\alpha} + WB''\mathrm{e}^{\mathrm{i}\alpha}$	$B''\mathrm{e}^{\mathrm{i}\alpha}$
4	$-\mathrm{i}B''\mathrm{e}^{\mathrm{i}\alpha}$	$-\mathrm{i}[(B' + VB'')\mathrm{e}^{-\mathrm{i}\alpha} + WB''\mathrm{e}^{\mathrm{i}\alpha}]$
5	$-\mathrm{i}\{B''\mathrm{e}^{\mathrm{i}\alpha} + X\{(B' + VB'')\mathrm{e}^{-\mathrm{i}\alpha} + WB''\mathrm{e}^{\mathrm{i}\alpha}]\}$ $\equiv A'$	$-\mathrm{i}[(B' + VB'')\mathrm{e}^{-\mathrm{i}\alpha} + WB''\mathrm{e}^{\mathrm{i}\alpha}]$ $\equiv A''$
6	A'	$A'' + YA'$
Region	B_+	B_-
6	$A'\mathrm{e}^{\mathrm{i}\alpha}$	$(A'' + YA')\mathrm{e}^{-\mathrm{i}\alpha}$
7	$A'\mathrm{e}^{\mathrm{i}\alpha}$	$(A'' + YA')\mathrm{e}^{-\mathrm{i}\alpha} + ZA'\mathrm{e}^{\mathrm{i}\alpha}$
8	$-\mathrm{i}[(A'' + YA')\mathrm{e}^{-\mathrm{i}\alpha} + ZA'\mathrm{e}^{\mathrm{i}\alpha}]$	$-\mathrm{i}A'\mathrm{e}^{\mathrm{i}\alpha}$
1	$-\mathrm{i}[(A'' + YA')\mathrm{e}^{-\mathrm{i}\alpha} + ZA'\mathrm{e}^{\mathrm{i}\alpha}]$	$-\mathrm{i}\{A'\mathrm{e}^{\mathrm{i}\alpha} + U[(A'' + YA')$ $\mathrm{e}^{-\mathrm{i}\alpha} + ZA'\mathrm{e}^{\mathrm{i}\alpha}]\}$

which means, in the light of the definition of A' in Table A.3, that

$$A'' = -\mathrm{i}B''G = 0, \tag{A.60}$$

and

$$A' = \mathrm{i}\mathrm{e}^{-\mathrm{i}\alpha}B''. \tag{A.61}$$

Equation (A.60) requires that the composite Stokes constant G should vanish for a bound state, which is consistent with (A.58) provided that

$$\alpha = \int_a^b k(z)\mathrm{d}z = \left(n + \frac{1}{2}\right)\pi, \tag{A.62}$$

while at the same time equation (A.61) ensures that

$$A' = (-1)^n B'', \tag{A.63}$$

which correctly expresses the fact that the wavefunction amplitudes at the two turning points have the same or opposite signs according to whether n is even or odd.

Equation (A.62) is of course the Bohr quantization condition, which is shown to be valid provided the asymptotic approximations (A.52) apply everywhere around the encircling contour. This means in practice that all other zeros of $k(z)$ must be well separated from the real turning points a and b. In the case of the harmonic oscillator, $k(z)$ has strictly two zeros; hence the circle may be made arbitrarily large and the quantization condition is exact.

The complications associated with more general types of oscillator are conveniently illustrated by the quartic case, for which

$$k(z) = (a^4 - z^4)^{1/2}, \tag{A.64}$$

giving rise to zeros at $z = \pm a$ and $\pm \mathrm{i}a$. The pattern of Stokes lines, shown in Fig. A.5(a), is inevitably more complicated than that in Fig. A.4, but it is still possible to follow a relatively simple path from region 1 to region 5 in Fig. A.5(b), provided that the parameter a is sufficiently large that the regions of non-asymptotic behaviour (shown as circles around the branch points) do not overlap. The choice of path in Fig. A.5(b) treats the two real branch points as isolated, and it is a relatively simple matter to recover the Bohr quantization condition by what is in effect a phase integral equivalent of the argument in Section 2.1.

The essential step is to set all the Stokes constants for anticlockwise crossings equal to i, according to the conclusions of Section A.2. In the notation of eqns (A.52), the resulting coefficient changes are summarized in Table A.4. Note that the final Stokes constant is taken as $-\mathrm{i}$ because the path from region 2 to region 5 is taken in a clockwise sense. Now B_- and A_+ are the dominant coefficients on the positive and negative real axes respectively, and both must vanish for a bound state; hence the Table A.4 entries in regions 1 and 5 impose the requirements that $B'' = 0$, $B'' \neq 0$ and $2B' \cos\alpha = 0$. In other words α must satisfy the Bohr quantization condition, $\alpha = (n + \frac{1}{2})\pi$.

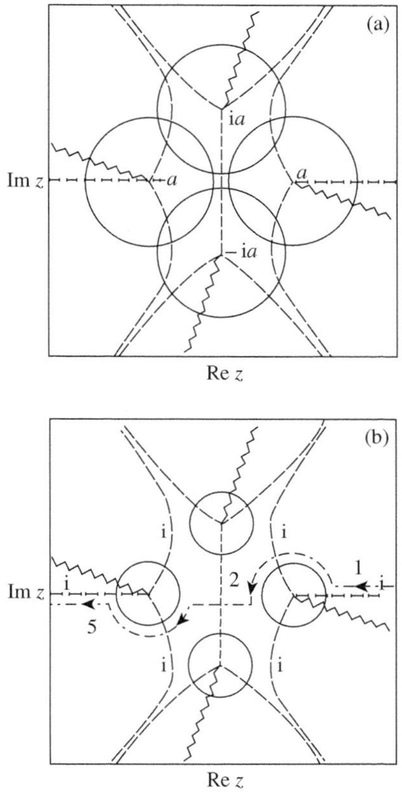

Fig. A.5 Stokes lines (dashed) and branch cuts (zig-zag) for the quartic potential, for which $k^2(z) = (a^4 - z^4)$, (a) for $a^4 < 6.27$ and (b) for $a^4 > 6.27$. Connections along the chain line in (b) with Stokes constants i yield the Bohr–Sommerfeld quantization condition.

Table A.4 Coefficient changes for the path in Fig. A.5(b).

Region	B_+	B_-
1	B'	B''
2	B'	$B'' + \mathrm{i}B'$
Region	A_+	A_-
5	$B'\mathrm{e}^{-\mathrm{i}\alpha} - \mathrm{i}(B'' + \mathrm{i}B')\mathrm{e}^{\mathrm{i}\alpha}$	$(B'' + \mathrm{i}B')\mathrm{e}^{\mathrm{i}\alpha}$

It remains to estimate the value of a at which this primitive quantization becomes valid, in the sense that coalescence of the non-asymptotic zones in Fig. A.5 is avoided. As an order of magnitude estimate, asymptotic behaviour may be assumed (see eqn (2.24)) when

$$\left| \int_a^z k(z)\mathrm{d}z \right| > \pi, \tag{A.65}$$

which translates roughly to

$$\frac{4}{3}a^{3/2}|z-a|^{3/2} > \pi, \tag{A.66}$$

in the case of a quartic oscillator. On the other hand overlap will be avoided if

$$|z-a| < a/\sqrt{2}. \tag{A.67}$$

Hence the condition for a valid path of the form shown in Fig. A.5(b) may be expressed as

$$a^4 > 2(3\pi/4)^{4/3} \simeq 6.27. \tag{A.68}$$

The results in Table 2.1 confirm that the first-order quantization condition is indeed valid within 5 per cent at $v \simeq 2$, when the inequality (A.68) is satisfied, because x and ε in eqn (2.112) scale to $2^{-1/4}z$ and $2^{1/2}a^4$ in the present notation.

A further interesting result, due to Vöros (1983), is that much higher accuracy may be obtained, for small a values, by employing the double minimum quantization eqn (3.72), in the limit that the energy is far above the barrier maximum. Of the two required phase integrals, one is taken from $-a$ to a in Fig. A.5 and the other from $-ia$ to ia (see problem 3.4). This suggests that the same result could be obtained by taking a path around all four branch points in Fig. A.5.

A.5 Curve crossing

As first shown by Stückelberg (1932) and later reanalysed by Thorsons et al. (1971), Crothers (1971) and Baranyi (1978, 1979), the phase integral method can also be applied to the coupled equations

$$\begin{aligned}\left(\hbar^2\frac{\mathrm{d}^2}{\mathrm{d}R^2} + 2m[E - V_1(R)]\right)\psi_1 &= 2mV_{12}(R)\psi_2 \\ \left(\hbar^2\frac{\mathrm{d}^2}{\mathrm{d}R^2} + 2m[E - V_2(R)]\right)\psi_2 &= 2mV_{12}(R)\psi_1,\end{aligned} \tag{A.69}$$

in a way which leads to the curve-crossing connection formulae given by eqn (3.59).

Stückelberg (1932) started by eliminating ψ_2 from eqns (A.69) and employing a JWKB expansion for the resulting fourth-order equation for ψ_1. The following approach, taken from Crothers (1971), is, however, both more direct and more revealing. It relies on the substitutions

$$\psi_{\mathrm{i}} = a_i \exp(\mathrm{i}S/\hbar) \tag{A.70}$$

and

$$S = S_0 + \hbar S_1 + \ldots, \tag{A.71}$$

where the a_i are treated initially as constants and the phase term S is taken as common to ψ_1 and ψ_2. The solution of eqns (A.69)–(A.71) to order $\hbar^0$ is then found to yield

$$2m\begin{pmatrix} E - V_1(R) & -V_{12}(R) \\ -V_{12}(R) & E - V_2(R) \end{pmatrix}\begin{pmatrix} a_1 \\ a_2 \end{pmatrix} = [S_0']^2\begin{pmatrix} a_1 \\ a_2 \end{pmatrix}. \tag{A.72}$$

Hence the solutions $[S_0']^2$ may be identified with the eigenvalues of the matrix in eqn (A.72), namely

$$[S_0']^2 = 2m[E - V_\pm(R)] = \hbar^2 k_\pm^2(R), \tag{A.73}$$

where

$$V_\pm(R) = \frac{1}{2}[V_1(R) + V_2(R)] \pm \frac{1}{2}\{[V_1(R) - V_2(R)]^2 + 4V_{12}^2(R)\}^{1/2}. \tag{A.74}$$

Similarly the coefficients a_i constitute the components of the eigenvectors

$$\begin{pmatrix} a_{1+} \\ a_{2+} \end{pmatrix} = \begin{pmatrix} \sin\theta \\ \cos\theta \end{pmatrix}, \quad \begin{pmatrix} a_{1-} \\ a_{2-} \end{pmatrix} = \begin{pmatrix} \cos\theta \\ -\sin\theta \end{pmatrix}, \tag{A.75}$$

where

$$\tan\, 2\theta = \frac{2V_{12}}{V_2 - V_1}. \tag{A.76}$$

It is assumed, to avoid ambiguity, that $V_2(R) > V_1(R)$ and $V_{12}(R) \to 0$ as $R \to \infty$; thus $\theta \to 0$ as $R \to \infty$.

Extension of the solution to order $\hbar$ yields, as usual,

$$S_1 = \frac{\mathrm{i}}{2}\ln S_0', \tag{A.77}$$

so that, overall,

$$\begin{pmatrix} \psi_1 \\ \psi_2 \end{pmatrix} = C\begin{pmatrix} \cos\theta \\ -\sin\theta \end{pmatrix} k_-^{-1/2}\exp\left(\pm\mathrm{i}\int k_-\mathrm{d}R\right) + D\begin{pmatrix} \sin\theta \\ \cos\theta \end{pmatrix} k_+^{-1/2}\exp\left(\pm\mathrm{i}\int k_+\mathrm{d}R\right), \tag{A.78}$$

which shows that the primitive semiclassical diabatic amplitudes ψ_1 and ψ_2 are, perhaps surprisingly, expressed in terms of quantities, $k_\pm(R)$, and the adiabatic mixing angle θ.

The validity of this approximation rests on the assumption that the a_i and hence θ are independent of R, whereas it is readily verified that θ varies from 0 to $\pi/2$ on passing through a curve crossing (at which $V_1(R) = V_2(R)$). Similarly, θ increases from 0 to $\pi/4$ in the Demkov (1964) or perturbed symmetric resonance (Crothers 1973) case, characterized by an increase from $V_{12}(R) \ll V_2(R) - V_1(R)$

to $V_{12}(R) \gg V_2(R) - V_1(R)$. The aim of the following analysis is to circumvent detailed solution of the coupled equations within these non-adiabatic coupling zones. It is assumed that complications due to the branch points at zeros of $k_\pm(R)$ can be ignored, although certain situations giving rise to such complications can be treated by similar methods (Ovchinnikova 1964; Nikitin 1968; Baranyi 1979; Coveney et al. 1985).

The crucial observation is that the non-adiabatic behaviour may be traced to the existence of a pair of complex conjugate transition points, R_c and R_c^* say, at which the two adiabatic potentials $V_\pm(R)$ intersect because the square root factor in eqn (A.74) vanishes. For example, the Landau–Zener model

$$\begin{aligned} V_1(R) - V_2(R) &= -\Delta F(R - R_x) \\ V_{12}(R) &= \text{constant} \end{aligned} \tag{A.79}$$

leads to

$$R_c, R_c^* = R_x \pm 2\mathrm{i}V_{12}/\Delta F. \tag{A.80}$$

The consequences of this complex intersection are best brought out by introducing a variable

$$t = [V_2(R) - V_1(R)]/2V_{12}(R) \tag{A.81}$$

in terms of which the transition points lie at $t = \pm\mathrm{i}$, and

$$\begin{aligned} k_-(R) - k_+(R) &= [k_-^2(R) - k_+^2(R)]/[k_-(R) + k_+(R)] \\ &= T_0(t)(1 + t^2)^{1/2}(\mathrm{d}t/\mathrm{d}R), \end{aligned} \tag{A.82}$$

where

$$T_0(t) = V_{12}(R)(\mathrm{d}R/\mathrm{d}t)/\hbar\bar{v}, \tag{A.83}$$

$$\bar{v} = [k_+(R) + k_-(R)]\hbar/2m. \tag{A.84}$$

It is assumed in what follows that the function $T_0(t)$ is slowly varying and single valued. An important consequence of this definition is that

$$\begin{aligned} \sin\theta &= \left[\frac{1}{2}\left(1 - \frac{t}{[1+t^2]^{1/2}}\right)\right]^{1/2} = -\frac{\mathrm{i}}{2}\left[\left(\frac{t+\mathrm{i}}{t-\mathrm{i}}\right)^{1/4} - \left(\frac{t-\mathrm{i}}{t+\mathrm{i}}\right)^{1/4}\right], \\ \cos\theta &= \left[\frac{1}{2}\left(1 + \frac{t}{[1+t^2]^{1/2}}\right)\right]^{1/2} = \frac{1}{2}\left[\left(\frac{t+\mathrm{i}}{t-\mathrm{i}}\right)^{1/4} + \left(\frac{t-\mathrm{i}}{t+\mathrm{i}}\right)^{1/4}\right]. \end{aligned} \tag{A.85}$$

Equations (A.85) show, as one might expect, that the coefficients $\sin\theta$ and $\cos\theta$ in eqn (A.78) diverge at the transition points $t = \pm\mathrm{i}$.

The consequent coefficient changes in eqn (A.78) are conveniently followed by factoring out an uninteresting common phase factor $\exp\{\frac{1}{2}\mathrm{i}\int[k_+(R) + k_-(R)]\mathrm{d}R\}$ which remains single valued in the transition zone. It proves sufficient to follow the behaviour of the outgoing part of a single amplitude, say ψ_1, because ψ_2 is related to it by the $\sin\theta$ and $\cos\theta$ factors in (A.78) and the incoming terms are complex conjugates of the outgoing ones. The reduced amplitude of interest may therefore be expressed as

$$\begin{aligned}\varphi_1 &= \psi_1 \exp\left(-\frac{\mathrm{i}}{2}\int_{R_c}^{R}[k_+(R)+k_-(R)]\mathrm{d}R\right)\\ &= C_+\cos\theta\exp\left(\frac{\mathrm{i}}{2}\int_{-\mathrm{i}}^{t}T_0(t)(1+t^2)^{1/2}\mathrm{d}t\right)\\ &\quad + C_-\sin\theta\exp\left(-\frac{\mathrm{i}}{2}\int_{-\mathrm{i}}^{t}T_0(t)(1+t^2)^{1/2}\mathrm{d}t\right) \qquad \text{(A.86a)}\\ &= D_+\cos\theta\exp\left(\frac{\mathrm{i}}{2}\int_{\mathrm{i}}^{t}T_0(t)(1+t^2)^{1/2}\mathrm{d}t\right)\\ &\quad + D_-\sin\theta\exp\left(-\frac{\mathrm{i}}{2}\int_{\mathrm{i}}^{t}T_0(t)(1+t^2)^{1/2}\mathrm{d}t\right), \qquad \text{(A.86b)}\end{aligned}$$

where the two sets of coefficients are related by

$$D_\pm = \mathrm{e}^{\mp\delta}C_\pm, \tag{A.87}$$

with

$$\delta = -\frac{\mathrm{i}}{2}\int_{-\mathrm{i}}^{\mathrm{i}}T_0(t)(1+t^2)^{1/2}\mathrm{d}t = -\frac{\mathrm{i}}{2}\int_{R_c^*}^{R_c}[k_-(R)-k_+(R)]\mathrm{d}R. \tag{A.88}$$

Note that eqn (A.82) has been used to express the integrals over $[k_+(R)-k_-(R)]$ in terms of t, and that the choice of branch cuts in Fig. A.6 makes δ a positive quantity because $\arg(t^2+1)=0$ and $\arg[T_0(t)\mathrm{d}t]=\pi/2$.

The changes in $C_\pm$ and $D_\pm$ on circumscribing the transition zone in an anticlockwise sense may now be deduced by noting that the first term in eqn (A.86a) is dominant on the Stokes lines T and U, while the second term in (A.86b) is dominant on the lines V and W. Secondly, $\arg(t-\mathrm{i})$ decreases by 2π on crossing the upper branch cut which causes the phase terms in (A.86b) to interchange and also leads, according to (A.85), to the following changes in the trigonometric factors: $\sin\theta \to -\cos\theta$ and $\cos\theta \to -\sin\theta$. A similar argument for the lower branch cut, where $\arg(t+\mathrm{i})$ decreases by 2π, indicates an interchange of phase terms accompanied by the trigonometric changes $\sin\theta\to\cos\theta$ and $\cos\theta\to\sin\theta$. The effects of crossing the branch cuts may therefore be summarized in the form

$$C_+\to -C_-, \qquad C_-\to C_+ \tag{A.89a}$$

at the lower branch cut, and

$$D_+\to D_-, \qquad D_-\to -D_+ \tag{A.89b}$$

at the upper one. Table A.5 summarizes the coefficient changes implied by the above considerations.

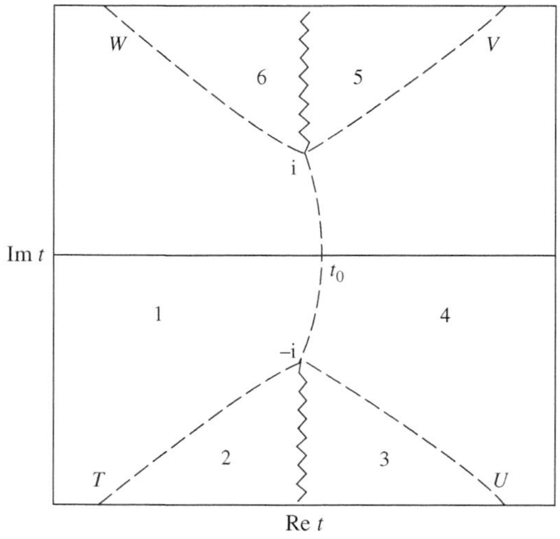

Fig. A.6 Stokes lines (dashed) and branch cuts (zig-zag) for curve crossing. T, U, V and W denote the Stokes constants. t_0 lies on the intersection of the Stokes line from i to $-$i and the real t axis.

Table A.5 Coefficient changes at a curve crossing (Fig. A.6).

Region	C_+	C_-
1	C'	C''
2	C'	$C'' + TC'$
3	$C'' + TC'$	$-C'$
4	$C'' + TC'$	$-C' + U(C'' + TC')$
Region	D_+	D_-
4	$(C'' + TC')\mathrm{e}^{-\delta}$ $\equiv D'$	$[-C' + U(C'' + TC')]\mathrm{e}^{\delta}$ $\equiv D''$
5	$D' + VD''$	D''
6	$-D''$	$D' + VD''$
1	$-D'' + W(D' + VD'')$	$D' + VD''$
Region	C_+	C_-
1	$[-D'' + W(D' + VD'')]\mathrm{e}^{\delta}$	$[D' + VD'']\mathrm{e}^{-\delta}$

The two descriptions in region 1 imply the following identities

$$\begin{aligned} 1 &= (1 - TU)(1 - VW)\mathrm{e}^{2\delta} + TW \\ 0 &= -(1 - VW)U\mathrm{e}^{2\delta} + W \\ 0 &= T\mathrm{e}^{-2\delta} - V(1 - TU) \\ 1 &= \mathrm{e}^{-2\delta} + UV, \end{aligned} \tag{A.90}$$

which rearrange to

$$T = U^* = V = W^* = (1 - e^{-2\delta})^{1/2} e^{-i\chi}, \tag{A.91}$$

where χ is an undetermined phase parameter. Taken in conjunction with the entries in Table A.5, this implies that

$$\begin{pmatrix} e^{-\delta/2} D'' \\ e^{\delta/2} D' \end{pmatrix} = \begin{pmatrix} (1 - e^{-2\delta})^{1/2} e^{i\chi} & -e^{-\delta} \\ e^{-\delta} & (1 - e^{-2\delta})^{1/2} e^{-i\chi} \end{pmatrix} \begin{pmatrix} e^{-\delta/2} C'' \\ e^{\delta/2} C' \end{pmatrix}. \tag{A.92}$$

The final step in the argument is to relate eqn (A.92) to the connection formula in eqn (3.59). To do this one notes that the standard representations in (3.57),

$$\begin{aligned} \psi_\pm(R) &\overset{R \ll R_0}{\approx} P'_\pm k_\pm^{-1/2}(R) \exp\left(i \int_{R_0}^{R} k_\pm(R) \mathrm{d}R \right) \\ &\overset{R \ll R_0}{\approx} Q'_\pm k_\pm^{-1/2}(R) \exp\left(i \int_{R_0}^{R} k_\pm(R) \mathrm{d}R \right), \end{aligned} \tag{A.93}$$

where R_0 is the radius corresponding to t_0 in Fig. A.6, are readily shown to imply that

$$\begin{aligned} P'_+ &= e^{\delta/2} C'', \quad P'_- = e^{-\delta/2} C' \\ Q'_+ &= e^{-\delta/2} D'', \quad Q'_- = e^{\delta/2} D', \end{aligned} \tag{A.94}$$

while in the notation of eqn (3.61)

$$\delta = -\frac{i}{2} \int_{R_c^*}^{R_c} [k_-(R_-) - k_+(R)] \mathrm{d}R = \pi v. \tag{A.95}$$

Thus

$$\begin{pmatrix} Q'_+ \\ Q'_- \end{pmatrix} = \begin{pmatrix} [1 - \exp(-2\pi v)]^{1/2} e^{i\chi} & -e^{-\pi v} \\ e^{-\pi v} & [1 - \exp(-2\pi v)]^{1/2} e^{-i\chi} \end{pmatrix} \begin{pmatrix} P'_+ \\ P'_- \end{pmatrix}. \tag{A.96}$$

Only the phase term χ remains to be determined.

The simplest way to find χ is to use the comparison equation technique. To do this, it is convenient to set

$$T_0(t) = 4\nu = \text{constant} \tag{A.97}$$

in eqns (A.86a) and (A.86b) and to interpret the function $\varphi_1(t)$ as a JWKB solution of

$$\left(\frac{d^2}{dt^2} + f^2(t) \right) \varphi = 0, \tag{A.98}$$

where

$$f^2(t) = 4\nu t^2 + 4\nu - \mathrm{i}/2\nu, \tag{A.99}$$

on the basis that

$$\begin{aligned}\int_0^t f(t)\mathrm{d}t &\overset{t\gg 1}{\approx} 2\nu \int_0^t (1+t^2)^{1/2}\mathrm{d}t - \frac{\mathrm{i}}{2}\int_0^t \frac{\mathrm{d}t}{(1+t^2)^{1/2}} \\ &= 2\nu \int_0^t (1+t^2)^{1/2}\mathrm{d}t - \frac{\mathrm{i}}{2}\ln[t + (1+t^2)^{1/2}].\end{aligned} \tag{A.100}$$

Consequently

$$\begin{aligned}&\nu^{1/4} f^{-1/2}(t) \exp\left(\pm \mathrm{i}\int_0^t f(t)\mathrm{d}t\right) \overset{t\gg 1}{\approx} \\ &\left(\frac{(1+t^2)^{1/2} \pm t}{2(1+t^2)^{1/2}}\right)^{1/2} \exp\left(2\mathrm{i}\nu \int_0^t (1+t^2)^{1/2}\mathrm{d}t\right)\end{aligned} \tag{A.101}$$

which agrees with the forms in eqn (A.86a) because the upper and lower sign choices in the pre-exponent coincide with the prescriptions for $\cos\theta$ and $\sin\theta$ given by (A.85). A similar construction clearly applies for any constant value of T_0, but (A.97) is convenient in ensuring that

$$\delta = 2\nu \int_{-\mathrm{i}}^{\mathrm{i}} (1+t^2)^{1/2}\mathrm{d}t = \pi\nu, \tag{A.102}$$

as required by the notation of (A.95)–(A.96).

One now proceeds to substitute

$$z = 2\nu^{1/2} t \exp(-\mathrm{i}\pi/4), \tag{A.103}$$

which reduces (A.98) to the parabolic cylinder form (Whittaker and Watson 1962)

$$\left(\frac{\mathrm{d}^2}{\mathrm{d}z^2} - \frac{1}{4}z^2 + \frac{1}{2} + \mathrm{i}\nu\right)\varphi = 0, \tag{A.104}$$

from which the known asymptotic properties of the solution

$$\begin{aligned}\varphi &= D_{\mathrm{i}v}(z) \\ &= \mathrm{e}^{-\pi v} D_{\mathrm{i}v}(-z) + \frac{\mathrm{i}(2\pi)^{1/2}\exp(-\pi\nu/2)}{\Gamma(-\mathrm{i}v)} D_{-1-\mathrm{i}v}(-\mathrm{i}z)\end{aligned} \tag{A.105}$$

may be used to determine $\chi(v)$. The first of eqns (A.105) is convenient as $t \to \infty$ and the second as $t \to -\infty$, because the following asymptotic approximation (Abramowitz and Stegun 1965; Whittaker and Watson 1962) is valid in the range $-\pi/4 < \arg z < \pi/4$:

$$
\begin{aligned}
D_{\mathrm{i}\nu}(z) &\approx z^{\mathrm{i}v}\exp(-z^2/4) \\
&\overset{t\gg 1}{\approx} \exp\left(\mathrm{i}\nu t^2 + \mathrm{i}\nu\ln 2t + \frac{\mathrm{i}}{2}\nu\ln\nu + \frac{\pi\nu}{4}\right).
\end{aligned} \tag{A.106}
$$

But

$$
\nu t^2 + \nu\ln 2t \overset{t\gg 1}{\approx} 2\nu\int_0^t (1+t^2)^{1/2}\mathrm{d}t - \nu/2, \tag{A.107}
$$

from which it follows that

$$
B\varphi(t) \overset{t\gg 1}{\approx} \cos\theta\exp\left(2\mathrm{i}\nu\int_0^t (1+t^2)^{1/2}\mathrm{d}t - \mathrm{i}\beta\right), \tag{A.108}
$$

where

$$
\begin{aligned}
B &= \exp\left(-\frac{\pi\nu}{4} - \frac{\mathrm{i}\pi}{8} - \frac{\mathrm{i}}{2}\arg\Gamma(\mathrm{i}\nu)\right) \\
\beta &= \frac{1}{2}[\arg(\mathrm{i}\nu) - \nu\ln\nu + \nu + \pi/4].
\end{aligned} \tag{A.109}
$$

Here the approximation $1 \simeq \cos\theta$ has been inserted for convenient comparison with (A.85). Similarly the identity (Abramowitz and Stegun 1965)

$$
\Gamma(-\mathrm{i}\nu) = (2\pi)^{1/2}\sinh\pi\nu\exp[-\mathrm{i}\arg\Gamma(\mathrm{i}\nu)] \tag{A.110}
$$

may be combined with (A.106) and the second of (A.105) to yield

$$
\begin{aligned}
B\varphi(t) \overset{t\ll -1}{\approx} &[1-\exp(-2\pi\nu)]^{1/2}\cos\theta\exp\left(2\mathrm{i}\nu\int_0^t (1+t^2)^{1/2}\mathrm{d}t + \mathrm{i}\beta\right) \\
&+ \mathrm{e}^{-\pi\nu}\sin\theta\exp\left(-2\mathrm{i}\nu\int_0^t (1+t^2)^{1/2}\mathrm{d}t - \mathrm{i}\beta\right),
\end{aligned} \tag{A.111}
$$

where in this case one has recognized that $\cos\theta \simeq 1$ and $\sin\theta \simeq (-2t)^{-1}$.

Finally, after factoring out the constant B and comparing with eqns (A.85) and (A.94), it follows that

$$
\begin{aligned}
Q'_- &= \mathrm{e}^{-\mathrm{i}\beta} \\
P'_- &= [1-\exp(-2\pi\nu)]^{1/2}\,\mathrm{e}^{\mathrm{i}\beta} \\
P'_+ &= \mathrm{e}^{-\pi\nu}\,\mathrm{e}^{-\mathrm{i}\beta}.
\end{aligned} \tag{A.112}
$$

On the other hand eqn (A.96) requires a choice of χ such that

$$
Q'_- = [1-\exp(1-2\pi\nu)]^{1/2}\,\mathrm{e}^{-\mathrm{i}\chi}P'_- + \mathrm{e}^{-\pi\nu}P'_+, \tag{A.113}
$$

which is consistent with (A.112) if

$$
\chi(\nu) = 2\beta = \arg\Gamma(\mathrm{i}\nu) - \nu\ln\nu + \nu + \pi/4. \tag{A.114}
$$

This completes the determination of $\chi(\nu)$.

This is, however, very far from the whole story because the argument tacitly assumes that the classical turning points on the two adiabatic potentials are sufficiently well separated in phase that inward and outward motions can be handled separately. In other words the non-adiabatic coupling depends only on the integral in eqn (A.102) between the transition points at $t = \pm\mathrm{i}$ in Fig. A.6. In practical terms this means that the energy must be chosen well above that of the classical turning point. Some extensions to lower energies are, however, available by reference to the Bykovskii et al. (1964) (BNO) model, by which the Schrödinger equation is solved in the momentum representation within the potential function approximations of eqns (A.79). The early results are discussed by Nikitin (1968), and details of the mathematical argument are given in Appendix D of Child (1974b). The problem is that there are in fact four transition points, which have so far been treated as two separate pairs, each closely grouped around the real t axis. Ovchinnikova treats the situation in which the energy is well below the crossing point, in which case the transition points regroup to lie in close pairs around the imaginary t axis, which requires separate treatments according to whether the forces F_1 and F_2 in eqn (A.79) have equal or opposite signs for inward and outward motion respectively.

More recently, Nikitin et al. (1976) and Baranyi (1978, 1979) have extended the phase integral argument for the BNO model involving potential slopes F_i with the same sign, to show that the transition amplitudes, interference terms and tunnelling integrals are characterized over the whole energy range by the real and imaginary parts of the single complex integral

$$\hat{\sigma} + \mathrm{i}\hat{\delta} = \frac{\beta}{2}\int_{-\varepsilon}^{\mathrm{i}} \left(\frac{1+t^2}{\varepsilon + t}\right)^{1/2} \mathrm{d}t, \tag{A.115}$$

where

$$\begin{aligned} &\varepsilon = E\Delta F/(2FV_{12}), \qquad F = (F_1F_2)^{1/2}, \qquad \Delta F = F_1 - F_2 \\ &\beta = (4V_{12}/\hbar)\left[mV_{12}/F\Delta F\right]^{1/2}. \end{aligned} \tag{A.116}$$

It is readily confirmed, in the high energy limit, $\varepsilon \gg 1$, that

$$\begin{aligned} \hat{\sigma} + \mathrm{i}\hat{\delta} &\simeq \frac{\beta}{2}\left[\int_{-\varepsilon}^{0} \frac{\varepsilon}{(\epsilon + t)^{1/2}}\mathrm{d}t + \int_{0}^{\mathrm{i}} \left(\frac{1+t^2}{\varepsilon}\right)^{1/2} \mathrm{d}t\right] \\ &= \beta\varepsilon^{3/2} + \mathrm{i}\frac{\pi\beta}{8\varepsilon^{1/2}} = \beta\varepsilon^{3/2} + \mathrm{i}\pi\nu, \end{aligned} \tag{A.117}$$

where $\beta\varepsilon^{3/2}$ is the Stückelberg interference term for the BNO model (Child 1974b) and ν is the non-adiabatic coupling parameter in eqn (A.95). The following interesting symmetry relation between positive and negative energies, relative to the crossing point, may be verified by substituting $\varepsilon = -\varepsilon'$ and $t = -t'$, with the branch cut taken such that $\arg(\varepsilon + t) = -\pi$ on the negative real axis;

$$\hat{\sigma}(-\varepsilon, \beta) = \hat{\delta}(\varepsilon, \beta), \qquad \hat{\delta}(-\varepsilon, \beta) = \hat{\sigma}(\varepsilon, \beta). \tag{A.118}$$

In addition, Baranyi (1978) gives an expression for $\hat{\sigma}$ in terms of elliptic integrals (Gradsteyn and Ryyzhik 1980).

Substantial further progress was obtained by Zhu and Nakamura (1995, 1996), who were able to determine and parameterize the behaviour of the Stokes constants at all four transition points over a wide range of energies and coupling strengths. Details of the resulting Zhu–Nakamura theory are beyond the scope of this appendix, but readers are referred to the book by Nakamura (2007) for a full account of the mathematical argument and a wide variety of applications.

It should also be mentioned that Crothers (1972) has extended the Stückelberg–Landau–Zener argument to the Demkov (1964) model with

$$\begin{aligned} V_2(R) - V_1(R) &= \Delta(R) = \text{constant} \\ V_{12}(R) &= A\mathrm{e}^{-\alpha R}, \end{aligned} \tag{A.119}$$

which has much in common with the curve-crossing problem because the discriminant equation

$$[V_2(R) - V_1(R)]^2 + 4V_{12}^2(R) = 0 \tag{A.120}$$

again has two complex roots close to the real axis, given this time by

$$R = R_0 \pm \mathrm{i}\pi/2\alpha, \quad R_0 = -\alpha^{-1}\ln(\Delta/2A). \tag{A.121}$$

There is, however, an important difference, in that the transition zone is now centred around $t_0 \simeq 1$, in the notation of eqn (A.81), and that t itself remains positive for all R, albeit tending to small values as $R \to 0$, This means that the asymptotic region $t \to -\infty$ is no longer physically accessible (Crothers 1971). Further generalizations have been given by Crothers (1972) and Nikitin (1979, 1980).

Appendix B
Uniform approximations and diffraction integrals

The theory of semiclassical interferences has close similarities with optical theory, the classical trajectory being the analogue of the geometrical light ray. In both cases the true picture—quantum mechanics on one hand or wave optics on the other—requires an integral over an infinite number of paths but in the short-wavelength limit the dominant contributions come from a few 'classical' paths, which appear as stationary phase points of the integrand. Such points are real if the event is classically allowed and complex if it is forbidden. The allowed events contribute to the bright part of the diffraction pattern and the forbidden ones to the shadow region.

The simplest way to handle such situations is by means of a stationary phase approximation (Jeffreys and Jeffreys 1956) to the integral, which inevitably leads to spurious divergences whenever two or more stationary phase points coalesce on a 'caustic' boundary between bright and shadow regions. The purpose of this appendix is to show how knowledge of the topologies of such coalescences may be used to construct accurate uniform approximations to the relevant integrals despite the divergent stationary phase properties of the integrand. As emphasized by Berry (1976) and Connor et al. (1976), the natural classification is according to Thom's (1975) catastrophe theory which deals with the local topology of the stationary points of particular polynomial functions (see also Postern and Stewart 1978). Each catastrophe type has its own diffraction integral and its own machinery for constructing the relevant uniform approximation.

The underlying ideas are introduced in Section B.1 by reference to the simplest example—technically a 'fold' catastrophe involving interference between two possibly coalescent contributions, for which the proper diffraction integral is the Airy function (Abramowitz and Stegun 1965). Section B.2 then expounds the connection between the catastrophe type and the form of the proper diffraction integral. Further details required for the construction of uniform approximations of 'cusp' and 'elliptic' or 'hyperbolic umbilic' types are described in Section B.3. Finally, Section B.4 contains an account of two non-catastrophe-based uniform approximations that are required elsewhere in the text—the Bessel and harmonic approximations.

B.1 The uniform Airy approximation

The uniform Airy approximation is designed as an approximation to the integral

$$I(\alpha) = \int_{-\infty}^{\infty} g(u) \exp[\mathrm{i}f(\alpha; u)]\mathrm{d}u \tag{B.1}$$

in situations such that the phase function $f(\alpha; u)$ has two stationary points, say u_a and u_b, whose positions and characters are controlled by a parameter α, which may be the energy, a final quantum number or a scattering angle according to the physical context. The notation is chosen such that u_a and u_b correspond to phase minima and phase maxima respectively when the two points are real; thus $f_a'' > 0$ and $f_b'' < 0$ and the stationary phase or primitive semiclassical (PSC) approximation yields (see Section 2.2)

$$I_{\mathrm{PSC}}(\alpha) = P_a^{1/2}(\alpha)\exp[\mathrm{i}f_a(\alpha) + \mathrm{i}\pi/4] + P_b^{1/2}(\alpha)\exp[\mathrm{i}f_b(\alpha) - \mathrm{i}\pi/4], \tag{B.2}$$

where

$$P_a(\alpha) = 2\pi g_a^2/f_a'', \quad P_b(\alpha) = -2\pi g_b^2/f_b''. \tag{B.3}$$

As usual $I_{\mathrm{PSC}}(\alpha)$ diverges wherever the two stationary points coalesce because $f(\alpha, u)$ then has a point of inflexion at which $f_a'' = f_b'' = 0$.

The uniform approximation technique due to Chester et al. (1957) removes such divergences by means of a variable transformation $u(x)$ which maps the phase function $f(\alpha; u)$ on to a canonical function $\phi(\zeta; x)$ with the same stationary point characteristics. The defining equation in the case of two stationary phase points is

$$f[\alpha; u(x)] = \phi(\zeta; x) = A(\alpha) + \frac{1}{3}x^3 - \zeta(\alpha)x, \tag{B.4}$$

in which the quantities $A(\alpha)$ and $\zeta(\alpha)$ remain to be determined by values of the phase function $f(\alpha; u)$ at the stationary points u_a and u_b. The immediate effect is that (B.1) transforms to

$$I(\alpha) = \int_{-\infty}^{\infty} g[u(x)](\mathrm{d}u/\mathrm{d}x)\exp[\mathrm{i}A(\alpha) + \mathrm{i}x^3/3 - \mathrm{i}\zeta(\alpha)x]\mathrm{d}x. \tag{B.5}$$

The next step is to introduce the approximation

$$g(u)(\mathrm{d}u/\mathrm{d}x) \simeq X + xY, \tag{B.6}$$

thereby reducing (B.5) to the form

$$I(\alpha) \simeq 2\pi\exp[\mathrm{i}A(\alpha)]\{X\,Ai[-\zeta(\alpha)] - \mathrm{i}Y\,Ai'[-\zeta(\alpha)]\}, \tag{B.7}$$

where $Ai(z)$ is the Airy function (Abramowitz and Stegun 1965)

$$Ai(z) = \frac{1}{2\pi}\int_{-\infty}^{\infty} \exp(\mathrm{i}x^3/3 + \mathrm{i}zx)\mathrm{d}x \tag{B.8}$$

and $Ai'(z)$ is its derivative.

It remains to relate the quantities $A(\alpha), \zeta(\alpha), X$ and Y in (B.7) to f_a, f_b, P_a and P_b in (B.2) by forcing a correspondence between the stationary phase points on the two sides of the identity (B.4):

$$u_a \longleftrightarrow x_a = \zeta^{1/2}(\alpha) \quad \text{and} \quad u_b \longleftrightarrow x_b = -\zeta^{1/2}(\alpha). \tag{B.9}$$

Consequently, on substitution in (B.4),

$$f_a = A(\alpha) - \frac{2}{3}\zeta^{2/3}(\alpha), \qquad f_b = A(\alpha) + \frac{2}{3}\zeta^{2/3}(\alpha), \tag{B.10}$$

from which

$$A(\alpha) = \frac{1}{2}[f_a(\alpha) + f_b(\alpha)], \qquad \zeta(\alpha) = \left\{\frac{3}{4}\left[f_b(\alpha) - f_a(\alpha)\right]\right\}^{2/3}, \tag{B.11}$$

showing that the coalescence point occurs at $\zeta(\alpha) = 0$. By this definition, with $f_b(\alpha)$ and $f_a(\alpha)$ taken as maximum and minimum values of $f(u)$ respectively, $\zeta(\alpha) > 0$ when the stationary points are real. If, on the other hand, they are complex conjugates, it is necessary to choose $x_a = \exp(\mathrm{i}\pi/2)|x_a|$ and $x_b = \exp(-\mathrm{i}\pi/2)|x_b|$ in order to ensure that $\arg(f_b - f_a) = 3\pi/2$ and hence that $\zeta(\alpha)$ is real and negative.

The final step is to take the second derivative of eqn (B.4) with respect to x, and to recall that $\mathrm{d}f/\mathrm{d}u = 0$ at the stationary phase points; hence

$$f_v''(\mathrm{d}u/\mathrm{d}x)_v^2 = 2x_v, \qquad v = a, b. \tag{B.12}$$

It then follows by substitution for $(\mathrm{d}u/\mathrm{d}x)_v$ and x_v in (B.6) that

$$\begin{aligned} g_a(2/f_a'')^{1/2}\zeta^{1/4}(\alpha) &= X + Y\zeta^{1/2}(\alpha) \\ g_b(-2/f_b'')^{1/2}\zeta^{1/4}(\alpha) &= X - Y\zeta^{1/2}(\alpha). \end{aligned} \tag{B.13}$$

The combination of (B.7) and (B.13) readily rearranges to

$$\begin{aligned} I(\alpha) = \sqrt{\pi}\exp[\mathrm{i}A(\alpha)]\{&(P_a^{1/2} + P_b^{1/2})\zeta^{1/4}(\alpha)Ai[-\zeta(\alpha)] \\ &-\mathrm{i}(P_a^{1/2} - P_b^{1/2})\zeta^{-1/4}(\alpha)Ai'[-\zeta(\alpha)]\}, \end{aligned} \tag{B.14}$$

where P_a and P_b are given by eqn (B.3).

Equation (B.14) is the desired uniform Airy approximation, which remains finite even at the coalescence points because the divergence of P_a and P_b is exactly cancelled by the vanishing of $\zeta^{1/4}(\alpha)$ provided that $f(u)$ varies quadratically in the stationary regions; similarly the divergence of $\zeta^{-1/4}(\alpha)$ at the coalescence point is balanced by the disappearance of the term $(P_a^{1/2} - P_b^{1/2})$. It is also readily verified by use of the following asymptotic forms (Abramowitz and Stegun 1965),

$$\begin{aligned} Ai(-\zeta) &\overset{\zeta\gg 1}{\approx} \pi^{-1/2}\zeta^{-1/4}\sin\left(\frac{2}{3}\zeta^{3/2} + \pi/4\right) \\ &\overset{\zeta\ll 1}{\approx} \frac{1}{2}\pi^{-1/2}|\zeta|^{-1/4}\exp\left(-\frac{2}{3}|\zeta|^{3/2}\right) \\ Ai'(-\zeta) &\overset{\zeta\gg 1}{\approx} -\pi^{-1/2}\zeta^{1/4}\cos\left(\frac{2}{3}\zeta^{3/2} + \pi/4\right) \\ &\overset{\zeta\ll 1}{\approx} -\frac{1}{2}\pi^{-1/2}|\zeta|^{1/4}\exp\left(-\frac{2}{3}|\zeta|^{3/2}\right), \end{aligned} \tag{B.15}$$

that eqn (B.14) goes over to the primitive semiclassical form (B.2) for $\zeta \gg 1$, while, in the classically forbidden case, $\zeta \ll -1$,

$$I(\alpha) \simeq |P_a|^{1/2} \exp[\mathrm{iRe}(f_a) - \mathrm{Im}(f_a)]. \tag{B.16}$$

The strength of this analysis is that the interference pattern close to any binary stationary phase coalescence always takes the form shown in Fig. B.1, regardless of the physical context; it is equally applicable to the form of a wavefunction close to a turning point (Section 2.2), to the predissociation linewidth variation with vibrational quantum number (Section 5.3) and to elastic and inelastic scattering (Sections 8.2, 9.2 and 9.3).

Moreover, the basic Airy form applies also to higher-dimensional integrals provided that the number of stationary phase points is limited to two. To see this, note that the integrand in the multidimensional case,

$$I(\alpha) = \int_{-\infty}^{\infty} \int_{-\infty}^{\infty} \cdots \int_{-\infty}^{\infty} g(\mathbf{u}) \exp[\mathrm{i}f(\boldsymbol{\alpha}; \mathbf{u})] du_1\, du_2 \ldots du_n, \tag{B.17}$$

is amenable to a variable transformation $(u_1, u_2, \ldots, u_n) \rightarrow (x_1, x_2, \ldots, x_n)$ such that the two stationary points lie on the x_1 axis, as implied by the following generalization of (B.4):

$$f(\boldsymbol{\alpha}; \mathbf{u}) = \phi[\boldsymbol{\zeta}(\boldsymbol{\alpha}); \mathbf{x}] \tag{B.18}$$

where

$$\phi[\boldsymbol{\zeta}(\boldsymbol{\alpha}); \mathbf{x}] = \zeta_0(\boldsymbol{\alpha}) + \frac{1}{3}x_1^3 - \zeta_1(\boldsymbol{\alpha})x_1 + \sum_{i=2}^{n} \frac{1}{2}\zeta_i(\boldsymbol{\alpha})x_i^2. \tag{B.19}$$

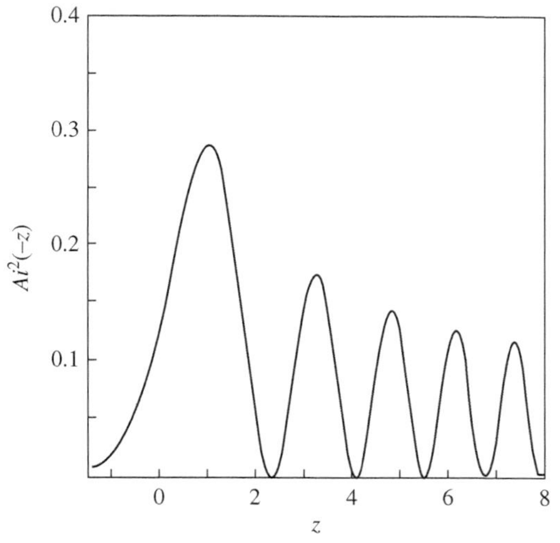

Fig. B.1 The function $|Ai(-z)|^2$.

Similarly, eqn (B.6) generalizes to

$$g(x)\left[\frac{\partial(u_1,u_2,\ldots,u_n)}{\partial(x_1,x_2,\ldots,x_n)}\right] = X + x_1 Y. \tag{B.20}$$

One also finds by taking second derivatives of eqn (B.18) with respect to the various x_i at the stationary points that

$$\left[\frac{\partial(u_1,u_2,\ldots,u_n)}{\partial(x_1,x_2,\ldots,x_n)}\right]_v = \left[\frac{\|\ \partial^2\phi/\partial x_i\partial x_j\ \|}{\|\ \partial^2 f/\partial u_i\partial u_j\ \|}\right]_v^{1/2}, \tag{B.21}$$

where $\|\ \partial^2\phi/\partial x_i\partial x_j\ \|$ is the Hessian determinant of the second derivative matrix and the subscript v implies evaluation of the stationary phase point $v = a$ or b. It is then a simple matter to substitute (B.20) into (B.18) and to integrate out the quadratic terms, after which the previous argument (with x_1 in place of x) yields a result identical with (B.14) except that $\zeta(\alpha) \to \xi_1(\boldsymbol{\alpha})$ and P_a and P_b are replaced by (Connor 1973)

$$\begin{aligned} P_a &= (2\pi)^n g_a^2/\ \|\ \partial^2 f/\partial u_i\partial u_j\ \|_a \\ P_b &= -(2\pi)^n g_b^2/\ \|\ \partial^2 f/\partial u_i\partial u_j\ \|_b. \end{aligned} \tag{B.22}$$

Similar reasoning applies to the higher interference patterns discussed below in the sense that the generic diffraction pattern is unaltered by extension of the dimensionality. The behaviour is always governed by the number of stationary phase points and the manner in which they coalesce.

B.2 Waves and catastrophes

The above discussion shows that the functional behaviour of the integral $I(\alpha)$ with respect to α depends on the disposition of the stationary phase points, u_a and u_b; the integral oscillates with respect to α if u_a and u_b are real and decreases exponentially as their imaginary parts increase into the complex domain. More complicated situations can be envisaged, involving more than two stationary phase points in higher-dimensional spaces. The Thom (1975) theory of catastrophes may be used to categorize the various possibilities and to devise appropriate diffraction integrals.

A catastrophe in this context corresponds to a coalescence between stationary points of an arbitrary function, which leads indeed to a catastrophic divergence of the primitive semiclassical or stationary phase approximation. Each catastrophe type is categorized by its own particular 'unfolding' function $\phi(\zeta; x)$, of which the right-hand side of (B.4) is the simplest example. The argument of $\phi(\zeta; x)$ depends on two types of parameter; the variables x define what is known as the 'behaviour space' in which the stationary character is observed while the parameters ζ or 'control variables' determine the relative positions and the real or complex characters of the stationary points in question. Hence the physical observation space is in effect the control space, and the surfaces in this space, on which coalescences occur, play an important role in the theory, as what are termed the 'caustics' or 'bifurcation sets' (Postern and

Stewart 1978). Finally, each catastrophe type has its own canonical diffraction integral determined by its unfolding function in the form

$$U(\zeta) = \left[\frac{1}{2\pi}\right]^n \int_{-\infty}^{\infty}\int_{-\infty}^{\infty}\cdots\int_{-\infty}^{\infty} \exp[\mathrm{i}\phi(\zeta;\mathbf{x})]\mathrm{d}x_1\mathrm{d}x_2\ldots\mathrm{d}x_n. \tag{B.23}$$

To see these ideas in their simplest context it is convenient to refer back to eqn (B.4) and to recognize the function

$$\phi^{\mathrm{I}}(\zeta;x) = \frac{1}{3}x^3 - \zeta x \tag{B.24}$$

as the unfolding of the simplest catastrophe type, the 'fold'. The control parameter ζ fixes the nature of the stationary points

$$x_v = \pm\zeta^{1/2}, \quad v = a \text{ or } b, \tag{B.25}$$

with the caustic at $\zeta = 0$. Finally, the diffraction integral

$$\begin{aligned} U_1(\zeta) &= \frac{1}{2\pi}\int_{-\infty}^{\infty} \exp[\mathrm{i}\phi^{\mathrm{I}}(\zeta;x)]\mathrm{d}x \\ &= \frac{1}{2\pi}\int_{-\infty}^{\infty} \exp[\mathrm{i}(x^3/3 - \zeta x)]\mathrm{d}x \end{aligned} \tag{B.26}$$

is the Airy function $Ai(-\zeta)$, which is seen in Fig. B.1 to oscillate on the bright (real root) side of the caustic, $\zeta > 0$, and to decrease exponentially into the shadow (complex root) region, $\zeta < 0$.

There are six other elementary catastrophe types, which cover the topologies of coalescences with up to five stationary points. Their names and unfoldings are listed as follows (Postern and Stewart 1978):

cusp:	$\phi^{\mathrm{II}} = x^4 + \zeta_1 x^2 + \zeta_2 x$	(B.27)
swallowtail:	$\phi^{\mathrm{III}} = x^5 + \zeta_1 x^3 + \zeta_2 x^2 + \zeta_3 x$	(B.28)
elliptic umbilic:	$\phi^{\mathrm{IV}} = 2x^3 - 6xy^2 + \zeta_1(x^2+y^2) + \zeta_2 x + \zeta_3 y$	(B.29)
hyperbolic umbilic:	$\phi^{\mathrm{V}} = x^3 + y^3 + \zeta_1 xy + \zeta_2 x + \zeta_3 y$	(B.30)
butterfly:	$\phi^{\mathrm{VI}} = x^6 + \zeta_1 x^4 + \zeta_2 x^3 + \zeta_3 x^2 + \zeta_4 x$	(B.31)
parabolic umbilic:	$\phi^{\mathrm{VII}} = y^4 + x^2 y + \zeta_1 x^2 + \zeta_2 y^2 + \zeta_3 x + \zeta_4 y.$	(B.32)

It is beyond the scope of this appendix to explore the consequences of this classification in every detail but aspects of the cusp and hyperbolic/elliptic umbilic situations relevant to the nature of possible diffraction patterns are discussed below. Their significance for the derivation of uniform approximations is examined in Section B.3. Connor (1990b) gives a comprehensive account of the fold, cusp and swallowtail cases, with particular reference to evaluation of the diffraction integrals.

Cusp catastrophes

The form of the diffraction pattern is always governed by the shape of the caustic or bifurcation set, on which the coalescences occur. In the cusp case, the stationary points of the unfolding $\phi^{\mathrm{II}}(\zeta_1, \zeta_2; x)$ are given by

$$\mathrm{d}\phi^{\mathrm{II}}/\mathrm{d}x = 4x^3 + 2\zeta_1 x + \zeta_2 = 0, \tag{B.33}$$

and repeated roots occur when

$$\mathrm{d}^2\phi^{\mathrm{II}}/\mathrm{d}x^2 = 12x^2 + 2\zeta_1 = 0. \tag{B.34}$$

Hence the caustic surface, obtained by eliminating x from (B.33) and (B.34), is defined by the equation

$$\Delta = 8\zeta_1^3 + 27\zeta_2^2 = 0. \tag{B.35}$$

The resulting division of the control space (ζ_l, ζ_2) is shown by the dashed line in Fig. B.2 (the full picture has mirror symmetry in the ζ_1 axis). Equation (B.33) has three real roots (for real ζ_1 and ζ_2) when $\Delta < 0$ (the bright side of the caustic) and one real and two complex conjugate roots for $\Delta > 0$. Finally, all three roots coalesce together at the cusp point $\zeta_1 = \zeta_2 = 0$.

The existence of real or complex roots is, of course, associated with classically allowed or classically forbidden events respectively and it should be no surprise that the appropriate diffraction integral

$$U^{\mathrm{II}}(\zeta_1, \zeta_2) = \frac{1}{2\pi} \int_{-\infty}^{\infty} \exp[\mathrm{i}(x^4 + \zeta_1 x^2 + \zeta_2 x)]\mathrm{d}x, \tag{B.36}$$

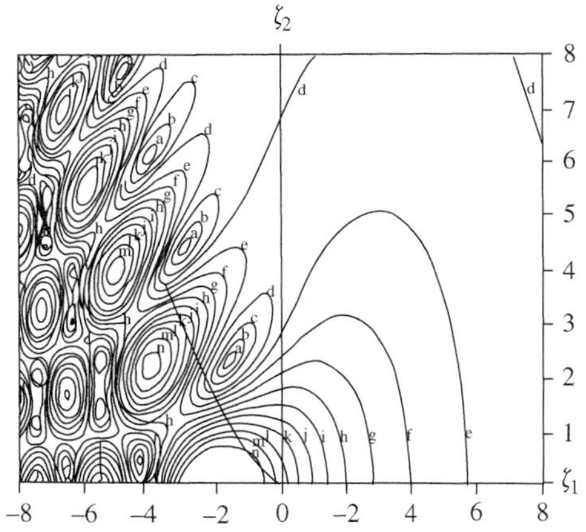

Fig. B.2 Contour plot of $|U^{\mathrm{II}}(\zeta_1, \zeta_2)|$ in the upper half (ζ_1, ζ_2) plane. The dashed line follows the cusp-shaped (Pearcey) caustic boundary (or bifurcation set) given by (B.35). (Taken from Connor (1990b) by courtesy of Marcel Dekker Inc.)

which is known as the Pearcey (1946) integral, shows a marked change of character on crossing the caustic. As displayed in Fig. B.2, $|U''(\zeta_1, \zeta_2)|$ undergoes marked oscillations in the bright (three real root) region where $\Delta < 0$, and an exponential fall-off into the shadow region (see also Trinkhaus and Drepper 1977).

Elliptic and umbilic catastrophes

The elliptic and hyperbolic umbilic cases are conveniently taken together because they are interrelated for complex values of the variables, as may be confirmed by noting that the hyperbolic unfolding ϕ^{V} goes over, under the substitutions

$$\begin{gathered} x = \tilde{x} + \mathrm{i}\tilde{y}, \quad y = \tilde{x} - \mathrm{i}\tilde{y} \\ \zeta_2 = \tfrac{1}{2}(\tilde{\zeta}_2 - \mathrm{i}\tilde{\zeta}_3), \quad \zeta_3 = \tfrac{1}{2}(\tilde{\zeta}_2 + \mathrm{i}\tilde{\zeta}_3), \end{gathered} \tag{B.37}$$

to

$$\phi^{\mathrm{V}} = 2\tilde{x}^3 - 6\tilde{x}\tilde{y}^2 + \zeta_1(\tilde{x}^2 + \tilde{y}^2) + \tilde{\zeta}_2\tilde{x} + \tilde{\zeta}_3\tilde{y}. \tag{B.38}$$

Equation (B.38) corresponds exactly to the elliptic umbilic form in (B.29), which is written in this way rather than with an initial term $x^3 - 3xy^2$ as given by Postern and Stewart (1978) in order to underline the connection. It should be emphasized, however, that the caustic structures and types of coexistent stationary points are quite distinct in the two cases, if the control parameters $\{\zeta_i\}$ in (B.29) and (B.30) are restricted to real values.

For example, in the elliptic case the stationary points are given by

$$\begin{aligned} \partial\phi^{\mathrm{IV}}/\partial x &= 6(x^2 - y^2) + 2\zeta_1 x + \zeta_2 = 0, \\ \partial\phi^{\mathrm{IV}}/\partial y &= -12xy + 2\zeta_1 y + \zeta_3 = 0, \end{aligned} \tag{B.39}$$

and their characters depend on the sign of the Hessian determinant

$$\Delta = \| \partial^2\phi^{\mathrm{IV}}/\partial x\partial y \| = 4[\zeta_1^2 - 36(x^2 + y^2)]. \tag{B.40}$$

Roots of (B.39) with $\Delta > 0$ are maxima or minima, those with $\Delta < 0$ are saddle points, and coalescences occur when $\Delta = 0$. The direct solution of (B.39) is quite awkward in the general case, but the forms of the caustics, at which

$$36(x^2 + y^2) = \zeta_1^2, \tag{B.41}$$

are conveniently described by means of the following parameterization (Postern and Stewart 1978), which is designed to be consistent with (B.39):

$$\begin{aligned} &x = \alpha\cos\theta, \quad y = \alpha\sin\theta \\ &\zeta_1 = 6\alpha, \\ &\zeta_2 = -6\alpha^2(\cos 2\theta + 2\cos\theta), \quad \zeta_3 = 6\alpha^2(\sin 2\theta - 2\sin\theta). \end{aligned} \tag{B.42}$$

The solid line in Fig. B.3 shows a (ζ_2, ζ_3) section of this bifurcation set at fixed ζ_1, with the three cusps parameterized by $\theta = 0, \pm 2\pi/3$.

The nature of the stationary points inside and outside this caustic boundary may be displayed by examining the solutions of (B.39) along the line $\zeta_3 = 0$. Their properties are summarized in Table B.1. The first two roots are real for $\zeta_2 > -18\alpha^2$ and the second two for $\zeta_2 < 6\alpha^2$, inequalities that may be verified to extend to points on the

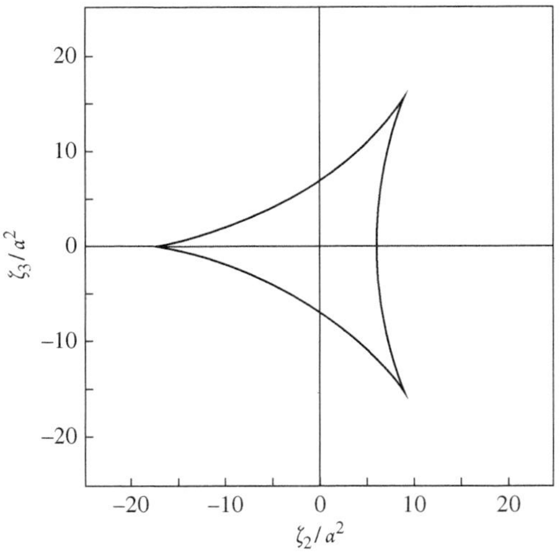

Fig. B.3 The elliptic umbilic bifurcation set at $\zeta_1 = 6\alpha$.

Table B.1 Stationary points of ϕ^{IV} for $\zeta_3 = 0$.

x	y	Δ
α	$[(36\alpha^2 + 2\zeta_2)/12]^{1/2}$	$-12(36\alpha^2 + 2\zeta_2)$
α	$-[(36\alpha^2 + 2\zeta_2)/12]^{1/2}$	$-12(36\alpha^2 + 2\zeta_2)$
$-\alpha + (\alpha^2 - \zeta_2/6)^{1/2}$	0	$-4[(36\alpha^2 - 6\zeta_2) - 2\zeta_1(36\alpha^2 - 6\zeta_2)^{1/2}]$
$-\alpha - (\alpha^2 - \zeta_2/6)^{1/2}$	0	$-4[(36\alpha^2 - 6\zeta_2) + 2\zeta_1(36\alpha^2 - 6\zeta_2)^{1/2}]$

bifurcation set parameterized by $\theta = 0$ and $\theta = \pi$ respectively. Hence all four stationary points are real inside the enclosed region in Fig. B.3, while the signs of Δ show that three are saddle points and that one is a maximum or a minimum (inspection of the eigenvalues of the Hessian matrix indicates a maximum for $\alpha < 0$ and a minimum for $\alpha > 0$). The first three roots in Table B.1 clearly coalesce at the cusp point, $(\zeta_2, \zeta_3) = (-18\alpha^2, 0)$, and the third and fourth coalesce at $(\zeta_2, \zeta_3) = (6\alpha^2, 0)$. Finally, two real roots remain outside the caustic boundary, both being saddle points.

The diffraction integral in this case is

$$U^{\mathrm{IV}} = \left[\frac{1}{2\pi}\right]^2 \int_{-\infty}^{\infty} \int_{-\infty}^{\infty} \exp\{\mathrm{i}[2x^3 - 6xy^2 + \zeta_1(x^2 + y^2) + \zeta_2 x + \zeta_3 y]\} \mathrm{d}x\mathrm{d}y, \quad \text{(B.43)}$$

and the contours of $|U^{\mathrm{IV}}|$ show an intricate oscillatory pattern in the bright internal region and a fall-off in intensity outside. Berry et al. (1979) show beautiful optically generated examples of this behaviour, which is also illustrated by Trinkhaus and Drepper (1977).

The hyperbolic case is treated in a similar manner. The stationary points of the unfolding ϕ^{V} are determined by

$$\begin{aligned} \partial\phi^{\mathrm{V}}/\partial x &= 3x^2 + \zeta_1 y + \zeta_2 = 0 \\ \partial\phi^{\mathrm{V}}/\partial y &= 3y^2 + \zeta_1 x + \zeta_3 = 0, \end{aligned} \qquad \text{(B.44)}$$

and the condition for repeated roots of (B.44) is

$$\Delta = 36xy - \zeta_1^2 = 0. \qquad \text{(B.45)}$$

The caustics or bifurcation sets are therefore conveniently parameterized by the equations (Postern and Stewart 1978)

$$\begin{aligned} &x = \beta t, \qquad y = \beta/t \\ &\zeta_1 = 6\beta \\ &\zeta_2 = -3\beta^2(t^2 + 2t^{-1}), \qquad \zeta_3 = -3\beta^2(t^{-2} + 2t); \end{aligned} \qquad \text{(B.46)}$$

an alternative is to replace t by $\pm e^{\theta}$, which makes a direct connection with the elliptic case. The solid lines in Fig. B.4(a) show that eqns (B.46) define two branches in the (ζ_2, ζ_3) plane at fixed ζ_1, the upper and lower of which correspond to positive and negative t values respectively.

The types of stationary point in the three regions may be determined by considering the four solutions of (B.44) along the line $\zeta_2 = \zeta_3$. Their properties are summarized in Table B.2. The first two roots are real for $\zeta_2 < 3\beta^2$ and the second two for $\zeta_2 < -9\beta^2$. Hence all four roots are real in region I but there is a vital difference from the elliptic case, in that now the first two Δ values are positive while the second two are negative; inspection of the eigenvalues of the Hessian matrix shows that of the first two, one is a maximum and the other a minimum while the second two are both saddle points. Similarly on crossing to region II (corresponding to $-9\beta^2 < \zeta_2 < 3\beta^2$ in Table B.2), one of the remaining two real stationary points is a saddle and the other may be shown to be a maximum if $\beta > 0$ and a minimum if $\beta < 0$. Finally, all four roots are complex in region III—a situation that cannot occur in the elliptic case.

Table B.3 summarizes the important differences with respect to the types of coexistent real stationary points between the elliptic and hyperbolic umbilic cases. It may also be noted in passing that the modified swallowtail unfolding

$$\phi^{\mathrm{III}'} = x^5 + \zeta_1 x^3 + \zeta_2 x^2 + \zeta_3 x + \lambda y^2 \qquad \text{(B.47)}$$

can also give rise to four real stationary points in a two-dimensional space, two of which are saddle points and two either both maxima or both minima. Such cases do not, however, constitute a further elementary catastrophe because, by the argument at the end of Section B.1, the nature of possible coalescences is entirely determined by the swallowtail part of $\phi^{\mathrm{III}'}$.

To complete this section, we note that the difference between the types of coexistent stationary points in the elliptic and hyperbolic umbilic cases carries with it a gross difference in the caustic structures and hence in the canonical diffraction patterns. The

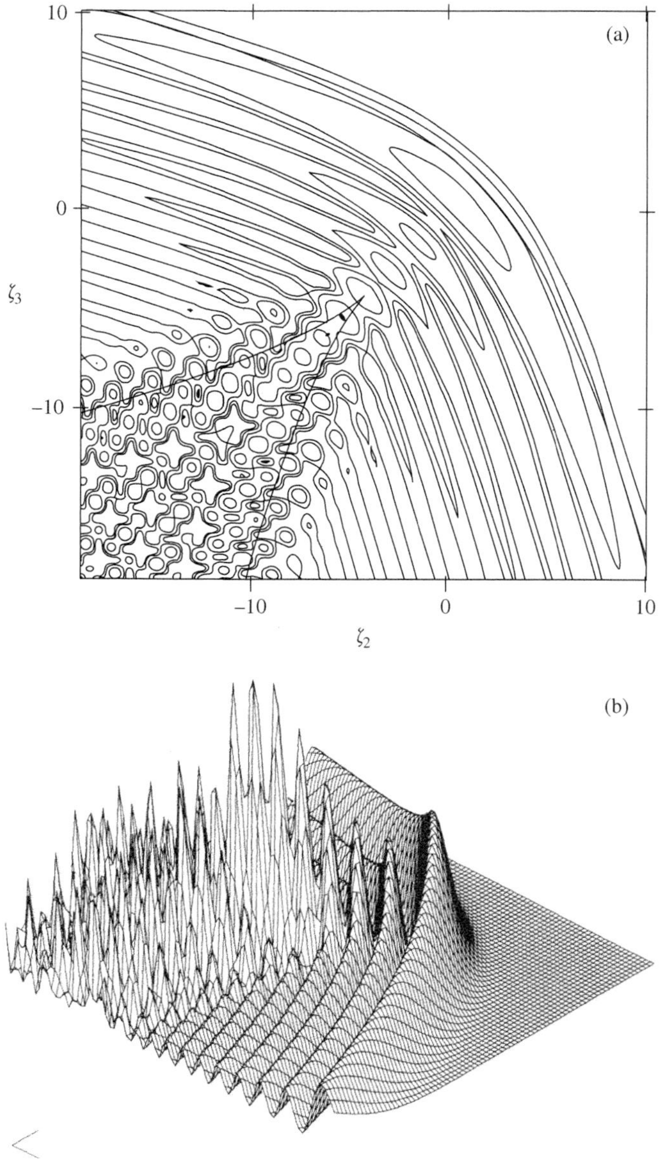

Fig. B.4 (a) Contour plot of $|U^{\text{V}}(\zeta_1, \zeta_2, \zeta_3)|^2$ in the (ζ_2, ζ_3) plane for $\zeta_1 = 4$. The solid lines show the hyperbolic umbilic bifurcation set. (b) A perspective view of (a). (Reprinted from Uzer et al. (1983a) by permission of Taylor and Francis.)

Table B.2 Hyperbolic umbilic stationary points when $\zeta_2 = \zeta_3$.

x	y	Δ
$-\beta + (\beta^2 - \zeta_2/3)^{1/2}$	$-\beta + (\beta^2 - \zeta_2/3)^{1/2}$	$\beta^2 - \zeta_2/3 - 2\beta(\beta^2 - \zeta_2/3)^{1/2}$
$-\beta - (\beta^2 - \zeta_2/3)^{1/2}$	$-\beta - (\beta^2 - \zeta_2/3)^{1/2}$	$\beta^2 - \zeta_2/3 + 2\beta(\beta^2 - \zeta_2/3)^{1/2}$
$\beta + (-3\beta^2 - \zeta_2/3)^{1/2}$	$\beta - (-3\beta^2 - \zeta_2/3)^{1/2}$	$3(\beta^2 + \zeta_2/9)$
$\beta - (-3\beta^2 - \zeta_2/3)^{1/2}$	$\beta + (-3\beta^2 - \zeta_2/3)^{1/2}$	$3(\beta^2 + \zeta_2/9)$

Table B.3 Real elliptic and hyperbolic umbilic stationary points.

Number	Elliptic	Hyperbolic
4	{ 3 saddles, 1 max ($\alpha < 0$); 3 saddles, 1 min ($\alpha > 0$)	2 saddles, 1 max, 1 min
2	2 saddles	{ 1 saddle, 1 max ($\beta > 0$); 1 saddle, 1 min ($\beta < 0$)

contours of the relevant hyperbolic umbilic diffraction integral, shown in Fig. B.4, are those of $|U^{\mathrm{V}}(\zeta_1, \zeta_2, \zeta_3)|$ where

$$U^{\mathrm{V}}(\zeta_1, \zeta_2, \zeta_3) = \left[\frac{1}{2\pi}\right]^2 \int_{-\infty}^{\infty} \int_{-\infty}^{\infty} \exp[\mathrm{i}(x^3 + y^3 + \zeta_1 xy + \zeta_2 x + \zeta_3 y)]\mathrm{d}x\mathrm{d}y. \qquad \text{(B.48)}$$

The numerical evaluation of (B.48) is discussed by Uzer et al. (1983a), and optically generated examples of the generic diffraction pattern may be seen in Figures 5 and 9 of Berry et al. (1979).

B.3 Higher catastrophe-based uniform approximations

Section B.1 gives the prototype uniform approximation for catastrophes of the simplest (fold) type. This section describes how the same ideas may be extended to more complicated situations. The illustrations below cover those with the topologies of the cusp and elliptic or hyperbolic umbilic types.

The cusp case

It is assumed that the required integral takes the form

$$I(\alpha) = \frac{1}{2\pi} \int_{-\infty}^{\infty} g(u) \exp[\mathrm{i}f(\alpha; u)]\mathrm{d}u, \qquad \text{(B.49)}$$

where $f(\alpha; u)$ is known to have three stationary points, u_v, whose dispositions are governed by the physical control parameters α, although reference to these parameters will henceforth be dropped to avoid obscuring the notation. The object is to express $I(\alpha)$ in terms of the cusp unfolding (B.36) and the stationary phase quantities

$$\left.\begin{aligned} P_v &= 2\pi g^2(u_v)/f''(u_v) \\ f_v &= f(u_v) \end{aligned}\right\} \quad v = a, b, c. \qquad \text{(B.50)}$$

It is further assumed for simplicity that $f(\alpha; u)$ has the same qualitative behaviour as ϕ^{II}, namely one maximum and two minima in the region of three real stationary points, going over to one maximum and two complex stationary points; cases in which $f(\alpha; u)$ has two real maxima and one minimum etc. are trivially handled by replacing ϕ^{II} by $-\phi^{\mathrm{II}}$ in what follows.

The analysis proceeds, as in Section B.1, by assuming a transformation $u(x)$, defined by the identity

$$\begin{aligned} f[u(x)] &= \phi^{\mathrm{II}}(\zeta_1, \zeta_2; x) + \zeta_0 \\ &= x^4 + \zeta_1 x^2 + \zeta_2 x + \zeta_0 \end{aligned} \tag{B.51}$$

in terms of which eqn (B.49) becomes

$$I(\alpha) - \frac{1}{2\pi}\int_{-\infty}^{\infty} g(u)(\mathrm{d}u/\mathrm{d}x)\exp[\mathrm{i}(x^4 + \zeta_1 x^2 + \zeta_2 x + \zeta_0)]\mathrm{d}x. \tag{B.52}$$

The next step is to approximate the pre-exponent as

$$g(u)(\mathrm{d}u/\mathrm{d}x) \simeq X_0 + X_1 x^2 + X_2 x, \tag{B.53}$$

by means of which (B.52) becomes

$$I(\alpha) = \exp(\mathrm{i}\zeta_0)[X_0 U^{\mathrm{II}}(\zeta_1, \zeta_2) - \mathrm{i}X_1 U_1^{\mathrm{II}}(\zeta_1, \zeta_2) - \mathrm{i}X_2 U_2^{\mathrm{II}}(\zeta_1, \zeta_2)], \tag{B.54}$$

where $U^{\mathrm{II}}(\zeta_1, \zeta_2)$ is the cusp diffraction integral (B.36) and U_i^{II} are the partial derivatives

$$U_i^{\mathrm{II}}(\zeta_1, \zeta_2) = (\partial U^{\mathrm{II}}/\partial \zeta_i)_{\zeta_j}, \quad i \neq j = 1, 2. \tag{B.55}$$

Techniques for the evaluation of $U^{\mathrm{II}}(\zeta_1, \zeta_2)$ and its various partial derivatives $U_i^{\mathrm{II}}(\zeta_1, \zeta_2)$ are discussed in a slightly different notation by Connor et al. (1983) and Connor (1990b).

Equation (B.54) gives the form of the desired uniform approximation but it remains to determine the parameters $\{\zeta_i\}$ and coefficients $\{X_i\}$ in terms of the assumedly known phases $\{f_v\}$ and classical probabilities $\{P_v\}$ given by (B.50). The connection between the $\{\zeta_i\}$ and $\{f_v\}$ is readily established by the requirement that the stationary points on the two sides of (B.51) should coincide; thus

$$x_v^4 + \zeta_1 x_v^2 + \zeta_2 x_v + \zeta_0 = f_v, \tag{B.56}$$

with the $x_v (v = a, b, c)$ given by the stationary conditions

$$4x_v^3 + 2\zeta_1 x_v + \zeta_2 = 0. \tag{B.57}$$

These six equations for the $\{\zeta_i\}$ and $\{x_v\}$ are the analogues of (B.9) and (B.10) in the simpler Airy case; but they are less amenable to direct solution. Their roots may be determined by a generalized Newton–Raphson approach (Connor and Farrelly 1981).

Starting values for the variables may be obtained by replacing the two closest phases, f_a and f_b say, by their mean $\bar{f}$. This simplification readily leads, via (B.33)–(B.35), to

$$(\zeta_1, \zeta_2) = (-6\bar{x}^2, 8\bar{x}^3), \tag{B.58}$$

where $\bar{x}$ is the as yet unknown coordinate of the repeated root. Furthermore, the vanishing of the sum of the roots of (B.57) requires that $x_c = -2\bar{x}$. Finally, the three equations (B.53) for $v = a, b$ and c readily rearrange to

$$\begin{aligned} \bar{x}^4 &= \frac{1}{27}(\bar{f} - f_c) = \frac{1}{54}(f_a + f_b - 2f_c), \\ \zeta_0 &= f_c + \frac{8}{9}(\bar{f} - f_c), \end{aligned} \tag{B.59}$$

which together with (B.58) provide convenient starting points for the iteration. Thus the solution of the first of (B.59) yields x, which then determines ζ_1, ζ_2 and x_c, while the second of (B.59) gives ζ_0. Note that the turning point disposition of a function with two minima and one maximum ensures that $\bar{f} - f_c > 0$ so that $\bar{x}$ may be taken as real; it is in fact immaterial whether it is taken as positive or negative.

The next step is to relate the coefficients $\{X_i\}$ to the classical probabilities $\{P_v\}$ by requiring equality between the second derivatives of the two sides of (B.51) at the stationary points x_v; thus

$$f''(\alpha; u_v)(\mathrm{d}u/\mathrm{d}x)^2 = 12x_v^2 + 2\zeta_1, \tag{B.60}$$

which means that (B.53) transforms to

$$\begin{aligned} g(u_v)(\mathrm{d}u/\mathrm{d}x)_v &= \left(\frac{(12x_v^2 + 2\zeta_1)P_v}{2\pi}\right)^{1/2} \\ &= X_0 + X_1 x_v^2 + X_2 x_v. \end{aligned} \tag{B.61}$$

Since the x_v are determined by (B.56) and (B.57) and the P_v are assumedly known, the three unknown coefficients X_i are readily derived by inversion of the three equations (B.61) for $v = a, b, c$; this therefore completes the approximation.

A key feature of the result, which arises from the forced coincidence of the stationary points of $f(u)$ and $\phi^{\mathrm{II}}(x)$, is that the divergence of the P_v associated with the vanishing of $\mathrm{d}^2 f/\mathrm{d}u^2$ is exactly cancelled by disappearance of the term $(12x_v^2 + 2\zeta_1)$ in (B.61), which is of course nothing but $\mathrm{d}^2\phi^{\mathrm{II}}/\mathrm{d}u^2$. Consequently the coefficients $\{X_i\}$ derived from (B.61) vary smoothly across the caustics in Fig. B.2 where two or more of the P_v go to infinity. The net result is that the resulting approximation to the integral is uniformly valid, regardless of the disposition of the points of stationary phase.

Hyperbolic and elliptic umbilic cases

The interrelation between the hyperbolic and elliptic unfoldings (see eqns (B.37) and (B.38)) implies an intimate connection between the related uniform approximations. Both aim to approximate a two-dimensional integral of the form

$$I(\alpha) = \left(\frac{1}{2\pi}\right)^2 \int_{-\infty}^{\infty} \int_{-\infty}^{\infty} g(u, v) \exp[\mathrm{i} f(\boldsymbol{\alpha}; u, v)] \mathrm{d}u \mathrm{d}v \tag{B.62}$$

where $f(\boldsymbol{\alpha}; u, v)$ has four stationary points, (u_v, v_v) for $v = a, b, c, d$ say, and the integral is to be expressed in terms of the four classical probabilities

$$P_v(\boldsymbol{\alpha}) = (2\pi)^2 g^2(u_v, \nu_v) / \parallel \partial^2 f / \partial u \partial \nu \parallel_v, \tag{B.63}$$

and the four stationary phases

$$f_v(\boldsymbol{\alpha}) = f(\boldsymbol{\alpha}; u_v, \nu_v), \tag{B.64}$$

in which $\|\partial^2 f / \partial u \partial v\|_v$ denotes the Hessian determinant of the second derivative matrix. The essential difference between the two cases lies in the types of stationary point. As discussed in Section B.2, if all the points are real, the elliptic case allows three saddles and one maximum or one minimum whereas the hyperbolic case gives rise to two saddles, one maximum and one minimum; other possibilities are listed in Table B.3. This means that the distinction between the two cases is carried by the probabilities P_v which identify the characters of the stationary points; hence analysis of the information carried by the phases f_v alone can be carried through for both cases at once.

One proceeds, by analogy with the previous cusp case, in terms of a mapping between the function $f(\boldsymbol{\alpha}; u, v)$ and one or other of the elliptic or hyperbolic unfoldings $\phi^{\mathrm{IV}}(\tilde{\zeta}; \tilde{x}, \tilde{y})$ or $\phi^{\mathrm{V}}(\zeta; x, y)$ in the knowledge that the two sets of variables are related by eqns (B.37). Following Uzer and Child (1982), who analyse the situation in detail, the hyperbolic mapping is adopted; that is

$$\begin{aligned} f[\boldsymbol{\alpha}; u(x, y), \nu(x, y)] &= \phi^{\mathrm{V}}(\zeta; x, y) + \zeta_0 \\ &= x^3 + y^3 + \zeta_1 xy + \zeta_2 x + \zeta_3 y + \zeta_0. \end{aligned} \tag{B.65}$$

The integral therefore transforms to

$$I(\alpha) = \left(\frac{1}{2\pi}\right)^2 \int_{-\infty}^{\infty} \int_{-\infty}^{\infty} g(u, \nu) \left(\frac{\partial(u, \nu)}{\partial(x, y)}\right) \exp[\mathrm{i}\phi^{\mathrm{V}}(\zeta, x, y) + \zeta_0] \mathrm{d}x \mathrm{d}y, \tag{B.66}$$

which reduces under approximation

$$g(u, \nu) \left(\frac{\partial(u, \nu)}{\partial(x, y)}\right) \simeq X_0 + X_1 xy + X_2 x + X_3 y, \tag{B.67}$$

to

$$I(\alpha) = \exp(\mathrm{i}\zeta_0)[X_0 U^{\mathrm{V}}(\zeta) - \mathrm{i}X_1 U_1^{\mathrm{V}}(\zeta) - \mathrm{i}X_2 U_2^{\mathrm{V}}(\zeta) - \mathrm{i}X_3 U_3^{\mathrm{V}}(\zeta)], \tag{B.68}$$

where $U^{\mathrm{V}}(\zeta_1, \zeta_2, \zeta_3)$ is the hyperbolic diffraction integral (see (B.48)) and the $U_i^{\mathrm{V}}(\zeta_1, \zeta_2, \zeta_3)$ are its partial derivatives

$$U_i^{\mathrm{V}} = \partial U^{\mathrm{V}} / \partial \zeta_i, \quad i = 1, 2, 3. \tag{B.69}$$

The equivalent form involving the elliptic umbilic diffraction integral is readily obtained by employing the substitutions (B.37) in eqns (B.66) and (B.67). Techniques for evaluating the various integrals are discussed by Uzer et al. (1983a) and by Berry et al. (1979) in the hyperbolic and elliptic cases respectively, and more generally by Connor (1990b).

Equation (B.68) merely establishes the structure of the approximation, which is clearly analogous to that of (B.54) in the cusp case. The main effort lies in relating the control parameters $\{\zeta_i\}$ and expansion coefficients $\{X_i\}$ to the phases $\{f_v\}$ and probabilities $\{P_v\}$ respectively. The first step is to solve the stationary phase identities

$$\begin{aligned} x_v^3 + y_v^3 + \zeta_1 x_v y_v + \zeta_2 x_v + \zeta_3 y_v + \zeta_0 &= f_v \\ 3x_v^2 + \zeta_1 y_v + \zeta_2 &= 0 \\ 3y_v^2 + \zeta_1 x_v + \zeta_3 &= 0 \end{aligned} \tag{B.70}$$

for the quantities $\{\zeta_i\}, i = 1, 2, 3, 4$, and $\{x_v, y_v\}, v = a, b, c, d$. Surprisingly Uzer and Child (1982) find that these 12 equations may be reduced to a single ninth-degree polynomial equation for ζ_1^3, which has three real solutions for any choice of real or complex conjugate f_v values.

This equation is most conveniently expressed in terms of the variable $\xi = 4\zeta_1^3/27$, by defining the quantities

$$\begin{aligned} \sigma_v &= f_v - \bar{f} \\ \bar{f} &= \frac{1}{4}\sum_{v=1}^{4} f_v \\ \Sigma &= \sigma_1 + \sigma_2 + \sigma_3 \\ \Pi &= \sigma_1\sigma_2 + \sigma_2\sigma_3 + \sigma_3\sigma_1 \\ R &= \sigma_1\sigma_2\sigma_3 \end{aligned} \tag{B.71}$$

and

$$X = \Pi - \Sigma^2, \quad Y = R - \Sigma\Pi, \quad Z = -R\Sigma, \tag{B.72}$$

in terms of which ξ must satisfy the equation (Uzer and Child 1983)

$$\begin{aligned} &\xi^9 + 24X\xi^7 + 168Y\xi^6 - 6(13X^2 - 180Z)\xi^5 - 336XY\xi^4 \\ &\quad + 4(20X^3 + 432XZ - 159Y^2)\xi^3 + 24(7X^2Y + 36YZ)\xi^2 \\ &\quad - 3(9X^4 - 72X^2Z + 40XY^2 + 144Z^2)\xi + 8Y^3 = 0. \end{aligned} \tag{B.73}$$

Each real solution for ξ (and hence ζ_1) gives rise to six physically equivalent (ζ_2, ζ_3) combinations related by the symmetry in Fig. B.5, in which the coordinates (a, b, c) are defined as

$$a = 36\,\mathrm{Re}(\zeta_2 - \zeta_3)/\zeta_1^2, \qquad b = 36(\zeta_2 + \zeta_3)/\zeta_1^2, \qquad c = 36\,\mathrm{Im}(\zeta_2 - \zeta_3)/\zeta_1^2. \tag{B.74}$$

The ambiguity between the physically distinct solutions with different real ζ_1 values may be removed by comparison with the topology of the real problem as revealed by the

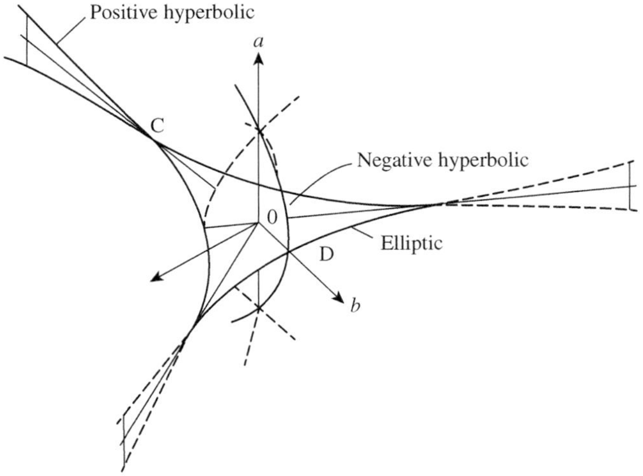

Fig. B.5 Symmetry of the umbilic bifurcation sets with respect to the axes (a, b, c) defined by eqn (B.74). Solid curves indicate real $(b+c)$ and dashed curves complex $(b+c)$. The points C and D denote junctions of the elliptic and hyperbolic sets.

nature of the contributing stationary points. For example, if the f_v are all real, Uzer and Child (1982) find that two of the solutions of (B.70) are of elliptic type (i.e. b given by (B.74) is imaginary), one with $\zeta_1 > 0$ (corresponding according to Section B.2 to three saddles plus a maximum), while the other has $\zeta_1 < 0$ (corresponding to three saddles plus a minimum). Finally, the third solution is of hyperbolic type (with b real) which implies two saddles, one maximum and one minimum. These are precisely the three possibilities listed for four real f_v values in Table B.3 and the choice between them may be resolved by reference to the signs of the P_v quantities, which together with the relative values of the f_v values determine the character of the stationary point. Uzer and Child (1982) explore the other possibilities, with complex f_v values, in a similar way.

The remaining problem is to determine the expansion coefficients $\{X_i\}$, which is done by examining the Hessian determinants on the two sides of (B.65) at the stationary points $\{x_v, y_v\}$ derived from (B.70); thus

$$\begin{aligned} \| \partial^2 f/\partial u \partial \nu \| \, [\partial(u,\nu)/\partial(x,y)]_v^2 &= \| \partial^2 \phi^{\mathrm{v}}/\partial x \partial y \| \\ &= 36 x_v y_v - \zeta_1^2. \end{aligned} \tag{B.75}$$

It follows on combining this result with (B.63) that

$$[(36 x_v y_v - \zeta_1^2) P_v]^{1/2} = 2\pi (X_0 + X_1 x_v y_v + X_2 x_v + X_3 y_v), \tag{B.76}$$

with x_v and y_v derived from (B.70) for the physically correct $(\zeta_l, \zeta_2, \zeta_3)$ combination. Equation (B.76), for $v = a, b, c, d$, are obviously readily inverted for the four coefficients $\{X_i\}$, and Uzer and Child (1982) find, as a practical confirmatory test, that the physically correct $(\zeta_1, \zeta_2, \zeta_3)$ combination invariably leads to real values of the X_i. Finally, Gray and Child (1984) show how possible ambiguities in contributions to the f_v of the form $(2n+1)\pi$ may be resolved. This latter ambiguity proves in fact to be the most troublesome problem in practical applications.

B.4 Non-generic uniform approximations: Bessel and harmonic approximations

The generic, catastrophe-based, uniform approximations were seen above to rely on reduction to integrals over an infinite range, whereas the physical context sometimes requires a finite range, normally from 0 to 2π. Frequently the phase variation outside the stationary phase region is sufficiently rapid to justify formal extension of the range, but in other situations it is vital to reflect the correct periodicity of the integrand (see, for example, Fig. 9.6, in which the uniform Airy approximation is clearly invalid). The following uniform Bessel approximation, due to Stine and Marcus (1973), is designed for just such situations. The uniform harmonic approximation, which follows, was developed by Hunt and Child (1978) to take advantage of flux conservation during scattering and spectroscopic processes.

To start with the Bessel approximation, suppose that the required integral takes the form

$$I(\alpha) = \int_0^{2\pi} g(\theta) \exp[\mathrm{i}f(\alpha;\theta]\mathrm{d}\theta, \tag{B.77}$$

where $f(\alpha;\theta)$ has two stationary points, taken when real as a minimum at θ_a and a maximum at θ_b. Suppose also that $f(\alpha;\theta)$ increases by $2m\pi$ over a cycle of θ and that the following classical probabilities are known:

$$P_a = 2\pi g_a^2/f_a'', \qquad P_b = -2\pi g_b^2/f_b''. \tag{B.78}$$

Stine and Marcus (1973) then approximate $I(\alpha)$ in terms of the $\{f_v\}$ and $\{P_v\}$ by using the identity

$$f[\alpha;\theta(t)] = A(\alpha) - \zeta(\alpha)\cos t - mt, \tag{B.79}$$

in order to transform from θ to t as the integration variable. Since coincidence between the stationary points on either side of (B.79) requires that

$$\begin{aligned} \theta_a &\longleftrightarrow t_a = \pi/2 - \arccos(m/\zeta) \\ \theta_b &\longleftrightarrow t_b = \pi/2 + \arccos(m/\zeta), \end{aligned} \tag{B.80}$$

it follows, on substitution in (B.79), that

$$\begin{aligned} f_b &= A - m\pi/2 - (\zeta^2 - m^2)^{1/2} + m\arccos(m/\zeta) \\ f_b &= A - m\pi/2 + (\zeta^2 - m^2)^{1/2} - m\arccos(m/\zeta). \end{aligned} \tag{B.81}$$

In other words

$$A = \frac{1}{2}(f_a + f_b) + m\pi/2,$$

while ζ is determined by the transcendental equation

$$(\zeta^2 - m^2)^{1/2} - m \arccos(m/\zeta) = \frac{1}{2}(f_b - f_a). \tag{B.82}$$

Equation (B.82) yields $\zeta > m$ if f_a and f_b are real (with $f_b > f_a$ by assumption), and the analytic continuation

$$(m^2 - \zeta^2)^{1/2} - m \operatorname{arccosh}(m/\zeta) = -\frac{1}{2}(f_b - f_a) \tag{B.83}$$

yields $\zeta < m$ when $f_a = f_b^*$.

The approximation is completed in the usual way, by introducing t as the integration variable in (B.74) to obtain

$$I(\alpha) = \int_0^{2\pi} g(\theta)(\mathrm{d}\theta/\mathrm{d}t) \exp[\mathrm{i}(A - \zeta \cos t - mt)]\mathrm{d}t, \tag{B.84}$$

with $\mathrm{d}\theta/\mathrm{d}t$ given at the stationary points t_v by the following second derivative identities derived from (B.79),

$$f_v''(\mathrm{d}\theta/\mathrm{d}t)^2 = \pm\,(\zeta^2 - m^2)^{1/2}, \tag{B.85}$$

with the upper and lower signs taken for $v = a$ and b respectively. The final step is to approximate

$$g[\theta(t)](\mathrm{d}\theta/\mathrm{d}t) \simeq X - Y \cos t \tag{B.86}$$

so that

$$I(\alpha) \simeq 2\pi \mathrm{e}^{\mathrm{i}\zeta}[X J_m(\zeta) - \mathrm{i}Y J_m'(\zeta)], \tag{B.87}$$

where $J_m(\zeta)$ is the Bessel function (Abramowitz and Stegun 1965) and $J_m'(\zeta)$ is its derivative. The required coefficients are obtained from (B.78), (B.80), (B.85) and (B.86) in the form

$$\begin{aligned} X &= (8\pi)^{-1/2}(P_a^{1/2} + P_b^{1/2})(\zeta^2 - m^2)^{1/4} \\ Y &= -(8\pi)^{-1/2}(P_a^{1/2} - P_b^{1/2})\zeta(\zeta^2 - m^2)^{-1/4}. \end{aligned} \tag{B.88}$$

In other words

$$\begin{aligned} I(\alpha) = (\pi/2)^{1/2}\mathrm{e}^{\mathrm{i}A}[&(P_a^{1/2} + P_b^{1/2})(\zeta^2 - m^2)^{1/4} J_m(\zeta) \\ &+ \mathrm{i}(P_a^{1/2} - P_b^{1/2})\zeta(\zeta^2 - m^2)^{-1/4} J_m'(\zeta)], \end{aligned} \tag{B.89}$$

which completes the Bessel approximation.

The following uniform harmonic form is applied in Sections 5.3 and 9.5 to situations in which the phase difference between two interfering trajectories is constrained by flux conservation to lie between 0 and $(2n+1)\pi$. The derivation makes use of a particular 'classical' integral representation for the nth harmonic oscillator wavefunction, given by eqns (C.62a) in the form

$$\psi_n(\zeta) = \frac{1}{2\pi}\int_{-\pi/2}^{\pi/2} \gamma_n(\zeta,\alpha)\exp[\mathrm{i}\phi_n(\alpha)]\mathrm{d}\alpha, \tag{B.90}$$

where

$$\gamma_n(\zeta,\alpha) = \begin{cases} C_n(2n+1)^{1/4}\sec^{1/2}\varphi & \text{for } n \text{ even} \\ C_n(2n+1)^{-1/4}\zeta\sec^{3/2}\varphi & \text{for } n \text{ odd} \end{cases} \tag{B.91}$$

$$\phi_n(\zeta,\alpha) = \frac{1}{2}\zeta^2\tan\alpha - \left(n+\frac{1}{2}\right)\alpha, \tag{B.92}$$

and the coefficients C_n differ by at most a few per cent from unity (see (C.66a)). Equation (B.90) is an exact representation, but it is convenient to recognize the equivalent form

$$\psi_n(\zeta) = \frac{1}{2\pi}\left(\int_{-\pi}^{-\pi/2} + \int_{\pi/2}^{\pi}\right)\gamma_n(\zeta,\alpha)\exp[\mathrm{i}\phi_n(\alpha)]\mathrm{d}\alpha, \tag{B.93}$$

which may be derived from (B.90) by the replacements $\alpha \to (\alpha \pm \pi)$ in the integrand. The reason for including this alternative lies in the form of the stationary phase condition

$$\zeta = \pm(2n+1)^{1/2}\cos\alpha, \tag{B.94}$$

which is of course common to both integrands, but it is convenient, for consistency with phase choices in the literature (Hunt and Child 1978; Child et al. 1983), to adopt the lower sign in (B.94), despite inconsistency with the convention in (4.21). In this case negative values of ζ are covered by the integration range in eqn (B.90) and positive values by that in (B.93). Given this sign choice, eqn (B.94) has two roots

$$\alpha_a = \arccos[-\zeta/(2n+1)^{1/2}] = -\alpha_b, \tag{B.95}$$

at which the phase term takes the values

$$\begin{aligned}\phi_a &= -\frac{1}{2}\zeta(2n+1-\zeta^2)^{1/2} - \left(n+\frac{1}{2}\right)\arccos[-\zeta/(2n+1)^{1/2}],\\ \phi_b &= \frac{1}{2}\zeta(2n+1-\zeta^2)^{1/2} + \left(n+\frac{1}{2}\right)\arccos[-\zeta/(2n+1)^{1/2}],\end{aligned} \tag{B.96}$$

respectively. Thus phase differences of 0 and $(2n+1)\pi$ imply ζ values of $-(2n+1)^{1/2}$ and $(2n+1)^{1/2}$ respectively.

The derivation of the uniform approximation proceeds in the usual way by comparing these phase properties with those in the reference integral. Thus, adopting the notation of (B.77) for the integral to be approximated, the first step is to use the identity

$$f(\alpha;\theta) = \phi_n(\zeta;\alpha) + A \tag{B.97}$$

to transform the integration variable from θ to α, with the parameters ζ and A determined by the validity of (B.97) at the stationary points; thus

$$\frac{1}{2}(f_b + f_a) = A, \tag{B.98}$$

and

$$\begin{aligned}\frac{1}{2}(f_b - f_a) &= \frac{1}{2}(\phi_b - \phi_a)\\ &= \frac{1}{2}\zeta(2n+1-\zeta^2)^{1/2} + \left(n+\frac{1}{2}\right)\arccos[-\zeta/(2n+1)^{1/2}],\end{aligned} \tag{B.99}$$

provided that $(f_b - f_a)$ is real and constrained to the range $[0, (2n+1)\pi]$. The resulting ζ value then lies between $\pm(2n+1)^{1/2}$. There are also two other possibilities: $(f_b - f_a)$ might be imaginary or complex with a real part of $(2n+1)\pi$. In the first case (B.99) goes over to

$$\frac{1}{2}(f_b - f_a) = -\frac{\mathrm{i}}{2}\zeta\left(\zeta^2 - 2n - 1\right)^{1/2} - \mathrm{i}\left(n+\frac{1}{2}\right)\ln\left(\frac{-\zeta + (\zeta^2 - 2n - 1)^{1/2}}{(2n+1)^{1/2}}\right), \tag{B.100}$$

with a solution in the range $\zeta < -(2n+1)^{1/2}$. The equivalent equation in the second case is

$$\frac{1}{2}(f_b - f_a) - \left(n+\frac{1}{2}\right)\pi = \frac{i}{2}\zeta(\zeta^2 - 2n - 1)^{1/2} + \mathrm{i}\left(n+\frac{1}{2}\right)\ln\left(\frac{\zeta - (\zeta^2 - 2n - 1)^{1/2}}{(2n+1)^{1/2}}\right), \tag{B.101}$$

with a solution in the range $\zeta > (2n+1)^{1/2}$.

Given these values of A and ζ, the integral in (B.74) reduces further, under the approximation

$$g(\theta)(\mathrm{d}\theta/\mathrm{d}\alpha) \simeq \frac{1}{2\pi}\gamma_n(\zeta,\alpha)(X + Y\zeta\tan\alpha) \tag{B.102}$$

to

$$\begin{aligned}I(\alpha) &= \int_{-\pi/2}^{\pi/2} g(\theta)(\mathrm{d}\theta/\mathrm{d}\alpha)\exp[\mathrm{i}\phi_n(\zeta,\alpha) + \mathrm{i}A]\mathrm{d}\alpha\\ &\simeq \frac{1}{2\pi}\int_{-\pi/2}^{\pi/2}\gamma_n(\zeta,\alpha)(X + Y\zeta\tan\alpha)\exp[\mathrm{i}\phi_n(\zeta,\alpha) + \mathrm{i}A]\mathrm{d}\alpha\\ &= X\psi_n(\zeta) - \mathrm{i}Y\{\psi_n'(\zeta) - \frac{1}{2}[1-(-1)^n]\zeta^{-1}\psi_n(\zeta)\},\end{aligned} \tag{B.103}$$

where the final term arises from the linear dependence of $\gamma_n(\zeta, \alpha)$ on ζ for odd n values.

The coefficients X and Y are determined in the usual way by inverting (B.102) at the stationary points with $\mathrm{d}\theta/\mathrm{d}x$ derived from the second derivative of (B.97), a procedure that may be verified to yield

$$\begin{aligned} X &= C_n(\pi/2)^{1/2}(P_a^{1/2} + P_b^{1/2})(2n + 1 - \zeta^2)^{1/4} \\ Y &= C_n(\pi/2)^{1/2}(P_a^{1/2} - P_b^{1/2})(2n + 1 - \zeta^2)^{-1/4}, \end{aligned} \tag{B.104}$$

where P_a and P_b are given by (B.78).

The final result therefore takes the form

$$\begin{aligned} I(\alpha) = C_n \mathrm{e}^{\mathrm{i}A}(\pi/2)^{1/2} &\Big[(P_a^{1/2} + P_b^{1/2}) \left(2n + 1 - \zeta^2\right)^{1/4} \psi_n(\zeta) \\ &-\mathrm{i}(P_a^{1/2} - P_b^{1/2}) \left(2n + 1 - \zeta^2\right)^{-1/4} \\ &\times\{\psi_n'(\zeta) - \frac{1}{2}[1 - (-1)^n]\zeta^{-1}\psi_n(\zeta)\}\Big]. \end{aligned} \tag{B.105}$$

Here $\psi_n(\zeta)$ is the harmonic oscillator wavefunction,

$$\psi_n(\zeta) = (\pi)^{-1/4}(2^n n!)^{-1/2} H_n(\zeta) \exp\left(-\frac{1}{2}\zeta^2\right), \tag{B.106}$$

where $H_n(\zeta)$ is the Hermite polynomial (Abramowitz and Stegun 1965). $\psi_n'(\zeta)$ is its derivative and the quantities A and ζ are given by eqns (B.98)–(B.101). Finally, the coefficients C_n, which are very close to unity, are given by (C.66a).

Several other non-generic uniform approximations are also available—in particular the Laguerre form due to Child and Hunt (1977) and a non-integer Bessel form due to Connor and Mayne (1979).

Appendix C
Transformations in classical and quantum mechanics

This appendix concerns the connection between classical canonical transformations and the corresponding unitary transformations in quantum mechanics, with particular emphasis on the role of the classical generator. The ideas, which stem from Van Vleck (1928), were later reviewed by Fock (1959) and Van der Waerden (1967), and most recently expounded in an elegant review by Miller (1974). The main ideas are introduced in Section C.1 and illustrated by reference to angle–action and energy–time systems in Section C.2, which includes certain results required elsewhere in the text. Section C.3 covers the theory of dynamical transformations along the lines of Miller's path integral approach to classical S matrix theory. Section C.4 applies similar ideas to the semiclassical Green's function. Finally, Section C.5 uses the theory to obtain uniform approximations to the Wigner $3j$ and $6j$ symbols, in a way that underlines their geometrical significance.

C.1 Classical and semiclassical transformations

A transformation $(q,\ p) \to (\tilde{q}, \tilde{p})$ is termed canonical in classical mechanics if the value of the Hamiltonian is preserved,

$$H(q,\ p) = \tilde{H}(\tilde{q}, \tilde{p}), \tag{C.1}$$

in such a way that Hamilton's equations apply in both systems:

$$\begin{aligned} \mathrm{d}q/\mathrm{d}t &= \partial H/\partial p, \quad \mathrm{d}p/\mathrm{d}t = -(\partial H/\partial q) \\ \mathrm{d}\tilde{q}/\mathrm{d}t &= \partial \tilde{H}/\partial \tilde{p}, \quad \mathrm{d}\tilde{p}/\mathrm{d}t = -(\partial \tilde{H}/\partial \tilde{q}), \end{aligned} \tag{C.2}$$

conditions that are readily verified to require a unit Jacobian,

$$\partial(\tilde{q}, \tilde{p})/\partial(q,\ p) = 1. \tag{C.3}$$

The systematic theory of such transformations (Goldstein 1980; Percival and Richards 1982) is developed in terms of one or other of four possible generating functions $F_1(q,\ \tilde{q})$, $F_2(q,\ \tilde{p})$, $F_3(p,\ \tilde{q})$ and $F_4(p,\ \tilde{p})$, each dependent on one old and one new variable, and with the remaining variables generated by the following partial derivative relations:

$$p = (\partial F_1/\partial q)_{\tilde{q}}, \qquad \tilde{p} = -(\partial F_1/\partial \tilde{q})_q, \tag{C.4a}$$
$$p = (\partial F_2/\partial q)_{\tilde{p}}, \qquad \tilde{q} = (\partial F_2/\partial \tilde{p})_q, \tag{C.4b}$$
$$q = -(\partial F_3/\partial p)_{\tilde{q}}, \qquad \tilde{p} = -(\partial F_3/\partial \tilde{q})_p, \tag{C.4c}$$
$$q = -(\partial F_4/\partial p)_{\tilde{p}}, \qquad \tilde{p} = (\partial F_4/\partial \tilde{p})_p. \tag{C.4d}$$

The equality between the mixed second derivatives of any of the F_i then automatically ensures the validity of (C.3), but the equations carry the awkward complication that, for example, (C.4a) yields mixed functions $p(q,\ \tilde{q})$ and $\tilde{p}(q,\ \tilde{q})$ that must be inverted in order to express $(\tilde{q},\ \tilde{p})$ in terms of $(q,\ p)$ or vice versa. The choice between the F_i in any context may be made for physical or mathematical convenience, because any one of the following interrelated set yields an equivalent transformation, as is readily confirmed by use of eqns (C.4a)–(C.4d):

$$F_2(q,\ \tilde{p}) = F_1(q,\ \tilde{q}) + \tilde{p}\tilde{q}, \tag{C.5a}$$

$$F_3(p,\ \tilde{q}) = F_1(q,\ \tilde{q}) - pq, \tag{C.5b}$$

$$F_4(p,\ \tilde{p}) = F_1(q,\ \tilde{q}) + \tilde{p}\tilde{q} - pq. \tag{C.5c}$$

For example, according to (C.4b) and (C.5a),

$$\begin{aligned} p &= (\partial F_2/\partial q)_{\tilde{p}} = (\partial F_1/\partial q)_{\tilde{q}} + (\partial F_1/\partial \tilde{q})_q(\partial \tilde{q}/\partial q)_p + \tilde{p}(\partial \tilde{q}/\partial q)_p \\ \tilde{q} &= (\partial F_2/\partial \tilde{p})_q = (\partial F_1/\partial \tilde{q})_q(\partial \tilde{q}/\partial \tilde{p})_q + \tilde{q} + \tilde{p}(\partial \tilde{q}/\partial \tilde{p})_q, \end{aligned} \tag{C.6}$$

which reduce to the identities

$$p = p, \quad \tilde{q} = \tilde{q} \tag{C.7}$$

after substitution from (C.4a) for the terms in F_1.

Equations (C.4a)–(C.4d) and (C.5a)–(C.5c) are the ingredients of classical canonical transformation theory. The next step is to show that the classical generators F_i may also provide foundations for the corresponding quantum mechanical unitary transformation functions U_i, at least in a semiclassical sense. Consider for example the transformation

$$\psi(q) = \int U_1(q,\ \tilde{q})\tilde{\psi}(\tilde{q})\mathrm{d}\tilde{q}, \tag{C.8}$$

where $U_1(q,\ \tilde{q})$ must be defined in such a way that (C.8) relates corresponding eigenfunctions of $H(q,\ -\mathrm{i}\hbar\partial/\partial q)$ and $\tilde{H}(\tilde{q},\ -\mathrm{i}\hbar\partial/\partial \tilde{q})$. (Note that if $\tilde{q}$ were an angle variable the operator $-\mathrm{i}\hbar\partial/\partial \tilde{q}$ would be modified by addition of a Maslov term, as in eqn (4.23)). Put in mathematical terms this requires that

$$H(q,\ -\mathrm{i}\hbar\partial/\partial q)\psi(q) = \int U_1(q,\ \tilde{q})\tilde{H}(\tilde{q},\ -\mathrm{i}\hbar\partial/\partial \tilde{q})\tilde{\psi}(\tilde{q})\mathrm{d}\tilde{q}, \tag{C.9}$$

which is the quantum mechanical analogue of (C.1). In other words, in the light of (C.8), $U_1(q,\ \tilde{q})$ must satisfy

$$\int H(q,\ -\mathrm{i}\hbar\partial/\partial q)U_1(q,\ \tilde{q})\tilde{\psi}(\tilde{q})\mathrm{d}\tilde{q} = \int U_1(q,\ \tilde{q})\tilde{H}(\tilde{q},\ -\mathrm{i}\hbar\partial/\partial\tilde{q})\tilde{\psi}(\tilde{q})\mathrm{d}\tilde{q}. \qquad \text{(C.10)}$$

It is shown below that a solution to order $\hbar^0$ may be expressed in the form

$$U_1(q,\ \tilde{q}) = A_1(q,\ \tilde{q})\exp[\mathrm{i}F_1(q,\ \tilde{q})/\hbar], \qquad \text{(C.11)}$$

where $F_1(q,\ \tilde{q})$ is the corresponding classical generator. To see this, note first, by virtue of (C.1), (C.4a) and (C.11), that, to order $\hbar^0$,

$$\begin{aligned} H(q,\ -\mathrm{i}\hbar\partial/\partial q)U_1(q,\ \tilde{q}) &= H(p,\ q)U_1(q,\ \tilde{q}) \\ &= \tilde{H}(\tilde{p},\ \tilde{q})U_1(q,\ \tilde{q}) \\ &= \tilde{H}(\tilde{q},\ \mathrm{i}\hbar\partial/\partial\tilde{q})U_1(q,\ \tilde{q}). \end{aligned} \qquad \text{(C.12)}$$

Secondly, provided the integrand takes the same value at both integration limits, it follows from (C.8), by repeated partial integration on the right-hand side of the equation below (again to order $\hbar^0$), that

$$\begin{aligned} H(q,\ -\mathrm{i}\hbar\partial/\partial q)\psi(q) &= \int [\tilde{H}(\tilde{q},\ \mathrm{i}\hbar\partial/\partial\tilde{q})U_1(q,\ \tilde{q})]\tilde{\psi}(\tilde{q})\mathrm{d}\tilde{q} \\ &= \int U_1(q,\ \tilde{q})\tilde{H}(\tilde{q},\ -\mathrm{i}\hbar\partial/\partial\tilde{q})\tilde{\psi}(\tilde{q})\mathrm{d}\tilde{q}, \end{aligned} \qquad \text{(C.13)}$$

which establishes the validity of (C.9).

Finally, the unitarity condition

$$\int_{-\infty}^{\infty} U_1(q,\ \tilde{q})U_1^*(q',\ \tilde{q})\mathrm{d}\tilde{q} = \delta(q - q') \qquad \text{(C.14)}$$

may be used to express the pre-exponent in $A_1(q,\ \tilde{q})$ in (C.11) in terms of partial derivatives of $F_1(q,\ \tilde{q})$. The existence of the delta function is well understood in the normal semiclassical sense that rapid oscillations in the full integrand will lead to complete cancellation except when $q' = q$; thus

$$\int_{-\infty}^{\infty} A_1(q,\ \tilde{q})A_1^*(q',\ \tilde{q})\exp\left(\frac{\mathrm{i}}{\hbar}[F_1(q,\ \tilde{q}) - F_1(q',\ \tilde{q})]\right)\mathrm{d}\tilde{q} = \delta(q' - q). \qquad \text{(C.15)}$$

All that remains is to fix the magnitude of $A_1(q,\ \tilde{q})$ by approximating

$$F_1(q,\ \tilde{q}) - F_1(q',\ \tilde{q}) \simeq (\partial F_1/\partial q)_{\tilde{q}}(q - q') \qquad \text{(C.16)}$$

and writing

$$p = (\partial F_1/\partial q)_{\tilde{q}}, \qquad \text{(C.17)}$$

as specified by (C.4a). The effect, provided that p varies monotonically with $\tilde{q}$, is that (C.15) may be expressed as

$$\int_{-\infty}^{\infty} |A_1(q,\ \tilde{q})|^2 (\partial\tilde{q}/\partial p)_{\tilde{q}} \exp\left(-\frac{\mathrm{i}}{\hbar} p(q - q')\right) \mathrm{d}p = \delta(q - q'). \tag{C.18}$$

It follows, by comparison with the standard form (Dirac 1958)

$$\int_{-\infty}^{\infty} \exp(\mathrm{i}xy)\mathrm{d}x = 2\pi\delta(y), \tag{C.19}$$

that

$$|A_1(q,\ \tilde{q})|^2 = \frac{1}{2\pi\hbar}\left(\frac{\partial\tilde{q}}{\partial p}\right)_q^{-1} = \frac{1}{2\pi\hbar}\left(\frac{\partial^2 F_1}{\partial q \partial\tilde{q}}\right). \tag{C.20}$$

Taken together with what will prove a convenient phase convention (Miller 1974), this means that the semiclassical unitary transform is given in terms of the corresponding classical generator by the equation

$$U_1(q,\ \tilde{q}) = \left[-\frac{1}{2\pi\mathrm{i}\hbar}\left(\frac{\partial^2 F_1}{\partial q\partial\tilde{q}}\right)\right]^{1/2} \exp\left(\frac{\mathrm{i}}{\hbar}F_1(q,\ \tilde{q})\right). \tag{C.21}$$

This assumes that $p(q,\ \tilde{q})$ is a monotonic function of $\tilde{q}$ at all values of q; otherwise one must sum over the various contributing branches.

Equation (C.21) applies to the transformation from one coordinate representation to another, but it is readily seen that the general structure is preserved even when one or both of the representations is a momentum one. To see this recall that $\tilde{\psi}(\tilde{q})$ is related to its momentum representative, $\tilde{\phi}(\tilde{p})$, by the Fourier transformation (Dirac 1958)

$$\tilde{\psi}(\tilde{q}) = \frac{1}{(2\pi\mathrm{i}\hbar)^{1/2}}\int_{-\infty}^{\infty} \exp(\mathrm{i}\tilde{p}\tilde{q}/\hbar)\tilde{\phi}(\tilde{p})\mathrm{d}\tilde{p}, \tag{C.22}$$

so that on substitution in (C.8)

$$\psi(q) = \int_{-\infty}^{\infty} U_2(q,\ \tilde{p})\tilde{\phi}(\tilde{p})\mathrm{d}\tilde{p}, \tag{C.23}$$

where

$$\begin{aligned} U_2(q,\ \tilde{p}) &= \frac{1}{(2\pi\mathrm{i}\hbar)^{1/2}}\int_{-\infty}^{\infty} U_1(q,\ \tilde{q})\exp(\mathrm{i}\tilde{p}\tilde{q}/\hbar)\mathrm{d}\tilde{q} \\ &= \frac{1}{(2\pi\mathrm{i}\hbar)^{1/2}}\int_{-\infty}^{\infty} A_1(q,\ \tilde{q})\exp\left(\frac{\mathrm{i}}{\hbar}[F_1(q,\ \tilde{q}) + \tilde{p}\tilde{q}]\right)\mathrm{d}\tilde{q}. \end{aligned} \tag{C.24}$$

Stationary phase reduction of the final integral now yields the analogue of (C.21), because by virtue of (C.4a) the stationarity condition

$$\partial F_1/\partial\tilde{q} = -\tilde{p} \tag{C.25}$$

yields a point $\tilde{q}(q,\ \tilde{p})$ consistent with the transformation previously generated by $F_1(q,\ \tilde{q})$. Moreover, the stationary value of the exponent

$$F_2(q,\ \tilde{p}) = F_1[q,\ \tilde{q}(q,\ \tilde{p})] + \tilde{p}\tilde{q}(q,\ \tilde{p}) \tag{C.26}$$

conforms exactly to the classical expression in (C.5a). Finally, on applying the standard quadratic approximation about $\tilde{q}(q, \tilde{p})$,

$$U_2(q,\ \tilde{p}) = A_2(q,\ \tilde{p}) \exp\left(\frac{\mathrm{i}}{\hbar} F_2(q,\ \tilde{p})\right), \tag{C.27}$$

where

$$\begin{aligned} A_2(q,\ \tilde{p}) &= (2\pi \mathrm{i}\hbar)^{-1/2} A_1[q,\ \tilde{q}(q,\ \tilde{p})][2\pi\hbar\mathrm{i}/(\partial^2 F_1/\partial\tilde{q}^2)]^{1/2} \\ &= \left(\frac{1}{2\pi\mathrm{i}\hbar}\frac{\partial^2 F_1/\partial q\partial\tilde{q}}{\partial^2 F_1/\partial\tilde{q}^2}\right)^{1/2} = \left[\frac{1}{2\pi\mathrm{i}\hbar}\left(\frac{\partial\tilde{q}}{\partial q}\right)_{\tilde{p}}\right]^{1/2} \\ &= \left(\frac{1}{2\pi\mathrm{i}\hbar}\frac{\partial^2 F_2}{\partial q\partial\tilde{p}}\right)^{1/2}. \end{aligned} \tag{C.28}$$

Equations (C.4a) and (C.4b) are required for this final reduction.

Similar analysis may be applied for any desired form of transformation, the general semiclassical result being that the unitary transform $U_i(x,\ \tilde{y})$, where x and y denote q or p, is related to the corresponding classical generator $F_i(x,\ \tilde{y})$ in the form

$$U_i(x,\ \tilde{y}) = \left[\pm\frac{1}{2\pi\mathrm{i}\hbar}\left(\frac{\partial^2 F_i}{\partial x\partial\tilde{y}}\right)\right]^{1/2} \exp\left(\frac{\mathrm{i}}{\hbar}F_i(x,\ \tilde{y})\right), \tag{C.29}$$

where the sign in the pre-exponent is positive for $i = 2,\ 3$ and negative for $i = 1, 4$ (Miller 1974). The generalization to f degrees of freedom takes the form

$$U_i(\mathbf{x},\ \tilde{\boldsymbol{y}}) = \left[\left(\frac{1}{2\pi\mathrm{i}\hbar}\right)^f \left\|\pm\frac{\partial^2 F_i}{\partial x\partial\tilde{y}}\right\|\right]^{1/2} \exp\left(\frac{\mathrm{i}}{\hbar}F_i(\mathbf{x},\ \tilde{\boldsymbol{y}})\right), \tag{C.30}$$

where $\|\partial^2 F_i/\partial x\partial\tilde{y}\|$ denotes the Van Vleck or Hessian determinant of the second derivative matrix.

C.2 Energy–time and angle–action representations

An illuminating, yet simple, illustration of the theory is provided by the transformation between Cartesian $(P,\ R)$ and energy–time $(E,\ t)$ descriptions of free motion. The starting point is the exact coordinate representation of the flux normalized state,

$$\psi(R) = (P/m)^{-1/2} \exp(\mathrm{i}PR/\hbar), \tag{C.31}$$

which is transformed to the energy–time picture by an $F_1(R,\ t)$ generator, which is itself obtained from an intermediate $F_2(R,\ E)$ generator determined by the Hamilton–Jacobi equation

$$H\left[\left(\frac{\partial F_2}{\partial R}\right)_E,\ R\right] = \frac{1}{2m}\left(\frac{\partial F_2}{\partial R}\right)_E^2 = E. \tag{C.32}$$

The solution

$$F_2(R,\ E) = (2mE)^{1/2}R \tag{C.33}$$

shows correctly that

$$\begin{aligned} P &= (\partial F_2/\partial R)_E = (2mE)^{1/2}, \\ t &= (\partial F_2/\partial E)_R = (m/2E)^{1/2}R = mR/P. \end{aligned} \tag{C.34}$$

The corresponding $F_1(R, t)$ generator is given according to (C.5a) by

$$F_1(R,\ t) = F_2[R,\ E(R,\ t)] - E(R,\ t)t = mR^2/2t, \tag{C.35}$$

after some rearrangement in the light of (C.34). The required semiclassical unitary transform is therefore given by

$$U_1(R,\ t) = \left(\frac{mR}{2\pi \mathrm{i}\hbar t^2}\right)^{1/2} \exp(\mathrm{i}mR^2/2\hbar t). \tag{C.36}$$

Consequently $\psi(R)$ transforms to

$$\varphi(t) = \int_{-\infty}^{\infty} \left(\frac{m^2R}{2\pi \mathrm{i}\hbar Pt^2}\right)^{1/2} \exp\left(\frac{\mathrm{i}mR^2}{2\hbar t} + \frac{\mathrm{i}PR}{\hbar}\right) \mathrm{d}R, \tag{C.37}$$

which reduces at the stationary phase level to

$$\varphi(t) = \exp(-\mathrm{i}Et/\hbar), \tag{C.38}$$

where $E = P^2/2m$.

The resultant form is exact in this simple case. It is also evident that the form for $\varphi(t)$ in (C.38) must apply to any system with one degree of freedom; hence it is relatively uninformative as it stands. However, the reverse transformation

$$\psi(R) = \int_{-\infty}^{\infty} U_1(R,\ t) \exp(-\mathrm{i}Et/\hbar)\mathrm{d}t \tag{C.39}$$

offers a convenient classically based integral representation for $\psi(R)$ in certain contexts. The harmonic oscillator representation given below is an angle–action (rather than energy–time) version of this idea and similar energy–time forms for the harmonic oscillator and for quadratic barrier passage may be found in Feynman and Hibbs (1965), where they are derived by path integral methods.

It is also useful to extend the above argument, for the free-motion case, to situations in which there is an additional internal degree of freedom, with Hamiltonian $H_0(I)$, because the resulting analogue of (C.38) is required in Chapter 10. One might, for example, start with the mixed representation

$$\Psi(R,\ n_0) = (P/m)^{-1/2}\exp(\mathrm{i}PR/\hbar)\delta(I - I_0) \tag{C.40}$$

which describes a state with unit translational flux and a particular action I_0 corresponding to the quantum number

$$n_0 = I_0/\hbar - \delta, \tag{C.41}$$

where δ is the usual Maslov index. Since the action operator is given by $\hat{I} = -\mathrm{i}\hbar(\partial/\partial\alpha) + \delta\hbar$ (see Sections 3.1 and 4.1), the Hamilton–Jacobi equation for the generator $F_2(R,\ \alpha;\ E,\ I)$ takes the form

$$\frac{1}{2m}\left(\frac{\partial F_2}{\partial R}\right)^2 + H_0\left[\frac{\partial F_2}{\partial \alpha} + \delta\hbar\right] = E, \tag{C.42}$$

and the solution is readily found to be

$$F_2(R,\ \alpha;\ E,\ I) = (I - \delta\hbar)\alpha + P(E,\ I)R, \tag{C.43}$$

where

$$P(E,\ I) = \{2m[E - H_0(I)]\}^{1/2}. \tag{C.44}$$

Notice, on applying (C.5a), that the variables conjugate to E and I in the transformed representation take the interesting forms

$$\begin{aligned}\tau &= \partial F_2/\partial E = mR/P(E,\ I)\\ \bar{\alpha} &= \partial F_2/\partial I = \alpha - (\partial H_0/\partial I)t = \alpha - \omega\tau,\end{aligned} \tag{C.45}$$

where ω is the frequency of the internal motion. The conjugate to E is of course the time (now denoted τ for later convenience), but the conjugate to I is no longer α but the modified variable $\bar{\alpha}$, which is a constant of the motion in this non-interacting picture because, by construction, α varies with time as $\alpha = \omega\tau +$ constant (see Section 4.1). The remaining step in the transformation of (C.40) is to define the mixed generator $F(R,\ I;\ \tau,\ \bar{\alpha})$ which may be reduced with the help of (C.45) to

$$\begin{aligned}F(R,\ I;\ \tau,\ \bar{\alpha}) &= F_2(R,\ \alpha;\ E,\ I) - (I - \delta\hbar)(\alpha - \bar{\alpha}) - E\tau\\ &= \frac{mR^2}{2\tau} + (I - \delta\hbar)\bar{\alpha} - H_0(I)\tau.\end{aligned} \tag{C.46}$$

The form of the associated unitary transform,

$$U(R,\ I;\ \tau,\ \bar{\alpha}) = \left(\frac{1}{2\pi\mathrm{i}\hbar}\right)\left(\left\|\frac{\partial^2 F}{\partial R\partial t}\right\|\right)^{1/2}\exp[\mathrm{i}F(R,\ I;\ \tau,\ \bar{\alpha})/\hbar], \tag{C.47}$$

then implies, after some manipulation, that

$$\varphi(\tau,\ \bar{\alpha}) = \int_{-\infty}^{\infty}\int_{-\infty}^{\infty} U(R,\ I;\ \tau,\ \bar{\alpha})\Psi(R,\ n_0)\mathrm{d}R\mathrm{d}I$$
$$= \frac{1}{(2\pi)^{1/2}}\exp(\mathrm{i}n_0\bar{\alpha} - \mathrm{i}E\tau/\hbar), \tag{C.48}$$

which is the result employed (as eqn (10.24)) in Chapter 10.

The scaled harmonic oscillator, with Hamiltonian

$$H = \frac{1}{2}(p^2 + q^2) = n + \frac{1}{2}, \tag{C.49}$$

offers further insights into the scope of the transformation theory. Here the quantum number n rather than the action, $I = n + \frac{1}{2}$, is conveniently taken as the variable conjugate to the angle α, because according to eqn (4.23) its operator equivalent,

$$\hat{n} = \hat{I} - \frac{1}{2} = -\mathrm{i}\partial/\partial\alpha, \tag{C.50}$$

has the form assumed in eqns (C.9)–(C.13) ($\hbar = 1$ in the present units).

Following the above Hamilton–Jacobi route (eqns (C.33)–(C.35)) one readily finds that

$$F_1(q,\ \alpha) = F_2[q,\ n(q,\ \alpha)] - n(q,\ \alpha)\alpha$$
$$= -\frac{1}{2}q^2\tan\alpha + \frac{1}{2}\alpha. \tag{C.51}$$

Hence, following (C.4a),

$$p = \partial F_1/\partial q = -q\tan\alpha \tag{C.52}$$
$$n = -(\partial F_1/\partial\alpha) = \frac{1}{2}q^2\sec^2\alpha - \frac{1}{2},$$

which rearrange to the canonical form (see (4.21))

$$q = (2n+1)^{1/2}\cos\alpha, \quad p = -(2n+1)^{1/2}\sin\alpha. \tag{C.53}$$

The analogue of eqns (C.8) and (C.10) therefore suggests a representation for the Cartesian wavefunction $\psi_n(q)$ of the form

$$\psi_n(q) = \int U_1(q,\ \alpha)\varphi_n(\alpha)\mathrm{d}\alpha, \tag{C.54}$$

where

$$U_1(q,\ \alpha) = A_1(q,\ \alpha)\exp\left(-\frac{\mathrm{i}}{2}q^2\tan\alpha + \frac{\mathrm{i}}{2}\alpha\right) \tag{C.55}$$

and, according to (4.25),

$$\varphi_n(\alpha) = (2\pi)^{-1/2}\exp(\mathrm{i}n\alpha). \tag{C.56}$$

The integration limits in (C.54) will be chosen later. Turning to the function $A_1(q,\ \alpha)$, the semiclassical form,

$$A_1^{\text{sc}} = \left(-\frac{1}{2\pi\mathrm{i}}\frac{\partial^2 F_1}{\partial q \partial \alpha}\right)^{1/2} = \left(\frac{q\sec^2\alpha}{2\pi\mathrm{i}}\right)^{1/2}, \tag{C.57}$$

cannot be exact because $\psi_n(q)$ must be either even or odd in q, for even or odd values of n. One can, however, obtain an exact representation in this special case because eqn (C.12), which takes the form

$$\left(-\frac{1}{2}\frac{\partial^2}{\partial q^2} + \frac{1}{2}q^2\right) U_1(q,\ \alpha) = \left(\mathrm{i}\frac{\partial}{\partial q} + \frac{1}{2}\right) U_1(q,\ \alpha), \tag{C.58}$$

can be solved exactly by substitution from (C.55) to yield

$$\frac{\mathrm{i}}{2}\tan\alpha A + \mathrm{i}q\tan\alpha\left(\frac{\partial A}{\partial q}\right) - \frac{1}{2}\left(\frac{\partial^2 A}{\partial q^2}\right) = \mathrm{i}\left(\frac{\partial A}{\partial \alpha}\right). \tag{C.59}$$

Two fundamental solutions may be recognized:

$$\begin{aligned} A(q,\ \alpha) &= B_n \sec^{1/2}\alpha \quad && \text{for } n = 2k \\ &= B_n q \sec^{3/2}\alpha \quad && \text{for } n = 2k+1; \end{aligned} \tag{C.60}$$

other forms involving higher powers of q may be reduced to these by integration by parts in (C.54). One interesting feature is that the coefficient choices

$$\begin{aligned} B_n &= (2\pi)^{-1/2}(2n+1)^{1/4} \quad && \text{for } n = 2k \\ &= (2\pi)^{-1/2}(2n+1)^{-1/4} \quad && \text{for } n = 2k+1 \end{aligned} \tag{C.61}$$

bring (C.60) into coincidence with (C.57) (apart from a phase factor) in the semiclassical sense that q and α are related by (C.53).

The following integral representations are therefore suggested for $\psi_n(q)$:

$$\psi_n(q) = C_n \frac{(2n+1)^{1/4}}{2\pi}\int_{-\pi/2}^{\pi/2} \sec^{1/2}\alpha \exp\left(-\frac{\mathrm{i}}{2}q^2\tan\alpha + \mathrm{i}\left(n+\frac{1}{2}\right)\alpha\right)\mathrm{d}\alpha \tag{C.62a}$$

for n even, and

$$\psi_n(q) = C_n \frac{q}{2\pi(2n+1)^{1/4}}\int_{-\pi/2}^{\pi/2} \sec^{3/2}\alpha \exp\left(-\frac{\mathrm{i}}{2}q^2\tan\alpha + \mathrm{i}\left(n+\frac{1}{2}\right)\alpha\right)\mathrm{d}\alpha \tag{C.62b}$$

for n odd. Here the integration limits have been chosen to ensure that $\psi_n(q)$ is real (as may be verified by the substitution $\alpha' = -\alpha$) and such that $\sec^{1/2}\alpha$ is real over the integration range.

Equations (C.62a) and (C.62b) have been expressed in this form to emphasize their semiclassical origin, and also because the as yet undetermined factors C_n are close to unity. To see the latter note that the standard form

$$\psi_n(q) = \pi^{-1/4}(2^n n!)^{-1/2} H_n(q) \exp\left(-\frac{1}{2}q^2\right) \tag{C.63}$$

for even n may be shown to take the following value at $q = 0$:

$$\psi_n(0) = (-1)^{n/2}\pi^{-1/4}2^{-n/2}[\Gamma(n+1)]^{1/2}/\Gamma\left(\frac{n}{2}+1\right), \tag{C.64}$$

where $\Gamma(x)$ is the gamma function (Abramowitz and Stegun 1965), while the integral in (C.62a) at $q = 0$ gives (Gradsteyn and Ryyzhik 1980)

$$2\int_0^{\pi/2} \sec^{1/2}\alpha \cos\left(n+\frac{1}{2}\right)\alpha \, d\alpha = 2^{3/2}\pi\Gamma\left(\frac{3}{2}\right)\Big/\Gamma\left(\frac{n}{2}+1\right)\Gamma\left(\frac{1-n}{2}\right). \tag{C.65}$$

It follows by use of the gamma function duplication formula (Abramowitz and Stegun 1965) that

$$C_n = (2n+1)^{-1/4}\left[2\Gamma\left(\frac{n}{2}+1\right)\Big/\Gamma\left(\frac{n+1}{2}\right)\right]^{1/2} \quad \text{for } n = 2k, \tag{C.66a}$$

with the numerical values $C_0 = 1.0623$, $C_2 = 1.0046$, $C_4 = 1.0015$ and $C_n \to 1$ as $n \to \infty$. A similar comparison for $\psi_n'(0)$ leads to

$$C_n = (2n+1)^{-1/4}\left[\Gamma\left(\frac{n+1}{2}\right)\Big/2\Gamma\left(\frac{n}{2}+1\right)\right]^{1/2} \quad \text{for } n = 2k+1, \tag{C.66b}$$

with the values $C_1 = 0.9885$, $C_3 = 0.9976$, $C_5 = 0.9980$, etc.

This classically based integral representation has close analogies with forms given by Feynman and Hibbs (1965), Ovchinnikova (1974) and Boyer and Wolf (1975). It also provides a direct route to the harmonic uniform approximation discussed in Section B.4.

C.3 Dynamical transformations and the classical *S* matrix

Miller (1974) and coworkers have made elegant use of the foregoing theory in deriving semiclassical transition amplitudes, or S matrix elements, from the classical limit of the Feynman propagator (Feynman and Hibbs 1965). The underlying idea is that passage from a phase point (p_1, q_1) at time t_1 to (p_2, q_2) at t_2 constitutes a dynamical transformation whose classical generator determines the quantum mechanical propagator.

The simplest illustration applies to the case of free motion, for which it is readily verified that the generator

$$F_1(q_2t_2;\, q_1t_1) = \frac{m(q_2-q_1)^2}{2(t_2-t_1)} \tag{C.67}$$

yields the correct momenta:

$$p_2 = \partial F_1/\partial q_2 = p_1 = -(\partial F_1/\partial q_1) = \frac{m(q_2 - q_1)}{(t_2 - t_1)}. \tag{C.68}$$

The corresponding unitary transform, denoted for consistency with what follows by

$$K(q_2t_2;\ q_1t_1) = \left(-\frac{1}{2\pi \mathrm{i}\hbar}\frac{\partial^2 F_1}{\partial q_1 \partial q_2}\right)^{1/2} \exp[\mathrm{i}F_1(q_2t_2;\ q_1t_1)/\hbar], \tag{C.69}$$

may be confirmed to propagate the wavefunction from time t_1 to t_2 in the sense that (Feynman and Hibbs 1965)

$$\psi(q_2t_2) = \int_{-\infty}^{\infty} K(q_2t_2;\ q_1t_1)\psi(q_1t_1)\mathrm{d}q_1. \tag{C.70}$$

To see this, note that the exact initial free-motion wavefunction, at energy $E = p^2/2m$, is

$$\psi(q_1t_1) = \exp[\mathrm{i}(pq_1 - Et_1)/\hbar], \tag{C.71}$$

so that on using the standard integral

$$\int_{-\infty}^{\infty} \exp(\mathrm{i}ax^2)\mathrm{d}x = (2\pi \mathrm{i}/a)^{1/2}, \tag{C.72}$$

(C.70) yields an identical form to (C.71) except that the subscript 2 appears in place of 1. Equations (C.67)–(C.71) are exact for this simple case.

The corresponding exact propagator in more general situations may be expressed as a path integral (Feynman and Hibbs 1965),

$$K(q_2t_2;\ q_1t_1) = \int \exp\{\mathrm{i}S[q(t)]/\hbar\}\mathrm{d}\text{path}, \tag{C.73}$$

where S is the action along a particular path,

$$S[q(t)] = \int_{-\infty}^{\infty} \{p(t)\dot{q}(t) - H[p(t),\ q(t)]\}\mathrm{d}t, \tag{C.74}$$

and the integral in (C.73) is taken over all possible phase space paths with end points q_1 at t_1 and q_2 at t_2. A convenient way to approximate S for a given path is to break it into short time segments of length Δt_i along which the momentum p_i consistent with H at a given energy may be taken as constant; hence, following (C.67)–(C.68) the ith action increment becomes

$$\begin{aligned} \Delta S_i &= F_1(q_2t_2;\ q_1t_1) - V_i\Delta t_i \\ &= m\Delta q_i^2/2\Delta t_i - V_i\Delta t_i, \end{aligned} \tag{C.75}$$

where $m\Delta q_i = p_i\Delta t_i$.

Such expedients are, however, unnecessary in the semiclassical treatment of small isolated species, because the fluctuations in $S[q(t)]$ from one path to another may be assumed to be so large that constructive interference occurs only around the paths of minimum action, which are by Hamilton's principle those traced out by the classical trajectories from q_1 at t_1 to q_2 at t_2 (Goldstein 1980; Percival and Richards 1982). Hence in the semiclassical limit (Feynman and Hibbs 1965)

$$K(q_2t_2;\ q_1t_1) \simeq \left(-\frac{1}{2\pi i\hbar}\frac{\partial^2 S_{\rm cl}}{\partial q_1 \partial q_2}\right)^{1/2} \exp[{\rm i}S_{\rm cl}(q_2t_2;\ q_1t_1)/\hbar], \tag{C.76}$$

where

$$S_{\rm cl}(q_2t_2;\ q_1t_1) = \int_{t_1}^{t_2} [p\dot{q} - H(p,\ q)]{\rm d}t, \tag{C.77}$$

with the integral taken along a classical trajectory. Note that the pre-exponent in (C.76), which includes contributions from the immediately neighbouring paths, is fixed by unitarity (compare eqns (C.14)–(C.20)). The sum of terms of the Cartesian form in eqn (C.76), taken over all 'root trajectories' from q_1 to q_2 in time $t_2 - t_1$, is known is the Van Vleck (1928) propagator.

Henceforth it is assumed that the system is conservative (i.e. that H contains no explicit time dependence), in which case terms in H (or E) may be factored out of the multidimensional analogue of (C.76) in the form

$$\begin{aligned}\kappa(\mathbf{q}_2;\ \mathbf{q}_1) &= K(\mathbf{q}_2t_2;\ \mathbf{q}_1t_1)\exp[{\rm i}E(t_2 - t_1)/\hbar] \\ &= a(\mathbf{q}_2;\ \mathbf{q}_1)\exp[{\rm i}W(\mathbf{q}_2;\ \mathbf{q}_1)/\hbar],\end{aligned} \tag{C.78}$$

where

$$W(\mathbf{q}_2;\ \mathbf{q}_1) = \int_{\mathbf{q}_1}^{\mathbf{q}_2} \mathbf{p}\cdot{\rm d}\mathbf{q}, \tag{C.79}$$

with the integral taken along a trajectory from $\mathbf{q}_1$ to $\mathbf{q}_2$. This reduced propagator governs the evolution of the spatial part of the wavefunction in the form

$$\psi(\mathbf{q}_2) = \int_{-\infty}^{\infty} \kappa(\mathbf{q}_2;\ \mathbf{q}_1)\psi(\mathbf{q}_1){\rm d}\mathbf{q}_1. \tag{C.80}$$

Since the trajectory used to determine $\kappa(\mathbf{q}_1;\ \mathbf{q}_2)$ assumedly depends on $\mathbf{q}_1$ and $\mathbf{q}_2$, the initial and final momenta, generated by the equations

$$p_{1i}(\mathbf{q}_2;\ \mathbf{q}_1) = -(\partial W/\partial q_{1i}), \quad p_{2i}(\mathbf{q}_2;\ \mathbf{q}_1) = (\partial W/\partial q_{2i}), \tag{C.81}$$

are also strictly dependent on $\mathbf{q}_2$ and $\mathbf{q}_1$. Note also that the double-ended nature of the boundary conditions may allow more than one classical path from $\mathbf{q}_1$ to $\mathbf{q}_2$, in which case (C.78) must be replaced by an appropriate sum.

We turn now to the relation between the propagator and the scattering matrix in collisional applications of the theory. Miller (1974) argues that an observable transition from one quantum state $\mathbf{n}_1$ to another $\mathbf{n}_2$ involves not a coordinate change, but a change of action from $\mathbf{I}_1$ to $\mathbf{I}_2$. Hence the S matrix depends on a propagator $\kappa(\mathbf{I}_2;\ \mathbf{I}_1)$ in an angle–action representation. Not surprisingly, if the transformations are followed through at the stationary phase level, the result is the primitive semiclassical approximation of eqn (9.37). Here, however, the aim is to obtain the integral representation of eqn (9.28), from which other results are deduced in Section 10.2.

It is assumed for simplicity that the system involves translational variables $(P,\ R)$ and a single set of internal variables given in either angle–action $(I,\ \alpha)$ or Cartesian $(p,\ x)$ form. It is also useful to remember, according to the discussion in Section C.2, that the semiclassical conjugate to α is not I but

$$N = I - \delta\hbar = n\hbar, \tag{C.82}$$

where δ is the Maslov index and n the quantum number.

Before proceeding to the details, it is important to establish the dependence of these variables upon one another. For example, in considering a trajectory from asymptotic variables $(P_1,\ R_1,\ N_1,\ \alpha_1)$ to $(P_2,\ R_2,\ N_2,\ \alpha_2)$ the final quantities P_2 and N_2 are dependent on the choice of all four initial variables, R_2 may be chosen arbitrarily, and α_2 depends on R_2 as well as the initial set, because (see eqn (4.5))

$$\alpha_i = \omega(N_i)t + \text{constant}, \tag{C.83}$$

where $\omega(N_i) = \partial H_0/\partial N_i$, and a different value of R_2 implies a different final time. Closer inspection shows that P_2 and N_2 are also interdependent because

$$E = \frac{1}{2m}P^2 + H_0(I), \tag{C.84}$$

and the same is true of P_1 and N_1 if the energy E is taken as an independent variable. Finally, it is convenient to accommodate the interdependence of α_i and R_i by defining the modified angle

$$\bar{\alpha}_i = \alpha_i + \omega(N_i)mR_i/P_i = \alpha_i + \omega(N_i)\tau_i, \tag{C.85}$$

which is in effect the constant in eqn (C.83). The upshot is that the variables $(N,\ \bar{\alpha}, E,\ \tau)$, which were introduced earlier in eqns (C.42)–(C.48), are the most convenient for conceptual purposes, because E is fixed and the outcome of the trajectory is independent of the τ_i, provided the motion starts and ends in an asymptotic region. The essential dependence is therefore either $(N_2, \bar{\alpha}_2)$ on $(N_1,\ \bar{\alpha}_1)$, or $(N_1,\ N_2)$ on $(\bar{\alpha}_1,\ \bar{\alpha}_2)$, or vice versa. The aim of what follows is to transform from the physically convenient practical variables, used to determine the trajectory, to the conceptual $(E,\ \tau,\ N,\ \bar{\alpha})$ system, and then to relate the S matrix to the propagator $\kappa(\bar{\alpha}_2;\ \bar{\alpha}_1)$. The necessary transformation steps will be followed via the classical generators.

Two possibilities are considered, according to whether the motion is followed in the $(P,\ R,\ N,\ \alpha)$ or $(P,\ R,\ p,\ x)$ system. In the first case the dynamical transformation itself is induced by the F_1 type generator

$$W(\alpha_2 R_2;\ \alpha_1 R_1) = \int_{\alpha_1}^{\alpha_2} N\mathrm{d}\alpha + \int_{R_1}^{R_2} P\mathrm{d}R, \tag{C.86}$$

and in the second case

$$W(x_2 R_2;\ x_1 R_1) = \int_{x_1}^{x_2} p\mathrm{d}x + \int_{R_1}^{R_2} P\mathrm{d}R, \tag{C.87}$$

with the integrals taken along classical trajectories.

The easiest route to the corresponding $(\tau_1,\ \bar{\alpha}_1)$ to $(\tau_2,\ \bar{\alpha}_2)$ generator involves an intermediate transformation to the $(N,\ R)$ representation, using as the generator either

$$\begin{aligned}\rho(N_2 R_2;\ N_1 R_1) &= W(\alpha_2 R_2; \alpha_1 R_1) - N_2\alpha_2 + N_1\alpha_1 \\ &= -\int_{N_1}^{N_2} \alpha\mathrm{d}N + \int_{R_1}^{R_2} P\mathrm{d}R\end{aligned} \tag{C.88}$$

or

$$\tilde{\rho}(N_2 R_2;\ N_1 R_1) = W(x_2 R_2; x_1 R_1) - F_2(x_2,\ N_2) + F_2(x_1,\ N_1), \tag{C.89}$$

where $F_2(x,\ N)$ is the type 2 generator from $(p,\ x)$ to $(N,\ \alpha)$, given according to eqn (4.12) by

$$\begin{aligned}F_2(x,\ N) &= \int_a^x \{2\mu[H_0(I) - V_{\text{int}}(x)]\}^{1/2}\mathrm{d}x, \qquad 0 < \alpha < \pi \\ &= 2\pi I - \int_a^x \{2\mu[H_0(I) - V_{\text{int}}(x)]\}^{1/2}\mathrm{d}x \quad \pi < \alpha < 2\pi.\end{aligned} \tag{C.90}$$

It is readily verified from (C.89) that

$$\begin{aligned}\partial\rho/\partial N_2 &= -\alpha_2, \quad \partial\rho/\partial N_1 = \alpha_1, \\ \partial\rho/\partial R_2 &= P_2, \quad \partial\rho/\partial R_1 = -P_1,\end{aligned} \tag{C.91}$$

and similarly for $\tilde{p}$ because by construction

$$\partial F_2/\partial x = p, \quad \partial F_2/\partial N = \alpha. \tag{C.92}$$

The next step is to transform to the $(\bar{\alpha},\ \tau)$ representation with the help of eqns (C.42)–(C.48). The resulting generator takes the form

$$\begin{aligned}\lambda(\bar{\alpha}_2,\ \tau_2;\ \bar{\alpha}_1\tau_1) = \rho(N_2 R_2;\ N_1 R_1) \\ -G(N_2 R_2;\ \bar{\alpha}_2\tau_2) + G(N_1 R_1;\ \bar{\alpha}_1\tau_1),\end{aligned} \tag{C.93}$$

where

$$G(NR;\ \bar{\alpha}\tau) = \frac{mR^2}{2\tau} - H_0(I)\tau - N\bar{\alpha}; \tag{C.94}$$

alternatively, $\tilde{\rho}(N_2R_2;\ N_1R_1)$ may be employed in place of $\rho(N_2R_2;\ N_1R_1)$. Finally, (C.93) may be cast into a more appealing form by noting that the variables α_i and P_i,

$$\begin{aligned} \alpha_i &= -(\partial G/\partial N_i) = \bar{\alpha}_i + (\partial H_0/\partial N_i)\tau_i = \bar{\alpha}_i + \omega(N_i)\tau_i \\ P_i &= \partial G/\partial R_i = mR_i/\tau_i, \end{aligned} \tag{C.95}$$

implied by (C.94) are necessarily consistent with those given by (C.91). Hence, on combining (C.83) with (C.88)–(C.95),

$$\lambda(\bar{\alpha}_2\tau_2;\ \bar{\alpha}_1\tau_1) = -\int_{N_1}^{N_2} \alpha \mathrm{d}N - \int_{P_1}^{P_2} R\mathrm{d}P + N_2\bar{\alpha}_2 - N_1\bar{\alpha}_1 + E(\tau_2 - \tau_1). \tag{C.96}$$

Note that although the N_i and P_i are dependent on the $\bar{\alpha}_i$, the function $\lambda(\bar{\alpha}_2\tau_2;\ \bar{\alpha}_1\tau_1)$ generates the proper conjugates to $\bar{\alpha}_i$ and τ_i in the form

$$\begin{aligned} \partial\lambda/\partial\bar{\alpha}_2 &= N_2, \quad \partial\lambda/\partial\bar{\alpha}_1 = -N_1, \\ \partial\lambda/\partial\tau_2 &= -(\partial\lambda/\partial\tau_1) = E, \end{aligned} \tag{C.97}$$

because, by virtue of (C.84) and (C.95), the terms in $\partial N_i/\partial\bar{\alpha}_j$ cancel exactly with those in $\partial P_i/\partial\bar{\alpha}_j$.

This function $\lambda(\bar{\alpha}_2\tau_2;\ \bar{\alpha}_1\tau_1)$ plays the role of $W(\mathbf{q}_2;\ \mathbf{q}_1)$ in (C.79); hence the semiclassical propagator takes the form

$$\kappa(\bar{\alpha}_2\tau_2;\ \bar{\alpha}_1\tau_1) = a(\bar{\alpha}_2\tau_2;\ \bar{\alpha}_1\tau_1)\exp[\mathrm{i}\lambda(\bar{\alpha}_2\tau_2;\ \bar{\alpha}_1\tau_1)/\hbar]. \tag{C.98}$$

Moreover, it is clear from the energy dependences of $\mathrm{i}\lambda(\bar{\alpha}_2\tau_2;\ \bar{\alpha}_1\tau_1)$ and of the typical asymptotic wavefunction given by (C.48)

$$\psi_{n_i}(\bar{\alpha}_i\tau_i) = (2\pi)^{-1/2}\exp[\mathrm{i}(N_i\bar{\alpha}_i - E\tau_i)/\hbar], \tag{C.99}$$

that propagation with respect to τ is purely multiplicative, as one might expect by analogy with (C.74). The propagation equation is therefore

$$\psi^{(n_1)}(\bar{\alpha}_2\tau_2) = \int_0^{2\pi} \kappa(\bar{\alpha}_2\tau_2;\ \bar{\alpha}_1\tau_1)\psi_{n_1}(\bar{\alpha}_1\tau_1)\mathrm{d}\bar{\alpha}_1, \tag{C.100}$$

where the superscript on the left-hand side implies propagation of ψ_{n_1}. It follows, by analogy with (C.76), that

$$\kappa(\bar{\alpha}_2\tau_2;\ \bar{\alpha}_1\tau_1) = \left(\frac{1}{2\pi\mathrm{i}\hbar}\frac{\partial^2\lambda}{\partial\bar{\alpha}_1\partial\bar{\alpha}_2}\right)^{1/2} \exp[\mathrm{i}\lambda(\bar{\alpha}_2\tau_2;\ \bar{\alpha}_1\tau_1)/\hbar]. \tag{C.101}$$

The relation between $\kappa(\bar{\alpha}_2\tau_2;\ \bar{\alpha}_1\tau_1)$ and the S matrix is now readily established by noting that $\psi^{(n_1)}(\bar{\alpha}_2\tau_2)$ must decompose as

$$\psi^{(n_1)}(\bar{\alpha}_2\tau_2) = \sum_{n_2} S_{n_1 n_2}\psi^{(n_2)}(\bar{\alpha}_2\tau_2), \tag{C.102}$$

so that on projecting out the n_2^0th term from (C.100)

$$\begin{aligned} S_{n_1^0 n_2^0} &= \int_0^{2\pi}\int_0^{2\pi} \psi^*_{n_2^0}(\bar{\alpha}_2\tau_2)\kappa(\bar{\alpha}_2\tau_2;\ \bar{\alpha}_1\tau_1)\psi_{n_1^0}(\bar{\alpha}_1\tau_1)\mathrm{d}\bar{\alpha}_1\mathrm{d}\bar{\alpha}_2 \\ &= \frac{1}{2\pi}\int_0^{2\pi}\int_0^{2\pi}\left[\frac{\mathrm{i}}{2\pi}\left(\frac{\partial n_1}{\partial \bar{\alpha}_2}\right)_{\bar{\alpha}_2}\right]^{1/2} \exp[\mathrm{i}\Lambda_{n_1^0 n_2^0}(\bar{\alpha}_1\bar{\alpha}_2)]\mathrm{d}\bar{\alpha}_1\mathrm{d}\bar{\alpha}_2, \end{aligned} \tag{C.103}$$

where

$$\Lambda_{n_1^0 n_2^0} = -\int_{n_1}^{n_2} \alpha \mathrm{d}n - \int_{k_1}^{k_2} R\mathrm{d}k + (n_2 - n_2^0)\bar{\alpha}_2 - (n_1 - n_1^0)\bar{\alpha}_1. \tag{C.104}$$

Note that, for ease of comparison with (9.29), N_i and P_i have been replaced by

$$n_i = N_i/\hbar \quad \text{and} \quad k_i = P_i/\hbar, \tag{C.105}$$

that the quantum numbers n_i^0 have been given the superscript to distinguish them from the n_i which are functions of the $\bar{\alpha}_i$ (as are the k_i), and that the mixed second derivative of λ in (C.101) has been evaluated with the help of (C.97).

Equation (C.104) is the most general integral form for the S matrix but it assumes knowledge of the trajectories from all initial to all final angles $\bar{\alpha}$ at the energy of interest. The following simpler form,

$$S_{n_1^0 n_2^0} = \frac{1}{2\pi}\int_0^{2\pi}\left(\frac{\partial \bar{\alpha}_1}{\partial \bar{\alpha}_2}\right)^{1/2}_{n_1} \exp[\mathrm{i}\Delta_{n_1^0 n_2^0}(\bar{\alpha}_2)]\mathrm{d}\bar{\alpha}_2, \tag{C.106}$$

where

$$\Delta_{n_1^0 n_2^0} = -\int_{n_1}^{n_2} \alpha \mathrm{d}n - \int_{k_1}^{k_2} R\mathrm{d}k + (n_2 - n_2^0)\bar{\alpha}_2, \tag{C.107}$$

may be obtained by stationary phase integration with the help of the identity

$$\left(\frac{\partial n_1}{\partial \bar{\alpha}_2}\right)_{\bar{\alpha}_1}\left(\frac{\partial n_1}{\partial \bar{\alpha}_1}\right)^{-1}_{\bar{\alpha}_2} = \left(\frac{\partial(n_1,\ \bar{\alpha}_1)}{\partial(\bar{\alpha}_2,\ \bar{\alpha}_1)}\right)\left(\frac{\partial(\bar{\alpha}_1,\ \bar{\alpha}_2)}{\partial(n_1,\ \bar{\alpha}_2)}\right) = -\left(\frac{\partial \bar{\alpha}_1}{\partial \bar{\alpha}_2}\right)_{n_1}, \tag{C.108}$$

where n_2, k_2 and $\bar{\alpha}_1$ are now to be taken as functions of n_1^0 and $\bar{\alpha}_2$, while k_1 is of course determined by E and n_1^0. Reversal of the functional dependence of $\bar{\alpha}_1$ on $\bar{\alpha}_2$ transforms (C.106) to the form

$$S_{n_1^0 n_2^0} = \frac{1}{2\pi}\int_0^{2\pi}\left(\frac{\partial \bar{\alpha}_2}{\partial \bar{\alpha}_1}\right)^{1/2}_{n_1^0} \exp[\mathrm{i}\Delta_{n_1^0 n_2^0}(\bar{\alpha}_1)]\mathrm{d}\bar{\alpha}_1, \tag{C.109}$$

which is identical with the initial value representation of eqn (9.28). This closes the present analysis except to note that eqns (C.104) and (C.107) employ expressions for $\Lambda_{n_1^0 n_2^0}$ and $\Delta_{n_1^0 n_2^0}$ derived from $W(\alpha_2 R_2;\ \alpha_1 R_1)$ on the assumption that the trajectory is followed in the (N, α, P, R) system. It is readily verified by comparison between (C.88) and (C.89) that $W(x_2,\ R_2;\ x_1 R_1)$ would yield an equivalent expression, obtained by substituting

$$-\int_{n_1}^{n_2} \alpha \mathrm{d}n = \frac{1}{\hbar}\left(\int_{x_1}^{x_2} p\mathrm{d}x - F_2(x_2,\ N_2) + F_1(x_1,\ N_1)\right) \tag{C.110}$$

in eqn (C.107), which is applicable in situations where the internal motion is most conveniently followed in Cartesian variables.

C.4 The semiclassical Green's function

The semiclassical Green's function is given by Gutzwiller (1990) as a half Fourier transform of the Van Vleck (1928) propagator in eqn (C.76):

$$\begin{aligned} G(\mathbf{q}''\mathbf{q}'; E) &= \left(\frac{1}{\mathrm{i}\hbar}\right)\int_0^\infty \mathrm{d}t K(\mathbf{q}''\mathbf{q}'; t)\mathrm{e}^{\mathrm{i}Et/\hbar} \\ &= \left(\frac{1}{\mathrm{i}\hbar}\right)\int_0^\infty \mathrm{d}t \left[\left(\frac{1}{2\pi\mathrm{i}\hbar}\right)^f \det\left(-\frac{\partial^2 S}{\partial\mathbf{q}''\partial\mathbf{q}'}\right)_t\right]^{1/2} \\ &\qquad \times \mathrm{e}^{\mathrm{i}S[\mathbf{q}'',\mathbf{q}',t]/\hbar + \mathrm{i}Et/\hbar + \mathrm{phase}}, \end{aligned} \tag{C.111}$$

where the subscript on the pre-exponent is a reminder that the derivatives are taken at fixed t. Moreover, $\partial S/\partial t = -E(\mathbf{q}'', \mathbf{q}', t)$—the negative of the energy of the trajectory from $\mathbf{q}'$ to $\mathbf{q}''$ in time t. The exponent is therefore stationary at times t such that $E(\mathbf{q}'', \mathbf{q}', t)$ coincides with the target energy E. Subsequent manipulations are simplified by the Legendre transformation

$$S[\mathbf{q}'', \mathbf{q}', t] = \int_{\mathbf{q}'}^{\mathbf{q}''} p(\mathbf{q}, E)\cdot \mathrm{d}\mathbf{q} - Et = W(\mathbf{q}'', \mathbf{q}', E) - Et, \tag{C.112}$$

where the reduced classical action integral

$$W(\mathbf{q}''\mathbf{q}'; E) = \int_{\mathbf{q}'}^{\mathbf{q}''} \mathbf{p}\cdot\mathrm{d}\mathbf{q} \tag{C.113}$$

is the stationary value of the exponent in (C.111). Consequently the stationary phase approximation to $G(\mathbf{q}''\mathbf{q}'; E)$ is given by

$$\begin{aligned} G_{sc}(\mathbf{q}''\mathbf{q}'; E) \simeq &\frac{2\pi}{(2\pi\mathrm{i}\hbar)^{(f+1)/2}}\sum_{\mathrm{traj}}\left[\left(\frac{\partial^2 S}{\partial t^2}\right)^{-1}\det\left(-\frac{\partial^2 S}{\partial\mathbf{q}''\partial\mathbf{q}'}\right)_t\right]^{1/2} \\ &\times \exp\left[\mathrm{i}W(\mathbf{q}''\mathbf{q}'; E)/\hbar - \mathrm{i}\mu\pi/2\right], \end{aligned} \tag{C.114}$$

where the sum is taken over all trajectories from $\mathbf{q}'$ to $\mathbf{q}''$ at energy E, and the Maslov index μ counts the number of sign changes of the determinant at the so-called conjugate points along the trajectory (Gutzwiller 1990).

Significant simplifications of eqn (C.114) may be obtained by recognizing that $\partial W/\partial E = t(E)$. In the first place it follows that $\partial^2 S/\partial t^2 = -\partial E/\partial t = -\left(\partial t/\partial E\right)^{-1} = -\left[\partial^2 W/\partial E^2\right]^{-1}$. Secondly, following Gutzwiller (1990), eqn (C.112) may be used to express the Van Vleck determinant in terms of $\det\left(\partial^2 W/\partial \mathbf{q}''\partial \mathbf{q}'\right)$. As a preliminary, note that

$$\left(\frac{\partial t}{\partial q_i'}\right)_t = \left(\frac{\partial^2 W}{\partial q_i'\partial E}\right) + \left(\frac{\partial^2 W}{\partial E^2}\right)\left(\frac{\partial E}{\partial q_i'}\right)_t = 0, \tag{C.115}$$

from which it follows that

$$\begin{aligned}
\left[\frac{\partial}{\partial q_j''}\left(\frac{\partial S}{\partial q_i'}\right)_t\right] &= \frac{\partial}{\partial q_j''}\left[\left(\frac{\partial W}{\partial q_i'}\right)_E + \left(\frac{\partial W}{\partial E}\right)\left(\frac{\partial E}{\partial q_i'}\right)_t\right] \\
&= \left(\frac{\partial^2 W}{\partial q_j''\partial q_i'}\right) + \left(\frac{\partial^2 W}{\partial q_j''\partial E}\right)\left(\frac{\partial E}{\partial q_i'}\right) \\
&= \left(\frac{\partial^2 W}{\partial q_j''\partial q_i'}\right) - \left(\frac{\partial^2 W}{\partial q_j''\partial E}\right)\left(\frac{\partial^2 W}{\partial q_i'\partial E}\right)\left(\frac{\partial^2 W}{\partial E^2}\right)^{-1}.
\end{aligned} \tag{C.116}$$

Terms like $\left(\partial^2 E/\partial q_j''\partial q_i'\right)$ vanish because $E = H(\mathbf{p}', \mathbf{q}')$ is independent of $\mathbf{q}''$. When expressed in terms of determinants, this means that

$$\det\left(-\frac{\partial^2 S}{\partial \mathbf{q}_j''\partial \mathbf{q}_i'}\right)_t = -\left(\frac{\partial^2 W}{\partial E^2}\right)^{-1}\det\begin{pmatrix} -\ \partial^2 W/\partial \mathbf{q}''\partial \mathbf{q}' & \partial^2 W/\partial \mathbf{q}''\partial E \\ \partial^2 W/\partial E\partial \mathbf{q}' & \partial^2 W/\partial E^2 \end{pmatrix}, \tag{C.117}$$

where the coordinate derivatives of W are taken at constant E.

The determinant on the right must, however, be evaluated with care (Gutzwiller 1990), because $\det(\partial^2 W/\partial \mathbf{q}''\partial \mathbf{q}') = 0$. To see this note by the Hamilton–Jacobi equation that

$$H(\mathbf{p}'', \mathbf{q}'') = H\left[\left(\partial W/\partial \mathbf{q}''\right), \mathbf{q}''\right] = E. \tag{C.118}$$

Hence, on differentiating with respect to q_i',

$$\sum_{p_j''}\left(\frac{\partial H}{\partial p_j''}\right)\left(\frac{\partial p_j''}{\partial q_i'}\right) = \sum_{q_j''}\dot{q}_j''\left(\frac{\partial^2 W}{\partial q_j''\partial q_i'}\right) = 0, \tag{C.119}$$

which implies that the matrix $(\partial^2 W/\partial \mathbf{q}''\partial \mathbf{q}')$ is singular. Moreover, the nature of the singularity may be understood by adopting a coordinate system such that q_1 runs along a particular classical trajectory, and that $q_2, q_3, \ldots, q_f$ are perpendicular to it. Consequently $\dot{\mathbf{q}} = (\dot{q}_1, 0, \ldots, 0)$, which means that $\left(\partial^2 W/\partial q_j''\partial q_i'\right) = 0$ whenever

$i = 1$ or $j = 1$. It also follows, by differentiating (C.118) with respect to E, that $\dot{q}_1 \left(\partial^2 W/\partial q_1 \partial E\right) = 1$. Consequently

$$\det\left(-\frac{\partial^2 S}{\partial \mathbf{q}'' \partial \mathbf{q}'}\right)_t = -\left(\dot{q}_1'\dot{q}_1''\frac{\partial^2 W}{\partial E^2}\right)^{-1} \det\left(-\frac{\partial^2 W}{\partial \tilde{\mathbf{q}}'' \partial \tilde{\mathbf{q}}'}\right)_E, \tag{C.120}$$

in which $(\partial^2 W/\partial \tilde{\mathbf{q}}'' \partial \tilde{\mathbf{q}}')$ denotes the reduced Hessian matrix with indices i or j restricted to $(2, 3, \ldots, f)$.

Taken in conjunction with eqn (C.114) this means that

$$G(\mathbf{q}''\mathbf{q}'; E) \simeq 2\pi \left(\frac{1}{2\pi i \hbar}\right)^{(f+1)/2} \sum_{\text{traj}} \left[\left(\dot{q}_1'\dot{q}_1''\right)^{-1} \det\left(-\frac{\partial^2 W}{\partial \tilde{\mathbf{q}}'' \partial \tilde{\mathbf{q}}'}\right)_E\right]^{1/2}$$
$$\times e^{iW(\mathbf{q}''\mathbf{q}';E)/\hbar - i\mu\pi/2}. \tag{C.121}$$

In conclusion it should be noted that eqns (C.115)–(C.117) may be extended to imaginary time propagation, with $t = -i\tau$, as required, for example in the instanton theory of chemical reactions in Section 11.4. The difference is that eqn (C.112) is replaced by

$$\bar{S}[\mathbf{q}'', \mathbf{q}', t] = \int_{\mathbf{q}'}^{\mathbf{q}''} \bar{\mathbf{p}}(\mathbf{q}, E) \cdot d\mathbf{q} - Et = \bar{W}\left(\mathbf{q}'', \mathbf{q}', E\right) + E\tau, \tag{C.122}$$

in which $\bar{\mathbf{p}}$ is the momentum on the upturned potential energy surface, $\bar{V}(\mathbf{q}) = -V(\mathbf{q})$ and $\tau(E) = -\partial \bar{W}/\partial E$. Thus

$$\det\left(-\frac{\partial^2 \bar{S}}{\partial \mathbf{q}'' \partial \mathbf{q}'}\right)_t = \left(\dot{q}_1'\dot{q}_1''\frac{\partial^2 \bar{W}}{\partial E^2}\right)^{-1} \det\left(-\frac{\partial^2 \bar{W}}{\partial \tilde{\mathbf{q}}'' \partial \tilde{\mathbf{q}}'}\right)_E, \tag{C.123}$$

or in the context of eqn (11.64)

$$\det\left(-\frac{\partial^2 \bar{S}}{\partial \mathbf{q}'' \partial \mathbf{q}'}\right)_t = \left(-\dot{q}_1'\dot{q}_1''\frac{\partial \tau}{\partial E}\right)^{-1} \det\left(-\frac{\partial^2 \bar{W}}{\partial \tilde{\mathbf{q}}'' \partial \tilde{\mathbf{q}}'}\right)_E. \tag{C.124}$$

C.5 Angular momentum coupling coefficients

The angular momentum vector coupling coefficients relate to transformations between one angular momentum coupling scheme and another. Semiclassical approximations for them have been obtained in a variety of ways (Racah 1951; Beidenharn 1953; Adler et al. 1956; Brussard and Tolhoek 1957; Ponzano and Regge 1968; Miller 1974; Schulten and Gordon 1975). Following Miller (1974), they are here derived by use of classical generators, in a way that also borrows illuminating geometrical insights from Ponzano and Regge (1968) and Schulten and Gordon (1975).

To start with the simplest case, the $3j$ symbol is a symmetrical version of the Clebsch–Gordan coefficient, which relates the uncoupled states $|j_2m_2\rangle|j_3m_3\rangle$ to the coupled states $|(j_2j_3)j_1-m_1\rangle$ by the equations (Brink and Satchler 1968; Zare 1988)

$$\begin{pmatrix} j_1 & j_2 & j_3 \\ m_1 & m_2 & m_3 \end{pmatrix} = (-1)^{j_2-j_3+m_1}(2j_1+1)^{-1/2}T_{j_1m_2}, \tag{C.125}$$

where the elements $T_{j_1m_2}$ define the unitary transformation

$$|(j_2j_3)j_1-m_1\rangle = \sum_{m_2} T_{j_1m_2}|j_2m_2\rangle|j_3m_3\rangle, \tag{C.126}$$

subject to

$$m_1+m_2+m_3=0. \tag{C.127}$$

Note, for future reference, that the sum over m_2 in (C.126) could be more symmetrically regarded as a sum over m_2-m_3 at fixed m_1.

The derivation that follows makes use of the classical analogue of eqn (C.126) depicted in Fig. C.1, in which the $\mathbf{J}_i$ have fixed magnitudes $j_i+\frac{1}{2}$ (in units of $\hbar$) with $\mathbf{J}_1$ fixed and $\mathbf{J}_2$ and $\mathbf{J}_3$ free to rotate around it, subject to the triangular constraint

$$\mathbf{J}_1+\mathbf{J}_2+\mathbf{J}_3=0, \tag{C.128}$$

consistent with (C.127). This choice of the senses of the $\mathbf{J}_i$ ensures the known symmetry of the $3j$ symbol (Brink and Satchler 1968; Zare 1988) under cyclic permutation of the indices. Seen in relation to the above equations, the triangle with a fixed edge $\mathbf{J}_1$ represents the left-hand side of (C.126), and the freedom of rotation about $\mathbf{J}_1$ corresponds to the uncertainty in m_2-m_3 represented by the sum over m_2 on the right, all other actions being fixed by $|\mathbf{J}_1|$, m_1 and m_2-m_3. The method of derivation involves constructing a semiclassical unitary transform equivalent to $T_{j_1m_2}$ in terms of an F_4 type generator between the magnitude $J_1=j_1+\frac{1}{2}$, on one hand, and the difference m_2-m_3 on the other, subject to the geometrical constraints in Fig. C.1. One vital observation, which may be confirmed by reference to Fig. 4.3, is that the angle α_1, measured around the vector $\mathbf{J}_1$, constitutes the conjugate variable to the magnitude J_1. Similarly, the azimuthal angles β_i in the component representation

$$\mathbf{J}_i=(P_i\cos\beta_i,\ P_i\sin\beta_i,\ m_i), \tag{C.129}$$

where

$$P_i=(J_i^2-m_i^2)^{1/2}, \tag{C.130}$$

are the angles conjugate to the z components m_i. Hence, bearing in mind the known symmetry of the $3j$ symbol under permutation of the indices, the natural choice for F_4 (obtained by setting $F_3=0$ in (C.5b)) may be written

$$F_4=\Omega^{(3)}=\sum_i(J_i\alpha_i+m_i\beta_i), \tag{C.131}$$

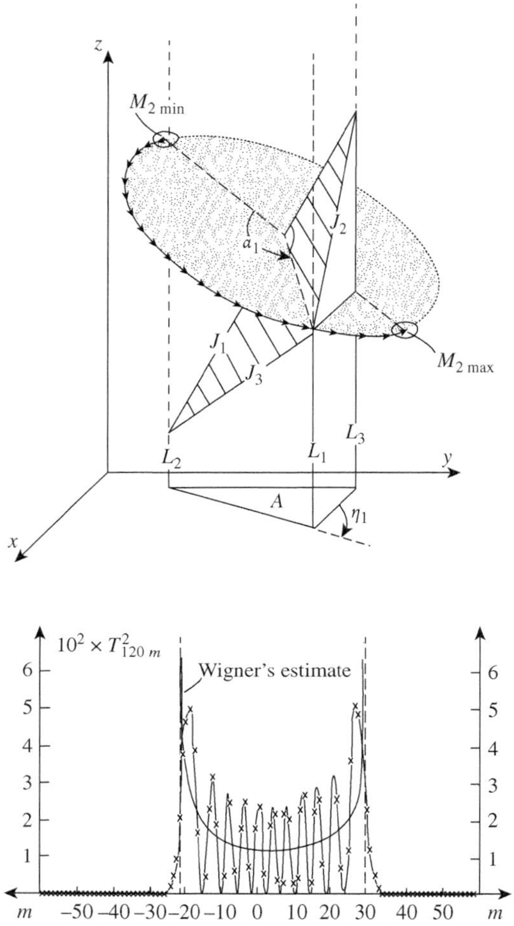

Fig. C.1 A graphical representation of eqn (C.125). The lower panel shows $|T_{j_1 m_2}|^2$ for $j_1 = 120$, as given by (C.146) (curves) and by exact recursion relations (points). Wigner's (1959) estimate is $[(2j_1 + 1)/4\pi A]$, where A is the area of the projected triangle. (Reprinted with permission from Schulten and Gordon (1975). Copyright 1975, AIP Publishing LLP.)

where the angles α_i and β_i are geometrically determined functions of the J_i and m_i. Moreover, the absolute orientation about the z axis is irrelevant, which makes it convenient to transform temporarily to new conjugate variables $\{L_i,\ \eta_i\}$ defined by the equations

$$\begin{aligned} m_1 &= L_2 - L_3, \quad & \eta_1 &= \beta_3 - \beta_2 \\ m_2 &= L_2 - L_1, \quad & \eta_2 &= \beta_1 - \beta_3 \\ m_3 &= L_1 - L_2, \quad & \eta_3 &= \beta_2 - \beta_1, \end{aligned} \tag{C.132}$$

a construction that also later aids extension of the theory to $6j$ symbols; note also that $T_{j_1 m_2}$, which determines the transformation between J_1 and the difference $m_2 - m_3$, relates more symmetrically to a transformation from J_1 to L_1 at fixed $L_1 + L_2 + L_3$.

In these new variables (C.131) transforms to

$$\Omega^{(3)} = \sum_i (J_i \alpha_i + L_i \eta_i), \tag{C.133}$$

which is the desired form for the F_4 generator, because the angles α_i and η_i are determined, in terms of the J_i and m_i, by the following geometrical identities (Ponzano and Regge 1968):

$$\begin{aligned} \cos\alpha_1 &= \frac{J_1^2(m_3 - m_2) + (J_3^2 - J_2^2)m_1}{4P_1 F(J_1,\ J_2,\ J_3)}, \\ \sin\alpha_1 &= \frac{J_1 A(P_1,\ P_2,\ P_3)}{P_1 F(J_1,\ J_2,\ J_3)}, \end{aligned} \tag{C.134}$$

and

$$\begin{aligned} \cos\eta_1 &= \frac{P_1^2 - P_2^2 - P_3^2}{2P_2P_3} = \frac{J_1^2 - J_2^2 - J_3^2 - 2m_2m_3}{2P_2P_3}, \\ \sin\eta_1 &= \frac{2A(P_1,\ P_2,\ P_3)}{P_2P_3}, \end{aligned} \tag{C.135}$$

where $F(J_1,\ J_2,\ J_3)$ and $A(P_1,\ P_2,\ P_3)$ are respectively the areas of the triangle $(J_1,\ J_2,\ J_3)$ and its projection $(P_1,\ P_2,\ P_3)$ in the xy plane:

$$\begin{aligned} F(J_1,\ J_2,\ J_3) &= \frac{1}{4}[(J_1 + J_2 + J_3)(-J_1 + J_2 + J_3)(J_1 - J_2 + J_3) \\ &\quad (J_1 + J_2 - J_3)]^{1/2}, \\ A(P_1,\ P_2,\ P_3) &= \frac{1}{4}[(P_1 + P_2 + P_3)(-P_1 + P_2 + P_3)(P_1 - P_2 + P_3) \\ &\quad (P_1 + P_2 - P_3)]^{1/2}. \end{aligned} \tag{C.136}$$

Similar expressions for $(\alpha_2,\ \eta_2)$ and $(\alpha_3,\ \eta_3)$ are obtained from (C.134) and (C.135) by cyclic permutation of the indices. It is also evident by inspection of Fig. C.1 that any angular momentum set $\{J_i,\ m_i\}$ is equally consistent with a mirror-image geometry (obtained by reflection in the $(J_1,\ z)$ plane), in which the signs of the α_i and η_i are reversed.

The validity of $\Omega^{(3)}$ as an F_4 generator, such that

$$\partial\Omega^{(3)}/\partial J_i = \alpha_i, \quad \partial\Omega^{(3)}/\partial L_i = \eta_i, \tag{C.137}$$

is most simply confirmed by direct but lengthy differentiation, while a more elegant but more abstract proof is outlined by Schulten and Gordon (1975). The important consequence is that the analogue of (C.29) yields the desired unitary transform

$$U_4^{(3)}(J_1,\ L_1) = \sum \left(\frac{1}{2\pi \mathrm{i}} \frac{\partial^2 \boldsymbol{\Omega}^{(3)}}{\partial J_1 \partial L_1} \right)^{1/2} \exp(\mathrm{i}\boldsymbol{\Omega}^{(3)})$$
$$= \left(\frac{J_1}{2A} \right)^{1/2} \cos(\boldsymbol{\Omega}^{(3)} + \pi/4), \tag{C.138}$$

where the sum that leads to the cosine is taken over the two equivalent orientations of the triangle, and the term $\pi/4$ arises because

$$\frac{\partial^2 \boldsymbol{\Omega}^{(3)}}{\partial J_1 \partial L_1} = \frac{\partial \alpha_1}{\partial L_1} = -\frac{J_1}{2A} \tag{C.139}$$

for the orientation specified by (C.134)–(C.135).

The final result for the semiclassical $3j$ symbol, obtained by substituting $U_4^{(3)}$ for $T_{j_1 m_2}$ in (C.125), is therefore

$$\begin{pmatrix} j_1 & j_2 & j_3 \\ m_1 & m_2 & m_3 \end{pmatrix} = (-1)^{\sigma} \frac{1}{(2\pi A)^{1/2}} \cos(\boldsymbol{\Omega}^{(3)} + \pi/4), \tag{C.140}$$

where $\Omega^{(3)}$ is most conveniently expressed for computational purposes in the following form derived from (C.127), (C.131) and (C.132):

$$\Omega^{(3)} = J_1\alpha_1 + J_2\alpha_2 + J_3\alpha_3 - m_2\eta_1 + m_1\eta_2. \tag{C.141}$$

Here $J_i = j_i + \frac{1}{2}$, the angles α_i and η_i are determined by (C.134) and (C.135) plus their cyclic permutations, and the area $A(P_1,\ P_2,\ P_3)$ is given by (C.130) and (C.136). The phase term σ may be determined by comparison between (C.142) and the identity (Brink and Satchler 1968)

$$\begin{pmatrix} j_1 & j_2 & j_3 \\ 0 & 0 & 0 \end{pmatrix} = \frac{(-1)^{(j_1+j_2+j_3)}}{2(2\pi A)^{1/2}} [1 + (-1)^{j_1+j_2+j_3}], \tag{C.142}$$

from which it may be verified that

$$\sigma = j_1 + j_2 - j_3 + 1. \tag{C.143}$$

Equation (C.140) was first given by Miller (1974). If two of the j_i are much larger than the other, say $j_2 \simeq j_3 \gg j_1$, it is simpler to employ the $3j$ equivalent of the earlier result of Brussard and Tolhoek (1957), namely

$$\begin{pmatrix} j_1 & j_2 & j_3 \\ m_1 & m_2 & m_3 \end{pmatrix} \simeq (-1)^{m_3 - j_3} (2j_3 + 1)^{-1/2} D^{j_1}_{m_1,\ j_3 - j_2}(0,\ \theta,\ 0), \tag{C.144}$$

where $D^{j}_{\mu\nu}(0,\ \theta,\ 0)$ is the rotation matrix (Brink and Satchler 1968) and

$$\cos\theta = \frac{m_3}{j_3 + 1/2}. \tag{C.145}$$

Schulten and Gordon (1975) obtained eqn (C.140) (albeit with an apparent sign error in the expression for $\cos\alpha_1$) by the interesting device of converting the $3j$ symbol recurrence relations to JWKB-type differential equations. This approach has the added benefit that it also covers the non-classical range of $\{j_i,\ m_i\}$ combinations for which the area $A(P_1,\ P_2,\ P_3)$ is imaginary and for which the angles given by eqns (C.134) and (C.135) are complex. The following uniform approximations cover both real and complex situations and, as usual, smooth out the spurious singularities at the boundaries (Schulten and Gordon 1975):

$$\begin{pmatrix} j_1 & j_2 & j_3 \\ m_1 & m_2 & m_3 \end{pmatrix} = (-1)^\sigma z^{1/4}(2A)^{-1/2} \times \begin{cases} \cos\boldsymbol{\Omega}_0 Ai(-z) - \sin\boldsymbol{\Omega}_0 Bi(-z) & \boldsymbol{\Omega} < \boldsymbol{\Omega}_0 \\ \cos\boldsymbol{\Omega}_0 Bi(-z) - \sin\boldsymbol{\Omega}_0 Ai(-z) & \boldsymbol{\Omega} > \boldsymbol{\Omega}_0, \end{cases} \tag{C.146}$$

where the Airy functions $Ai(-z)$ and $Bi(-z)$ (Abramowitz and Stegun 1965) have argument

$$z = \pm\left(\frac{3}{2}|\boldsymbol{\Omega} - \boldsymbol{\Omega}_0|\right)^{2/3}, \tag{C.147}$$

with the positive and negative signs taken in the real and complex cases respectively. Finally,

$$\boldsymbol{\Omega}_0^{(3)} = J_1\alpha_1^0 + J_2\alpha_2^0 + J_3\alpha_3^0 - m_2\eta_1^0 + m_1\eta_2^0, \tag{C.148}$$

α_i^0 and η_i^0 being the angles on the nearest boundary:

$$\alpha_i^0 = \begin{cases} 0 \text{ if } & 0 \leqslant \text{Re}\ \alpha_i \leqslant \pi/2 \\ \pi \text{ if } & \pi/2 \leqslant \text{Re}\ \alpha_i \leqslant \pi \end{cases} \tag{C.149}$$

$$\eta_i^0 = \begin{cases} 0 \text{ if } & 0 \leqslant \text{Re}\ \eta_i \leqslant \pi/2 \\ \pi \text{ if } & \pi/2 \leqslant \text{Re}\ \eta_i \leqslant \pi. \end{cases}$$

It is evident from Fig. C.2 that this uniform result provides an excellent approximation to the exact $3j$ symbol, regarded as a function of j_1 at fixed m_2 or vice versa. One also clearly sees the familiar oscillatory patterns in the bright classical ranges flanked by exponential decay into the shadow regions.

The theory of the $6j$ symbol follows similar lines. In this case

$$\begin{Bmatrix} j_1 & j_2 & j_3 \\ l_1 & l_2 & l_3 \end{Bmatrix} = [(2j_1+1)(2l_1+1)]^{-1/2} T_{j_1 l_1}, \tag{C.150}$$

where the transformation elements $T_{j_1 l_1}$ connect states with common internal labels, l_2, l_3 and j_3, and a common resultant j_2 but with different intermediate coupling schemes. In the j_1 set, labelled $|(l_2,\ l_3)j_1,\ j_3;\ j_2\rangle$, l_2 and l_3 have a resultant j_1, which couples to j_3 to form j_2, while in the l_1 set, labelled $|(l_2,\ j_3)l_1,\ l_3;\ j_2\rangle$, the initial coupling is between l_2 and j_3 to form l_1, which then combines with l_3 to reach the same resultant j_2. Thus $T_{j_1 l_1}$ defined by the equation

$$|(l_2,\ l_3)j_1,\ j_3;\ j_2\rangle = \sum_{l_1} T_{j_1 l_1}|(l_2,\ j_3)l_1,\ l_3;\ j_2\rangle. \tag{C.151}$$

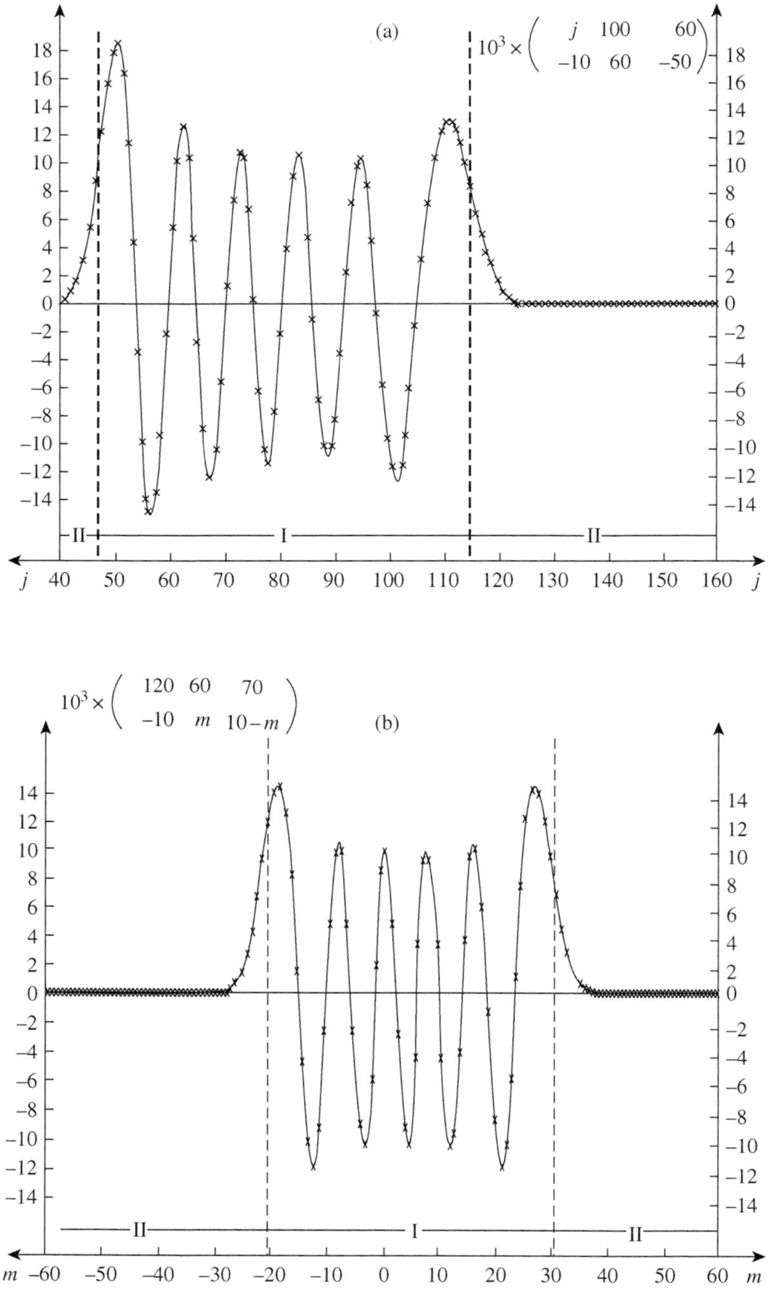

Fig. C.2 Semiclassical (curves) and exact (points) $3j$ symbols, (a) $\begin{pmatrix} j & 100 & 60 \\ -10 & 60 & -50 \end{pmatrix}$ and (b) $\begin{pmatrix} 120 & 60 & 70 \\ -10 & m & 10-m \end{pmatrix}$. (Reprinted with permission from Schulten and Gordon (1975). Copyright 1975, AIP Publishing LLP.)

The corresponding classical situation is illustrated in Fig. C.3, in which $J_i = j_i + \frac{1}{2}$ and $L_i = l_i + \frac{1}{2}$. Thus, in geometrical terms, the left-hand side of (C.151) corresponds to the fixed triangles $(J_1,\ L_2,\ L_3)$ and $(J_1,\ J_2,\ J_3)$ with a common edge J_1, while variation of the arbitrary dihedral angle $\tilde{\alpha}_1$ about J_1 gives rise to a range of L_1 values, consistent with the triangles $(L_1,\ L_2,\ J_3)$ and $(L_1,\ J_2,\ L_3)$, which is the classical analogue of the uncertainty in l_1 implied by the sum on the right-hand side of (C.151).

This geometrical picture also shows that the choice of the symbol L_i in eqns (C.128)–(C.143) was far from accidental because, as underlined by Ponzano and Regge (1968), the $3j$ symbol may be regarded as a large l_i approximation to its $6j$ counterpart;

$$\left\{ \begin{array}{ccc} j_1 & j_2 & j_3 \\ l_1 & l_2 & l_3 \end{array} \right\} \simeq (-1)^{j_1+j_2+j_3+2(l_1+l_2+l_3)}(2R)^{-1/2} \left(\begin{array}{ccc} j_1 & j_2 & j_3 \\ m_1 & m_2 & m_3 \end{array} \right), \tag{C.152}$$

where

$$m_1 = l_2 - l_3, \qquad m_2 = l_3 - l_1, \qquad m_3 = l_1 - l_2 \tag{C.153}$$
$$R = \frac{1}{3}(l_1 + l_2 + l_3).$$

The geometrical analogue is that the edges L_i of the $6j$ tetrahedron in Fig. C.3 go over, as $L_i \to \infty$, to those of a vertical prism, with an upper triangular face (J_1, J_2, J_3), as in Fig. C.3, and a mean height R. Not surprisingly, therefore, the dihedral angles $\tilde{\alpha}_i$ and $\tilde{\eta}_i$ in Fig. C.3 play precisely the same role in the $6j$ context as that played by α_i and η_i in the theory of the $3j$ symbol. In particular the function

$$\Omega^{(6)} = \sum_i (J_i\tilde{\alpha}_i + L_i\tilde{\eta}_i) \tag{C.154}$$

again acts as the proper F_4 type generator, in the sense that

$$\partial\Omega^{(6)}/\partial J_i = \tilde{\alpha}_i \quad \text{and} \quad \partial\Omega^{(6)}/\partial L_i = \tilde{\eta}_i, \tag{C.155}$$

but the expressions for $\tilde{\alpha}_i$ and $\tilde{\eta}_i$ in terms of J_i and L_i are much more complicated, namely (Ponzano and Regge 1968; Schulten and Gordon 1975):

$$\begin{aligned}
\cos\tilde{\alpha}_1 &= \\
&\frac{2J_1^2L_1^2 - J_1^2(-J_1^2 + L_2^2 + L_3^2) - J_2^2(J_1^2 + L_2^2 - L_3^2) - J_3^2(J_1^2 - L_2^2 + L_3^2)}{16F(J_1J_2J_3)F(J_1L_2L_3)} \\
\sin\tilde{\alpha}_1 &= \frac{3J_1V}{2F(J_1J_2J_3)F(J_1L_2L_3)} \\
\cos\tilde{\eta}_1 &= \\
&\frac{2J_1^2L_1^2 - L_1^2(-L_1^2 + L_2^2 + L_3^2) - J_2^2(-L_1^2 + L_2^2 - J_3^2) - L_3^2(L_1^2 - L_2^2 + J_3^2)}{16F(L_1J_2L_3)F(L_1L_2J_3)} \\
\sin\tilde{\eta}_1 &= \frac{3L_1V}{2F(L_1J_2L_3)F(L_1L_2J_3)}.
\end{aligned} \tag{C.156}$$

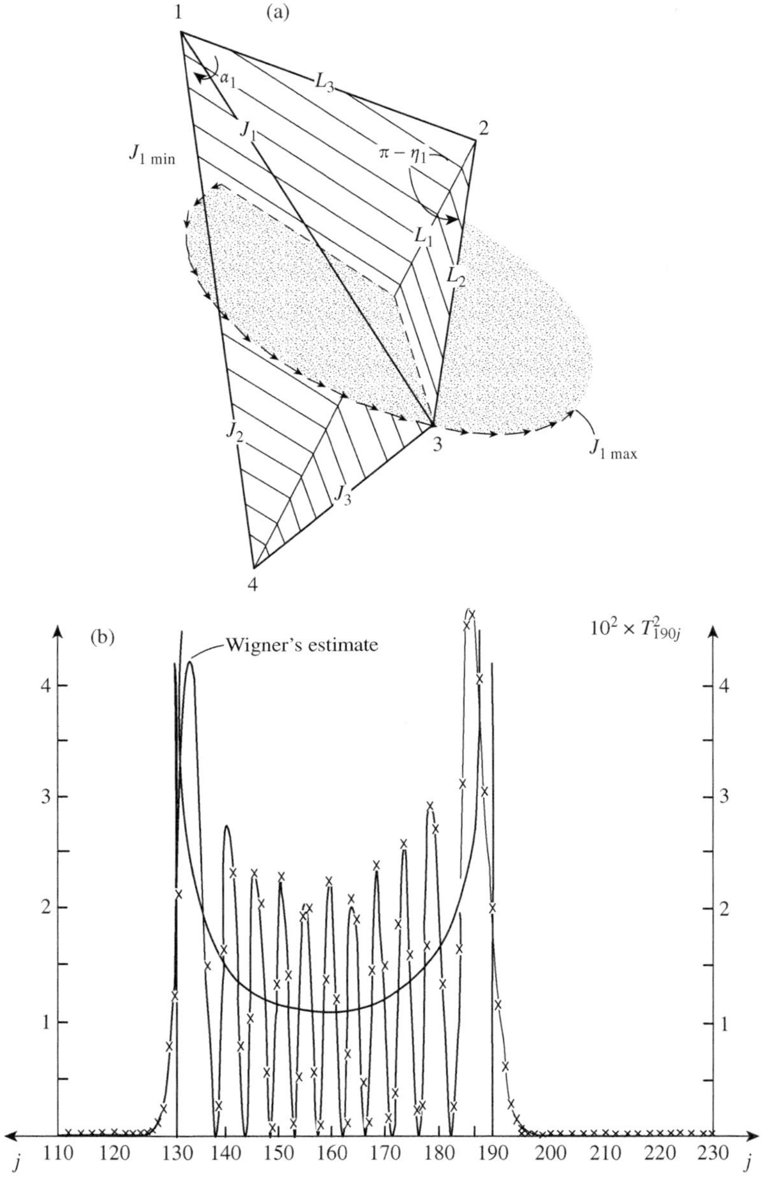

Fig. C.3 A graphical representation of eqn (C.151). The lower panel shows $|T_{j_1 l_1}|^2$ for $l_1 = 190$, as given by (C.161) (curves) and by exact recursion relations (points). Wigner's (1959) estimate is $[(2l_1 + 1)/4\pi V]$, where V is the volume of the tetrahedron. (Reprinted with permission from Schulten and Gordon (1975). Copyright 1975, AIP Publishing LLP.)

Here the various area terms $F(J_1J_2J_3)$ etc. are given by (C.136) and V is the volume of the tetrahedron, which is conveniently calculated from the Caley determinant

$$V^2 = \frac{1}{288}\begin{vmatrix} 0 & L_1^2 & L_2^2 & L_3^2 & 1 \\ L_1^2 & 0 & J_3^2 & J_2^2 & 1 \\ L_2^2 & J_3^2 & 0 & J_1^2 & 1 \\ L_3^2 & J_2^2 & J_1^2 & 0 & 1 \\ 1 & 1 & 1 & 1 & 0 \end{vmatrix}. \tag{C.157}$$

Similar expressions for the remaining $\tilde{\alpha}_i$ and $\tilde{\eta}_i$ are again obtained by cyclic permutation. Furthermore each angular momentum set $\{J_i,\ L_i\}$ is again realizable in two ways, one given by (C.156) and the other in which the signs of all the angles are reversed. The validity of (C.155) may be verified by direct differentiation, or by the more abstract argument outlined by Ponzano and Regge (1968).

The final step in the primitive semiclassical argument is to note, by virtue of (C.155) and (C.156), that

$$\frac{\partial^2\Omega^{(6)}}{\partial J_1 \partial L_1} = \left(\frac{\partial \tilde{\alpha}_1}{\partial L_1}\right)_{J_1} = -\frac{J_1L_1}{6V}, \tag{C.158}$$

so that the semiclassical unitary transform analogous to (C.142) becomes

$$U_4(J_1,\ L_1) = (J_1L_1/3\pi V)^{1/2}\cos(\Omega^{(6)} + \pi/4). \tag{C.159}$$

Alternatively, on replacing $T_{j_1 l_1}$ in (C.150) by $U_4(J_1,\ L_1)$,

$$\left\{\begin{matrix} j_1 & j_2 & j_3 \\ l_1 & l_2 & l_3 \end{matrix}\right\} = \left(\frac{1}{12\pi V}\right)^{1/2}\cos(\Omega^{(6)} + \pi/4), \tag{C.160}$$

where $\Omega^{(6)}$ and V are defined by (C.154)–(C.157) with $J_i = j_i + \frac{1}{2}$ and $L_i = l_i + \frac{1}{2}$.

Equation (C.160) was first suggested on heuristic grounds by Ponzano and Regge (1968) and later derived from the recurrence relations by Schulten and Gordon (1975). The latter also give the following uniform Airy approximations which extend the approximation to angular momentum combinations for which the volume V is negative and the angles $\tilde{\alpha}_i$ and $\tilde{\eta}_i$ complex:

$$\left\{\begin{matrix} j_1 & j_2 & j_3 \\ l_1 & l_2 & l_3 \end{matrix}\right\} = \left(\frac{Z^2}{12|V|}\right)^{1/2}\left\{\begin{matrix} \cos\Omega_0^{(6)}Ai(-Z) - \sin\Omega_0^{(6)}Bi(-Z), & \mathbf{\Omega}^{(6)} < \mathbf{\Omega}_0^{(6)} \\ \cos\Omega_0^{(6)}Bi(-Z) - \sin\Omega_0^{(6)}Ai(-Z), & \mathbf{\Omega}^{(6)} > \mathbf{\Omega}_0^{(6)} \end{matrix}\right. \tag{C.161}$$

where

$$\Omega_0^{(6)} = \sum_i (J_i\tilde{\alpha}_i^0 + L_i\tilde{\eta}_i^0), \tag{C.162}$$

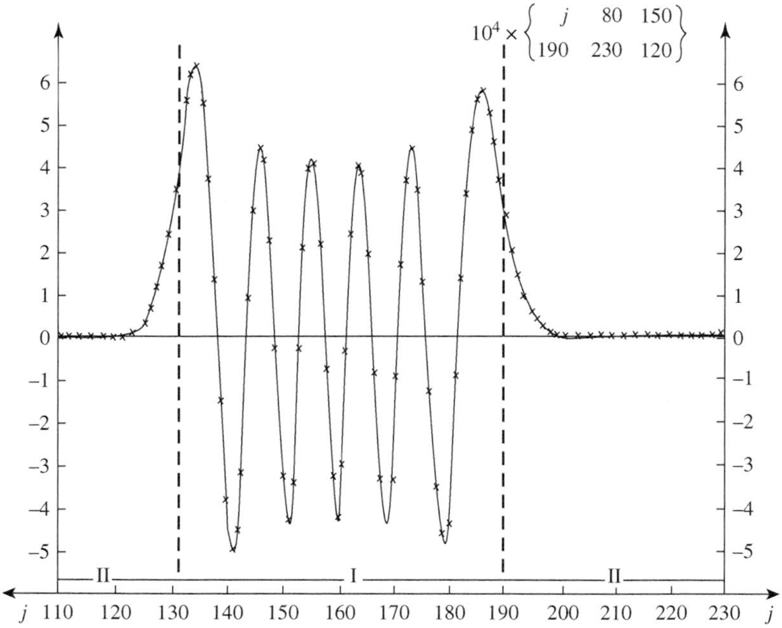

Fig. C.4 Semiclassical (curves) and exact (points) $6j$ symbols $\left\{\begin{matrix} j & 80 & 150 \\ 190 & 230 & 120 \end{matrix}\right\}$. (Reprinted with permission from Schulten and Gordon (1975). Copyright 1975, AIP Publishing LLP.)

Table C.1 Quantum mechanical and semiclassical 6j symbols[a] taken from Schulten and Gordon (1975).

j_1		$\lambda = 1$	$\lambda = 4$	$\lambda = 6$	$\lambda = 32$
1	QM	0.2789(00)	0.9692(−01)	0.0977(−01)	0.0806(−02)
	SC	0.3034(00)	0.9520(−01)	0.0926(−01)	0.0758(−02)
2	QM	−0.9535(−01)	0.0457(−01)	0.1288(−01)	0.0297(−02)
	SC	−0.9303(−01)	0.0523(−01)	0.1285(−01)	0.0285(−02)
3	QM	−0.6742(−01)	−0.0141(−01)	0.1378(−01)	0.6773(−02)
	SC	−0.6975(−01)	−0.0143(−01)	0.1378(−01)	0.6773(−02)
4	QM	0.1533(00)	−0.2083(−01)	−0.0825(−01)	−0.0275(−02)
	SC	0.1558(00)	−0.2158(−01)	−0.0823(−01)	−0.0281(−02)
5	QM	−0.1564(00)	0.5313(−01)	0.1021(−01)	−0.4379(−02)
	SC	−0.1566(00)	0.5315(−01)	0.1023(−01)	−0.4376(−02)
6	QM	0.1099(00)	0.1711(−01)	0.4149(−03)	0.8876(−05)
	SC	0.1090(00)	0.1704(−01)	0.4145(−03)	0.8870(−05)
7	QM	−0.5536(−01)	0.9524(−03)	0.3496(−08)	0.6071(−15)
	SC	−0.5441(−01)	0.9452(−03)	0.3488(−08)	0.6064(−15)
8	QM	0.1800(−01)	0.4072(−05)	0.1804(−18)	0.6568(−36)
	SC	0.1727(−01)	0.3915(−05)	0.1738(−18)	0.6332(−36)

[a] $\left\{\begin{matrix} j_1\lambda & 4.5\lambda & 3.5\lambda \\ \lambda & -3.5\lambda & 2.5\lambda \end{matrix}\right\}$

Table C.2 Quantum mechanical and semiclassical 6j symbols[b] taken from Schulten and Gordon (1975).

j_1		$\lambda = 1$	$\lambda = 4$	$\lambda = 8$	$\lambda = 16$
1	QM	0.3491(−01)	0.3226(−02)	0.0513(−02)	0.0302(−03)
	SC	0.3482(−01)	0.3155(−02)	0.0499(−02)	0.0292(−03)
3	QM	0.1891(−01)	0.0185(−02)	−0.1458(−02)	0.2156(−03)
	SC	0.1905(−01)	0.0180(−02)	−0.1458(−02)	0.2157(−03)
5	QM	−0.2359(−01)	0.2973(−02)	0.0887(−02)	0.1358(−03)
	SC	−0.2382(−01)	0.2982(−02)	0.0889(−02)	0.1362(−03)
7	QM	0.0129(−01)	0.1603(−02)	−0.0698(−02)	0.0315(−03)
	SC	0.0152(−01)	0.1569(−02)	−0.0699(−02)	0.0315(−03)
9	QM	0.1677(−01)	−0.2800(−02)	0.0854(−02)	0.0697(−03)
	SC	0.1671(−01)	−0.2800(−02)	0.0854(−02)	0.0696(−03)
11	QM	−0.2135(−01)	0.2264(−02)	−0.0020(−02)	−0.3562(−03)
	SC	−0.2147(−01)	0.2259(−02)	0.0022(−02)	−0.3561(−03)
13	QM	−0.2521(−01)	0.1724(−02)	0.1184(−02)	0.4040(−03)
	SC	0.2527(−01)	0.1731(−02)	−0.1183(−02)	0.4039(−03)
15	QM	0.0271(−01)	0.1171(−06)	0.5415(−12)	0.2293(−22)
	SC	0.0257(−01)	0.1095(−06)	0.5051(−12)	0.2136(−22)

[b] $\begin{Bmatrix} j_1\lambda & 8\lambda & 7\lambda \\ 6.5\lambda & 7.5\lambda & 7.5\lambda \end{Bmatrix}$

with $\tilde{\alpha}_i^0$ and $\tilde{\eta}_i^0$ given by the analogues of (C.148), and

$$Z = \pm\left(\frac{3}{2}|\Omega - \Omega_0|\right)^{1/2}. \tag{C.163}$$

As usual the upper and lower signs in (C.163) apply in the classical and non-classical cases respectively.

The comparison in Fig. C.4 shows that eqn (C.161) gives an excellent approximation, at least for moderately large quantum numbers, and the numerical accuracy is found to be equally good for integer and half-integer quantum numbers (Schulten and Gordon 1975). Further information on the accuracy of the approximation in relation to the magnitudes of the angular momenta is contained in Tables C.1 and C.2, which are taken from scaling tests by Schulten and Gordon (1975). It is evident that the errors are typically of order 10 per cent or less even for quite small quantum numbers.

Appendix D
The onset of chaos

Inspection of the phase space structures in Section 7.1 shows that the onset of chaos, as signalled by a scatter of points in the Poincaré section, is localized around the separatrices between different types of motion (e.g. between local and normal modes in Fig. 7.4 or librational and precessional ones in Fig. 7.5). The origin of this scatter lies in the influence of two or more Chirikov resonances, which is related below to the way in which the separatrix structure is lost. Consequences relevant to the quantization question, to intramolecular dynamics (Davis 1985; Davis and Gray 1986) and to chemical reactions (Davis 1987; Skodje and Davis 1988) are also discussed. The reader is referred for more details to the papers by MacKay et al. (1984) and Bensimon and Kadanoff (1984), and to the literature on chaos in dynamical systems for other aspects of the situation (Casati 1985; Berry et al. 1987; Tabor 1989; Ozorio de Almeida 1988; Gutzwiller 1990).

D.1 Breaking the separatrix

The existence of a separatrix derives from the presence of an unstable periodic trajectory, whose intersection with the Poincaré section constitutes a hyperbolic or unstable fixed point. The example shown in Fig. D.1 derives from the Hamiltonian

$$H = \frac{1}{2}(p_x^2 + p_y^2) + D(1 - \mathrm{e}^{-ax})^2 + \frac{1}{2}\omega^2 y^2 + \lambda x y^2, \qquad \text{(D.1)}$$

with parameter values $D = 12.75, a = (2D)^{-1/2}, \omega = 0.45$ and $E = 5$. Inward motion towards the hyperbolic point, P, occurs in directions marked H^+, which are termed stable manifolds (Tabor 1989), while outward flow occurs along the unstable manifolds, H^-. As first recognized by Poincaré (1892), the regularity or otherwise of motions in the neighbourhood depends on the fate of H^- as it is integrated outwards from P, or equivalently of H^+ as it is integrated backwards in time. One possibility is that H^- joins smoothly with H^+, as shown in Fig. D.1(a), which is a diagnostic for regularity; no internally generated trajectory could pass outside or vice versa, because to do so it would cross the union of H^+ and H^-, which is itself a trajectory, and no two trajectories can cross in phase space. (Because ambiguity in the path from the intersection would make it undeterministic.)

Suppose next that H^+ fails to join with H^- in a weakly chaotic situation with islands of regularity around the separatrix, as in Fig. D.1(b). Since H^- cannot cross either of the bounding regular orbits, it must eventually tangle up with itself, and similarly for the backward integration of H^+. Moreover, the nature of this tangling

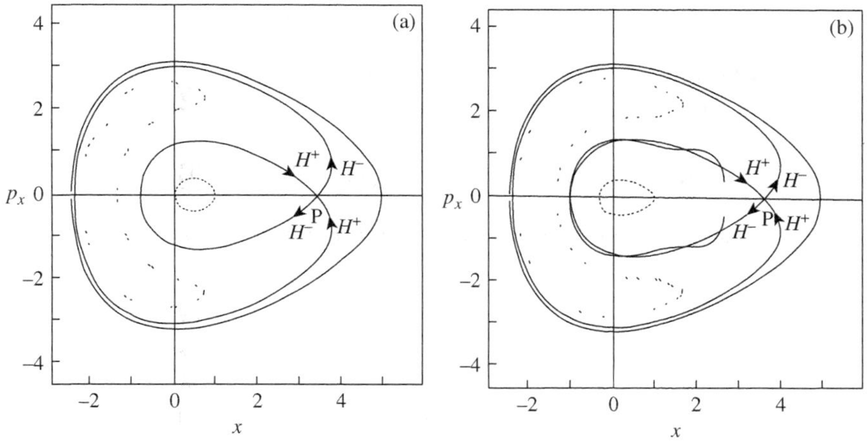

Fig. D.1 A surface of section at $y = 0, p_y > 0$ for the Hamiltonian (D.1), showing an unstable fixed point P, and stable and unstable manifolds, H^+ and H^- respectively. Two cases are considered: (a) $\lambda = 0.015$ and (b) $\lambda = 0.025$. In the first the manifolds meet to complete the separatrix. In the second the separatrix is broken.

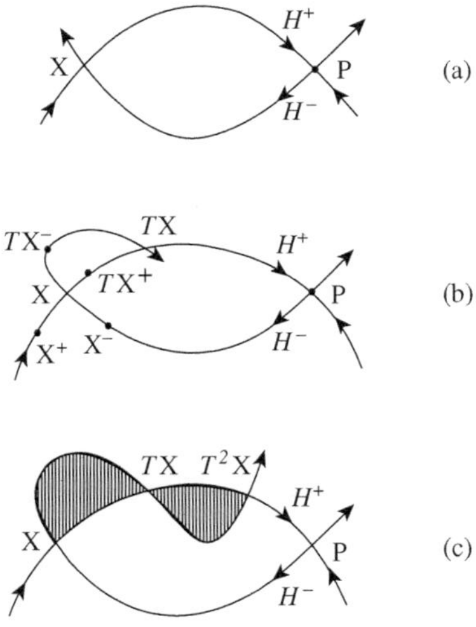

Fig. D.2 Steps towards the homoclinic tangle. The equivalent tangle of H^+ is omitted to avoid confusing the diagram.

is quite specific. At some time the unstable manifold H^+ must cross H^-, at what is termed a homoclinic point X, as shown in Fig. D.2(a)—a crossing that is allowed by the deterministic principle because $H^{\pm}$ are defined by separate equations of motion. Poincaré (1892) was the first to recognize that such homoclinic intersections have remarkable consequences. To see this, consider the fate of neighbouring points X^+ and

X^-, when integrated in the appropriate time directions, indicated by the arrows in Fig. D.2(a). X^- will appear at TX^- on H^- and X^+ at TX^+ on H^+, but what happens to the homoclinic point X? It can only appear both on H^- ahead of TX^- (in the arrow direction) and on H^+ ahead of TX^+ if there is a second intersection at TX, as shown in Fig. D.2(b). Repetition of the argument implies the existence of an infinite family of further homoclinic intersections $T^2X, T^3X, T^4X, \ldots$, etc., of which the first is sketched in Fig. D.2(c). The story is not finished, however, because the area-preserving nature of the dynamics ensures that the two shaded areas are equal to each other and to the areas of all succeeding loops. On the other hand the density of homoclinic intersections on H^+ increases without limit towards the fixed point; hence, by area preservation, the tendrils of H^- must become ever longer and thinner. Similar arguments apply to the loops and tendrils of the stable manifold H^+.

The nature of this homoclinic tangle and some consequences for the chaotic dynamics are illustrated in Figs. D.3 and D.4, by use of a two-dimensional area-preserving map, which offers a computationally efficient alternative to a full trajectory study for pedagogic purposes (Tabor 1989). The map employed is due to Henon (1969), and details of the algorithm employed are given in Section D.2. Figure D.3 illustrates an overview of the map for a particular parameter value, $\cos\alpha = 0.2114$. The five islands are interconverted, one to the next, by each iteration of the map, so that each island has a central elliptic fixed point equivalent to the intersection of a $5\colon n$ stable periodic orbit in a normal Poincaré section. Similarly the five hyperbolic points belong to a single unstable periodic orbit.

The magnified view of one of the islands in Fig. D.4 shows the beginnings of the homoclinic tangle and charts the steps involved in a typical chaotic escape from the central region. One unusual feature is that the two hyperbolic points, A and B, are directly linked by one branch of the separatrix because the unstable manifold of A

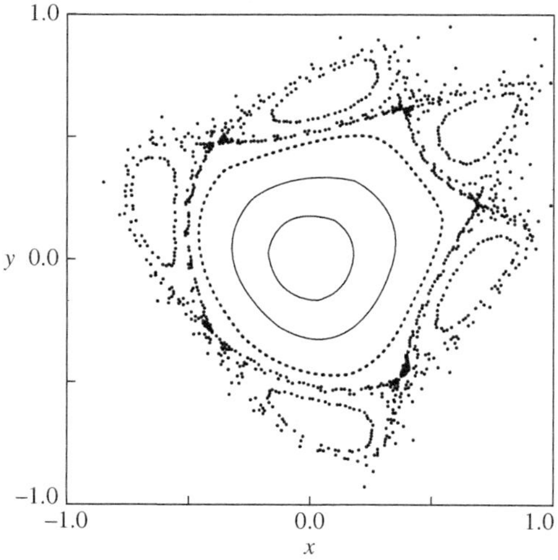

Fig. D.3 Overview of the Poincaré (1892) map for $\cos\alpha = 0.2114$.

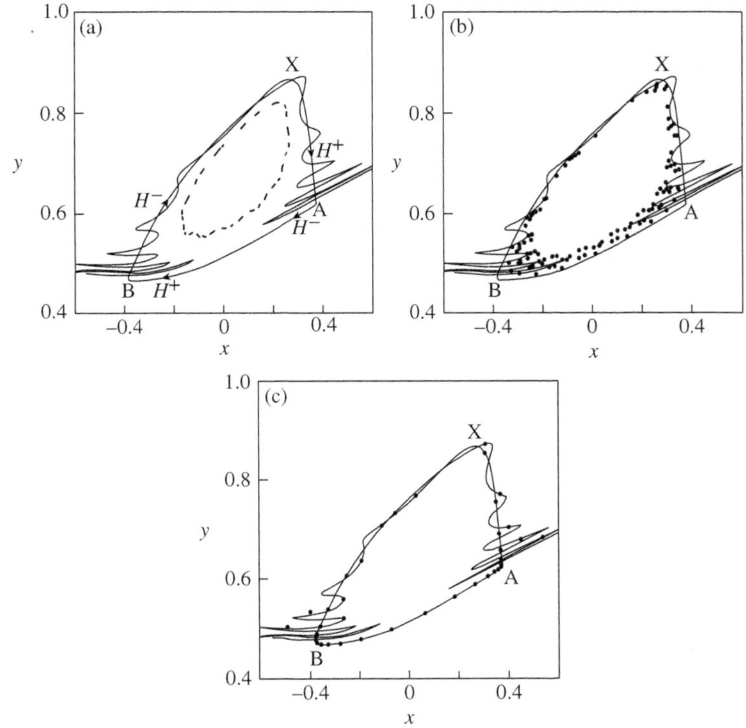

Fig. D.4 (a) First steps in the homoclinic tangle, in the uppermost island of Fig. D.3. X denotes the first homoclinic intersection. The central island indicates regular motion. (b) Successive points obtained by about a hundred five-fold iterations of the map from a central starting point. (c) Steps in the final escape initiated from the arrowed point.

joins smoothly with the stable manifold of B. Consequently the inner region of the complete map (Fig. D.3) contains only regular orbits because the smooth manifolds cut off any connection with the islands. The more common situation is that A and B would both generate stable and unstable homoclinic tangles, such as those shown in Fig. D.4(a). A central island is also included to indicate the bounds of clearly regular motion.

The next diagram, Fig. D.4(b), shows the first hundred or so iterations of a trajectory initiated at the point $(x, y) = (-0.3, 0.5)$. Notice how the trajectory avoids the visible inner loops of H^+ and H^- during this internal phase of its motion. The next stage, in Fig. D.4(c), which shows the escape phase of the same trajectory, is very illuminating. The motion is carried round clockwise and the first five steps in the lower left-hand part of the figure are seen to be captured by successive inner loops of the stable manifold, H^+ from A, the final loop of which initiated the tangle as it propagated backwards in time. Thereafter the trajectory follows the smooth part of H^+ until it collides with the fixed point A, and emerges on H^- of A. The smooth join between H^- of A and H^+ of B leads to a further collision, now with B, and a switch to H^- of B. Finally it appears in successive outer loops of H^- until apparently 'ejected'. An alternative scenario, in the opposite direction, would involve capture by an outside loop of the H^+ followed by ultimate transfer to the inner loops of H^-.

The unresolvable overlap between the inner tendrils of H^+ and H^- is immensely complicated, but it is clear by continuity with the above discussion that successive iterations of the map (equivalent to successive points in a Poincaré section) lie in successive closed loops of $H^\pm$, wherever such loops are defined; there are, for example, no closed loops of H^+ (or H^-) in the forward (or backward) time direction of the primary homoclinic point X in Fig. D.4(a). This means that motion in the inner region proceeds by passage from one tendril of H^+ to the next, and simultaneously for H^- wherever the two sets of tendrils overlap, until escape occurs from the last loop of H^+. Hence the first hundred or more iterations shown in Fig. D.4(b) were devoted to patiently unravelling the loops of H^+, after which the trajectory was under the control of H^- alone. Hence even the final escape would be illusory if the external motion was bounded, because the tendrils of H^- would eventually overlap with those of H^+, or with those of the stable manifold of another hyperbolic point (a heteroclinic intersection), and the whole process would go into reverse. The Henon (1969) map is, however, dissociative in the external region so that many, but not all, of the trajectories reach infinity.

Two insights from this picture into the discussion in Chapter 7 may be noted. The 'vague tori' of Jaffé and Reinhardt (1982) mentioned in Section 7.1 are clearly a manifestation of the long-time trapping illustrated in Fig. D.4(b). Secondly, the tendrils attached to the adiabatically switched torus of Skodje et al. (1985), shown in Fig. 7.11, must correspond to incipient trapping by some stable manifold which may bend and distort the figure out of all recognition, without affecting the area enclosed. In other words there may be a class of chaotic-looking but in fact invariant tori, whose Poincaré section areas cannot be estimated by any direct method, but which can be reached by adiabatic switching from a much simpler invariant torus.

The above picture also casts light on the nature of bottlenecks to molecular dissociation and chemical reaction, in ways that have been described by Davis (1985), Davis and Gray (1986) and Skodje and Davis (1988).

D.2 Henon map and the separatrix algorithm

The Henon (1969) map is a twist map (Tabor 1989) defined by the scheme

$$T: \qquad \begin{aligned} x_{i+1} &= x_i \cos\alpha - (y_i - x_i^2)\sin\alpha \\ y_{i+1} &= x_i \sin\alpha - (y_i - x_i^2)\cos\alpha. \end{aligned} \tag{D.2}$$

It has the inverse

$$T^{-1}: \qquad \begin{aligned} x_i &= x_{i+1}\cos\alpha + y_{i+1}\sin\alpha \\ y_i &= -x_{i+1} + y_{i+1}\cos\alpha + (x_{i+1}\cos\alpha + y_{i+1}\sin\alpha)^2, \end{aligned} \tag{D.3}$$

as may be confirmed by using (D.2) and (D.3) to demonstrate the identity $(x_i, y_i) = (x_i, y_i)$. The area-preserving property is readily verified by confirming that

$$\frac{\partial(x_{i+1}, y_{i+1})}{\partial(x_i, y_i)} = 1. \tag{D.4}$$

Henon (1969) gives a systematic study of the structure of the map for different values of the parameter cos α. Attention was restricted in the previous section to the value $\cos\alpha = 0.2114$.

The steps in determining the evolution of the broken separatrix are first to locate the hyperbolic fixed points and then to propagate the manifolds H^- and H^+ by use of eqns (D.2) and (D.3) respectively. Since the five islands in Fig. D.3 are interconverted by successive iterations, five iterations of (D.2) or (D.3) are required to generate a new point in any given island. Hence, if $\mathbf{r_i}$ denotes (x_i, y_i), it is convenient to denote the outcome of five iterations as

$$\mathbf{R_i} = T^5\mathbf{r_i} = (x_{i+5}, y_{i+5}) = (X_i, Y_i). \tag{D.5}$$

The fixed point algorithm then takes the form

$$\mathbf{R} - \mathbf{r} = 0, \tag{D.6}$$

which is readily solved by the Newton–Raphson scheme

$$\mathbf{r}^{(\mathbf{k}+1)} = \mathbf{r}^{(\mathbf{k})} - (\mathbf{A} - \mathbf{I})^{-1}(\mathbf{R}^{(\mathbf{k})} - \mathbf{r}^{(\mathbf{k})}), \tag{D.7}$$

where $\mathbf{I}$ is the unit matrix and the stability matrix $\mathbf{A}$ has typical elements

$$A_{Xy} = \partial X/\partial y. \tag{D.8}$$

Thus A_{Xy} may be approximated as

$$\frac{\partial X}{\partial y} \simeq \frac{\Delta X}{\Delta y} = \frac{X(x^{(k)}, y^{(k)} + \Delta y/2) - X(x^{(k)}, y^{(k)} - \Delta y/2)}{\Delta y}, \tag{D.9}$$

and similarly for A_{Xx}, A_{Yx} and A_{Yy}. As a test of the step size it may be noted that (D.3) requires that $\det(\mathbf{A}) = 1$. Convergence to six significant figures was obtained in the present application with a step size of 0.01.

The character of the resulting point is indicated by the eigenvalues of $\mathbf{A}$, which are real for a hyperbolic and complex for an elliptic point, while in both cases their product must equal unity (Percival and Richards 1982). Moreover, the eigenvectors of $\mathbf{A}$ provide a linearization of the manifolds $H^{\pm}$ in the neighbourhood of the fixed points, in the Poincaré section, with the larger eigenvalue belonging to the unstable manifold H^-. Propagation of H^- therefore involves construction of a short ladder, of say N points, spaced along this eigenvector so as to lie on a straight line between some initial point $\mathbf{r}_1$ and its image $\mathbf{R}_1$ after five iterations of (D.2). Successive five-fold iterations of this manifold map out the tangles of H^- shown in Fig. D.4. A similar construction for the stable manifold H^+ starts with a ladder along the second eigenvector and proceeds by five-fold iterations of the inverse map defined by (D.3). Note that the number of points in each loop or tendril is equal to N, the number of points in the ladder, because each five-fold iteration takes one loop to the next. The larger the initial displacement from the fixed point, the more rapidly but the less accurately will the tangles be drawn. A step size of the order of 0.001 was found convenient for the construction of Fig. D.4.

Appendix E
Angle–action transformations

E.1 Harmonic oscillator

$$H = \frac{1}{2m}p^2 + \frac{1}{2}m\omega^2 q^2$$

$$E = I\omega$$

$$q = (2I/m\omega)^{1/2}\cos\alpha$$

$$p = -(2Im\omega)^{1/2}\sin\alpha.$$

Quantum restriction

$$I = \left(v + \frac{1}{2}\right)\hbar, \quad v = 0, 1, 2, \ldots.$$

E.2 Morse oscillator

$$H = \frac{1}{2m}p^2 + D\left\{1 - \exp\left[-a(x - x_e)\right]\right\}^2$$

$$E = a\left(\frac{2D}{m}\right)^{1/2} I - \frac{a^2}{2m}I^2$$

$$x = x_e + a^{-1}\ln[(1 + \sqrt{\varepsilon}\cos\alpha)/(1 - \varepsilon)]$$

$$p = -(2mD)^{1/2}[\varepsilon(1 - \varepsilon)]^{1/2}\sin\alpha/(1 + \sqrt{\varepsilon}\cos\alpha)$$

$$\varepsilon = E/D.$$

Quantum restriction

$$I = \left(v + \frac{1}{2}\right)\hbar, \quad v = 0, 1, 2, \ldots.$$

E.3 Degenerate harmonic oscillator

(i) Cartesian form

$$
\begin{aligned}
H &= \frac{1}{2\mu}\left(p_x^2 + p_y^2\right) + \frac{1}{2}\mu\omega^2(x^2 + y^2) \\
E &= I\omega \\
x &= [(I + J)/\mu\omega]^{1/2}\cos(\alpha_I + \alpha_J) \\
y &= [(I - J)/\mu\omega]^{1/2}\cos(\alpha_I - \alpha_J) \\
p_x &= -[(I + J)\mu\omega]^{1/2}\sin(\alpha_I + \alpha_J) \\
p_y &= -[(I - J)\mu\omega]^{1/2}\sin(\alpha_I - \alpha_J).
\end{aligned}
$$

Quantum restrictions

$$
\begin{aligned}
I &= (v+1)\hbar, \qquad && v = 0, 1, 2, \ldots \\
J &= j\hbar, && j = v,\ v-2, \ldots, -v.
\end{aligned}
$$

(ii) Polar form

$$
\begin{aligned}
H &= \frac{1}{2\mu}\left(p_r^2 + p_\phi^2/r^2\right) + \frac{1}{2}\mu\omega^2 r^2 \\
E &= I\omega \\
r &= \left\{\left[I + \left(I^2 - I_\phi^2\right)^{1/2}\cos 2\alpha_I\right]\Big/\mu\omega\right\}^{1/2} \\
\phi &= \alpha_\phi + \arctan\left\{I_\phi \tan\alpha_I\Big/\left[I + \left(I^2 - I_\phi^2\right)^{1/2}\right]\right\} \\
p_r &= -\frac{\left[\mu\omega\left(I^2 - I_\phi^2\right)\right]^{1/2}\sin 2\alpha_I}{\left[I + \left(I^2 - I_\phi^2\right)^{1/2}\cos 2\alpha_I\right]^{1/2}} \\
p_\phi &= I_\phi.
\end{aligned}
$$

Quantum restrictions

$$
\begin{aligned}
I &= (v+1)\hbar, \qquad && v = 0, 1, 2, \ldots \\
I_\phi &= m\hbar, && m = v, v-2, \ldots, -v.
\end{aligned}
$$

E.4 Angular momentum

(i) Of a vector with spherical polar angles (θ, ϕ)

$$
\begin{aligned}
J^2 &= p_\theta^2 + p_\phi^2/\sin^2\theta \\
\cos\theta &= (1 - M^2/J^2)^{1/2}\cos\alpha_J \\
\sin\theta &= (J^2\sin^2\alpha_J + M^2\cos^2\alpha_J)^{1/2}/J
\end{aligned}
$$

$$\phi = \alpha_M + \arctan[(J/M)\tan\alpha_J] - \pi$$
$$\sin\theta\cos\phi = -\cos\beta\cos\alpha_M\cos\alpha_J + \sin\alpha_M\sin\alpha_J$$
$$\sin\theta\sin\phi = -\cos\beta\sin\alpha_M\cos\alpha_J - \cos\alpha_M\sin\alpha_J$$
$$\cos\beta = M/J$$
$$p_\theta = \frac{J(J^2 - M^2)^{1/2}\sin\alpha_J}{(J^2\sin^2\alpha_J + M^2\cos^2\alpha_J)^{1/2}}$$
$$p_\phi = M$$
$$J_x = -\cos\phi\cot\theta\, p_\phi - \sin\phi\, p_\theta = (J^2 - M^2)^{1/2}\cos\alpha_M$$
$$J_y = -\sin\phi\cot\theta\, p_\phi + \cos\phi\, p_\theta = (J^2 - M^2)^{1/2}\sin\alpha_M$$
$$J_z = M.$$

Quantum restrictions

$$J = \left(j + \tfrac{1}{2}\right)\hbar, \qquad j = 0, 1, 2, \ldots$$
$$M = m\hbar, \qquad m = j, j-1, \ldots, -j.$$

(ii) Of a body with Euler angles (ϕ, θ, ψ) (see Fig. 4.6)

$$J^2 = p_\theta^2 + \frac{1}{\sin^2\theta}[p_\phi^2 + p_\psi^2 - 2p_\phi p_\psi\cos\theta]$$
$$\cos\theta = \frac{1}{J^2}\{[(J^2 - M^2)(J^2 - K^2)]^{1/2}\cos\alpha_J + MK\}$$
$$\sin\theta\cos\phi = \frac{1}{J^2}\{-[M(J^2 - K^2)^{1/2}\cos\alpha_J - K(J^2 - M^2)^{1/2}]\cos\alpha_M + J(J^2 - K^2)^{1/2}\sin\alpha_J\sin\alpha_M\}$$
$$\sin\theta\sin\phi = \frac{1}{J^2}\{-[M(J^2 - K^2)^{1/2}\cos\alpha_J - K(J^2 - M^2)^{1/2}]\sin\alpha_M - J(J^2 - K^2)^{1/2}\sin\alpha_J\cos\alpha_M\}$$
$$\sin\theta\cos\psi = \frac{1}{J^2}\{[K(J^2 - M^2)^{1/2}\cos\alpha_J - M(J^2 - K^2)^{1/2}]\cos\alpha_K - J(J^2 - M^2)^{1/2}\sin\alpha_J\sin\alpha_K\}$$
$$\sin\theta\sin\psi = \frac{1}{J^2}\{[K(J^2 - M^2)^{1/2}\cos\alpha_J - M(J^2 - K^2)^{1/2}]\sin\alpha_K + J(J^2 - M^2)^{1/2}\sin\alpha_J\cos\alpha_K\}$$
$$p_\theta = \frac{[(J^2 - M^2)(J^2 - K^2)]^{1/2}\sin\alpha_J}{J\sin\theta}$$
$$p_\phi = M$$
$$p_\psi = K.$$

Space fixed

$$J_x = -\cos\phi\left(\cot\theta\, p_\phi - \frac{1}{\sin\theta}\, p_\psi\right) - \sin\phi\; p_\theta = (J^2 - M^2)^{1/2}\cos\alpha_M$$

$$J_y = -\sin\phi\left(\cot\theta\, p_\phi - \frac{1}{\sin\theta}\, p_\psi\right) + \cos\phi\; p_\theta = (J^2 - M^2)^{1/2}\sin\alpha_M$$

$$J_z = M.$$

Body fixed

$$J_{x'} = \cos\psi\left(\cot\theta\, p_\psi - \frac{1}{\sin\theta}\, p_\phi\right) + \sin\psi\; p_\theta = (J^2 - K^2)^{1/2}\cos\alpha_K$$

$$J_{y'} = -\sin\psi\left(\cot\theta\, p_\psi - \frac{1}{\sin\theta}\, p_\phi\right) + \cos\psi\; p_\theta = (J^2 - K^2)^{1/2}\sin\alpha_K$$

$$J_{z'} = K.$$

Quantum restrictions

$$\begin{aligned} J &= \left(j + \tfrac{1}{2}\right)\hbar, & j &= 0, 1, 2, \ldots \\ M &= m\hbar, & m &= j, j-1, \ldots, -j \\ K &= k\hbar, & k &= j, j-1, \ldots, -j. \end{aligned}$$

E.5 Hydrogen atom

$$H = \frac{1}{2\mu}(p_r^2 + p_\theta^2/r^2 + p_\phi^2/r^2\sin^2\theta) - \frac{e^2}{r}$$

$$E = \tilde{H}(N) = -\mu e^4/2N^2$$

$$r = (N^2/\mu e^2)(1 + \varepsilon\cos u)$$

$$\cos\theta = \sin\beta\cos(\alpha_L + \psi)$$

$$\sin\theta\exp(\pm \mathrm{i}\phi) = -[\cos\beta\cos(\alpha_L + \psi) \pm \mathrm{i}\sin(\alpha_L + \psi)]\exp(\pm \mathrm{i}\alpha_M)$$

$$p_r = \frac{\mu e^2(N^2 - L^2)^{1/2}\sin u}{N^2(1 + \varepsilon\cos u)}$$

$$p_\theta = \frac{L(L^2 - M^2)^{1/2}\sin(\alpha_L + \psi)}{[L^2\sin^2(\alpha_L + \psi) + M^2\cos^2(\alpha_L + \psi)]^{1/2}}$$

$$p_\phi = M,$$

where

$$\cos\beta = M/L$$

$$\varepsilon = (N^2 - L^2)^{1/2}/N.$$

Auxiliary relations

$$\alpha_N = u + \varepsilon \sin u$$

$$\cos\psi = \frac{\cos u + \varepsilon}{1 + \varepsilon \cos u}$$

$$\sin\psi = \frac{(1-\varepsilon^2)^{1/2} \sin u}{1 + \varepsilon \cos u}.$$

Quantum restrictions

$$\begin{aligned} N &= n\hbar, & n &= 1, 2, 3, \ldots \\ L &= \left(l + \tfrac{1}{2}\right)\hbar, & l &= n-1, n-2, \ldots, 0 \\ M &= m\hbar, & m &= l, l-1, \ldots, -l. \end{aligned}$$

Appendix F
The monodromy matrix

It is convenient to introduce the monodromy matrix

$$M = \begin{pmatrix} M_{qq} & M_{qp} \\ M_{pq} & M_{pp} \end{pmatrix} \tag{F.1}$$

in a one-dimensional context, although, as explained below, the results readily generalize to n dimensions. Its purpose is to measure the local spreading of the classical motion around the path of a reference trajectory $[q_{\rm ref}(t), p_{\rm ref}(t)]$. Thus the displacements $[\Delta q(t), \Delta p(t)] = [q(t) - q_{\rm ref}(t),\ p(t) - p_{\rm ref}(t)]$ at time t are related to those at $t = 0$ by

$$\begin{pmatrix} \Delta q(t) \\ \Delta p(t) \end{pmatrix} = \begin{pmatrix} M_{qq}(t) & M_{qp}(t) \\ M_{pq}(t) & M_{pp}(t) \end{pmatrix} \begin{pmatrix} \Delta q(0) \\ \Delta p(0) \end{pmatrix} \tag{F.2}$$

or

$$z(t) = M(t)\ z(0), \tag{F.3}$$

where $z(t)$ is the column vector $[\Delta q(t), \Delta p(t)]^T$. It is obvious from (F.2) that $M(t)$ reduces to the unit matrix at time $t = 0$, and that its subsequent evolution follows from Hamilton's equations for z. The notation $H_q = \partial H/\partial q$, $H_{qp} = \partial^2 H/\partial q \partial p$, etc. will be employed, while the symbol Δ indicates a difference from the reference trajectory. Thus

$$\begin{aligned} \frac{\mathrm{d}}{\mathrm{d}t} \begin{pmatrix} \Delta q \\ \Delta p \end{pmatrix} &= \begin{pmatrix} \Delta H_p \\ -\Delta H_q \end{pmatrix} = \begin{pmatrix} 0 & 1 \\ -1 & 0 \end{pmatrix} \begin{pmatrix} \Delta H_q \\ \Delta H_p \end{pmatrix} \\ &\simeq \begin{pmatrix} 0 & 1 \\ -1 & 0 \end{pmatrix} \begin{pmatrix} H_{qq} & H_{qp} \\ H_{pq} & H_{pp} \end{pmatrix} \begin{pmatrix} \Delta q \\ \Delta p \end{pmatrix}, \end{aligned} \tag{F.4}$$

or

$$\frac{\mathrm{d}z}{\mathrm{d}t} \simeq J\ \left[\partial^2 H\right]\ z, \tag{F.5}$$

where J denotes the *symplectic matrix* to the right of the two final forms in (F.4), which has the important properties

$$J^T = -J \qquad \text{and} \qquad J^2 = -I, \tag{F.6}$$

where I is the unit matrix. The approximation in (F.4) and (F.5) assumes a quadratic approximation to $H(q, p)$ around the reference trajectory.

The corresponding equation for $M(t)$ follows from (F.2) and (F.3), namely

$$\frac{\mathrm{d}M}{\mathrm{d}t} = J\ \left[\partial^2 H\right]\ M, \tag{F.7}$$

which leads to a variety of important properties. Note first, by combining (F.7) with its transpose and employing the result in (F.6), that

$$\frac{\mathrm{d}}{\mathrm{d}t}(M^T JM) = \frac{\mathrm{d}M^T}{\mathrm{d}t} JM + M^T J\frac{\mathrm{d}M}{\mathrm{d}t} = 0. \tag{F.8}$$

Moreover, since $M(0) = I$, the constant matrix $M^T JM$, implied by (F.8) is given by

$$M^T JM = J. \tag{F.9}$$

Now the determinant of a matrix product is the product of the individual determinants, and $M(0) = I$, which means that

$$\det(M) = 1. \tag{F.10}$$

Finally, it is readily verified that eqns (F.7)–(F.10) generalize to n dimensions, simply by replacing individual elements by block matrices. Thus, for example,

$$M^T = \begin{pmatrix} M_{qq}^T & M_{pq}^T \\ M_{qp}^T & M_{pp}^T \end{pmatrix}, \tag{F.11}$$

where M_{qp}^T denotes the transpose of the M_{qp} block of M. It follows, with the help of (F.9), that

$$M_{qq}^T M_{pp} - M_{pq}^T M_{qp} = M_{pp}^T M_{qq} - M_{qp}^T M_{pq} = I, \tag{F.12}$$

$$M_{qq}^T M_{pq} - M_{pq}^T M_{qq} = M_{pp}^T M_{qp} - M_{qp}^T M_{pp} = 0 \tag{F.13}$$

and, by transposing (F.13),

$$M_{pq}^T M_{qq} - M_{qq}^T M_{pq} = M_{qp}^T M_{pp} - M_{pp}^T M_{qp} = 0. \tag{F.14}$$

Similar connections between the blocks of M^T and those of M^{-1} may be obtained by rearranging (F.9) to the form

$$M^{-1} = JM^T J. \tag{F.15}$$

A symplectic algorithm for the numerical solution of eqn (F.7) has been given by Brewer et al. (1997)

Eigenvalues, eigenvectors and residues

Given a set of eigenvectors z_i, such that $Mz_i = \lambda_i z_i$, and $z_j^T M^T = \lambda_j z_j^T$, it follows from eqn (F.9) that

$$z_j^T M^T JMz_i = z_j^T Jz_i = \lambda_i\lambda_j z_j^T Jz_i. \tag{F.16}$$

Thus either $\lambda_i\lambda_j = 1$ or $z_j^T J z_i = 0$. The first of these alternatives means that the eigenvalues appear as inversely related pairs $\left(\lambda_i, \lambda_i^{-1}\right)$. Moreover, since the elements of M are real, the complex conjugate, λ_i^*, is also an eigenvalue. The possibilities arising from these conditions are listed in Table F.1, together with associated properties of M.

The second alternative above, which is termed the symplectic orthogonality condition, also applies when $\lambda_i\lambda_j = 1$, because, according to eqns (F.6) and (F.15),

$$z_j^T J^{-1} M^{-1} z_i = z_j^T M^T J z_i = -\lambda_i^{-1} z_j^T J z_i = \lambda_j z_j^T J z_i, \tag{F.17}$$

from which $z_j^T J z_i = 0$. The obvious normalization is chosen such that $z_j^T J z_i = \delta_{ij}$, or $Z^T J Z = I$, where the columns of Z are the eigenvectors z_i. It follows that M may be diagonalized by the transformation

$$Z^T J M Z = \Lambda, \tag{F.18}$$

where Λ and Z are the eigenvalue and eigenvalue matrices respectively. Finally, it is convenient to define the 'residue' of M by the equation

$$R = \det(M - I) = \det(\Lambda - I), \tag{F.19}$$

after recognizing that

$$\det[Z^T J(M - I)Z] = \det\left(Z^T J Z\right)\det(M - I) = \det(M - I). \tag{F.20}$$

There is then a convenient factorization

$$R = \prod_\nu R^{(\nu)} = \prod_\nu \det(\Lambda^{(\nu)} - I^{(\nu)}), \tag{F.21}$$

where ν denotes one of the pairs or quadruplets in Table F.1.

Table F.1 Types and properties of possible eigenvalue pairs or quadruplets.

name	$\{\lambda_i\}$	$\mathrm{Tr}(M)$	R
elliptic	e^{iv}, e^{-iv}	$-1 < 2\cos v < 1$	$0 < 4\sin^2(v/2) < 1$
parabolic	1, 1	2	0
inverse parabolic	-1, -1	-2	4
hyperbolic	e^u, e^{-u}	$2 < 2\cosh u < \infty$	$-\infty < -4\sinh^2(u/2) < 0$
inverse hyperbolic	$-e^u$, $-e^{-u}$	$-\infty < 2\cosh u < -2$	$1 < 4\cosh^2(u/2) < \infty$
loxodromic	e^{u+iv}, e^{u-iv}, e^{-u+iv}, e^{-u-iv}	$-\infty < \cosh u \cos v < \infty$	$0 < 4\left[\cosh(u) - \cos(v)\right]^2 < \infty$

Appendix G
Solutions to problems

The initials AS, DW and GR are abbreviations for Abramowitz and Stegun (1965), Dwight (1961) and Gradsteyn and Ryyzhik (1980) respectively.

G.1 Introduction

No problems.

G.2 Phase integral approximations

2.1 Compare the first-order condition $\frac{\hbar^2}{2m}\left[\left(\frac{\partial S^0}{\partial x}\right)_t\right]^2 + V(x) = -\left(\frac{\partial S^0}{\partial t}\right)_x$ with the third equation, to deduce that $(\partial S^0/\partial x)_t = p(x,t)$ and $(\partial S^0/\partial t)_x = -H(p,x,t)$. Finally Hamilton's equations follow from equality between mixed second derivatives of $S_0(x,t)$, with $p(x,t)$ taken as a dependent variable. Remember that $\mathrm{d}p/\mathrm{d}t = (\partial p/\partial t)_x + (\partial p/\partial x)_t \mathrm{d}x/\mathrm{d}t$.

2.2

(i) The result follows by careful differentiation.

(ii) With $k^2(\theta) = (l+\frac{1}{2})^2\sin^2\theta + \frac{1}{4}$, the validity criterion becomes $k^{-2}(\theta)\,|\mathrm{d}k/\mathrm{d}\theta| = |\cos\theta|/\left[2k^3(\theta)\right] \ll 1$, which will be satisfied if $(l+\frac{1}{2})\sin\theta \gg 1$, because $|\cos\theta| < 1$. Moreover, it is justified, under these conditions, to ignore the term in $\sin\theta$ in the equation for χ, so that $\chi \simeq A\cos\left[(l+\frac{1}{2})\theta + \alpha\right]$.

(iii) The first identity requires that

$$\cos\left[\left(l+\frac{1}{2}\right)\theta + \alpha\right] = \cos\left[\left(l+\frac{1}{2}\right)(\pi-\theta) + \alpha - l\pi\right],$$

from which $\alpha = -\pi/4$. In addition, the second identity and the hint imply that $A^2 \simeq 2\left[1+\pi(l+1/2)\right]^{-1}$. The stated result follows because the inequality $(l+\frac{1}{2})\sin\theta \gg 1$ implies $(l+\frac{1}{2})\pi \gg 1$.

2.3 The second equation follows from the first, via the identity $\mathrm{d}^2\psi/\mathrm{d}R^2 = \mathrm{e}^{-3x/2}\left[\mathrm{d}^2\phi/\mathrm{d}x^2 - 1/4\right]$. The JWKB solution to the second equation then implies

$$\begin{aligned}\psi = \mathrm{e}^{x/2}\Phi &\simeq \left[k_l(\mathrm{e}^x)\right]^{-1/2}\sin\left[\int_\alpha^x \mathrm{e}^x k_l(\mathrm{e}^x)\mathrm{d}x + \frac{\pi}{4}\right]\\ &= \left[k_l(\mathrm{r})\right]^{-1/2}\sin\left[\int_a^{\mathrm{r}} k_l(\mathrm{r})\mathrm{d}\mathrm{r} + \frac{\pi}{4}\right].\end{aligned}$$

2.4 The exponent $f(\theta) = z\sin\theta - n\theta$ is stationary at $\cos\theta_0 = n/z \ll 1$ so that $\theta_0 \simeq 0$ and $f''(\theta_0) = -z\sin\theta_0 \simeq -z$. Hence, using eqn (2.50),

$$J_n(z) \simeq \mathrm{Re}\,\frac{1}{\pi}\sqrt{2\pi\mathrm{i}/f''(\theta_0)}\exp\left[\mathrm{i}f(\theta_0)\right] = \sqrt{\frac{2}{\pi z}}\cos\left[z - \left(n + \frac{1}{2}\right)\frac{\pi}{2}\right].$$

2.5 The substitutions $a = \alpha^{-1}\ln(A/E)$ and $u = 1 + \sqrt{1 - \exp[\alpha(R-a)]}$ yield the desired result.

2.6 The substitution $k_l^2(R) = k^2\left(1 - a^2/R^2\right)$ is employed with $a = (l + 1/2)/k$, and the required integrals are given in § 2.274.3 of GS. The result for $R \to \infty$ is given by

$$j_l(kR) \simeq AR^{-1}k_l^{-1/2}(R)\sin\left(k\int_a^R \frac{\sqrt{R^2 - a^2}}{R}\mathrm{d}R + \frac{\pi}{4}\right) \sim (C/kR)\sin\left(kR - l\pi/2\right).$$

Similarly, in the limit $R \to 0$,

$$j_l(kR) \simeq A'\left|k_l(R)\right|^{-1/2}\exp\left(-k\int_R^a \frac{\sqrt{a^2 - R^2}}{R}\mathrm{d}R\right) \sim C'(kR)^l.$$

2.7 The substitution $P(\cos\theta) = A\left[z(\theta)\right]J\left[z(\theta)\right]$ in the Legendre equation requires that

$$\frac{\mathrm{d}^2J}{\mathrm{d}z^2} + \frac{1}{k}\left(\frac{2}{A}\frac{\mathrm{d}A}{\mathrm{d}\theta} + \cot\theta\right)\frac{\mathrm{d}J}{\mathrm{d}z} + \frac{1}{k^2}\left(l(l+1) + \frac{1}{A}\frac{\mathrm{d}^2A}{\mathrm{d}\theta^2} + \frac{\cot\theta}{A}\frac{\mathrm{d}A}{\mathrm{d}\theta}\right)J = 0.$$

Consistency with the first derivative term in the Bessel equation requires that

$$\frac{1}{k}\left(\frac{2}{A}\frac{\mathrm{d}A}{\mathrm{d}\theta} + \cot\theta\right) = \frac{1}{z(\theta)} = \frac{1}{k\theta},$$

which is a differential equation for $A(\theta) = (\theta/\sin\theta)^{1/2}$. The integration constant has been taken as unity, because both $P_l(\cos\theta)$ and $J_0(k\theta)$ tend to unity as $\theta \to 0$. One also finds after some manipulation that

$$\frac{1}{k^2}\left(l(l+1) + \frac{1}{A}\frac{\mathrm{d}^2A}{\mathrm{d}\theta^2} + \frac{\cot\theta}{A}\frac{\mathrm{d}A}{\mathrm{d}\theta}\right) = \frac{1}{k^2}\left((l+1/2)^2 + \frac{1}{4\sin^2\theta} - \frac{1}{4\theta^2}\right).$$

Moreover, § 1.411.6 shows that $\mathrm{cosec}^2\theta - \theta^{-2}$ vanishes at $\theta = 0$ and increases monotonically to the value $\left(\pi^2 - 4\right)/\pi^2$ at $\theta = \pi/2$. Consistency with the Bessel equation therefore requires that $k = (l+1/2)$ and $(\pi^2-4)/4\pi^2 \ll (l+1/2)^2$. Finally, the symmetry

$$P_l(\cos\theta) = (-1)^l P_l(-\cos\theta) = (-1)^l P_l[\cos(\pi-\theta)]$$

suggests the approximation

$$P_l(\cos\theta) \simeq (-1)^l \sqrt{\frac{(\pi-\theta)}{\sin\theta}} J_0\left[(l+1/2)(\pi-\theta)\right]$$

for $\pi/2 < \theta < \pi$.

G.3 Quantization

3.1 The substitution leads to § 858.553 of DW, from which $(v+1/2) = \sqrt{2mD/\alpha^2\hbar^2}(1-\sqrt{\varepsilon})$, which rearranges to the desired expression. The exact form includes a contribution from the second-order integral in eqn (2.111).

3.2

(i) Within the stated approximations,

$$\begin{aligned} F &\simeq \frac{1}{16}B_e A_1 A_3 + B_e^{1/2} A_1 (v+1/2) - \frac{1}{4}B_2 F^2 \\ &\simeq \frac{1}{16}B_e A_1 A_3 + B_e^{1/2} A_1 (v+1/2) - \frac{1}{4}B_e A_1^2 B_2 (v+1/2)^2. \end{aligned}$$

(ii) Expansions for U' and the required powers of $U^{1/2}$ lead to

$$\begin{aligned} a_0\left[2\xi + 3a_1\xi^2 + 4a_2\xi^3 + \ldots\right] &\simeq A_1 a_0^{1/2}\xi + \left[\frac{1}{2}A_1 a_0^{1/2} a_1 + A_2 a_0\right]\xi^2 \\ &+ \left[4a_0 a_2 - \frac{1}{2}A_1 a_0^{1/2}\left(a_2 - \frac{1}{4}a_1^2\right) - A_2 a_0 a_1 - A_3 a_0^{3/2}\right]\xi^3 + \ldots, \end{aligned}$$

while inversion of the A_ν expansion leads to the expressions for B_ν. The resulting formulae for A_ν and B_ν lead to the expressions for Y_{i0} for $i = 0, \ldots, 2$.

(iii) The expression for Y_{11} follows from the centrifugally modified coefficient $A_1^J = 2\left[a_0 + 3(1+a_1)J(J+1)B_e\right]^{1/2} \simeq B_e^{-1/2}[\omega_e + 6(1+a_1)J(J+1)B_e^2/\omega_e]$.

(iv) The formula for Y_{00} follows by substituting for a_1 and a_2 in terms of Y_{20} and Y_{11}.

3.3 Substitute $y = \mathrm{i}x$ in the second integral. The phase integrals in eqn (3.72) are then given by $2\alpha = 2\delta = \pi\epsilon = aE^{3/4}$ and $\kappa = \exp(-\pi\epsilon)$. Following Senn (1987) the phase correction ϕ is evaluated with the help of the algorithm, with $N = 50$:

$$\begin{aligned} \arg\Gamma(1/2 + \mathrm{i}x) \simeq &\frac{1}{2}x\ln\left[(N+1/2)^2 + x^2\right] + N\arctan\left[x/(N+1/2)\right] \\ &+ \arg\left[z^{-1}/12 - z^{-3}/360 + z^{-5}/1260\right] \\ &- \sum_{r=1}^{N}\arctan\left[x/(r+1/2)\right], \end{aligned}$$

where $z = N + 1/2 + \mathrm{i}x$. Eight-figure accuracy is obtained with $N = 50$. Solutions to the quantization equation, which is abbreviated as $f(E) = 0$, are obtained by looking for a sign change in $f(E)$ over a coarse grid and refining the solution by cycles of the algorithm $E_{r+1} = (f_{r-1}E_r - f_r E_{r-1})/(f_{r-1} - f_r)$, where $f_r = f(E_r)$.

The resulting eigenvalues, which are given in Table G.1, are more accurate than the first-order values in Table 2.1, but less accurate than the third-order results, except for $v = 0$.

3.4 Eqn (3.72) rearranges to

$$\left[\sqrt{1+\kappa^2}+\kappa\right]\cos\alpha'\cos\delta' = \left[\sqrt{1+\kappa^2}-\kappa\right]\sin\alpha'\sin\delta'.$$

Hence, by use of eqn (3.81), $(E_a - E)(E_d - E_d) = V_{ad}^2$, with V_{ad} in the given form.

The eigenvector coefficient ratio $(c_a/c_d) = V_{ad}/(E - E_a)$ may be compared with

$$(D/A) \simeq \left[\sqrt{1+\kappa^2}+\kappa\right]\cos\alpha'\sin\delta' \simeq (-1)^{v_a+v_d}\left[\sqrt{1+\kappa^2}+\kappa\right](E-E_a)\pi/\hbar\bar{\omega}_a.$$

The stated result follows by comparison with the formula for V_{ad}.

3.5 Eqn (3.92) applies, except with the final factor $\cos r\pi = \pm 1$ for $r = 0, 1$. Since the required energies lie below the barrier, the phase integrals are given by

$$\alpha = \int_a^b \sqrt{E - 2q\cos 2x}\,\mathrm{d}x \quad \text{and} \quad \pi\epsilon = \int_{-a}^a \sqrt{2q\cos 2x - E}\,\mathrm{d}x\ ,$$

where $a = \pi - b = \frac{1}{2}\arccos(E/2q)$. They are conveniently evaluated by the Gaussian quadrature formula AS § 25.4.41, in which

$$f(x) = \sqrt{(E - 2q\cos 2x)/[(b-x)(x-a)]}$$

for the α integral. The phase correction is evaluated as in problem 3.3. The resulting energy levels for $r = 1$, namely -5.717, 2.181 and 7.547, are systematically higher than the exact values by 0.092–0.098 units, while those for $r = 0$, namely -5.708 and 1.946, are higher by 0.082–0.088 units, discrepancies that point to the need for third-order quantization corrections, such as Y_{00} in problem 3.2.

3.6 Eqn (3.59) implies that

$$\begin{aligned} B_+ &= -\mathrm{i}\sqrt{1-\lambda^2}\mathrm{e}^{\mathrm{i}\theta'_+}A_+ + \mathrm{i}\lambda\mathrm{e}^{\mathrm{i}\theta'_2}A_- \\ \mathrm{e}^{\mathrm{i}\gamma} &= \lambda\mathrm{e}^{\mathrm{i}\alpha_+ - \mathrm{i}\pi/4}A_+ + \sqrt{1-\lambda^2}\mathrm{e}^{\mathrm{i}(\alpha_- - \chi) - \mathrm{i}\pi/4}A_-. \end{aligned}$$

Table G.1 Estimated quartic oscillator eigenvalues.

v	E	v	E	v	E
0	1.05182	2	7.43623	4	16.24913
1	3.77318	3	11.62928	5	21.22774

The real part of the first equation and the moduli on the two sides of the second yield

$$0 = \sqrt{1-\lambda^2}\cos\theta'_+ A_+ - \lambda\cos\theta'_2 A_-,$$
$$1 = \lambda^2 A_+^2 + 2\lambda\sqrt{1-\lambda^2}\cos(\theta'_+ - \theta'_2)A_+A_- + (1-\lambda^2)A_-^2.$$

The required expression for A_-^2 follows by eliminating A_+ and substituting $\lambda^{-2} = 1+u$.

The final reduction requires the approximations $\cos\theta'_2 \simeq (-)^{v_2}(E-E_2)\pi/\hbar\omega_2$ and $\cos(\theta'_+ - \theta'_2) \simeq (-1)^{v_2+1}\sin\theta'_+$, so that, after completing the square, the denominator reads

$$\begin{aligned} D &= \left[(E-E_2)\pi/\hbar\omega_2 - u\sin\theta'_+\cos\theta'_+\right]^2 + u^2\cos^2\theta'_+(1-\sin^2\theta'_+) \\ &= \left(\frac{\pi}{\hbar\omega_2}\right)^2\left\{\left[(E-E_2)-\Delta\right]^2 + \Gamma_2^2/4\right\} \end{aligned}$$

and the numerator becomes $N \simeq u\cos^2\theta'_+ = \pi\Gamma_2/2\hbar\omega_2$. The ratio N/D takes the desired Lorentzian form.

G.4 Angle–action variables

4.1 The substitution casts eqn (4.12) into the following form with the help of DW§ 446.00 or GR§ 2.553.3:

$$t = \tan\frac{\alpha}{2} = \sqrt{\frac{1+\sqrt{\varepsilon}}{1-\sqrt{\varepsilon}}}\tan\frac{u}{2}.$$

Thus, after some manipulation

$$\cos\alpha = \frac{1-t^2}{1+t^2} = \frac{\cos u - \varepsilon}{1-\sqrt{\varepsilon}\cos u},$$

from which the expression for $\exp(-aq)$ is obtained by back-substituting for $\cos u$.

The corresponding expression for p may be derived (albeit with a sign ambiguity) by substituting for $\exp(-aq)$ in the Hamiltonian. It is, however, more illuminating to employ the identity

$$\begin{aligned} p = m\dot{q} &= m\left(\frac{\partial q}{\partial\alpha}\right)\dot{\alpha} = m\left(\frac{\partial q}{\partial\alpha}\right)\left(\frac{\partial H}{\partial I}\right) \\ &= -\frac{m\sqrt{\varepsilon}\sin\alpha}{a(1+\sqrt{\varepsilon}\cos\alpha)}\sqrt{\frac{2D(1-\varepsilon)}{m}} = -\frac{\sqrt{2mD}}{a}\frac{\sqrt{\varepsilon(1-\varepsilon)}\sin\alpha}{(1+\sqrt{\varepsilon}\cos\alpha)}. \end{aligned}$$

4.2 Inversion of the linear relations $p_\xi = -(\partial F_3/\partial\xi) = \xi p_x + \eta p_y$ and $p_\eta = -(\partial F_3/\partial\eta) = -\eta p_x + \xi p_y$ provides the desired forms for p_x and p_y and hence for the transformed Hamiltonian. The identity

$$\frac{1}{2}\left[p_\xi^2 + p_\eta^2 + \omega^2(\xi^6 + \eta^6)\right] - \left(\xi^2 + \eta^2\right)H = 0$$

allows the separation

$$p_\xi^2 = 2E\xi^2 - \omega^2\xi^6 + K, \qquad p_\eta^2 = 2E\eta^2 - \omega^2\eta^6 - K,$$

in terms of which $J_\xi = \oint p_\xi \mathrm{d}\xi$ and $J_\eta = \oint p_\eta \mathrm{d}\eta$.

4.3

(i) It is convenient to define polar variables R and $\theta_\pm$, such that

$$r + \mathrm{i}p_r \pm p_\phi/r = \frac{1}{r}\left[I + \sqrt{I^2 - L^2}\cos 2\alpha_I - \mathrm{i}\sqrt{I^2 - L^2}\sin 2\alpha_I \pm L\right]$$
$$= \frac{\sqrt{I \pm L}}{r}\left[\sqrt{I \pm L}\mathrm{e}^{\mathrm{i}\alpha_I} + \sqrt{I \mp L}\mathrm{e}^{-\mathrm{i}\alpha_I}\right]\mathrm{e}^{-\mathrm{i}\alpha_I} = \frac{\sqrt{I \pm L}}{r}R\,\mathrm{e}^{\mathrm{i}\theta_\pm}.$$

Hence

$$R^2 = 2\left[I + \sqrt{I^2 - L^2}\cos 2\alpha_I\right] = 2r^2$$

and

$$\tan\theta_\pm = \frac{\left[\sqrt{I \pm L} - \sqrt{I \mp L}\right]\sin\alpha_I}{\left[\sqrt{I \pm L} + \sqrt{I \mp L}\right]\cos\alpha_I} = \pm\frac{L\tan\alpha_I}{I + \sqrt{I^2 - L^2}} = \pm\tan(\phi - \alpha_L),$$

where $\alpha_L = \alpha_\phi$. Thus $R\,\exp(\mathrm{i}\theta_\pm) = \sqrt{2}r\exp[\pm\mathrm{i}(\phi - \alpha_l)]$.

(ii) It follows, by combining these results, and taking their complex conjugates, that

$$a_d = \frac{1}{2}(r + \mathrm{i}p_r + p_\phi/r)\mathrm{e}^{-\mathrm{i}\phi} = \sqrt{(I + L)/2}\mathrm{e}^{-\mathrm{i}\alpha_I - \mathrm{i}\alpha_L}$$

$$a_g = \frac{1}{2}(r + \mathrm{i}p_r - p_\phi/r)\mathrm{e}^{-\mathrm{i}\phi} = \sqrt{(I - L)/2}\mathrm{e}^{-\mathrm{i}\alpha_I + \mathrm{i}\alpha_L}$$

$$a_d^\dagger = (a_d)^* = \frac{1}{2}(r - \mathrm{i}p_r + p_\phi/r)\mathrm{e}^{\mathrm{i}\phi} = \sqrt{(I + L)/2}\mathrm{e}^{+\mathrm{i}\alpha_I + \mathrm{i}\alpha_L}$$

$$a_g^\dagger = (a_g)^* = \frac{1}{2}(r - \mathrm{i}p_r - p_\phi/r)\mathrm{e}^{\mathrm{i}\phi} = \sqrt{(I + L)/2}\mathrm{e}^{+\mathrm{i}\alpha_I - \mathrm{i}\alpha_L}$$

4.4 The angle between $\mathbf{j}_1$ and $\mathbf{j}_2$ is the complement to the angle opposite $\mathbf{J}$ in the triangle that defines the resultant. Hence

$$J^2 = j_1^2 + j_2^2 - 2j_1j_2\cos(\pi - \theta),$$

so that

$$\cos\theta = -\frac{J^2 - j_1^2 - j_2^2}{2j_1j_2}.$$

The condition that α_1 and α_2 are measured from the common normal to $\mathbf{j}_1$ and $\mathbf{j}_2$ means, in the notation of Fig. 4.3, that $\mathbf{y}_1' = \mathbf{y}_2'$. Hence, from eqn (4.76),

$$\cos\chi = \mathbf{r}_1 \cdot \boldsymbol{r}_2 = \sin\alpha_1\sin\alpha_2\mathbf{x}_1' \cdot \boldsymbol{x}_2' + \cos\alpha_1\cos\alpha_2,$$

because $\mathbf{x}_1' \cdot \boldsymbol{y}' = \mathbf{x}_2' \cdot \boldsymbol{y}' = 0$. In addition, $\mathbf{x}_1'$, $\mathbf{x}_2'$, $\mathbf{j}_1$ and $\mathbf{j}_2$ all lie in the plane normal to $\mathbf{y}'$ with $\mathbf{x}_1' \cdot \boldsymbol{j}_1 = \mathbf{x}_2' \cdot \boldsymbol{j}_2 = 0$. Consequently

$$\mathbf{x}_1' \cdot \boldsymbol{x}_2' = -\cos\theta,$$

which means that

$$\cos\chi = \cos\alpha_1\cos\alpha_2 + \left[\left(J^2 - j_1^2 - j_2^2\right)/2j_1j_2\right]\sin\alpha_1\sin\alpha_2.$$

4.5 The transformed expression for α_N is conveniently expressed in the form

$$\alpha_N = \frac{2\pi}{D}\int_0^{\psi}\frac{1}{2}r^2(\psi)\mathrm{d}\psi,$$

where $r(\psi) = \rho(1-\varepsilon^2)/(1-\varepsilon\cos\psi)$ and $D = \pi\sqrt{1-\varepsilon^2}\rho^2$. Also note by reference to Fig. 4.4 that $\frac{1}{2}r^2(\psi)\mathrm{d}\psi$ is an element of the area of the ellipse. Hence the area swept out increases linearly with α_N and therefore with time. In addition, $D = \pi ab$ is the total area of the ellipse with principal radii a and b. The major radius $a = \rho$ is the average of the maximum and minimum values of r, while b is the largest value of $r\sin\psi$, which is given, according to eqns (4.104) and (4.107), by $b = \sqrt{1-\varepsilon^2}\rho$.

4.6 The separation follows by writing $H = E$ and multiplying through by $\left(\xi^2+\eta^2\right)$ to obtain

$$p_\xi^2 - 2x_1 - 2E\xi^2 + \xi^{-2}J_\varphi^2 + \mathcal{E}\xi^4 = -\left[p_\eta^2 - 2x_1 - 2E\eta^2 + \eta^{-2}J_\varphi^2 - \mathcal{E}\eta^4\right] = 0.$$

Zeroth-order solutions for x_1 and x_2 yield

$$x_1 = (2J_\xi + J_\varphi)\sqrt{-2E^{(0)}}, \qquad x_2 = (2J_\eta + J_\varphi)\sqrt{-2E^{(0)}},$$

from which

$$E^{(0)} = -\frac{(x_1+x_2)^2}{8(J_\xi + J_\eta + J_\varphi)^2} = -\frac{1}{2J^2}.$$

The first-order correction is obtained from the sum

$$\begin{aligned} J = (J_\xi + J_\eta + J_\varphi) &= \frac{(x_1 - x_2)}{\sqrt{-2E^{(1)}}} + \frac{3}{4}\mathcal{E}(-2E^{(0)})^{-3/2}\frac{(x_1^2 - x_2^2)}{2E^{(0)}} \\ &= \frac{1}{\sqrt{-2E^{(1)}}} - \frac{3}{2}\mathcal{E}J^4\left(J_\xi - J_\eta\right) = \frac{1}{\sqrt{-2E^{(1)}}} - \frac{3}{2}\mathcal{E}J^4 J_e. \end{aligned}$$

Thus

$$E^{(1)} = -\frac{1}{2}\left[J - \frac{3}{2}\mathcal{E}J^4 J_e\right]^{-2} \simeq -\frac{1}{2J^2} - \frac{3}{2}\mathcal{E}JJ_e.$$

G.5 Matrix elements

5.1

(i) Note that $(\mathrm{d}q/\mathrm{d}t) = (\mathrm{d}q/\mathrm{d}\alpha)(\mathrm{d}\alpha/\mathrm{d}t)$, where $(\mathrm{d}\alpha/\mathrm{d}t)$ is the frequency, ω, of the classical motion, the average value of which must be taken in order to ensure the Hermitian character of the matrix. Hence

$$\begin{aligned}\langle n'|p|n\rangle &= m\omega(\bar{n})\langle n'|\frac{\mathrm{d}q}{\mathrm{d}\alpha}|n\rangle = \frac{m\omega(\bar{n})}{2\pi}\int_0^{2\pi}\frac{\mathrm{d}q}{\mathrm{d}\alpha}\mathrm{e}^{\mathrm{i}(n-n')\alpha}\mathrm{d}\alpha \\ &= \frac{-\mathrm{i}(n-n')m\omega(\bar{n})}{2\pi}\int_0^{2\pi} q(\alpha,\bar{n})\mathrm{e}^{\mathrm{i}(n-n')\alpha}\mathrm{d}\alpha \\ &= -\mathrm{i}(n-n')m\omega(\bar{n})\langle n'|q|n\rangle.\end{aligned}$$

(ii)

$$\langle n'|\hat{p}|n\rangle = \frac{\mathrm{i}m}{\hbar}\langle n'|\hat{H}\hat{q}-\hat{q}\hat{H}|n\rangle = \frac{-\mathrm{i}(E_n - E_{n'})m}{\hbar}\langle n'|\,\hat{q}\,|n\rangle\,.$$

The two expressions therefore correspond to the extent that $(E_n - E_{n'}) = (n-n')\hbar\omega(\bar{n})$.

5.2 The formula for x in Appendix E.2 implies that

$$\begin{aligned}\langle n'|\mathrm{e}^{-ax}|n\rangle &= \frac{1}{2\pi}\int_0^{2\pi}\frac{(1-\varepsilon)\mathrm{e}^{\mathrm{i}(n-n')\alpha}\mathrm{d}\alpha}{1+\sqrt{\varepsilon}\cos\alpha} = \frac{(1-\varepsilon)}{\pi}\int_0^{\pi}\frac{\cos\Delta n\alpha\mathrm{d}\alpha}{1+\sqrt{\varepsilon}\cos\alpha} \\ &= \sqrt{1-\varepsilon}f(\Delta n,\varepsilon).\end{aligned}$$

The corresponding formula for p gives

$$\begin{aligned}\langle n'|p|n\rangle &= -\frac{\sqrt{2mD\varepsilon(1-\varepsilon)}}{2\pi}\int_0^{2\pi}\frac{\sin\alpha\mathrm{e}^{\mathrm{i}(n-n')\alpha}\mathrm{d}\alpha}{1+\sqrt{\varepsilon}\cos\alpha} \\ &= -\frac{\mathrm{i}\sqrt{2mD\varepsilon(1-\varepsilon)}}{\pi}\int_0^{\pi}\frac{\sin\alpha\sin(\Delta n\alpha)\mathrm{d}\alpha}{1+\sqrt{\varepsilon}\cos\alpha} \\ &= -\frac{\mathrm{i}\sqrt{2mD\varepsilon(1-\varepsilon)}}{2\pi}\int_0^{\pi}\frac{[\cos(\Delta n-1)\alpha-\cos(\Delta n+1)\alpha]\,\mathrm{d}\alpha}{1+\sqrt{\varepsilon}\cos\alpha} \\ &= -\frac{\mathrm{i}\sqrt{2mD\varepsilon(1-\varepsilon)}}{2\pi}[f(\Delta n-1,\varepsilon)-f(\Delta n+1,\varepsilon)] \\ &= -\mathrm{i}\sqrt{2mD}(1-\varepsilon)f(\Delta n,\varepsilon).\end{aligned}$$

The matrix elements of x for $n \neq n'$ are obtained from the result in problem 5.1, together with the identity $\omega = (\partial H/\partial I) = a\sqrt{2D(1-\varepsilon)/m}$ implied by eqn (3.9). Thus

$$\begin{aligned}\langle n'|x|n\rangle &= \frac{\mathrm{i}}{(n-n')m\omega}\langle n'|p|n\rangle = \frac{\sqrt{2mD}(1-\varepsilon)}{\Delta n m\omega}f(\Delta n,\varepsilon) \\ &= -\frac{\sqrt{1-\varepsilon}}{a\Delta n}f(\Delta n,\varepsilon) \quad \text{for} \quad n\neq n'.\end{aligned}$$

Finally, the diagonal matrix element is given by DW§ 865.44, or after integrating by parts by GR§ 3.792.5, in the form

$$\langle n|x|n\rangle = \frac{1}{2\pi a}\int_0^{2\pi} \ln\left(\frac{1+\sqrt{\varepsilon}\cos\alpha}{1-\varepsilon}\right) d\alpha = \frac{1}{a}\ln\left(\frac{1+\sqrt{1-\varepsilon}}{2(1-\varepsilon)}\right).$$

5.3 The required angle–action expressions are

$$\cos\theta = \frac{1}{J}\sqrt{(J^2-M^2)}\cos\alpha_J$$

and

$$\sin\theta e^{i\phi} = -\frac{1}{2J}\left[(J+M)\,e^{i\alpha_J} - (J-M)e^{-i\alpha_J}\right]e^{i\alpha_M}.$$

Thus

$$\begin{aligned}&\langle j'm'|\cos\theta|jm\rangle\\ &= \frac{1}{4\pi^2}\int_0^{2\pi}\int_0^{2\pi}\frac{\sqrt{J^2-M^2}}{J}\cos\alpha_J e^{i[(j-j')\alpha_J+(m-m')\alpha_M}d\alpha_J d\alpha_M\\ &\simeq \frac{1}{2}\left[\frac{\sqrt{\bar{j}_+^2-m^2}}{\bar{j}_+}\delta_{j'j+1}\delta_{m'm} + \frac{\sqrt{\bar{j}_-^2-m^2}}{\bar{j}_-}\delta_{j'j-1}\delta_{m'm}\right],\end{aligned}$$

with the average j values given by $2\bar{j}_+ = j+1/2+j+3/2 = 2(j+1)$ and $2j_- = j+1/2+j-1/2 = 2j$. Consequently

$$\langle j'm'|\cos\theta|jm\rangle = \frac{\sqrt{(j+1)^2-m^2}}{2(j+1)}\delta_{j'j+1}\delta_{m'm} + \frac{\sqrt{j^2-m^2}}{2j}\delta_{j'j-1}\delta_{m'm}.$$

The numerators are seen to agree with the exact quantum mechanical results, while squares of the denominators are in error by a factor of $(4j^2)^{-1}$.

A similar argument applies to the matrix elements of $\sin\theta e^{i\phi}$, with $\bar{m} = m+1/2$, for which

$$\langle j'm'|\sin\theta e^{i\phi}|jm\rangle = -\frac{(j+m+3/2)}{2(j+1)}\delta_{j'j+1}\delta_{m'm+1} + \frac{(j-m-1/2)}{2j}\delta_{j'j-1}\delta_{m'm+1}.$$

The denominators have the same fractional error as for the previous matrix elements, while the squares of the numerators have fractional errors of order $\left[4(j\pm m)^2\right]^{-1}$.

5.4

(i) The transition point condition $x^2+\alpha x-(1+\alpha x_0)=0$ has roots

$$x_\pm = \frac{-\alpha\pm\sqrt{\alpha^2+4(1+\alpha x_0)}}{2},$$

which are real for $\alpha^2+4(1+\alpha x_0)>0$.

The conditions for classical accessibility in the harmonic potential are that $x_+ < 1$ and $x_- > -1$. Hence only x_- is inaccessible for $\alpha > 2$. Turning to the doubly accessible root case, any such pairs must coalesce with $x_+ = x_- > -1$, which requires that $\alpha < 2$. Finally, the real root condition may be combined with a condition on the product, namely $x_+ x_- = 1 + \alpha x_0 < 1$, which together require that $\alpha^2 + 4 > -4\alpha x_0 > 0$.

(ii) The required double integral

$$S_{01} = \frac{\pi^{-1/4}}{2\pi} \int_{-\infty}^{\infty} \int_{-\infty}^{\infty} \exp\left[\frac{\mathrm{i}u^3}{3} - \mathrm{i}\alpha u(x - x_0) - \frac{x^2}{2}\right] \mathrm{d}u\mathrm{d}x$$

is evaluated by reversing the order of integration, performing the Gaussian integral over x and shifting the origin of u to eliminate the quadratic term. Thus

$$\int_{-\infty}^{\infty} \exp\left[-\frac{x^2}{2} - \mathrm{i}\alpha u x\right] \mathrm{d}x = \sqrt{2\pi} \exp\left(-\frac{\alpha^2 u^2}{2}\right)$$

because the vanishing of the integrand as $x \to \pm\infty$ allows a shift of the integration contour to the line $x' = x + \mathrm{i}\alpha u$. In addition, the substitution $u = v - \mathrm{i}\alpha^2/2$ may be verified to cast the resulting exponent, in the u integral, to the form

$$\frac{\mathrm{i}u^3}{3} - \frac{\alpha^2 u^2}{2} + \mathrm{i}\alpha x_0 u = \frac{\mathrm{i}v^3}{3} + \mathrm{i}\left(\frac{\alpha^4}{4} + \alpha x_0\right) v + \frac{\alpha^6}{12} + \frac{\alpha^3 x_0}{2}.$$

The integral over v is now in the correct Airy function form. Thus

$$S_{01} = (4\pi)^{1/4} \exp\left(\frac{\alpha^6}{12} + \frac{\alpha^3 x_0}{2}\right) Ai\left(\frac{\alpha^4}{4} + \alpha x_0\right).$$

It remains to note that the terms of the inequalities allow the approximations

$$\begin{aligned} Ai\left(\frac{\alpha^4}{4} + \alpha x_0\right) &\simeq \frac{\pi^{-1/4}}{2\left(\alpha^4/4 + \alpha x_0\right)^{1/4}} \exp\left[-\frac{2}{3}\left(\frac{\alpha^4}{4} + \alpha x_0\right)^{3/2}\right] \\ &\simeq \frac{\pi^{-1/4}}{\sqrt{2\alpha}} \exp\left[-\frac{\alpha^6}{12}\left(1 + \frac{6x_0}{\alpha^3} + \frac{6x_0^2}{\alpha^6}\right)\right]. \end{aligned}$$

Hence, after collecting terms in the above expression for S_{01} allowed by the stated inequalities,

$$S_{01} \simeq \frac{\pi^{-1/4}}{\alpha} \exp\left(-\frac{x_0^2}{2}\right) = \alpha^{-1}\psi_0(x_0).$$

(iii) The Airy function $Ai(z)$ may be evaluated by the series expansion (Abramowitz and Stegun 1965) for $|z| < 2$ and by the asymptotic approximations for larger $|z|$. Alternatively, Gordon (1969) gives accurate quadrature expressions, valid for all z.

G.6 Semiclassical inversion methods

6.1 The RKR integral transforms, under the suggested substitution, to the standard form (DW§ 260.01 or GR§ 2.261)

$$\begin{aligned} f(U) &= -\frac{2}{a}\int_1^{z(U)} \frac{dz}{\sqrt{z^2 - z^2(U)}} = -\frac{2}{a}\ln\left(\frac{z(U)}{1-\sqrt{1-z^2(U)}}\right) \\ &= \frac{2}{a}\ln\left(\frac{\sqrt{1-U/D}}{1-\sqrt{U/D}}\right) = \frac{1}{a}\ln\left(\frac{\sqrt{D}+\sqrt{U}}{\sqrt{D}-\sqrt{U}}\right). \end{aligned}$$

6.2 The necessary integrals are conveniently taken over n', with the lower limit at which $v = -0.5$, taken as $n' = l + 1/2$. Thus

$$f(n) = r_+ - r_- = 2\int_{l+1/2}^{n} \frac{nn'dn'}{\sqrt{n^2 - n'^2}} = 2n\sqrt{n^2 - (l+1/2)^2}$$

and

$$\begin{aligned} g(n) &= \frac{1}{r_-} - \frac{1}{r_+} = \frac{2}{(l+1/2)}\int_{l+1/2}^{n} \frac{(n')^{-3}\,dn'}{\sqrt{(n')^{-2} - (n)^{-2}}} \\ &= \frac{2}{(l+1/2)}\int_{n^{-1}}^{(l+1/2)^{-1}} \frac{zdz}{\sqrt{z^2 - n^{-2}}} = \frac{2\sqrt{n^2 - (l+1/2)^2}}{n(l+1/2)^2}. \end{aligned}$$

The quadratic equation $r^2 + br + c = 0$, with roots $r_\pm$, has $c = r_+r_- = f/g$ and $b = \sqrt{f^2 + 4f/g}$. Thus, after straightforward manipulation, the turning points are consistent with

$$E(n) = -\frac{1}{2n^2} = -\frac{1}{r} + \frac{(l+1/2)^2}{2r^2} = V_l(r).$$

6.3 Eqn (6.44) provides successive estimates of the dissociation energy, which increase from 2942 cm^{-1} from $\Delta E(0.5)$ to 3355 cm^{-1} at $\Delta E(3.5)$. A quadratic fit to these values predicts a maximum value, $D = 3948$ cm^{-1} at $v_D = 13.5$. However, the assumed quadratic convergence on the dissociation limit ignores the proper limiting behaviour indicated by eqns (6.44)–(6.48). LeRoy (1980) corrects the picture by employing modified versions of eqn (6.46) as specified by one or other of the following $v_D - v$ expansions,

$$\begin{aligned} D - E_v &= X_n(v_D - v)^{(n-1)/2n}\left[1 + a_1(v_D - v) + a_2(v_D - v)^2 + \cdots\right] \\ &= X_n(v_D - v)^{(n-1)/2n}\left[1 + b_1(v_D - v) + b_2(v_D - v)^2 + \cdots\right]^{(n-1)/2n}, \end{aligned}$$

both of which ensure the correct limiting behaviour as $v \to v_D$. Least squares fits to D, v_D and one or other of the coefficient sets yield $D = 4999 \pm 71$ cm^{-1} and $v_D = 40.78$ by the first expansion and $D = 4501 \pm 51$ cm^{-1} and $v_D = 40.82$ by the second. Knowledge of the long-range coefficient C_4 therefore allows a reliable extrapolation from $v = 3.5$ to $v = v_D = 41$.

G.7 Non-separable bound motion

7.1

(i) and (ii) The three representative Lissajous figures shown on the left of Fig. G.1 were run for 2000 time units with a step length of 0.002 units. In addition, 13 trajectories were initiated on the contour in the (x, y) plane at angles $\theta_k = (k\pi/14)$ for $k = 1, \ldots, 13$. Their intersections in the (x, p_x), Poincaré map, at $y = 0$ and $p_y > 0$, are shown after 4000 time steps on the right of the diagram. The trajectories labelled A, B, C indicate different types of trajectory, which occupy the corresponding regions in the Poincaré section.

(iii) Two periodic orbits which intersect the Poincaré section at the centres of regions A and B are stable, while a third unstable orbit intersects at the origin of the separatrix between regions A and C. Their properties are listed in Table G.2. Since $\det A = 1$ the eigenvalues are given by $\lambda = 0.5\left[trA \pm \sqrt{(trA)^2 - 4}\right]$. Thus the orbits are stable for $|trA| < 2$, with $\lambda = e^{\pm iu}$ where $2\cos u = trA$; and unstable for $|trA| > 2$, with $\lambda = e^{\pm v}$ where $2\cosh v = trA$. The properties of the three periodic orbits listed in table G.2 include the intersection point, x_0 at $y = 0$, the time period T and trA.

7.2

(i) The perturbation

$$H' = \frac{1}{2}\left(\frac{\delta}{\omega}\right)^2 \left[(I + J)\cos^2(\alpha_I + \alpha_J) + (I - J)\cos^2(\alpha_I - \alpha_J)\right]$$

averages over a cycle of α_I to the required form.

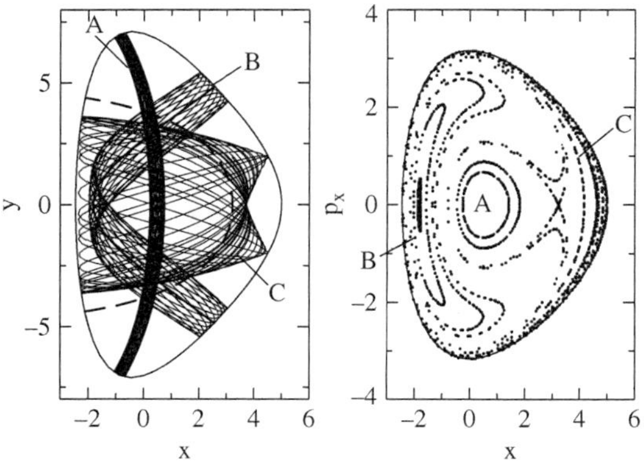

Fig. G.1 Left: Lissajous figures for three representative trajectories at $E = 5$. Right: The Poincaré surface of section. The labels A, B and C indicate the areas occupied by the different trajectory types.

Table G.2 Properties of the periodic orbits at $E = 5$.

orbit	x_0	T	$tr A$
1	−1.83043	12.9344	0.25672
2	0.47444	14.3210	−1.71764
3	3.22744	14.2090	3.32509

(ii) The transformation to polar variables involves lengthy manipulation via the steps

$$c = \cos(\phi - \alpha_\phi) = \frac{\left[I + \sqrt{I^2 - I_\phi^2}\right]^{1/2} \cos\alpha_I}{\left[I + \sqrt{I^2 - I_\phi^2 \cos 2\alpha_I}\right]^{1/2}}$$

$$s = \sin(\phi - \alpha_\phi) = \frac{I_\phi \sin\alpha_I}{\left[I + \sqrt{I^2 - I_\phi^2}\right]^{1/2} \left[I + \sqrt{I^2 - I_\phi^2 \cos 2\alpha_I}\right]^{1/2}}$$

and

$$\begin{aligned} x^2 - y^2 &= r^2(\cos^2\phi - \sin^2\phi) \\ &= r^2\left[(c^2 - s^2)\cos 2\alpha_\phi - 2cs \sin 2\alpha_\phi\right] \\ &= (\mu\omega)^{-1}\left[\left(\sqrt{I^2 - I_\phi^2} + I\cos 2\alpha_I\right)\cos 2\alpha_\phi - I_\phi \sin 2\alpha_I \sin 2\alpha_\phi\right]. \end{aligned}$$

Thus, after averaging over a cycle of α_I the residual slow motion evolves over the surface

$$H' = \frac{1}{2}\left(\frac{\delta^2}{\omega}\right)\sqrt{I^2 - I_\phi^2 \cos 2\alpha_\phi} = \text{const.}$$

7.3

(ii) The separation in problem 4.2 implies that $K = p_\xi^2 - 2E\xi^2 + \omega^2\xi^6 = -\left[p_\eta^2 - 2E\eta^2 + \omega^2\eta^6\right]$. Thus $\partial K/\partial p_\xi = \partial K/\partial \xi = 0$ at $\xi_0^2 = \sqrt{2E/3\omega^2}$ and $K = -\sqrt{32E^3/3\omega^2}$ and similarly for η_0.

(iii) The forward transformation implies $x \pm iy = re^{\pm i\phi} = (\xi \pm i\eta)^2/2$. Thus $\xi \pm i\eta = \sqrt{2r}e^{\pm i\phi/2}$. The corresponding expressions for p_ξ and p_η are given in the solution to problem 4.2.

(iv) The trajectories are conveniently run in Cartesian coordinates and transformed to $(\xi, \eta, p_\xi, p_\eta)$. Sections for the $K < 0$ trajectories are then taken by interpolating for (η, p_η) at a sign change of $\xi - \xi_0$, with $p_\xi > 0$, and similarly for (ξ, p_ξ) at a sign change of η. An enclosed area such as $\oint p_\xi d\xi$ is conveniently estimated via the angle–action transformation $\xi = \sqrt{2I}\cos\alpha$, $p_\xi = -\sqrt{2I}\sin\alpha$.

7.4

(i) It is seen by reference to Appendices E.1 and E.2 that the frequencies of the uncoupled Morse and harmonic oscillator modes are given by $\omega_x = \partial E/\partial I_x = 1-(v_x+1/2)/2D$ and $\omega_y = 0.45$. Hence the modes satisfy the Fermi resonance condition, $\omega_x = 2\omega_y$ at $v_x + 1/2 = 0.2D = 2.5$.

(ii) The coupling term in eqn (7.102) is taken as $H_1 = \lambda xy^2$ and the time period T of the switching term $\xi(t)$ given by (7.97) is taken as 10^3 or 10^4 multiples of the harmonic time period $2\pi/\omega_y$.

(iii) The initial trajectory under H_{Ferm} takes the form of a 2 : 1 periodic orbit, which evolves as a modulated Lissajous figure as the coupling term $H_1 = H - H_{\mathrm{Ferm}}$ is switched on.

G.8 Wavepackets

8.1 After use of the identity $(\partial\phi/\partial t)_x = (\partial\phi/\partial t)_\xi + (\partial\phi/\partial\xi)_t(\partial\xi/\partial t)_x$ the three conditions on the solution arise by comparing coefficients of terms in the Schrödinger equation for Ψ that are respectively linear in $\xi+u$, linear in $(\partial\phi/\partial\xi)_t$ and independent of the latter quantities.

8.2 Eqn (8.50) shows that

$$\begin{aligned}&\int_{-\infty}^{\infty} \mathrm{e}^{\left[ik\xi - z_1^2 + 2(x+\frac{1}{2}\xi)z_1 - \frac{1}{2}(x+\frac{1}{2}\xi)^2 - z_2^2 + 2(x-\frac{1}{2}\xi)z_2 - \frac{1}{2}(x-\frac{1}{2}\xi)^2\right]}\mathrm{d}\xi \\ &= \sum_{n_1 n_2} C_1 C_2 z_1^{n_1} z_2^{n_2} \int_{-\infty}^{\infty} \mathrm{e}^{\mathrm{i}k\xi}\psi_{n_1}(x+\xi/2)\psi_{n_2}(x-\xi/2)\mathrm{d}\xi.\end{aligned}$$

Hence $P_n(x,k)$ arises from terms in $(z_1z_2)^n$ in a series expansion for the left-hand integral, which reduces to

$$I = \sqrt{\pi}\exp\left[-(x^2+k^2) - 2z_1z_2 + 2(x+\mathrm{i}k)z_1 + 2(x-\mathrm{i}k)z_2\right].$$

Contributing terms are therefore of the form

$$\left[(2z_1z_2)^r/r!\right]\left[4\left(x^2+k^2\right)z_1z_2\right]^{n-r}/(n-r)!^2,$$

which depend only on (x^2+k^2).

8.3 The substitution $t' = t + \tau$ restores the form in eqn (11.52) apart from a phase factor $\exp[-\mathrm{i}(n+1/2)\tau]$, which has no effect on $|\psi_n(x)|^2$.

8.4. Solving Hamilton's equations, $q_t = q_0\cos\omega t + (p_0/m\omega)\sin\omega t$ and $p_t = -q_0 m\omega\sin\omega t + p_0\cos\omega t$. Moreover, with the help of the first hint,

$$\begin{aligned}S_t(q_0,p_0) &= \int_0^t [p\dot{q} - H]\,\mathrm{d}t = \frac{1}{2}[p_tq_t - p_0q_0] \\ &= \frac{m\omega}{2}\left\{\left[\left(\frac{p_0}{m\omega}\right)^2 - q_0^2\right]\sin\omega t\cos\omega t - \frac{2q_0p_0}{m\omega}\sin^2\omega t\right\}.\end{aligned}$$

Thus the phase term $\phi_t(q_0, p_0)$ is strictly quadratic in (q_0, p_0) with a single stationary point at which $q_0 = x_0$ and $q_t = x_t$, namely $p_0 = m\omega(q_t - q_0 \cos\omega t)/\sin\omega t$, at which

$$S_t = \frac{m\omega}{2\sin\omega t}\left\{\left[x_t^2 + x_0^2\right]\cos\omega t - 2x_0x_t\right\}.$$

Moreover, $\left(\partial^2 S/\partial q_0 \partial q_t\right) = \left(\partial p_0/\partial q_t\right) = \left(m\omega/\sin\omega t\right)$. Consequently

$$\left\langle x_t|\mathrm{e}^{-\mathrm{i}Ht/\hbar}|x_0\right\rangle = \left(\frac{m\omega}{2\pi\mathrm{i}\hbar\sin\omega t}\right)^{1/2}\exp\left\{\frac{\mathrm{i}m\omega}{2\hbar\sin\omega t}\left[(x_0^2 + x_t^2)\cos\omega t - 2x_0x_t\right]\right\},$$

which is the required result.

8.5

(i) By use of the chain rule

$$\left(\frac{\partial S}{\partial p_{1i}}\right)_{q_1} = \sum_j \left(\frac{\partial S}{\partial q_{2j}}\right)_{q_1}\left(\frac{\partial q_{2j}}{\partial p_{1i}}\right)_{q_1} = \sum_j p_{2j}M_{q_jp_i}$$

$$\left(\frac{\partial S}{\partial q_{1i}}\right)_{p_1} = \left(\frac{\partial S}{\partial q_{1i}}\right)_{q_2} + \sum_j \left(\frac{\partial S}{\partial q_{2j}}\right)_{q_1}\left(\frac{\partial q_{2j}}{\partial q_{1i}}\right)_{p_1} = -p_{1_i} + \sum_j p_{2j}M_{q_jq_i}.$$

The required second derivatives are given by

$$\left(\frac{\partial^2 S}{\partial p_{1k}\partial p_{1i}}\right) = \sum_j M_{p_jp_k}M_{q_jp_i} = \sum_j \left(M_{pp}^T\right)_{kj}\left(M_{qp}\right)_{ji}$$

$$\left(\frac{\partial^2 S}{\partial p_{1k}\partial q_{1i}}\right) = -\delta_{ki} + \sum_j M_{p_jp_k}M_{q_jq_i} = -I_{ki} + \sum_j \left(M_{pp}^T\right)_{kj}\left(M_{qq}\right)_{ji},$$

in which the final expressions are written as the appropriate matrix products.

G.9 Atom–atom scattering

9.1

(i) The integrals in eqns (9.7) and (9.25) are given according to DW § § 281.01 and 301.01 by

$$\chi(\lambda) = \pi - 2\lambda\int_d^\infty \frac{\mathrm{d}R}{R^2\sqrt{k^2 - \lambda^2/R^2}} = 2\cos^{-1}\left(\frac{\lambda}{kd}\right).$$

$$\eta(\lambda) = \lim_{R\to\infty}\left\{\int_d^\infty \sqrt{k^2 - \lambda^2/R^2}\mathrm{d}R - kR + \lambda\frac{\pi}{2}\right\}$$

$$= -\sqrt{k^2d^2 - \lambda^2} + \lambda\cos^{-1}\left(\frac{\lambda}{kd}\right).$$

(ii) The expressions for $(\mathrm{d}\sigma/\mathrm{d}\Omega)_{\mathrm{cl}}$, σ_{cl} and σ_{semi} follow directly from eqns (9.13), (9.15) and (9.21).

(iii) The partial-wave sum for $(d\sigma/d\Omega)$ shows a pronounced small angle shadow peak, over the range $0 < \theta < \pi/kd$, with oscillations around $d^2/4$ at larger angles. The estimated integrated value is approximately $2.16\pi d^2$.

9.2 The random phase approximations imply $\sin^2\eta \simeq 1/2$ and $\sum \sin 2\eta \simeq 0$, after which the quoted results follow from the hints. In addition,

$$\sigma_{\text{shad}} = 2\pi \int_0^{\theta_0} \left(\frac{\mathrm{d}\sigma}{\mathrm{d}\theta}\right)_{\text{shad}} \sin\theta \mathrm{d}\theta \simeq \pi d^2 \left(1 - \mathrm{e}^{-(kd\theta_0/2)^2}\right) = 0.92\pi d^2.$$

9.3 The integral representation AS § 9.1.22 is required for the first result. The second is obtained by recognizing that $2\eta''(\lambda) = \chi'(\lambda) < 0$ at λ_b and λ_c in Fig. 9.5. Thus

$$\int_0^\infty \lambda \mathrm{e}^{2\mathrm{i}\eta(\lambda) - \mathrm{i}\lambda\theta\cos\varphi} \mathrm{d}\lambda \simeq \mathrm{e}^{-\mathrm{i}\pi/4} \sqrt{\frac{2\pi}{|\chi'(\lambda)|}} \lambda(\theta,\varphi) \mathrm{e}^{2\mathrm{i}\eta[\lambda(\theta,\varphi)] - \lambda(\theta,\varphi)\theta\cos\varphi},$$

where $\lambda(\theta,\varphi)$ is defined such that $\chi[\lambda(\theta,\varphi)] = 2\mathrm{d}\eta/\mathrm{d}\lambda = \theta\cos\varphi$.

9.4 The expression for $f_g(\theta)$ follows by transforming to ψ as the integration variable, and recognizing the resulting integrals are representations for $J_0[\zeta(\theta)]$ and $J_0'[\zeta(\theta)]$. The expressions for $A(\theta)$, $\zeta(\theta)$, $p(\theta)$ and $q(\theta)$ are obtained by establishing a correspondence between the stationary points on the two sides of the mapping equation. Bearing in mind that $\lambda(\theta,\varphi)$ is defined such that derivatives with respect to λ cancel on the left-hand side, the derivative of the mapping equation reduces to

$$\lambda(\theta,\varphi)\theta\sin\varphi\,(\mathrm{d}\varphi/\mathrm{d}\psi) = \zeta(\theta)\sin\psi.$$

Properties of the roots, which are chosen to ensure that $\chi[\lambda_b(\theta,\pi)] = -\theta$ and $\chi[\lambda_c(\theta,0)] = \theta$, are given in Table G.3. The expressions for $F(\theta,\psi)$ follow from the mapping equation. Those of the pre-exponent $G(\theta,\psi)$ follow from the second derivative expression

$$\mathrm{d}^2F/\mathrm{d}\psi^2 = \lambda\theta\cos\varphi\,(\mathrm{d}\varphi/\mathrm{d}\psi)^2 = \zeta(\theta)\cos\psi,$$

from which

$$G(\theta,\psi) = p(\theta) + q(\theta)\cos\psi = \sqrt{(\lambda\zeta(\theta)\cos\psi)\,/\,(\theta|\chi'|\cos\varphi)}.$$

Finally, after substituting the appropriate forms from Table G.3,

$$A(\theta) = [F_c(\theta) + F_b(\theta)]/2 \quad \text{and} \quad \zeta(\theta) = [F_c(\theta) - F_b(\theta)]/2$$
$$p(\theta) = [G_c(\theta) + G_b(\theta)]/2 \quad \text{and} \quad q(\theta) = [G_c(\theta) - G_b(\theta)]/2.$$

9.5 The derivative relation $\chi(\lambda) = 2\eta'(\lambda)$, together with the identities $\chi_b = -\theta$ and $\chi_c = \theta$, ensure that $\lambda_{b,c} = \lambda_{\mathrm{g}} \pm \theta/\eta_{\mathrm{g}}''$. Hence $\eta_b = \eta_c$ and

$$\zeta(\theta) = \frac{1}{2}[2\eta_c - 2\eta_b + (\lambda_c + \lambda_c)\theta] = \lambda_{\mathrm{g}}\theta.$$

In addition, $\lambda_b \to \lambda_c \to \lambda_{\mathrm{g}}$ as $\theta \to 0$, which means that $p(\theta) \to \sqrt{\lambda_{\mathrm{g}}^2/|\chi_{\mathrm{g}}'|}$ and $q(\theta) \to 0$.

Table G.3 Properties of the stationary points of the uniform Bessel integrand.

λ	φ	ψ	χ	$F(\theta)$	$G(\theta)$
λ_b	π	π	$-\theta$	$2\eta_b+\lambda_b\theta$	$\sqrt{\lambda_c\zeta(\theta)/\theta\lvert\chi_c'\rvert}$
λ_c	0	0	θ	$2\eta_c-\lambda_c\theta$	$\sqrt{\lambda_b\zeta(\theta)/\theta\lvert\chi_b'\rvert}$

G.10 The classical *S* matrix

10.1

(i) Since $(\partial\bar{\alpha}/\partial\bar{\alpha}_1)=1$, and the integrand in eqn (10.28) is symmetric under the substitution $\bar{\alpha}_1\rightarrow 2\pi-\bar{\alpha}_1$,

$$S_{n_1^0n_2^0}=\frac{1}{\pi}\int_0^{\pi}\cos\left[(n_1^0-n_2^0)\bar{\alpha}_1-a\sin\bar{\alpha}_1\right]\mathrm{d}\bar{\alpha}_1=J_m(a).$$

(ii) In the notation of Section 10.2, the stationary phase condition, $m=a\cos\bar{\alpha}_1$, yields

$$f_a=-f_b=\begin{cases} m\cos^{-1}(m/a)-\sqrt{a^2-m^2} & \text{for}\quad a\geqslant m\\ \mathrm{i}\left[m\cosh^{-1}(m/a)-\sqrt{m^2-a^2}\,\right] & \text{for}\quad m>a\end{cases}$$

and

$$P_a=P_b=\begin{cases}\frac{1}{2\pi\sqrt{1-m^2/a^2}} & \text{for}\quad a\geqslant m\\ \frac{1}{2\pi\mathrm{i}\sqrt{m^2/a^2-1}} & \text{for}\quad m>a\end{cases}.$$

If the exponentially large term is neglected in the classically forbidden case, $m>a$, the primitive approximation is given by

$$S_{n_1^0n_2^0}^{\mathrm{PSC}}\simeq\begin{cases}2P_a^{1/2}\cos(f_b-\pi/4) & \text{for}\quad a\geqslant m\\ \lvert P_a^{1/2}\rvert\exp(-\lvert f_a\rvert) & \text{for}\quad m>a\end{cases}.$$

Note that $f_b>0$ for $a\geqslant m$. The corresponding uniform Airy approximation is

$$S_{n_1^0n_2^0}^{\mathrm{Airy}}\simeq 2\left(\pi P_a\right)^{1/2}\zeta^{1/4}Ai(-\zeta),\qquad \zeta=\left[3f_b/2\right]^{2/3}.$$

10.2 Note that $\cos\gamma$ is invariant under the simultaneous substitutions $\alpha_l\rightarrow\alpha_l\pm\pi$ and $\alpha_j\rightarrow\alpha_j\pm\pi$. Consequently the values of $l(\alpha_l^0,\alpha_j^0)$ and $j(\alpha_l^0,\alpha_j^0)$ on a trajectory initiated at (α_l^0,α_j^0) are invariant to the substitution $\alpha_l(\alpha_l^0,\alpha_j^0)\rightarrow\alpha_l(\alpha_l^0\pm\pi,\alpha_j^0\pm\pi)$, while $\alpha_l(\alpha_l^0,\alpha_j^0)$ and $\alpha_j(\alpha_l^0,\alpha_j^0)$ increase or decrease by the change in α_l^0 and α_j^0 respectively. Consequently, in the notation of eqn (10.28),

$$\Delta(\alpha_l^0+\pi,\alpha_j^0\pm\pi)=\Delta(\alpha_l^0,\alpha_j^0)+(l_1\pm j_1-l_2\mp j_2)\pi.$$

The implications of these symmetries for the S matrix is that

$$\begin{aligned}
&S(l_1 j_1, l_2 j_2) \\
&= \frac{1}{2\pi}\int_0^{2\pi}\int_0^{2\pi}\frac{\partial(\alpha_l,\alpha_j)}{\partial(\alpha_l^0,\alpha_j^0)}\mathrm{e}^{\mathrm{i}\Delta(\alpha_l^0,\alpha_j^0)}\mathrm{d}\alpha_l^0\mathrm{d}\alpha_j^0 \\
&= \frac{1}{2\pi}\left[1+e^{\mathrm{i}(l_1+j_1-l_2-j_2)\pi}\right]\int_0^{\pi}\int_0^{\pi}\frac{\partial(\alpha_l,\alpha_j)}{\partial(\alpha_l^0,\alpha_j^0)}\mathrm{e}^{\mathrm{i}\Delta(\alpha_l^0,\alpha_j^0)}\mathrm{d}\alpha_l^0\mathrm{d}\alpha_j^0 \\
&+ \frac{1}{2\pi}\left[1+\mathrm{e}^{\mathrm{i}(l_1-j_1-l_2+j_2)\pi}\right]\int_0^{\pi}\int_{\pi}^{2\pi}\frac{\partial(\alpha_l,\alpha_j)}{\partial(\alpha_l^0,\alpha_j^0)}\mathrm{e}^{\mathrm{i}\Delta(\alpha_l^0,\alpha_j^0)}\mathrm{d}\alpha_l^0\mathrm{d}\alpha_j^0,
\end{aligned}$$

in which the terms in square brackets carry the selection rules that $S(l_1 j_1, l_2 j_2)$ vanishes unless $(l_1 \pm j_1)$ has the same parity as $(l_2 \pm j_2)$.

A similar argument may be applied separately, either to α_l or to α_j, in the homonuclear case, because, for example, the substitution $\alpha_j \to \alpha_j + \pi$ causes a sign change in $\cos\gamma$ but no change in $V(R, \cos\gamma)$. Consequently

$$S(l_1 j_1, l_2 j_2) = \frac{1}{2\pi}\left[1+\mathrm{e}^{i(j_1-j_2)\pi}\right]\int_0^{\pi}\int_0^{2\pi}\frac{\partial(\alpha_l,\alpha_j)}{\partial(\alpha_l^0,\alpha_j^0)}\mathrm{e}^{\mathrm{i}\Delta(\alpha_l^0,\alpha_j^0)}\mathrm{d}\alpha_l^0\mathrm{d}\alpha_j^0,$$

where the term in square brackets carries the selection rule that $(j_1 - j_2)$ must be even.

10.3

(i) Figure G.2 shows the relevant geometry, with the point of impact at $\mathbf{r} = r[\cos(\gamma+\phi), -\sin(\gamma+\phi)]$, and initial and final velocities $\mathbf{v}_i = v(1,0,0)$ and $\mathbf{v}_f = v(\cos 2\psi, -\sin 2\psi, 0)$. In addition, r and ϵ depend on γ according to the equations

$$r^2\left[\frac{\cos^2\gamma}{a^2}+\frac{\sin^2\gamma}{b^2}\right] = 1, \qquad a^2\tan\epsilon\tan\gamma = b^2.$$

Moreover, $\psi = \phi - \epsilon$ and

$$\begin{aligned}
l_i &= m\left[\mathbf{r}\times\mathbf{v}_i\right]_z = mvr\sin(\gamma+\epsilon+\psi) \\
l_f &= m\left[\mathbf{r}\times\mathbf{v}_f\right]_z = mvr\sin(\gamma+\epsilon-\psi) \\
\bar{l} &= (l_i + l_f)/2 = mvr\sin(\gamma+\epsilon)\cos\psi \\
\Delta j &= l_i - l_f = 2mvr\cos(\gamma+\epsilon)\sin\psi = 2L\xi\sin\psi.
\end{aligned}$$

(ii) Note that ξ is the projection of $\rho = r/a$ onto the tangent, which passes through a maximum between vanishing values at the poles, where $\gamma + \epsilon = \pi/2$. This maximum determines either the minimum value ψ at given Δj or the maximum value of Δj at given ψ.

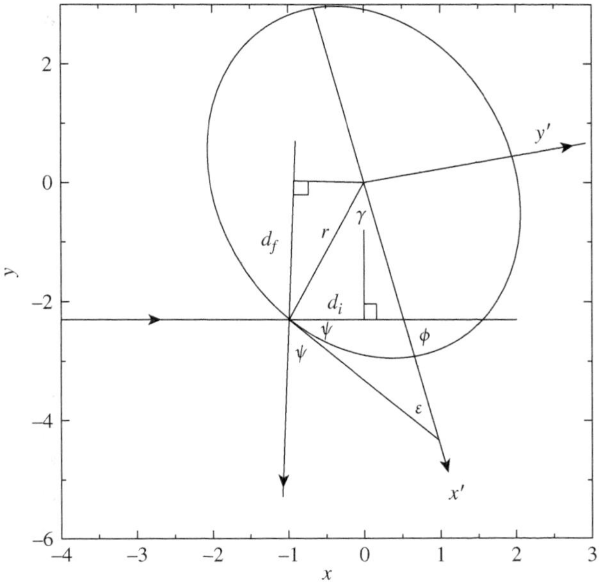

Fig. G.2 Scattering by a hard ellipse. The segments d_i and d_f are initial and final excluded free motion paths arising from deflections at the point of impact.

(iii) By analogy with problem 9.1, the phase shift for incident motion is given by $\eta_i \hbar = l_i \psi - mvd_i$, where, according to Fig. G.2, $d_i = r\cos(\gamma + \epsilon - \psi)$, with a similar formula for η_f except that $d_f = -r\cos(\gamma + \epsilon + \psi)$. Consequently

$$S\left(\bar{l}, \alpha\right) = (\eta_i + \eta_f)\,\hbar = (l_i + l_f)\,\psi - mv(d_i + d_f)$$
$$= 2[\bar{l}\psi - mvr\sin(\gamma + \epsilon)\sin\psi] = 2\left[\bar{l}\arccos\left(\bar{l}/L\xi_\alpha\right) - \sqrt{\left(L\xi_\alpha\right)^2 - \bar{l}^2}\right].$$

(iv) The numerical determination of $(\partial S/\partial \alpha)_{\bar{l}}$ involves varying ψ and γ [plus $r(\gamma)$ and $\epsilon(\gamma)$] to obtain $\xi_\alpha = (r/a)\cos(\gamma + \epsilon)$ and $\alpha = -(\epsilon + \psi)$ at the chosen fixed value of $\bar{l}$. Numerical differentiation of the resulting form for $S\left(\bar{l}, \alpha\right)$ yields the desired approximation for Δj.

G.11 Reactive scattering

11.1

(i) The solution follows from

$$(1 + \beta x)\sum a_J^{(0)} P_J(x)$$
$$= \sum_J a_J^{(0)} \left\{P_J(x) + \beta\,[J|x|J+1]\,P_{J+1}(x) + \beta\,[J|x|J-1]\,P_{J-1}(x)\right\}$$
$$= \sum_J \left\{a_J^{(0)} + \beta a_{J-1}^{(0)}\,[J-1|x|J] + \beta a_{J+1}^{(0)}\,[J+1|x|J]\right\} P_J(x).$$

(ii) It is convenient to write $(1+\beta_1 x)(1+\beta_2 x) = Ax^2 + Bx + 1$, where $A = \beta_1\beta_2$ and $B = \beta_1 + \beta_2$, which gives rise to the additional quadratic term

$$\begin{aligned} Ax^2 \sum_J a_J^{(0)} P_J(x) &\\ = \sum_J A a_J^{(0)} \left\{ \left[J|x^2|J+2\right] P_{J+2}(x) + \left[J|x^2|J\right] P_J(x) + \left[J|x|J-2\right] P_{J-2}(x) \right\} &\\ = \sum_J A \left\{ a_{J-2}^{(0)} \left[J-2|x^2|J\right] + a_J^{(0)} \left[J|x^2|J\right] + a_{J+2}^{(0)} \left[J+2|x^2|J\right] \right\} P_J(x). & \end{aligned}$$

Consequently the conditions $a_0^{(2)} = a_1^{(2)} = 0$, with β in part (i) written as B, require that

$$\begin{pmatrix} a_0^{(0)}\left[0|x^2|0\right] + a_2^{(0)}\left[2|x^2|0\right] & a_1^{(0)}\left[1|x|0\right] \\ a_1^{(0)}\left[1|x^2|1\right] + a_3^{(0)}\left[3|x^2|1\right] & a_0^{(0)}\left[0|x|1\right] + a_2^{(0)}\left[2|x|1\right] \end{pmatrix} \begin{pmatrix} A \\ B \end{pmatrix} = -\begin{pmatrix} a_0^{(0)} \\ a_1^0 \end{pmatrix}.$$

β_1 and β_2 are then given as the solutions of $\beta^2 - B\beta + A = 0$.

11.2

(i) $u_m(\phi) = (2\pi)^{-1/2} \exp\left[i(m+\gamma)\phi\right]$ evolves as $u_m(\phi + 2\pi) = \exp(i\alpha) u_m(\phi)$ with $\alpha = 2\pi\gamma$. Also, $E_m u_m(\phi) = -u_m''(\phi)/2I = \left[(m+\gamma)^2/2I\right] u_m(\phi)$.

(ii)

$$\begin{aligned} K(\phi, T) &= \frac{1}{2\pi} \sum_{m=-\infty}^{\infty} \exp\left[i\left((m+\gamma)\phi - (m+\gamma)\right)^2 T/2I\right] \\ &= \frac{1}{2\pi} \exp\left[i\gamma\phi - \frac{i\gamma^2 T}{2I}\right] \sum_{m=-\infty}^{\infty} \exp\left[-i\frac{m^2 T}{2I} + im\left(\phi - \frac{\gamma T}{I}\right)\right]. \end{aligned}$$

Thus, in the notation of $\theta_3(z,\tau)$, $z = (\phi - \gamma T/I)/2$ and $\tau = -T/2\pi I$. Similarly $z' = z/\tau = \gamma\pi - (\pi I\phi/T)$ and $\tau' = 2\pi I/T$.

(iii) It follows from the theta transformation that

$$\begin{aligned} K(\phi, T) &= \left(\frac{I}{2i\pi T}\right)^{1/2} \exp\left[\frac{iI\phi^2}{2T}\right] \theta_3(z', t') \\ &= \sum_{n=-\infty}^{\infty} e^{in\alpha} \left(\frac{I}{2i\pi T}\right)^{1/2} \exp\left[\frac{iI(\phi - 2n\pi)^2}{2T}\right]. \end{aligned}$$

11.3

(i)

$$\begin{aligned} \delta\tilde{\mathbf{p}}' &= \left(\frac{\partial \tilde{\mathbf{p}}'}{\partial \tilde{\mathbf{q}}'}\right) \delta\tilde{\mathbf{q}}' + \left(\frac{\partial \tilde{\mathbf{p}}'}{\partial \tilde{\mathbf{q}}}\right) \delta\tilde{\mathbf{q}} = \mathbf{W}_{\tilde{\mathbf{q}}'\tilde{\mathbf{q}}'} \delta\tilde{\mathbf{q}}' + \mathbf{W}_{\tilde{\mathbf{q}}'\tilde{\mathbf{q}}} \delta\tilde{\mathbf{q}} \\ \delta\tilde{\mathbf{p}} &= \left(\frac{\partial \tilde{\mathbf{p}}}{\partial \tilde{\mathbf{q}}'}\right) \delta\tilde{\mathbf{q}}' + \left(\frac{\partial \tilde{\mathbf{p}}}{\partial \tilde{\mathbf{q}}}\right) \delta\tilde{\mathbf{q}} = -\mathbf{W}_{\tilde{\mathbf{q}}\tilde{\mathbf{q}}'} \delta\tilde{\mathbf{q}}' - \mathbf{W}_{\tilde{\mathbf{q}}\tilde{\mathbf{q}}} \delta\tilde{\mathbf{q}}. \end{aligned}$$

(ii) The appropriate rearrangement yields

$$\delta\tilde{\mathbf{q}}' = -\mathbf{W}^{-1}_{\tilde{\mathbf{q}}\tilde{\mathbf{q}}'}\mathbf{W}_{\tilde{\mathbf{q}}\tilde{\mathbf{q}}}\delta\tilde{\mathbf{q}} - \mathbf{W}^{-1}_{\tilde{\mathbf{q}}\tilde{\mathbf{q}}'}\delta\tilde{\mathbf{p}}$$

$$\delta\tilde{\mathbf{p}}' = \left[-\mathbf{W}_{\tilde{\mathbf{q}}'\tilde{\mathbf{q}}'}\mathbf{W}^{-1}_{\tilde{\mathbf{q}}\tilde{\mathbf{q}}'}\mathbf{W}_{\tilde{\mathbf{q}}\tilde{\mathbf{q}}} + \mathbf{W}_{\tilde{\mathbf{q}}'\tilde{\mathbf{q}}}\right]\delta\tilde{\mathbf{q}} - \mathbf{W}_{\tilde{\mathbf{q}}'\tilde{\mathbf{q}}'}\mathbf{W}^{-1}_{\tilde{\mathbf{q}}\tilde{\mathbf{q}}'}\delta\tilde{\mathbf{p}}.$$

Hence, following Gutzwiller (1971),

$$\begin{aligned} R &= \det\begin{pmatrix} \mathbf{M}_{\tilde{\mathbf{q}}'\tilde{\mathbf{q}}} - \mathbf{I} & \mathbf{M}_{\tilde{\mathbf{q}}'\tilde{\mathbf{p}}} \\ \mathbf{M}_{\tilde{\mathbf{p}}'\tilde{\mathbf{q}}} & \mathbf{M}_{\tilde{\mathbf{p}}'\tilde{\mathbf{p}}} - \mathbf{I} \end{pmatrix} \\ &= \det\begin{pmatrix} -\mathbf{W}^{-1}_{\tilde{\mathbf{q}}\tilde{\mathbf{q}}'}\mathbf{W}_{\tilde{\mathbf{q}}\tilde{\mathbf{q}}} - \mathbf{I} & -\mathbf{W}^{-1}_{\tilde{\mathbf{q}}\tilde{\mathbf{q}}'} \\ -\mathbf{W}_{\tilde{\mathbf{q}}'\tilde{\mathbf{q}}'}\mathbf{W}^{-1}_{\tilde{\mathbf{q}}\tilde{\mathbf{q}}'}\mathbf{W}_{\tilde{\mathbf{q}}\tilde{\mathbf{q}}} + \mathbf{W}_{\tilde{\mathbf{q}}'\tilde{\mathbf{q}}} & -\mathbf{W}_{\tilde{\mathbf{q}}'\tilde{\mathbf{q}}'}\mathbf{W}^{-1}_{\tilde{\mathbf{q}}\tilde{\mathbf{q}}'} - \mathbf{I} \end{pmatrix} \\ &= \det\left(\mathbf{W}_{\tilde{\mathbf{q}}'\tilde{\mathbf{q}}'} + \mathbf{W}_{\tilde{\mathbf{q}}'\tilde{\mathbf{q}}} + \mathbf{W}_{\tilde{\mathbf{q}}\tilde{\mathbf{q}}'} + \mathbf{W}_{\tilde{\mathbf{q}}\tilde{\mathbf{q}}}\right) / \det\mathbf{W}_{\tilde{\mathbf{q}}\tilde{\mathbf{q}}'}. \end{aligned}$$

The steps in the reduction involve (a) subtracting $\mathbf{W}_{\tilde{\mathbf{q}}'\tilde{\mathbf{q}}'}$ times the first row from the second; (b) factoring $\mathbf{W}^{-1}_{\tilde{\mathbf{q}}\tilde{\mathbf{q}}'}$ from the resulting first row; and (c) subtracting the resulting first row from the second.

References

Abramowitz, M. and Stegun, I. A. (1965). *Handbook of mathematical functions.* New York: Dover.

Adler, K., Bohr, A., Huss, T., Mottelson, B. and Winther, A. (1956). *Rev. Mod. Phys.* **28**, 432.

Airy, G. B. (1838). *Trans. Philos. Cambridge Soc.* **6**, 379.

Alfaro, V. and Regge, T. (1965). *Potential scattering.* Amsterdam: North-Holland.

Allison, A. C. and Dalgarno, A. (1971). *J. Chem. Phys.* **55**, 4342.

Althorpe, S. C. (2006). *J. Chem. Phys.* **124**, 084105.

Althorpe, S. C. (2011). *J. Chem. Phys.* **134**, 114104.

Arnol'd, V. I. (1961). *Izv. Akad. Nauk. SSSR. Ser. Mat.* **25**, 21.

Arnol'd, V. I. (1963). *Usp. Mat. Nauk.* **18**, 13.

Atkins, P. and de Paula, J. (2010). *Physical chemistry* (9th ed.). Oxford: Oxford University Press.

Augustin, S. D. and Miller, W. H. (1974). *J. Chem. Phys.* **61**, 3155.

Augustin, S. D. and Rabitz, H. (1979). *J. Chem. Phys.* **71**, 4956.

Baer, M. (1973). *J. Chem. Phys.* **60**, 1057.

Balian, R. and Bloch, C. (1972). *Ann. Phys.* **69**, 76.

Balian, R., Parisi, G. and Vöros, A. (1978). *Phys. Rev. Lett.* **41**, 1141.

Bandrauk, A. D. and Child, M. S. (1970). *Mol. Phys.* **19**, 95.

Baranyi, A. (1978). *J. Phys. B: At. Mol. Phys.* **11**, L399.

Baranyi, A. (1979). *J. Phys. B: At. Mol. Phys.* **12**, 2841.

Barwell, M. G., LeRoy, R. J., Pajunen, P. and Child, M. S. (1979). *J. Chem. Phys.* **71**, 2618.

Bates, L. M. (1991). *J. App. Phys. (ZAMP)* **42**, 837.

Beidenharn, L. C. (1953). *J. Math. Phys.* **31**, 287.

Bender, C. M., Olaussen, K. and Wang, P. S. (1977). *Phys. Rev. D* **16**, 1740.

Benderskii, V. A., Makarov, D. E. and Wright, C. A. (1994). *Adv. Chem. Phys.* **88**, 55.

Bensimon, D. and Kadanoff, L. P. (1984). *Physica D* **13**, 82.

Bernstein, R. B. (1962). *J. Chem. Phys.* **37**, 1880.

Bernstein, R. B. (1963). *J. Chem. Phys.* **38**, 2599.

Bernstein, R. B. (1966). *Adv. Chem. Phys.* **10**, 75.

Berry, M. V. (1966). *Proc. Phys. Soc. (London)* **89**, 479.

Berry, M. V. (1975). *J. Phys. A: Math. Gen.* **8**, 566.

Berry, M. V. (1976). *Ann. Phys.* **25**, 1.

Berry, M. V. (1983). *J. Phys. A: Math. Gen.* **17**, 1225.

Berry, M. V. (1984). *Proc. Roy. Soc. A* **392**, 45.

Berry, M. V. and Mount, K. E. (1972). *Rep. Prog. Phys.* **35**, 315.

Berry, M. V., Nye, N. F. and Wright, F. J. (1979). *Phil. Trans. R. Soc. A* **291**, 453.

Berry, M. V., Percival, I. C. and Weiss, N. (1987). Dynamical chaos. In *Royal Society London Discussion Papers*, pp. 179–183. Princeton: Princeton University Press.
Berry, M. V. and Tabor, M. (1976). *Proc. Roy. Soc. A* **349**, 101.
Berry, M. V. and Tabor, M. (1977). *J. Phys. A: Math. Gen.* **10**, 371.
Bersuker, I. S. (2006). *The Jahn–Teller effect.* Cambridge: Cambridge University Press.
Bieniek, R. (1987). *Phys. Rev. A* **15**, 1513.
Birkhoff, G. D. (1927). Dynamical systems. In *A.M.S. Colloquium Publications.* New York: A. M. S. Publications.
Boas, R. P. and Stutz, C. (1971). *Am. J. Phys.* **39**, 745.
Bohr, N. (1913). *Philos. Mag.* **26**, 476.
Bordé, J. and Bordé, C. J. (1982). *Chem. Phys.* **71**, 417.
Born, M. (1960). *Mechanics of the atom.* London: Bell.
Bouakline, F., Althorpe, S. C. and Ruiz, D. P. (2008). *J. Chem. Phys.* **128**, 124322.
Boyer, C. P. and Wolf, K. B. (1975). *J. Math. Phys.* **16**, 1493.
Breford, E. J. and Engleke, F. (1978). *Chem. Phys. Lett.* **53**, 282.
Brewer, M. L., Hulme, J. S. and Manolopoulos, D. E. (1997). *J. Chem. Phys.* **106**, 4832.
Brillouin, L. (1926a). *CR Acad. Sci., Paris* **183**, 24.
Brillouin, L. (1926b). *J. Phys.* **7**, 353.
Brink, D. M. (1985). *Semiclassical methods in nucleus–nucleus scattering.* Cambridge: Cambridge University Press.
Brink, D. M. and Satchler, G. R. (1968). *Angular momentum.* Oxford: Oxford University Press.
Brussard, P. J. and Tolhoek, H. A. (1957). *Physica* **23**, 955.
Buck, U. (1975). *Adv. Chem. Phys.* **30**, 313.
Bykovskii, V., Nikitin, E. E. and Ovchinnikova, M. Y. (1964). *Sov. Phys.–JETP* **47**, 750; trans. **20**, 502.
Campbell, J. A. (1972). *J. Comput. Phys.* **10**, 308.
Cao, J. and Voth, G. A. (1996). *J. Chem. Phys.* **105**, 2710.
Carruthers, P. and Nieto, M. M. (1968). *Rev. Mod. Phys.* **40**, 411.
Casati, G. (1985). *Chaotic behaviour in quantum systems.* New York: Plenum.
Chapman, S., Garrett, B. C. and Miller, W. H. (1975). *J. Chem. Phys.* **63**, 2710.
Chapman, S., Garrett, B. C. and Miller, W. H. (1976). *J. Chem. Phys.* **64**, 502.
Chester, C., Friedman, B. and Ursell, F. (1957). *Proc. Comb. Phil. Soc.* **53**, 599.
Child, M. S. (1972). *Mol. Phys.* **23**, 269.
Child, M. S. (1974a). *J. Mol. Spectrosc.* **53**, 280.
Child, M. S. (1974b). *Molecular collision theory* (2nd ed.). New York: Dover.
Child, M. S. (1974c). In R. F. Barrow, D. A. Long and D. J. Miller (Eds), *Molecular Spectroscopy*, Vol. 2. Specialist Periodical Report. London: Chemical Society.
Child, M. S. (1975). *Mol. Phys.* **29**, 1421.
Child, M. S. (1976a). *Mol. Phys.* **32**, 1495.
Child, M. S. (1976b). In W. H. Miller (Ed.), *Dynamics of molecular collisions.* New York: Plenum Press.
Child, M. S. (1978). *Mol. Phys.* **35**, 759.
Child, M. S. (1979). In R. B. Bernstein (Ed.), *Atom molecule collisions: a guide for the experimentalist.* New York: Plenum.

Child, M. S. (1980). In M. S. Child (Ed.), *Semiclassical methods in molecular scattering and spectroscopy*, NATO ASI Series C, Chapter 4, p. 127. Dordrecht: Reidel.
Child, M. S. (1998). *J. Math. Phys.* **31**, 657.
Child, M. S. (2007). *Adv. Chem. Phys.* **136**, 657.
Child, M. S., Essen, H. and LeRoy, R. J. (1983). *J. Chem. Phys.* **78**, 6732.
Child, M. S. and Gerber, R. B. (1979). *Mol. Phys.* **38**, 421.
Child, M. S. and Halonen, L. (1984). *Adv. Chem. Phys.* **57**, 1.
Child, M. S. and Hunt, P. M. (1977). *Mol. Phys.* **34**, 261.
Child, M. S. and Lefebvre, R. (1978). *Chem. Phys. Lett.* **55**, 213.
Child, M. S. and Nesbitt, D. J. (1988). *Chem. Phys. Lett.* **149**, 404.
Child, M. S. and Shapiro, M. (1983). *Mol. Phys.* **48**, 111.
Child, M. S., Weston, T. and Tennyson, J. (1999). *Mol. Phys.* **96**, 371.
Chirikov, B. V. (1979). *Phys. Rep.* **52**, 265.
Clarke, A. P. and Dickinson, A. S. (1973). *J. Phys. B: At. Mol. Phys.* **6**, 164.
Cohen-Tannoudji, C., Diu, B. and Laloë, F. (1977). *Quantum mechanics.* New York: Wiley.
Colwell, S. M. and Handy, N. C. (1978). *Mol. Phys.* **35**, 1183.
Condon, E. U. and Shortley, G. H. (1957). *Theory of atomic spectra.* Cambridge: Cambridge University Press.
Connor, J. N. L. (1968). *Mol. Phys.* **15**, 621.
Connor, J. N. L. (1973). *Mol. Phys.* **26**, 1371.
Connor, J. N. L. (1976). *Mol. Phys.* **31**, 33.
Connor, J. N. L. (1980). In M. S. Child (Ed.), *Semiclassical methods in molecular scattering and spectroscopy*, NATO ASI Series C, Chapter 2, p. 45. Dordrecht: Reidel.
Connor, J. N. L. (1981). *J. Chem. Phys.* **74**, 1047.
Connor, J. N. L. (1990a). *Faraday Trans. Chem. Soc.* **86**, 1627.
Connor, J. N. L. (1990b). In R. Wong (Ed.), *Asymptotics and computational analysis.* New York: Marcel Dekker.
Connor, J. N. L. (2004). *Phys. Chem. Chem. Phys.* **6**, 377.
Connor, J. N. L. and Farrelly, D. (1981). *J. Chem. Phys.* **75**, 2831.
Connor, J. N. L. and Jakubetz, W. (1978). *Mol. Phys.* **35**, 949.
Connor, J. N. L., Jakubetz, W., Mackay, D. C. and Sukumar, C. V. (1980). *J. Phys. B: At. Mol. Phys.* **13**, 1823.
Connor, J. N. L., Jakubetz, W. and Sukumar, C. V. (1976). *J. Phys. B: At. Mol. Phys.* **9**, 1783.
Connor, J. N. L., Jakubetz, W. and Sukumar, C. V. (1979). *J. Phys. B: At. Mol. Phys.* **12**, L515.
Connor, J. N. L. and Mackay, D. C. (1978). *Mol. Phys.* **37**, 1703.
Connor, J. N. L. and Marcus, R. A. (1971). *J. Chem. Phys.* **55**, 5636.
Connor, J. N. L. and Mayne, H. R. (1979). *Mol. Phys.* **37**, 15.
Connor, J. N. L., Curtis, P. R. and Farrelly, D. (1983). *Mol. Phys.* **48**, 1305.
Contopoulos, G. (1971). *Astron. J.* **76**, 147.
Cotton, F. A. (1971). *Chemical applications of group theory* (2nd ed.). New York: Wiley.

Coveney, P. V., Child, M. S. and Baranyi, A. (1985). *J. Phys. B: At. Mol. Phys.* **18**, 4557.

Crothers, D. S. F. (1971). *Adv. Phys.* **20**, 405.

Crothers, D. S. F. (1972). *J. Phys. A: Gen. Phys.* **5**, 1680.

Crothers, D. S. F. (1973). *J. Phys. B: At. Mol. Phys.* **6**, 1418.

Cushman, R. H. and Bates, L. M. (1997). *Global aspects of classical integrable systems.* Basel: Birkhäuser.

Davis, M. J. (1985). *J. Chem. Phys.* **83**, 1016.

Davis, M. J. (1987). *J. Chem. Phys.* **86**, 3978.

Davis, M. J. and Gray, S. K. (1986). *J. Chem. Phys.* **84**, 5389.

Davis, M. J. and Heller, E. J. (1981). *J. Chem. Phys.* **75**, 3916.

Delande, D. and Gay, J. C. (1986). *Phys. Rev. A* **57**, 2006.

Delande, D. and Gay, J. C. (1987). *Phys. Rev. A* **59**, 1809.

DeLeon, N. and Heller, E. J. (1983). *J. Chem. Phys.* **78**, 4005.

DeLeon, N. and Heller, E. J. (1984). *J. Chem. Phys.* **81**, 5957.

DeLeon, N. and Mehta, M. A. (1988). *Comput. Phys. Rep.* **8**, 295.

Delos, J. B. (1986). *Adv. Chem. Phys.* **65**, 161.

Delos, J. B. and Carlson, C. E. (1975). *Phys. Rev. A* **11**, 210.

Delos, J. B., Knudson, S. K. and Noid, D. W. (1984). *Phys. Rev. A* **30**, 1208.

Delvigne, G. A. L. and Los, J. (1973). *Physica* **67**, 166.

Demkov, Y. N. (1964). *Sov. Phys.-JETP.* **18**, 138.

Dickinson, A. S. (1970). *Mol. Phys.* **18**, 441.

Dickinson, A. S. (1972). *J. Computational Phys.* **10**, 162.

Dickinson, A. S. (1980). In M. S. Child (Ed.), *Semiclassical methods in molecular scattering and spectroscopy,* NATO ASI Series C, Chapter 7, p. 263. Dordrecht: Reidel.

Dickinson, A. S. and Richards, D. (1982). *Adv. At. Mol. Phys.* **18**, 165.

Dingle, R. B. (1973). *Asymptotic expansions: their derivation and interpretation.* New York: Academic Press.

Dirac, P. A. M. (1958). *Principles of quantum mechanics* (4th ed.). Oxford: Oxford University Press.

Dixon, R. N. (1964). *Trans. Far. Soc.* **60**, 1363.

Dobbyn, A. J., McCabe, P., Connor, J. N. L. and Castillo, J. F. (1999). *Phys. Chem. Chem. Phys.* **1**, 1115.

Domcke, W., Yarkony, D. R. and Köppel, H. (2003). *Conical intersections: electronic structure, dynamics and spectroscopy.* New Jersey: World Scientific.

Duff, J. W. and Truhlar, D. G. (1975). *Chem. Phys.* **9**, 243.

Dunham, J. L. (1932). *Phys. Rev.* **41**, 713, 721.

Dwight, H. B. (1961). *Tables of integrals and other mathematical data* (4th ed.). Oxford: Oxford University Press.

Eaker, C. W. and Schatz, G. C. (1984). *J. Chem. Phys.* **81**, 2394.

Eaker, C. W., Schatz, G. C., DeLeon, N. and Heller, E. J. (1984). *J. Chem. Phys.* **81**, 5913.

Edmonds, A. R. (1974). *Angular momentum in quantum mechanics* (2nd ed.). Princeton: Princeton University Press.

Ehrenfest, P. (1916). *Versl. Kon. Akad. Amsterdam* **25**, 412.

Einstein, A. (1917). *Verh. Dtsch. Phys. Ges.* **19**, 82.

Englman, R. (1972). *The Jahn–Teller effect in molecules and crystals.* New York: Wiley.

Ezra, G. S., Martens, C. C. and Fried, L. E. (1987). *J. Phys. Chem.* **91**, 3721.

Faist, M. B. and Bernstein, R. B. (1976). *J. Chem. Phys.* **64**, 2971.

Faist, M. B. and Levine, R. D. (1976). *J. Chem. Phys.* **64**, 2953.

Feltgen, R., Meyer, W. and Reinsch, E. A. (1978). *Verh. Dtsch. Phys. Ges.* **2**, 419.

Feynman, R. P. and Hibbs, A. R. (1965). *Quantum mechanics and path integrals.* New York: McGraw-Hill.

Filinov, V. S. (1976). *Nucl. Phys. B* **271**, 717.

Firsov, O. B. (1953). *Zh. Eksp. Teor. Phys.* **24**, 279.

Fock, V. A. (1959). *Vestn. Leningr. Univ. Ser. Mat. Fiz. Khim.* **16**, 67. [Tech. transl. (1960), 60–17464, **4**, 53].

Ford, K. W. and Wheeler, J. A. (1959). *Ann. Phys.* **7**, 259, 287.

Fraser, S. J., Gottdiener, L. and Murrell, J. N. (1975). *Mol. Phys.* **29**, 415.

Fried, L. E. and Ezra, G. S. (1988). *Comput. Phys. Commun.* **51**, 103.

Froman, N. (1966). *Ark. Fys.* **32**, 541.

Froman, N. (1970). *Ann. Phys.* **61**, 451.

Froman, N. (1980). In M. S. Child (Ed.), *Semiclassical methods in molecular scattering and spectroscopy*, NATO ASI Series C, Chapter 1, p. 1. Dordrecht: Reidel.

Froman, N. and Froman, P. O. (1967). *JWKB-approximation, contributions to the theory.* Amsterdam: North-Holland.

Froman, N. and Froman, P. O. (1974). *Ann. Phys.* **83**, 103.

Froman, N. and Froman, P. O. (1978). *J. Math. Phys.* **19**, 1823.

Froman, N. and Froman, P. O. (1981). *J. Math. Phys.* **22**, 1190.

Froman, N., Froman, P. O., Myhrmann, U. and Paulsson, R. (1972). *Ann. Phys.* **74**, 314.

Garrett, B. C. and Truhlar, D. G. (1983). *J. Chem. Phys.* **79**, 4931.

Gel'fand, L. M. and Yaglom, A. M. (1960). *J. Math. Phys.* **1**, 48.

George, T. F. and Miller, W. H. (1972). *J. Chem. Phys.* **57**, 2458, 5722.

Gerber, R. B. and Shapiro, M. (1976). *Chem. Phys.* **13**, 227.

Gerber, R. B., Shapiro, M., Buck, U. and Schleusener, J. (1978). *Phys. Rev. Lett.* **41**, 236.

Goldstein, H. (1980). *Classical mechanics* (2nd ed.). New York: Addison-Wesley.

Gordon, R. G. (1969). *J. Chem. Phys.* **51**, 14.

Gradsteyn, I. S. and Ryyzhik, I. M. (1980). *Tables of integrals, series and products.* New York: Academic Press.

Gray, S. K. and Child, M. S. (1984). *Mol. Phys.* **51**, 189.

Gray, S. K., Child, M. S. and Noid, D. W. (1985). *Mol. Phys.* **54**, 573.

Gray, S. K. and Davis, M. J. (1989). *J. Chem. Phys.* **90**, 5420.

Grozdanov, T. P. and Solov'ev, E. A. (1982). *J. Phys. B: At. Mol. Phys.* **15**, 1195.

Gustavson, F. G. (1966). *Astron. J.* **71**, 670.

Gutzwiller, M. C. (1967). *J. Math. Phys.* **8**, 1979.

Gutzwiller, M. C. (1969). *J. Math. Phys.* **10**, 1004.

Gutzwiller, M. C. (1970). *J. Math. Phys.* **11**, 1791.

Gutzwiller, M. C. (1971). *J. Math. Phys.* **12**, 343.

Gutzwiller, M. C. (1990). *Chaos in classical and quantum mechanics.* New York: Springer Verlag.

Handy, N. C. (1980). In Child, M. S. (Ed.), *Semiclassical methods in molecular scattering and spectroscopy*, NATO ASI Series C, Chapter 8, p. 297. Dordrecht: Reidel.

Harter, W. G. (1986). *J. Chem. Phys.* **85**, 5560.

Harter, W. G. (1988). *Comput. Phys. Rep.* **8**, 321.

Harter, W. G. and Patterson, C. W. (1977). *J. Chem. Phys.* **66**, 4872, 4886.

Harter, W. G. and Patterson, C. W. (1984). *J. Chem. Phys.* **80**, 4241.

Heading, J. (1962). *Phase integral methods.* London: Methuen.

Hecht, K. T. (1960). *J. Mol. Spectrosc.* **5**, 355.

Heller, E. J. (1975). *J. Chem. Phys.* **62**, 1544.

Heller, E. J. (1981). *J. Chem. Phys.* **75**, 2923.

Heller, E. J. (1984). *Phys. Rev. Lett.* **53**, 1515.

Heller, E. J. (1987). *Phys. Rev. A* **35**, 1360.

Heller, E. J. (1991). *J. Chem. Phys.* **94**, 2723.

Henon, M. (1969). *Q. Appl. Math.* **27**, 291.

Henon, M. and Heiles, C. (1964). *Astron. J.* **69**, 73.

Herman, M. F. and Kluk, E. (1984). *Chem. Phys.* **91**, 27.

Herzberg, G. (1945). *Infra-red and Raman spectra.* New York: Van Nostrand.

Herzberg, G. (1950). *Spectra of diatomic molecules* (2nd ed.). New York: Van Nostrand.

Huber, D. and Heller, E. J. (1987). *J. Chem. Phys.* **87**, 5302.

Huber, D., Heller, E. J. and Littlejohn, R. G. (1988). *J. Chem. Phys.* **89**, 2003.

Hunt, P. M. and Child, M. S. (1978). *Chem. Phys. Lett.* **58**, 202.

Ishiwata, T., Otoshi, H., Sakaki, M. and Tanaka, I. (1984). *J. Chem. Phys.* **80**, 1411.

Jackson, J. M. and Mott, N. F. (1932). *Proc. Roy. Soc. A* **137**, 703.

Jacobson, M. P. and Child, M. S. (2001). *J. Phys. Chem. A* **105**, 2834.

Jaffé, C. and Brumer, P. (1980). *J. Chem. Phys.* **73**, 5646.

Jaffé, C. and Reinhardt, W. P. (1982). *J. Chem. Phys.* **77**, 5191.

Jeffreys, H. (1925). *Proc. London Math. Soc.* **23**, 428.

Jeffreys, H. (1962). *Asymptotic approximations.* Cambridge: Cambridge University Press.

Jeffreys, H. and Jeffreys, B. S. (1956). *Methods of mathematical physics* (3rd ed.). Cambridge: Cambridge University Press.

Johns, J. W. C. (1967). *Can. J. Phys.* **45**, 2369.

Johnson, R. B. (1985). *J. Chem. Phys.* **83**, 1204.

Juanes-Marcos, J. C., Althorpe, S. C. and Wrede, E. (2007). *J. Chem. Phys.* **126**, 044317.

Kay, K. G. (1994a). *J. Chem. Phys.* **100**, 4377.

Kay, K. G. (1994b). *J. Chem. Phys.* **101**, 2250.

Kay, K. G. (2005). *Ann. Rev. Phys. Chem.* **56**, 255.

Kay, K. G. (2006). *Chem. Phys.* **322**, 3.

Keller, J. B. (1958). *Ann. Phys.* **4**, 180.

Kendrick, B. K. (2003). *J. Chem. Phys.* **118**, 10502.

Kerner, E. (1958). *Can. J. Phys.* **36**, 371.

Klein, O. (1932). *Z. Phys.* **76**, 226.
Kokooline, V. and Greene, C. H. (2003). *Phys. Rev. A* **68**, 012703.
Kolmogorov, A. N. (1954). *Dokl. Akad. Nauk. SSSR* **98**, 527.
Kolmogorov, A. N. (1957). *Proc. Int. Congr. of Mathematicians* **1**, 315.
Kolos, W. and Wolniewicz, L. (1968). *J. Chem. Phys.* **49**, 404.
Kouri, D. J. (1979). In R. B. Bernstein (Ed.), *Atom molecule collisions: a guide for the experimentalist.* New York: Plenum.
Kramers, H. (1926). *Z. Phys.* **39**, 828.
Kreek, H., Ellis, R. and Marcus, R. A. (1974). *J. Chem. Phys.* **61**, 4540.
Kreek, H., Ellis, R. and Marcus, R. A. (1975). *J. Chem. Phys.* **62**, 913.
Kryvohuz, M. (2011). *J. Chem. Phys.* **134**, 114103.
Kuppermann, A. and Wu, Y. M. (2001). *Chem. Phys. Lett.* **349**, 537.
Lam, K. S. and George, T. F. (1980). In M. S. Child (Ed.), *Semiclassical methods in molecular scattering and spectroscopy,* NATO ASI Series C, Chapter 6, p. 179. Dordrecht: Reidel.
Landau, L. D. (1932). *Phys. Z. Sow.* **1**, 46.
Landau, L. D. and Lifshitz, E. M. (1965). *Quantum mechanics. Non-relativistic theory* (2nd ed.). Oxford: Pergamon Press.
Landau, L. D. and Lifshitz, E. M. (1976). *Mechanics* (3rd ed.). Oxford: Pergamon Press.
Langer, R. E. (1937). *Phys. Rev.* **51**, 669.
Leaf, B. (1969). *J. Math. Phys.* **10**, 1980.
Lee, S.-Y. and Heller, E. J. (1979). *J. Chem. Phys.* **71**, 4777.
Lee, S.-Y. and Heller, E. J. (1982). *J. Chem. Phys.* **76**, 3035.
Lefebvre-Brion, H. and Field, R. (1985). *Perturbations in spectra of diatomic molecules.* New York: Academic Press.
Leopold, J. G. and Percival, I. C. (1980). *J. Phys. B: At. Mol. Phys.* **13**, 1037.
LeRoy, R. J. (1972). *Can. J. Phys.* **50**, 953.
LeRoy, R. J. (1973). In R. F. Barrow, Long, D. A. and D. J. Millen (Eds.), *Molecular Spectroscopy,* Number 1 in Specialist periodical report. London: Royal Society of Chemistry.
LeRoy, R. J. (1980). In M. S. Child (Ed.), *Semiclassical methods in molecular scattering and spectroscopy,* NATO ASI Series C, Chapter 3, p. 109. Dordrecht: Reidel.
LeRoy, R. J. and Bernstein, R. B. (1970). *J. Chem. Phys.* **52**, 3869.
LeRoy, R. J. and Bernstein, R. B. (1971a). *J. Chem. Phys.* **54**, 5114.
LeRoy, R. J. and Bernstein, R. B. (1971b). *J. Mol. Spectrosc.* **37**, 109.
LeRoy, R. J., Keough, W. and Child, M. S. (1988). *J. Chem. Phys.* **89**, 4564.
LeRoy, R. J. and Schwartz, C. (1987). *J. Mol. Spectrosc.* **121**, 420.
Levine, R. D. and Bernstein, R. (1985). *Molecular reaction dynamics and chemical reactivity.* Oxford: Oxford University Press.
Levinson, N. (1949). *Danske Vidensk. Selsk. Mat-Fys. Meddr.* **25**, 9.
Lin, Y. W., George, T. F. and Morokuma, K. (1975). *J. Phys. B: At. Mol. Phys.* **8**, 265.
Littlejohn, R. G. (1975). *Phys. Rep.* **138**, 193.
Littlejohn, R. G. (1990). *J. Math. Phys.* **31**, 2952.
Littlejohn, R. G. and Robbins, J. M. (1987). *Phys. Rev. A* **36**, 2953.

Lombardi, M., Labastie, P., Bordas, M. C. and Broyer, M. (1988). *J. Chem. Phys.* **89**, 3479.
Longuet-Higgins, H. C. (1975). *Proc. Roy. Soc. A* **344**, 247.
Lundborg, B. and Froman, P. O. (1988). *Math. Proc. Cambridge Philos. Soc.* **104**, 581.
Luppi, J. and Pajunen, P. (1982). *J. Chem. Phys.* **76**, 4117.
Luppi, J. and Pajunen, P. (1983). *J. Chem. Phys.* **78**, 4551.
Luppi, J. and Pajunen, P. (1984). *J. Chem. Phys.* **81**, 1836.
Lynch, G. C., Truhlar, D. G. and Garrett, B. C. (1989). *J. Chem. Phys.* **90**, 3102.
MacKay, M. S., Meiss, J. D. and Percival, I. C. (1984). *Physica D* **13**, 55.
Mantz, A. W., Watson, J. K. G., Rao, K. N., Albritton, D. L., Schmetekopf, A. L. and Zare, R. N. (1971). *J. Mol. Spectrosc.* **39**, 180.
Marcus, R. A. (1970). *Chem. Phys. Lett.* **7**, 525.
Marcus, R. A. (1973). *J. Chem. Phys.* **59**, 5135.
Marcus, R. A. and Coltrin, M. E. (1977). *J. Chem. Phys.* **67**, 2607.
Martens, C. C. and Ezra, G. S. (1985). *J. Chem. Phys.* **83**, 2990.
Martens, C. C. and Ezra, G. S. (1987). *J. Chem. Phys.* **86**, 279.
Maslov, V. P. (1972). *Theorie des perturbations et methodes asymptotiques.* Paris: Dunod, Gautier Villars.
Maslov, V. P. and Fedoriouk, M. V. (1981). *Semiclassical approximations in quantum mechanics.* New York: Reidel.
Massey, H. S. W. and Mohr, C. B. (1934). *Proc. Phys. Soc. A* **144**, 188.
McCabe, P., Connor, J. N. L. and Sokolovski, D. (2001). *J. Chem. Phys.* **114**, 5194.
McCurdy, C. W. and Miller, W. H. (1977). *J. Chem. Phys.* **67**, 463.
Mead, C. A. and Truhlar, D. G. (1979). *J. Chem. Phys.* **70**, 2284.
Messiah, A. (1961). *Quantum mechanics.* Amsterdam: North-Holland.
Miller, S. C. and Good, R. H. (1953). *Phys. Rev.* **91**, 174.
Miller, W. H. and George, T. F. (1972). *J. Chem. Phys.* **56**, 5637, 5668.
Miller, W. H. (1968). *J. Chem. Phys.* **48**, 464.
Miller, W. H. (1969). *J. Chem. Phys.* **50**, 407.
Miller, W. H. (1970). *J. Chem. Phys.* **53**, 1949, 3578.
Miller, W. H. (1971). *J. Chem. Phys.* **54**, 4174.
Miller, W. H. (1974). *Adv. Chem. Phys.* **25**, 69.
Miller, W. H. (1975). *J. Chem. Phys.* **63**, 996.
Miller, W. H. (1976). *Adv. Chem. Phys.* **30**, 77.
Miller, W. H. (1977). *Far. Disc. Chem. Soc.* **62**, 40.
Miller, W. H. (2002). *Mol. Phys.* **100**, 397.
Monks, P. D. D., Xiahou, C. and Connor, J. N. L. (2006). *J. Chem. Phys.* **125**, 133504.
Moser, J. (1973). In *Stable and random motions in dynamical systems*, Ann. Math. Studies. Princeton: Princeton University Press.
Mott, N. F. and Massey, H. S. W. (1965). *Theory of atomic collisions* (3rd ed.). Oxford: Oxford University Press.
Nakamura, H. (2007). *Non-adabatic transition: concepts, basic theories and applications* (2nd ed.). Singapore: World Scientific.

Nikitin, E. E. (1968). In H. Hartmann (Ed.), *Chemische Elementarprozesse.* New York: Springer.
Nikitin, E. E. (1979). *Comm. At. Mol. Phys.* **1**, 111.
Nikitin, E. E. (1980). *Comm. At. Mol. Phys.* **2**, 118.
Nikitin, E. E., Ovchinnikova, M. Y., Andresen, B. and de Vries, A. E. (1976). *Chem. Phys.* **14**, 121.
Noid, D. W., Kosykowski, M. L. and Marcus, R. A. (1977). *J. Chem. Phys.* **67**, 404.
Noid, D. W., Kosykowski, M. L. and Marcus, R. A. (1979). *J. Chem. Phys.* **71**, 2864.
Noid, D. W., Kosykowski, M. L. and Marcus, R. A. (1980). *J. Chem. Phys.* **73**, 391.
Noid, D. W., Kosykowski, M. L. and Marcus, R. A. (1981). *Ann. Rev. Phys. Chem.* **31**, 267.
Noid, D. W. and Marcus, R. A. (1975). *J. Chem. Phys.* **62**, 2119.
Noid, D. W. and Marcus, R. A. (1977). *J. Chem. Phys.* **67**, 559.
Noid, D. W. and Marcus, R. A. (1986). *J. Chem. Phys.* **85**, 3305.
Noid, D. W. and Sumpter, B. G. (1985). *Chem. Phys. Lett.* **121**, 187.
Olsen, R. E. (1970). *Phys. Rev. A* **2**, 121.
Ovchinnikova, M. Y. (1964). *Opt. Spectrosk. [trans]* **17**, 882. trans. **17**, 447.
Ovchinnikova, M. Y. (1974). *Zh. é ksp. teor. Fiz.* **67**, 1276; transl. (1975), *Sov. Phys.-JETP* **40**, 635.
Ozorio de Almeida, A. M. (1988). *Hamiltonian systems: chaos and quantization.* Cambridge: Cambridge University Press.
Pack, R. T. and Dahler, J. S. (1969). *J. Chem. Phys.* **50**, 2397.
Pack, R. T. and Parker, G. A. (1987). *J. Chem. Phys.* **87**, 3888.
Pajunen, P. (1981). *Mol. Phys.* **43**, 753.
Pajunen, P. (1985). *J. Chem. Phys.* **83**, 2363.
Pajunen, P. and Child, M. S. (1980). *Mol. Phys.* **40**, 597.
Pauling, L. and Wilson, E. B. (1935). *Introduction to quantum mechanics.* New York: McGraw-Hill.
Paulsson, R., Karlsson, F. and LeRoy, R. J. (1983). *J. Chem. Phys.* **79**, 4346.
Pauly, H. (1979). In R. B. Bernstein (Ed.), *Atom molecule collisions: a guide for the experimentalist.* New York: Plenum.
Pearcey, T. (1946). *London, Edinburgh, Dublin Philos. Mag.* **37**, 311.
Pechukas, P. (1969). *Phys. Rev.* **181**, 166, 174.
Pechukas, P. (1976). In W. H. Miller (Ed.), *Dynamics of molecular collisions.* New York: Plenum Press.
Pechukas, P. and Child, M. S. (1976). *Mol. Phys.* **31**, 973.
Pechukas, P. and Light, J. C. (1966). *J. Chem. Phys.* **44**, 3897.
Percival, I. C. (1977). *Adv. Chem. Phys.* **36**, 1.
Percival, I. C. and Pomphrey, N. (1973). *J. Phys. B: At. Mol. Phys.* **6**, L229.
Percival, I. C. and Pomphrey, N. (1976). *Mol. Phys.* **31**, 97.
Percival, I. C. and Richards, D. (1982). *Introduction to dynamics.* Cambridge: Cambridge University Press.
Phillips, E. G. (1957). *Functions of a complex variable.* Edinburgh: Oliver and Boyd.
Poincaré, H. (1892). *Les methodes nouvelles de la mechanique celeste.* Paris: Gauthier-Villars.

Ponzano, G. and Regge, T. (1968). In F. Bloch (Ed.), *Spectroscopic and group theoretical methods in physics.* Amsterdam: North-Holland.
Postern, T. and Stewart, I. (1978). *Catastrophe theory and its applications.* London: Pitman.
Press, W. H., Teukolsky, S. A., Vetterling, W. T. and Flannery, B. P. (1992). *Numerical recipes. The art of scientific computing* (2nd ed.). Cambridge: Cambridge University Press.
Racah, C. (1951). *Phys. Rev.* **84**, 910.
Rankin, C. C. and Miller, W. H. (1971). *J. Chem. Phys.* **55**, 3150.
Rees, A. L. G. (1946). *Proc. Phys. Soc.* **59**, 998.
Reinhardt, W. P. and Jaffé, C. (1981). In K. E. Gustavson and W. P. Reinhardt (Eds.), *Quantum mechanics in mathematics.* New York: Plenum.
Robbins, M. J., Creagh, S. C. and Littlejohn, R. G. (1989). *Phys. Rev. A* **39**, 2838.
Robbins, M. J., Creagh, S. C. and Littlejohn, R. G. (1990). *Phys. Rev. A* **41**, 6052.
Robnik, M. (1981). *J. Phys. A: Math. Gen.* **14**, 3195.
Rydberg, R. (1931). *Z. Phys.* **73**, 326.
Sadovskii, D. A. and Zhilinskii, B. I. (2006). *Mol. Phys.* **104**, 2595.
Sage, M. L. and Child, M. S. (1989). *J. Chem. Phys.* **90**, 7257.
Schinke, R. (1989). In F. A. Gianturco (Ed.), *Collision theory for atoms and molecules.* New York: Plenum.
Schinke, R., Korsch, H. J. and Poppe, D. (1982). *J. Chem. Phys.* **77**, 6005.
Schinke, R. and McGuire, P. (1979). *J. Chem. Phys.* **71**, 4201.
Schmidt, I., Meyer, W., Kruger, B. and Engelke, F. (1988). *Chem. Phys. Lett.* **143**, 353.
Schulman, L. S. (1981). *Techniques and applications of path integrals.* New York: Wiley.
Schulten, K. and Gordon, R. G. (1975). *J. Math. Phys.* **16**, 1961, 1971.
Secrest, D. and Johnson, R. B. (1966). *J. Chem. Phys.* **45**, 4556.
Seiderman, T. and Miller, W. H. (1992). *J. Chem. Phys.* **98**, 4412.
Selin, L. A. (1962). *Ark. Phys.* **21**, 529.
Senn, P. (1987). *J. Comput. Chem.* **8**, 965.
Shapiro, M. and Gerber, R. B. (1976). *Chem. Phys.* **13**, 235.
Sibert, E. L., Hynes, J. T. and Reinhardt, W. (1982). *J. Chem. Phys.* **77**, 3595.
Siegert, A. (1939). *Phys. Rev.* **56**, 750.
Simons, G., Parr, R. G. and Finlan, J. M. (1973). *J. Chem. Phys.* **59**, 3229.
Skodje, R. T., Borondo, F. and Reinhardt, W. P. (1985). *J. Chem. Phys.* **82**, 4611.
Skodje, R. T. and Cary, J. R. (1988). *Comput. Phys. Rep.* **8**, 221.
Skodje, R. T. and Davis, M. J. (1988). *J. Chem. Phys.* **88**, 2429.
Skodje, R. T., Truhlar, D. G. and Garrett, B. (1982). *J. Chem. Phys.* **77**, 5955.
Skouteris, D., Castillo, J. F. and Manolopoulos, D. E. (2000). *Comp. Phys. Comm.* **133**, 128.
Smith, A. L. (1971). *J. Chem. Phys.* **55**, 4344.
Solov'ev, E. A. (1978). *Sov. Phys.–JETP.* **48**, 635.
Sorbie, K. S. and Handy, N. C. (1976). *Mol. Phys.* **31**, 97.
Stine, J. R. and Marcus, R. A. (1972). *Chem. Phys. Lett.* **15**, 536.
Stine, J. R. and Marcus, R. A. (1973). *J. Chem. Phys.* **59**, 5145.

Stine, J. R. and Noid, D. W. (1983). *J. Chem. Phys.* **78**, 3647.
Strand, M. P. and Reinhardt, W. P. (1979). *J. Chem. Phys.* **70**, 3812.
Stückelberg, E. C. G. (1932). *Helv. Phys. Acta.* **5**, 369.
Stwalley, W. C. (1973). *Chem. Phys. Lett.* **19**, 337.
Stwalley, W. C. (1975). *J. Chem. Phys.* **63**, 3062.
Stwalley, W. C. (1978). *Contemp. Phys.* **1**, 65.
Sukumar, C. V. and Bardsley, J. (1975). *J. Phys. B: At. Mol. Phys.* **8**, 568.
Sumpter, B. G. and Noid, D. W. (1986). *Chem. Phys. Lett.* **126**, 181.
Swimm, R. T. and Delos, J. B. (1979). *J. Chem. Phys.* **71**, 1706.
Szegö, G. (1975). *Orthogonal polynomials* (4th ed.). Providence: American Mathematical Society.
Tabor, M. (1989). *Chaos and integrability in non-linear dynamics.* New York: Wiley.
Tannor, D. J. (2007). *Introduction to quantum mechanics: a time-dependent perspective.* Herndon, VA: University Science Books.
Tannor, D. J. and Garaschuk, S. (2000). *Ann. Rev. Phys. Chem.* **51**, 553.
Taylor, J. R. (1972). *Scattering theory. The quantum theory of non-relativistic collisions.* Reading, Mass: Addison-Wesley; transl. D. H. Fowler.
Tellinghuisen, J. (1972). *J. Mol. Spectrosc.* **44**, 194.
Tellinghuisen, J. (1985). *Adv. Chem. Phys.* **60**, 299.
Thakker, A. (1975). *J. Chem. Phys.* **62**, 1693.
Thom, R. (1975). *Structural stability and morphogenesis.* New York: Wiley.
Thorson, W. R., Delos, J. B. and Boorstein, S. A. (1971). *Phys. Rev. A* **4**, 1052.
Thylwe, K. E. and Connor, J. N. L. (1988). *J. Chem. Phys.* **91**, 1668.
Totenhofer, A. J., Noli, C. and Connor, J. N. L. (2010). *Phys. Chem. Chem. Phys.* **12**, 8772.
Trinkhaus, H. and Drepper, F. (1977). *J. Phys. A: Math. Gen.* **10**, L11.
Tully, J. C. (1976). In W. H. Miller (Ed.), *Dynamics of molecular collisions.* New York: Plenum Press.
Tully, J. C. and Preston, R. K. (1971). *J. Chem. Phys.* **55**, 562.
Untch, A., Henning, S. and Schinke, R. (1989). *Chem. Phys.* **126**, 181.
Uzer, T. and Child, M. S. (1980). *Mol. Phys.* **41**, 1177.
Uzer, T. and Child, M. S. (1982). *Mol. Phys.* **46**, 1371.
Uzer, T. and Child, M. S. (1983). *Mol. Phys.* **50**, 247.
Uzer, T., Muckerman, J. T. and Child, M. S. (1983a). *Mol. Phys.* **50**, 1215.
Uzer, T., Noid, D. W. and Marcus, R. A. (1983b). *J. Chem. Phys.* **79**, 4412.
Van der Waerden, B. L. (1967). *Sources of quantum mechanics.* New York: Dover.
Van Vleck, J. H. (1928). *Proc. Natl. Acad. Sci.* **14**, 178.
Van Vleck, J. H. (1951). *Rev. Mod. Phys.* **23**, 213.
Vollmer, G. (1969). *Z. Phys.* **226**, 423.
Vöros, A. (1983). *Ann. Inst. H. Poincaré Phys. Theor.* **39**, 211.
Voth, G. A., Chandler, D. and Miller, W. H. (1989). *J. Chem. Phys.* **91**, 7749.
Walton, A. R. and Manolopoulos, D. E. (1996). *Mol. Phys.* **87**, 961.
Warnock, R. L. and Ruth, R. D. (1987). *Physica D* **26**, 1.
Wentzel, G. (1926). *Z. Phys.* **38**, 518.

Whittaker, E. T. and Watson, G. N. (1962). *Modern analysis* (4th ed.). Cambridge: Cambridge University Press.
Wigner, E. (1932a). *Phys. Rev.* **40**, 749.
Wigner, E. (1932b). *Z. Phys. Chem. B* **19**, 203.
Wigner, E. (1959). *Group Theory and its Application to the Quantum Mechanics of Atomic Spectra.* New York: Academic Press.
Wilson, D. C., Decius, J. C. and Cross, P. C. (1954). *Molecular vibrations: the theory of infrared and Raman spectra.* New York: McGraw-Mill.
Winnewisser, M., Winnewisser, B. P., Medvedev, I. R., De Lucia, F. C., Ross, S. C. and Bates, L. M. (2006). *J. Mol. Struct.* **798**, 1.
Wintgen, D. and Friedrich, H. (1986). *J. Phys. B: At. Mol. Phys.* **19**, 991.
Wolf, F. and Korsch, H. J. (1984). *J. Chem. Phys.* **81**, 3127.
Wolniewicz, L. (1966). *J. Chem. Phys.* **45**, 515.
Wong, W. H. and Marcus, R. A. (1971). *J. Chem. Phys.* **55**, 5663.
Xiahou, C., Connor, J. N. L. and Zhang, D. H. (2011). *Phys. Chem. Chem. Phys.* **13**, 12981.
Yourshaw, I., Zhao, Y. and Neumark, D. M. (1996). *J. Chem. Phys.* **105**, 351.
Zahr, G. E. and Miller, W. H. (1975). *Mol. Phys.* **30**, 951.
Zare, R. N. (1988). *Angular momentum.* New York: Wiley.
Zener, C. (1932). *Proc. Roy. Soc. A* **137**, 69.
Zhang, J. Z. H. and Miller, W. H. (1989). *J. Chem. Phys.* **91**, 1528.
Zhu, C. and Nakamura, H. (1995). *J. Chem. Phys.* **102**, 7448.
Zhu, C. and Nakamura, H. (1996). *J. Chem. Phys.* **106**, 2599.

Author index

Subject index

D